ENTOMOLOGY
AND
PEST MANAGEMENT

ENTOMOLOGY

AND

PEST MANAGEMENT

LARRY P. PEDIGO
IOWA STATE UNIVERSITY

MACMILLAN PUBLISHING COMPANY
NEW YORK
COLLIER MACMILLAN PUBLISHERS
LONDON

Macmillan Publishing Company
866 Third Avenue, New York, New York 10022

Collier Macmillan Canada, Inc.

Library of Congress Cataloging in Publication Data

Pedigo, Larry P.
    Entomology and pest management / Larry P. Pedigo.
       p.   cm.
    Includes index.
    0-02-393310-0
    1. Insect pests—Control. 2. Insect pests. 3. Insects.
  4. Entomology. I. Title.
  SB931.P39 1988   88-5341
  632'.7—dc19   CIP

Printing: 1 2 3 4 5 6 7 8   Year: 9 0 1 2 3 4 5 6 7 8

Special acknowledgments
   I wish to give special thanks and credit to Scott
H. Hutchins and Robert K. D. Peterson for their
library research and writing of the insect diagnostic
boxes. This important feature could not have been achieved
without their help.
   Special acknowledgment is also due Laura L. Karr for her
very thorough compilation of the glossary. Her efforts have added
a very important educational dimension to this work.

This book is dedicated to my graduate students,
past and present, whose enthusiasm, talent,
and labor has been an inspiration.

Preface

THE PAST THREE DECADES HAVE provided many crossroads for entomology where a turn this way or that significantly altered the final course of the science. During this time probably no events have had more impact on direction than the publication of *Silent Spring* by Rachel Carson in 1962 and the Environmental Movement in the United States during the late 1960s and early 1970s. These events fostered the transformation of a discipline narrowly preoccupied with pesticide technology to one of broad-based principles emphasizing sustainable solutions to pest problems and conservation of environmental quality. These principles are embodied in the concept of pest management, an extraordinarily successful approach that has offered new ways of thinking about and dealing with insect pests.

Although several books have been written on pest management of insects, few have emphasized practice and fewer still have included basic information about the pests themselves. In writing this book, I have combined elements of general entomology and modern principles of insect pest management with the hope of promoting improved pest control skills and a better appreciation of insects and their role in our environment.

TARGET AUDIENCE

Entomology and Pest Management can be used as an introduction to applied entomology for undergraduates or beginning graduate students. For undergraduates with only an elementary biology background, early chapters need emphasis to provide a basis for understanding the remaining content on insect ecology, surveillance, and management. Students with at least one course in entomology may wish to omit early chapters and emphasize the strategy and tactics of management found in later chapters. Omitting Chapters 1 through 4 for graduate courses or Chapters 13 and 14 for undergraduate courses will not result in a loss of continuity.

Content and Organization

The book consists of sixteen chapters, an epilogue, three appendices, and a glossary. Concepts and principles are emphasized and supported by factual detail and specific examples. Beginning chapters, 1 through 3, concentrate on general entomology for the novice. Chapters 4 and 5 synthesize the elements of insect biology and ecology required for understanding insect pest management. Chapter 6 covers techniques and principles of sampling for problem assessment, and Chapter 7 builds on this knowledge by outlining types of reactions of crops to insect densities and develops the current concepts of decision making for management. The ideas and history of insect pest control are reviewed in Chapter 8, which also introduces the concept and philosophy of modern pest management.

With this basic information presented in the first eight chapters, the student is introduced, chapter by chapter, to the individual tactics used as elements in pest management programs. The order of tactics presented is based on their relative importance in existing pest management programs. Consequently, natural enemies and ecological management of the environment, primarily preventive tactics, are mentioned first, followed by conventional insecticides, the premier element of curative tactics. The remaining tactics discussed in Chapters 12 through 14 are more specialized but, nevertheless, convey some of the newest and most innovative ideas in the science.

Chapter 15 is unique in that it attempts to integrate the problems of resistance, resurgence, and replacement, as well as such recent phenomena as enhanced microbial degradation of insecticides, into a single concept, ecological backlash. The chapter suggests to students that applying the tactics discussed does not always result in sustainable pest management and recommends ways of reducing or avoiding such problems.

The book ends by discussing the principles of integrating management tactics and gives examples of current insect pest management programs in the context of diverse commodities. Finally, the Epilogue presents the role of insect pest management in the broader science of integrated pest management and discusses some of the challenges for pest sciences in the years ahead.

Special Features

Basic and applied entomology. The primary purpose of the book is to promote an understanding of major elements of general entomology and relate them to modern principles of insect pest management. Both theory and practice are emphasized in a conceptual approach to the topics, and numerous examples are presented to facilitate learning.

Ecological approach. Pest management topics are discussed as aspects of applied ecology, and solutions to pest problems are presented with regard to environmental quality, profitability, and sustainment.

Insect diagnostic boxes. Over 50 insect diagnostic boxes are presented throughout the text. Each box contains detailed information on distribution, importance, appearance, and life cycle of a species or species group. Insects chosen are from examples mentioned in the text. Grouping specific data in boxes keeps the text from being overburdened with background information about a species when examples and case histories are given. Students not familiar with the species being discussed can consult the boxes to better understand and appreciate the examples in the text. Information in the boxes is readily referenced from the index. Additional information about major pests is given in Chapter 3, as insect families of major economic importance were included in presenting information on insect classification.

Boldface type. This feature allows the student to recognize new terms and important comments quickly and serves as a basis for topic review.

Appendices. Three appendices facilitate learning and serve as reference material. Appendix 1 presents a key to the orders of insects, allowing identification of both adult and immature insects by the novice. Appendix 2 contains a list of insect common names along with the accompanying scientific names and classifications. Appendix 3 presents lists of common insecticides by common and trade names and gives manufacturer and toxicity information.

Glossary. A glossary at the end of the book provides quick reference for students. Words in the text that appear in bold type are included, as well as many others.

Supplementary Material

An accompanying manual, *Manual of Entomology and Pest Management,* is available for use with this book. The manual contains up-to-date, practical information about insects, as well as details on pests and crops. This companion publication is particularly appropriate for courses with formal laboratory sections and can complement lecture courses having a pest-by-commodity orientation.

Acknowledgments

Many people deserve thanks for their assistance in completing this book. First and foremost are my current graduate students who read drafts as chapters were completed and made many useful suggestions: Joseph Browde, Paula Davis, Leon Higley, Scott Hutchins, Tom Klubertanz, and Rick Smelser. Also, I wish to express my appreciation to Scott Hutchins and Bob Peterson for writing the insect diagnostic boxes and Laura Karr for her work on the glossary.

My colleagues in the Department of Entomology at Iowa State University deserve many thanks for their consultation, draft reviews, and contribution of figures and photographs for reproduction: Joel Coats, Ferdinand Dicke, David Foster, Wilbur Guthrie, Ken Holscher, Elliot Krafsur, Donald Lewis, Les Lewis, John Mutchmor, John Obrycki, Ken Racke, Wayne Rowley, Bill Showers, Barbara Spike, Harold Stockdale, Jon Tollefson, and Wendy Wintersteen.

I also owe gratitude to my entomology colleagues outside Iowa State University who reviewed drafts of the manuscript and are responsible for many of the improvements but none of the mistakes. Among these colleagues are the following: Ronald Hammond, O.A.R.D.C., Ohio State University; David Hogg, University of Wisconsin; Armon Keaster, University of Missouri; Gary Lentz, University of Tennessee; James Mertins, Ames, Iowa; John Meyer, North Carolina State University; Edward Radcliffe, University of Minnesota; Thomas Reagan, Louisiana State University; John Reese, Kansas State University; Ted Schulz, North Dakota State University; Robert Snetsinger, Pennsylvania State University; and Alan York, Purdue University.

Special thanks are due Macmillan Publishing Company in general and Beth Anderson and Bob Rogers in particular for their help in processing the manuscript. In addition, Avis Morse of Ames, Iowa, was indispensable in facilitating several aspects of the work. Thanks are also extended to Carol Kromminga of Des Moines, Iowa, for her excellent compilation of the index.

Finally, I am grateful to my wife, Lois, and children, Brenda and Bruce, for their help and encouragement during the writing. My task was made much easier and enjoyable because of their support.

L. P. P.

Brief Contents

Detailed Contents

List of Diagnostic Boxes

1 INTRODUCTION

LONG BEFORE THERE WERE human beings there were insects. These creeping, crawling, and flying creatures that are easily obliterated by your hand or foot are estimated to have existed for over 350 million years! It may seem paradoxical that this is so. After all, how could such small, dumb, defenseless objects have faced the natural disasters of flood, drought, volcanoes, hurricanes, blizzards, earthquakes, and disease and yet survived so long and so well? After all, humans have narrowly escaped such disasters, and think how strong and clever we are. Yet, our family tree probably took root only about 2 million years ago.

INSECT ABUNDANCE

The fact is, today's human population is adrift in a sea of insects. If we look at numbers alone, it has been estimated that humans count only one to insects' 200 million and that there are about 40 million insects for each acre of land. But being much larger than insects, we might be tempted to argue that humans are ecologically more successful, making up in mass for our relatively lower numbers. However, analysts have estimated that in the United States there are some 400 pounds of insect biomass per acre, compared with our 14 pounds of flesh and bone. Based solely on numbers and biomass, insects are the most successful animals on earth! This surprising achievement disproves the commonly held, biased belief that intelligence is a necessary component of evolutionary success.

INSECT DIVERSITY

Along with humans, insects live in almost every habitable place on the earth, except the ocean depths. According to the distinguished entomologists T. Eisner and E. Wilson, insects all but own the land. They are the chief consumers of plants; they are the major predators of the plant eaters; they play a major role in decay of organic matter; and they serve as food for other kinds of animals.

Knowing these ecological facts, we might expect these organisms to be diverse and adaptable—and they are. Today, over 900,000 kinds (species) have been described, and it is widely believed that there are two to ten times as many not yet discovered.

1

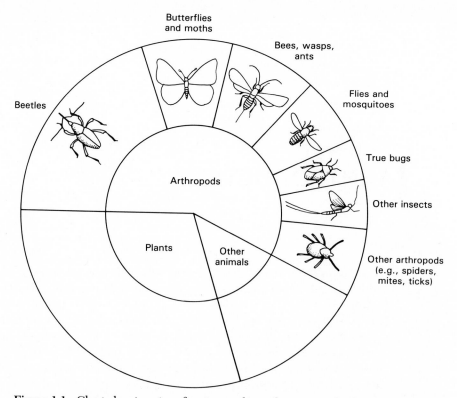

Figure 1.1 Chart showing size of various arthropod groups and other life forms. Pie slices are proportional to the number of species in the group. (Redrawn from Daly, et al., 1978, *Introduction to Insect Biology and Diversity*, McGraw-Hill Book Company)

This figure makes other animal groups small by comparison. For instance, species in the class Mammalia, of which humans are members, count less than 1 percent of those in the class Insecta. Furthermore, there are more species of insects in the world than any other kind of organism, making up about three-fourths of the total number known. Indeed, it has been estimated that every fifth animal is a beetle, only one group in the class Insecta. Without a doubt, this extraordinary diversity is yet another mark of insects' ecological success.

WHAT AN INSECT IS

Insects are grouped with other animals sharing similar characteristics in the phylum Arthropoda. Arthropods, as they are known, are characterized as having a body divided by grooves to form **segments** and a well-developed covering, the **integument**, that makes up the outer shell, or **exoskeleton**. Some of the arthropod body segments possess one or more pairs of jointed appendages from which the phylum name was derived (*arthro*, joint; *poda*, foot). In addition, arthropods have a heart at the top and a nerve cord at the bottom of their body cavity. Examples of arthropods other than insects include spiders, shrimp, and centipedes.

More specifically, insects belong to the class Insecta (from *insectum*, cut into), forming the most important category of the Arthropoda. The class Insecta has specific characteristics that set its members apart from other arthropod classes.

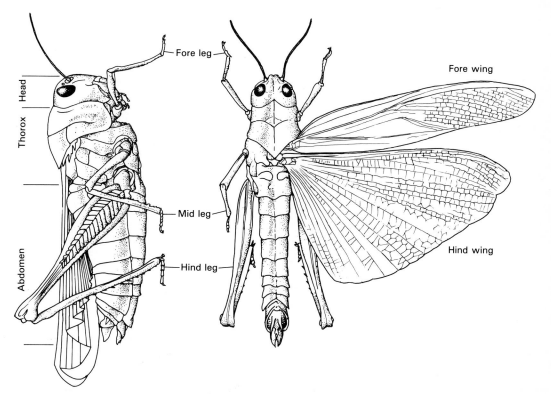

Figure 1.2 Side (left) and top aspects of a grasshopper, showing important external body features of insects. (Redrawn from T. Nolan, 1970, Commonwealth Scientific and Industrial Research Organization, *The Insects of Australia*, Melbourne University Press)

These characteristics include

1. A body divided into three distinct regions: head, thorax, and abdomen
2. A middle region, the thorax, bearing three pairs of legs and, most often, two pairs of wings
3. A system for breathing composed of air tubes.

OTHER ARTHROPODS

Other arthropod classes have fewer total numbers and species than insects. Although a detailed survey of all classes is beyond the scope of this book, a brief overview of the five most important classes may help clarify the uniqueness of insects. These other important arthropods include those in the classes Crustacea, Arachnida, Diplopoda, and Chilopoda. A complete classification of living arthropods is given in Table 1.1.

Class Crustacea. This class includes many common animals such as crayfish, lobsters, pillbugs, crabs, and shrimp. Crustaceans as a group are mostly aquatic, and they play major ecological roles in marine habitats that are absolutely devoid of insects. There are also a few species of terrestrial crustaceans such as pillbugs and sowbugs that are commonly found in basements and other humid places of the home. When disturbed, pillbugs curl up into a compact ball.

Table 1.1. Classification Scheme of Living Arthropods (Following Kaestner, 1968)

Phylum Arthropoda	arthropods
Subphylum Chelicerata	
Class Merostomata	
Order Xiphosura	horseshoe crabs
Class Arachnida	
Order Scorpiones	scorpions
Order Uropygi	whipscorpions
Order Amblypygi	tailless whipscorpions
Order Palpigradi	micro-whipscorpions
Order Araneae	spiders
Order Ricinulei	ricinuleids
Order Pseudoscorpiones	pseudoscorpions
Order Solifugae	windscorpions
Order Opiliones	harvestmen
Order Acari	mites and ticks
Class Pycnogonida	sea spiders
Subphylum Mandibulata	
Class Chilopoda	centipedes
Class Diplopoda	millipedes
Class Pauropoda	pauropods
Class Symphyla	greenhouse centipedes
Class Insecta	insects
(orders listed in Chapter 3)	

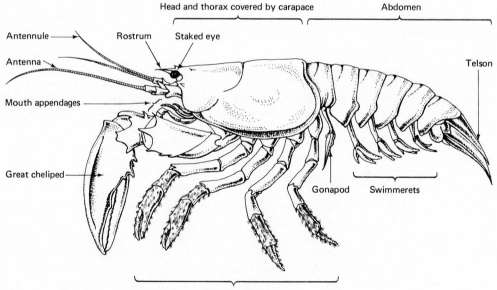

Figure 1.3 Side aspect of the American lobster, *Homaris americana*, showing the various modifications in appendages and other body structures. (Reprinted with permission of Macmillan Publishing Company from *Insects in Perspective* by Michael D. Atkins. Copyright ©1978 by Michael D. Atkins)

Most crustaceans breathe by using gills, and they are covered by a hard shell from which their name is derived (from *crusta*, shell). In a number of instances, they have two pairs of antennae, a number of pairs of legs modified for swimming, and a number of body segments fused with the head to form a **cephalothorax** (head-body).

As a group, crustaceans are mostly beneficial to humans, serving as food directly or as food for fish and other aquatic animals. Harmful species include barnacles, sessile marine forms that attach to vessels and destroy shore installations, and sowbugs and pillbugs, which may injure greenhouse and garden crops.

Class Diplopoda. The common name for members of this class is millipede (thousand legs). They are cylindrical with 25–100 segments, and most segments bear two pairs of legs. These arthropods are found in dark, humid environments such as under leaves and bark, in woodlands, and in basements of homes. When disturbed they curl up into a characteristic spiral.

Millipedes feed mostly on decaying organic matter and, therefore, are beneficial to humans. However, they are considered pests by most people because of their mere presence in homes.

Class Chilopoda. These arthropods are the sometimes feared centipedes. Some people fear them because certain ones are venomous and can inflict a painful bite. However, most times they are secretive and run away when approached.

Most centipedes have a flattened body with many segments and one pair of legs on each segment. A pair of claws behind the head is used to inject venom and paralyze insects and other invertebrates that serve as food for the centipede.

Class Arachnida. Next to the Insecta, this is the most diverse class of terrestrial arthropods. Members of this class possess a cephalothorax, as do crustaceans, but they lack antennae. Respiration is by means of bellowslike structures called **book lungs** in some groups (spiders), but others breath through the skin or with air tubes. Most arachnids live on land.

Based on their relationship to humans the most significant of the Arachnida are the spiders, mites, ticks, and scorpions. Other orders that contribute to the diversity of the class Arachnida include the Opiliones (harvestmen), Pseudoscorpiones (pseudoscorpions), and Uropygi (whipscorpions).

Spiders belong to the order Araneae and are represented by thousands of species. They are distinguished by their unsegmented abdomen, which is attached to the cephalothorax by a slender stalk or pedicel. Spiders feed mainly on insects by using specially modified mouthparts that crush their prey and allow these predators to suck out the body fluids. Silk-spinning organs are located on the underside of the abdomen, permitting spiders to build webs. Spider webs may be orb-shaped, funnel-shaped, triangle-shaped, or irregular. Insects and other prey become snared in the web and are killed outright or are paralyzed by venom from the spider's bite. As a group, spiders are very beneficial to humans by serving as natural enemies of insect pests. Some, however, are medically important pests because of their dangerous bite. In North America, these include the black widow (*Latrodectus mactans*) and the brown recluse (*Loxosceles reclusa*).

From a harmful standpoint, the most important order of the Arachnida is the Acari, comprised of the mites and ticks. Here can be found pests of humans, other animals, and plants. The Acari are distinct because of their saclike body and unsegmented abdomen broadly joined to the cephalothorax. Mouthparts of mites and ticks are specialized for piercing tissues and sucking out the contents.

Figure 1.4 Representatives of important groups in the class Crustacea. A. Freshwater mysid shrimp. B. Sand flea (amphipod). C. Shore crab (decapod). D. Freshwater isopod. E. Sowbug or pillbug (terrestrial isopod). (Reprinted with permission of Macmillan Publishing Company from *Insects in Perspective* by Michael D. Atkins. Copyright ©1978 by Michael D. Atkins)

Figure 1.5 Class Diplopoda, a millipede. (Reprinted with permission of Macmillan Publishing Company from *Insects in Perspective* by Michael D. Atkins. Copyright ©1978 by Michael D. Atkins)

Figure 1.6 Class Chilopoda, a centipede. (Reprinted with permission of Macmillan Publishing Company from *Insects in Perspective* by Michael D. Atkins. Copyright ©1978 by Michael D. Atkins)

Mites are usually very small arthropods (1–3 mm long) that feed on plants, animals, and organic debris. Some mites are very important predators of small insects and, particularly, other plant-feeding mites. One of the most important plant-feeding mites is the twospotted spider mite (*Tetranychus urticae*), which injures many crops in dry climates or during droughts in wetter regions. Chiggers (larvae of *Trombicula alfreddugesi* and others) infest humans, causing intense itching when they inject enzymes to dissolve skin tissue on which they feed. Mange mites (*Sarcoptes scabiei*) are examples of mites that feed on many animals including hogs, horses, dogs, and humans. These mites feed in the skin and produce burrows in which eggs are laid and young develop. In humans, these mites cause scabies, a skin condition often noted in the elderly.

Ticks are acarines larger than mites (10–20 mm long), with leathery integument or skin. Ticks feed only on animals (humans and other mammals, birds, and reptiles), sucking blood from and sometimes transmitting disease-causing organisms to their hosts. Examples of important tick pests in North America include the lone star tick (*Amblyomma americanum*) and the American dog tick (*Dermacentor variabilis*). Both these species attack humans, dogs, and livestock and are transmitters of the causal agent of Rocky Mountain spotted fever.

Scorpions are other arachnids belonging to the order Scorpiones. They can be found in tropical and temperate regions but probably are most abundant in

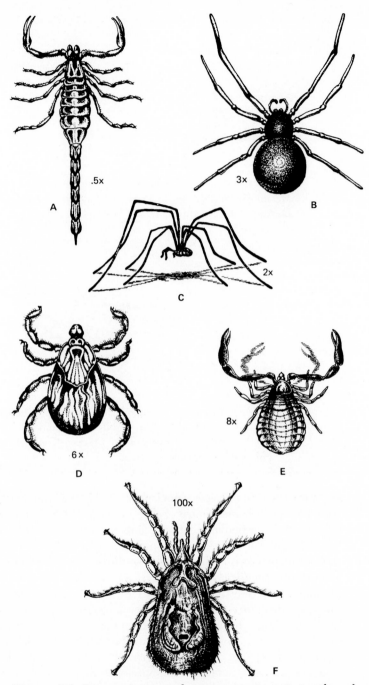

Figure 1.7 Representatives of important groups in the class Arachnida. A. Scorpion. B. Spider. C. Harvestman (daddy longlegs). D. Tick. E. Pseudoscorpion, F. Mite. (Reprinted with permission of Macmillan Publishing Company from *Insects in Perspective* by Michael D. Atkins. Copyright ©1978 by Michael D. Atkins)

warm, dry climates. These arthropods are distinguished by having a segmented abdomen broadly joined to a cephalothorax and long front appendages with strong pincers at the tip. Scorpions have a well-known stinging structure at the end of the abdomen used in immobilizing and killing prey, usually an insect, before feeding. When humans are stung by most species of scorpions, the effect of the poison, a neurotoxin, is painful but not usually dangerous. However, the sting of some species found in Mexico, for example, the Durango scorpion (*Centruroides suffusus*), can be fatal.

WHAT ENTOMOLOGY IS

Entomology is the science of insects. It is an organized study to obtain knowledge of all phases of insect life and to understand insects' roles in nature. As in any science, entomology assumes that elements of the universe interact with one another in a predictable way, and that insects, being one of these elements, can be understood relative to these interactions. From the knowledge obtained through **science**, insect events can be predicted and, if desired, modified with **technology**. Therefore, we can learn about insect feeding behavior and developmental rate from scientific study and apply this knowledge to alter these processes for plant and animal protection with technology.

Entomology is a biological science. It is the study of life, with particular focus on one life form, the insect form. More specifically, entomology is treated as a division of zoology, the study of animal life. In this context, we can think of entomology's position as one in a hierarchy of disciplines that become increasingly specific in their subject matter.

SCIENCE—study of the universe
BIOLOGY—study of life
ZOOLOGY—study of animals
ENTOMOLOGY—study of insects

Following this hierarchy, entomology is usually taught as a separate course(s) in universities, being based in departments of biology, zoology, or, in many instances, separate departments of entomology.

Although in the strictest sense, entomology is the study of insects, other arthropods usually are treated by the science. In particular, these include the terrestrial arthropods that share habitats and interact with insects and many that cause similar pest problems.

Entomologists. Entomologists are people who specialize in entomology. They have background and skills in biology and have particular interests in and knowledge of zoology.

The first entomologists were mostly general biologists who made significant contributions to entomology and laid the groundwork for its development as a separate discipline. Today, many people make a living working with insects, and some simply study them for fun. According to British entomologist Sir Richard Southwood, insects are of interest to people in one of three main ways: as a branch of zoology, as the scientific basis for pest control, or as a pastime. The first two ways are of concern to professional entomologists who teach, conduct research, and deal with pest problems. The last way is mainly of concern to amateur entomologists but also to many professionals, who enjoy observing,

Figure 1.8 Entomologist sampling insects with a sweep net. In addition to their special knowledge of insects, entomologists have basic backgrounds in zoology and other biological sciences.

photographing, and collecting insects because of their beauty and fascinating nature. Consequently, both professionals and amateurs contribute to the overall body of knowledge we call entomology, and all can be considered entomologists.

Producers of entomological information and services. Professional entomologists supply the greatest share of entomological information and services. They conduct their activities either through public or private agencies.

Public sources of information and services include universities, institutes, and governmental agencies. In the United States, most land-grant universities (those established by the Morrill Act of 1862) have a department of entomology that acquires information through research and disseminates it by teaching students, publishing scientific and popular articles, and using mass media. An important aspect of this educational system is the Cooperative Extension Service, a state and federally supported organization attached to the university, with a mission of providing up-to-date information and advice to farmers and the general public. In addition to the universities in the United States and other countries, research organizations like the Boyce Thompson Institute for Plant Research (Ithaca, N.Y.), the Rothamsted Experiment Station (Harpenden, Hertfordshire, United Kingdom), and the Institute of Animal Resource Ecology (Vancouver, British Columbia, Canada) supply information on insects that adds significantly to our current body of knowledge.

Major governmental agencies supplying entomological information in the United States include the Agricultural Research Service, the Forest Service, and the Animal and Plant Health Inspection Service of the U.S. Department of Agriculture. Equivalents of these federal organizations exist also in many other parts of the world, for example, the Canadian Department of Agriculture in

Figure 1.9 An extension entomologist (right) showing a farmer how to sample alfalfa weevils. The Cooperative Extension Service located in each state of the United States is one of the most important sources of entomological information.

Canada and the Commonwealth Scientific and Industrial Research Organization (C.S.I.R.O.) in Australia.

Not to be overlooked is entomological information and services available from private, mainly profit-oriented companies. Insect control services for households and businesses are obtained from structural pest control firms, which rid buildings of termites and other insect pests. Some of these companies also provide local mosquito control services and deal with pests of turf and ornamentals. Insecticide-manufacturing firms also conduct entomological research and provide a good deal of information about insect control to their customers and the general public. Other private firms who consult in the management of crop pests contract to assess insect and other pest conditions on growers' property and give expert advice for dealing with pests.

Users of entomological information and services. Directly or indirectly all of us use information about insects and benefit from the services of entomologists. On a personal basis, the general public utilizes entomological information and services to save dwellings, ward off mosquitoes, protect woolens in the closet, keep flies in check, and help grow vegetable gardens, among other ways.

Other direct beneficiaries, of course, are the producers of raw agricultural products. These individuals depend on a wide background of knowledge and techniques to avoid or reduce economic losses from insects. For some crops, particularly those of high value like apples, managing insect pests can mean the difference between profit or loss for the season.

Likewise, food processing firms greatly benefit from entomological information. In this industry, contamination from insects in processed foods can necessitate discarding huge quantities of an otherwise marketable product. The processor then is required to absorb the loss. Knowing about insect behavior and sanitation procedures can minimize or avoid such losses.

Because insects affect human health and the health of livestock, both physicians and veterinarians are primary users of entomological information and techniques. Problems ranging from head louse epidemics in schools to transmission of disease organisms by mosquitoes are addressed by physicians who depend on knowledge of insects. These professionals often consult medical entomologists before making a diagnosis and recommending therapy. Veterinarians presented with such problems as hog mange or cattle grubs rely on entomological research and often seek advice from livestock entomologists in suggesting cures.

Recently, entomological information has been employed even by the police to solve crimes of homicide. In solving these crimes, estimating the time of death becomes crucial to determine circumstances and suspects. For example, knowledge of behavior and developmental rates of carrion fly maggots inhabiting corpses allows estimates of egg-laying time and therefore the time of death of the victim. Such information developed through **forensic entomology** is expected to contribute significantly to future advancements in criminology.

These are only a few examples of the users and uses of entomological information and services. Certainly, the quality of human life has been and will continue to be enhanced through our knowledge of insects.

RELATIONSHIPS BETWEEN INSECTS AND PEOPLE

It is usually with a great deal of hesitation that human beings share their habitats with insects. Having had no choice in the matter, we have adjusted our existence because of insects, and many customs and life-styles reflect the intimacy of our relationship with them.

Brief History of Relationships

Insects in antiquity. Insects have lived unhampered by human contrivances for most of their existence on earth. They were probably the first small animals to inhabit the land with complete success, having evolved from creatures that probably looked like our present-day earthworms.

The land had been colonized by low-growing plants some 100 million years before insects became prominent and had witnessed a succession of vegetative changes that ultimately resulted in the great coal forests. It was during this time, the Carboniferous Period (about 350 million years ago), when amphibians and reptiles flourished that insects gained a firm foothold on the land. During this early period, change in insect form and degree of species diversity accelerated rapidly. It was a time of large dragonflylike insects, gigantic forms with wingspans as wide as 29 inches (75 cm)!

By the Permian Period (280 million years ago) most major insect characteristics had evolved, including wings; ability to fold wings over the body, which allowed winged insects to move through vegetation and soil; and complete metamorphosis, a kind of body change during growth that occurs in the most advanced insects. Indeed, many of the common insect groups we see today, for example, silverfish, cockroaches, and beetles, were present by the end of the Paleozoic Era (220 million years ago).

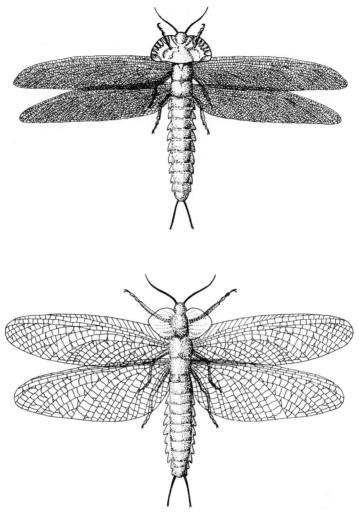

Figure 1.10 Reconstructions of Carboniferous fossil insects. (Redrawn from Jeannel, 1960, *Introduction to Entomology*, Hutchinson and Company)

Success of insects. From these early beginnings, the class Insecta developed into one of the dominant life forms on earth. Using our evolutionary perspective, we might question this great explosion of insects. Appropriately having been called the "conquerors of the land," what reasons can we advance for their great success? What characteristics can account for their tremendous diversity and numbers? Most specialists on this topic believe that insect success is the result of several biological features acting in combination and the rather unstable environment of the land that insects invaded.

Probably first and foremost of the biological features is arthropod **body architecture**. This architecture emphasizes an integument that is light and strong, forming a shell, for the protection of inner tissues and the attachment of muscles. Moreover, this shell, which usually includes an outermost wax layer, helps to prevent water loss from evaporation, an extremely important problem for small animals living on the land. Arthropod body architecture also includes jointed appendages that, in insects, have been profusely adapted into legs for locomotion, mouthparts for feeding, and structures of reproduction, among others.

Insects also are animals of relatively **small size**. Most vary from about 1/16 inch (about 2 mm) to 1 inch (about 3 cm). Some may be smaller, however, and a few, such as the 6-inch (about 15 cm) goliath beetle of Africa and the 13-inch (about 35 cm) walkingstick of Malaysia, definitely fall outside this normal range. Nevertheless, the small size of most insects facilitates dispersal, allows them to escape from predators (for example, birds), and enables them to use food present only in small amounts.

Differing from all other arthropods and invertebrates, most insects can fly. This **ability to fly** is one of the most important reasons for success of the class as a whole. Flying aids further in escaping predators and, perhaps more important, enables widespread dispersal of species. This dispersal promotes colonization of new habitats, which, in turn, promotes the evolution of new species.

Finally, the great **reproductive capacity** of insects and features of their growth and development have enhanced their ability to persist even in unfavorable environments. The ability to lay large numbers of eggs, combined with a relatively short generation time, produces a great amount of genetic variability that can be tested against the environment. The end result is rapid adaptation of populations to changing environmental conditions and formation of new species.

These major features, collectively unique to insects, were combined with great changes in physical conditions and habitat on the land. Together, they caused tremendous radiation of species and subsequent persistence of these through geological time.

Prehistoric times. At the dawn of human existence, a split second ago on the

Figure 1.11 Dragonflies (Odonata) in flight. Ability to fly is one of the most important reasons for success of the class Insecta. (Redrawn from a drawing by P. Eades)

Figure 1.12 Aphids (Homoptera: Aphididae) on a plant stem. The great reproductive capacity in these and other insects has enhanced their ability to persist even in unfavorable environments.

clock of geological time, insects were already entrenched in every conceivable land habitat. They fed on plants, all sorts of animals (including themselves), and all forms of nonliving organic debris. Therefore, it might be imagined that they already fed, as parasites, on the closest relatives of humankind, the apes and prehumans, when *Homo sapiens* finally arrived.

Human hairlessness seems to point to a tropical environment where protohumans probably originated, and where they fed on nuts and fruits, vegetable shoots, and probably **insects**. On the basis of abundance in the tropical habitat, insects are believed to have been one of the first animals that humans ate on a regular basis!

Because prehistoric humans probably coexisted with monkeys and arboreal apes, it is not unrealistic to assume that they also shared the same parasites. Additionally, some authorities believe that human associations with bed bugs (species closely related to bat parasites) arose out of humans sharing caves with bats during the ice ages. Therefore, it is supposed that early humans were infested with the array of parasites known from modern primates, including mites, fleas, ticks, flies, and worms. Humans also were subject to diseases caused by many microorganisms, including the so-called **arboviruses**, or arthropod-borne viruses transmitted by mosquitoes and other blood-sucking arthropods. Further, plasmodia, transmitted by mosquitoes to humans, causing epidemics of **malaria**, must have had horrendous effects on those early populations, as they still have today. Indeed, malaria has undoubtedly altered the entire pattern of human history and probably contributed to the decline of some civilizations.

As culture evolved, humans developed skills and weapons for killing large-bodied herbivores like the antelopes found on the grasslands of the African savanna and perhaps in Asia. The development of this new technology led to human language and cooperation, which fostered humans' invasion of new habitats and geographical regions. As this invasion occurred humans were exposed to greater and greater contact with insects species previously not encountered. This was the case when humans began to venture into uninhabited regions of Africa in search of antelope and encountered the tsetse fly. This fly was (and still is) capable of transmitting a trypanosome from antelopes (and other ungulates) to humans, causing African sleeping sickness, a debilitating disease that can be lethal in a few weeks.

Therefore, *Homo sapiens*, as a scavenger and later as a hunter, had pervasive encounters with insects, some good and some bad. Certainly, it would be difficult to overstate the impact of arthropod-borne disease not only in terms of human suffering but also in terms of retarding the development of civilization. In total, the relationships that developed left an indelible mark on human development.

The rise of agriculture and civilization. With the numerous extinctions of large-bodied game animals 20,000 to 50,000 years ago, human populations again became omnivorous, feeding on an expanded number of plant and animal species. In particular, vegetable matter was made more palatable by grinding, cooking, and

Figure 1.13 A tick (Acari: Ixodidae) on human skin. Ticks, mites, fleas, flies, and worms have parasitized humans since prehistoric times. (Courtesy USDA)

fermenting. These changes gave rise to the **technology of food production**—the cultivation of plants and the domestication of animals.

As agriculture developed, the landscape was remodeled significantly when humans caused some animals and plants to multiply and others to be displaced. The result was local areas with reduced biological diversity and, as might be expected, a greater confrontation of humans with insects. Now, not only were humans to contend with insects feeding on their bodies and transmitting diseases, they also had to concern themselves with insects competing for a desired resource, their crops. Subsequently, as agriculture developed and greater areas of land were utilized for farming, pressure from insect populations (and other pests like weeds) increased disproportionately, making pest control (particularly weed control) a major preoccupation.

In addition to the competition with insects fostered by agricultural technology came an added intensity of parasite problems. For instance, in western Africa slash-and-burn methods of cultivation created more intensive epidemics of malaria. This phenomenon occurred when clearings were made for cultivation that in turn provided additional breeding habitats (more standing water) for the mosquito, *Anopheles gambiae*, a transmitter of malarial plasmodia. This mosquito displaced other mosquito species that fed on animals, and it became a menace to every human who ventured into the new forest clearings.

Problems with parasitic insects also became more prevalent with urbanization, the congregation of people to form cities that was the direct result of agriculture, when one person began to supply more food than the family could consume. This urbanization created greater opportunities for insects to be transferred from one person to the next, as with lice and fleas, as well as from one household to the next, as with cockroaches and bed bugs.

So it was that insects in recorded history have affected culture and life-style but not only as adversaries. Insects have figured positively in products they provided like honey and silk. Others have been central figures in religion, such as

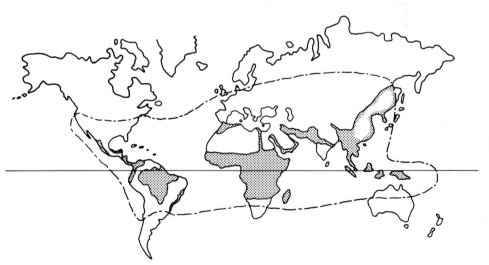

Figure 1.14 Map of the world showing the present distribution of significant numbers of cases of malaria (shaded) and estimated extent of its distribution in the past (dashed line). (Reprinted with permission of Macmillan Publishing Company from *Insects in Perspective* by Michael D. Atkins. Copyright ©1978 by Michael D. Atkins)

the scarab beetle in ancient Egypt and butterflies which migrate to the sacred mountain, Samanala Kanda, in Sri Lanka. Insects also have been used in ornamental jewelry and wall decorations of many kinds, as well as in predicting the severity of winter (according to the width of a woollybear's band).

Modern times. Today, human relationships with insects are the sum of those that have been acquired since our existence together began. This accumulation of relationships resulted primarily from our own activities to tame nature. In the history of this relationship, there seems to be a correlation with technology and insect encounters. Because technology is an overpowering force today, we can expect to see continued modifications in the natural landscape. Consequently, it seems likely that new associations and perhaps confrontations with insects are in the offing.

The Ledger

In analyzing the history of human and insect relationships, it might be tempting to believe that insects are our enemies incarnate and that their elimination would result in a better world. But insects have redeeming values both for human existence and the overall ecology of our planet. In an analysis, therefore, it may be advantageous for us to picture associations as profits and losses, as would a business enterprise. To do this, we might prepare a ledger showing black ink, the profits, and red ink, the losses, then form a balance sheet to arrive at our conclusions about insects.

Black ink: the benefits. Insects can be considered of benefit to humans by providing a product desired for human consumption, (**a primary resource**), or by interacting with elements of our environment to yield a benefit (**an intermediate resource**). Probably the most valued primary resources insects provide today are **silk, honey, wax**, and their **bodies** for human consumption and experimentation. Among the intermediate resources are insect activities as **pollinators, natural enemies of pests, food for wildlife**, and **scavengers**.

Silk, the queen of fabrics, is woven from thread of fine strands secreted from the salivary glands of silkworms, caterpillars of the moth *Bombyx mori*. The prop-

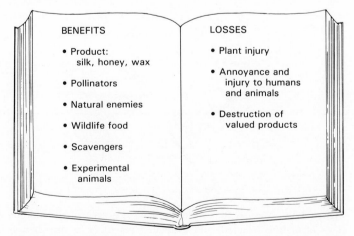

Figure 1.15 The ledger of insect relationships with humans, indicating benefits and losses.

agation of silkworms for production of this luxurious fabric is termed **sericulture** and is believed to have been developed in China as far back as 2500 B.C. With over 45,000 metric tons of silk produced on an annual basis in Asia and Europe, it has been estimated that at least 2,000 billion silkworms are reared on more than 90,000,000 tons of mulberry leaves annually! These statistics and the price of silk on the world market argue that the silkworm is one of our most significant insect species and is truly a valued resource.

Honey and wax, of course, are produced by the honeybee, *Apis mellifera*. This species is probably even more important than the silkworm. Wild honeybee hives are believed to have been raided by primitive humans as long ago as 7000 B.C., and management of the species to obtain honey and beeswax, **beekeeping** (or **apiculture**), has developed into a major agricultural industry in much of the world. Honey is produced by **worker bees** which feed on minute quantities of nectar from flower blossoms. The nectar is mixed with a bee's saliva, and this mixture, after reacting with certain enzymes, is regurgitated into individual hexagonal wax cells in the hive. These cells remain open until much of the water in the mixture has evaporated, and the cells with cured honey are sealed with a thin capping of wax by the workers. This stored honey serves as food for young and developing bees. An important by-product of extracting honey from a hive is beeswax, which is melted down and sold for use in furniture polish, candles, car wax, cosmetics, and other products.

The remaining products of insects, although small by comparison with silk and honey, have nonetheless been important in certain instances. They include raw material from lac scales (*Lacifer lacca*) used in making shellac and the coloring of other scale insects for making red and purple dyes.

As mentioned earlier, whole insects, in addition to their products, have been eaten by humans since the beginning of human life. Today, insects continue to serve as food directly, and edible insects are a source of protein and fat in otherwise deficient diets of people in many parts of the world. Among the edible insects, termites and grasshoppers are probably the most widely consumed, with caloric value sometimes exceeding 500 calories/100 insect grams. Europeans and their descendants are the only major group of people not consuming insects in any appreciable amount. This may be surprising when considering that the Greeks and Romans, who laid the foundations of modern European culture, ate a wide variety of insects and reared some as specialty foods.

Lastly as a primary resource, insects play an important role today as experimental animals. Many scientists use these six-legged guinea pigs to test theories and solve biological problems. For laboratory work, insects are small, require little space, and many, such as large cockroaches, demand little care and maintenance. Indeed, insects used as experimental animals have been indispensable in such fields as genetics, toxicology, and neurobiology, having allowed advances that would have been prohibitively costly or otherwise impossible. The science of genetics, in particular, has relied on insects, and probably no other species has been more important than the fruit, or pomace fly, *Drosophila melanogaster*, in unraveling the complexities of inheritance. In applied ecology, scientists have relied on the presence and absence of certain insect species as indicators of pollution. This is possible with insects like immature mayflies, which are particularly sensitive to chemical changes in water.

Of the ways insects serve as intermediate resources, probably none is more important than their plant pollination activity. Many groups of insects, for exam-

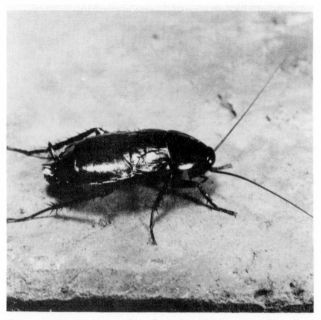

Figure 1.16 An oriental cockroach, *Blatta orientalis*. Such large cockroaches are often used as experimental subjects for laboratory research. (Courtesy USDA)

ple, butterflies, bees, and flies, inadvertently pollinate plants as they feed on pollen and nectar of flowers. Moving from flower to flower to collect nectar, insects like bumble bees come into contact with pollen grains that adhere to their legs and body hair. With each visitation, some grains are left on the sticky female stigmata, thereby cross-fertilizing the plant. In the process, other pollen grains are picked up that subsequently are transferred to yet other plants. Such relationships are believed to have played an important role in the evolution of both insects and plants and have been used to account for some of the specialized forms, colors, and fragrances in flowers and behavior patterns of insects. Although some of our important crops like wheat, corn and other cereals are wind-pollinated, most fruits and vegetable crops are pollinated by insects. Therefore, berries, pears, apples, citrus, melons, peas, beans, and tomatoes are but a few of our important crops that would not exist in their present form, if at all, without insects. In fact, species like alkali bees, *Nomia melanderi*, and leafcutter bees, *Megachile rotundata*, are managed so as to improve alfalfa pollination and, consequently, seed production in the western United States and Canada. These wild bees, particularly leafcutter bees, are cultured by individual alfalfa seed producers and provided through mobile field nesting sites. Larger and more successful producers may produce bee populations of over 30 million females!

Equally important to human welfare is the activity of some insects that kill and eat (or vice versa) other kinds of insects. These insects, known as natural enemies, function as predators or parasites in reducing numbers of harmful insects. For example, ladybird beetles feed voraciously on many species of aphids, sometimes averting the need for any pest management action. Parasites such as tiny *Trichogramma* wasps lay eggs inside the eggs of many moth pests, producing larvae that consume pest-egg contents and sometimes prevent pest populations

Figure 1.17 A whitelined sphinx moth feeding on a flower. Insect feeding on flower nectar is an important means of transferring pollen from flower to flower. (Photo by J. Mertins)

Figure 1.18 A *Trichogramma* wasp (Hymenoptera: Trichogrammatidae) laying an egg inside a moth egg. (Courtesy USDA)

from developing. Although the exact magnitude of natural enemy effects on all insect pests and potential pests cannot be known, it is believed momentous, because probably 30 to 40 percent of all insect species are natural enemies of other insects species.

Yet another activity of insects that benefits humans is scavenging. When animals and plants die, the organic matter that makes up their bodies possesses energy in the form of chemical bonds. That energy is released via a step-by-step breakdown of body tissue constituents. Insects, in feeding on dead plant and animal tissue, often carry out the first stage in this decomposition process by predisposing matter for enhanced decay and ultimate breakdown by microorganisms. Some prominent examples of insect decomposers include termites that break down wood, springtails that assist in the decomposition of dead leaves, and carrion beetles and many fly maggots that feed on dead animals.

The role of insects as food for wildlife and, in some instances, for domesticated animals is not to be overlooked. Insects are also consumed by an array of fish, amphibians, reptiles, birds, and small mammals, in addition to other insects. Indeed, insects are the major source of food for many small-bodied vertebrates, and effective colonization of certain areas by these animals depends on the insect food source. For example, mayflies are thought to be the single most important dietary component of trout, and the skilled angler would not expect good fishing without these insects. In other instances, insects are reared for sale as fish bait and pet food and are sometimes collected during mass outbreaks for swine and poultry feed, for example, crickets and locusts.

Red ink: the losses. Although insects are usually thought of as harmful to humans, probably fewer than 1 percent of all insect species fall into the pest category. From this group probably 3,500 or so species require regular attention, and in the United States, as few as 600 species can be considered significant.

Nevertheless, these species we call pests have imposed and continue to impose burdens on human populations by causing

1. Injury to crop plants, forests, and ornamentals
2. Annoyance, injury, and death to humans and domesticated animals
3. Destruction or value depreciation of stored products and possessions

Injury to crop plants, forests, and ornamentals is one of the most conspicuous ways insects cause economic losses to humans. Because approximately half of all known insects are plant feeders, as might be expected, our greatest number of insect problems involve plants.

Insects injure plants by feeding on them and laying eggs in plant tissues. By far, the more important of these injuries is feeding. Insects consume plant tissues with various types of chewing mouthparts and remove plant juices with piercing-sucking mouthparts. In the process of feeding, some insects, for example, many aphids and leafhoppers, also transmit (**vector**) hundreds of kinds of plant pathogens, including bacteria, fungi, viruses, and mycoplasmas that subsequently cause losses from diseases. Other insects such as the potato leafhopper, *Empoasca fabae*, inject toxins that influence the plant's physiology, resulting in yield and quality changes.

Losses from insect feeding may be **direct** or **indirect**. Direct losses occur when insects feed on potentially harvestable produce, thereby eliminating it or causing devaluation. Significant examples of this injury include that of the boll weevil, *Anthonomus grandis*, which destroys cotton squares and bolls, the codling moth, *Cydia pomonella*, which feeds inside apples, and bark beetles, *Scolytus* species and others, which bore into bark and cambium of trees. Direct losses also may occur merely from the presence of insects on produce, causing value depreciation (**dockage**), for example, oystershell scales, *Lepidosaphes ulmi*, on the rind of grapefruit. Indirect losses arise when insects feed on roots, stems, and leaves, thereby causing a decline in quantity or quality of potentially harvestable produce, usually seeds, fruits, or tubers. However, because of the recuperative potential of most plants, large populations of indirect feeders may be required to cause a significant loss. Examples of these pests include the Colorado potato beetle, *Leptinotarsa decemlineata*, consuming leaf tissue of potatoes; northern corn rootworms, *Diabrotica barberi*, eating roots of corn; and Hessian fly, *Mayetiola destructor*, boring in wheat stems.

Annoyance and injury to humans and domesticated animals is an equally important way that insects affect humans. However, probably fewer insect species are involved in causing these kinds of problems than are involved with plants. Insects of medical importance may affect humans directly or indirectly. Directly, a wide variety of species are parasitic on humans, primarily feeding on blood, for example, mosquitoes, lice, fleas, and biting flies; others cause annoyance and discomfort simply by their presence. Some insects and other arthropods inject venoms when they bite or sting, often causing a local reaction to the wound, but sometimes causing death. Although rare, death from envenomation occurs most frequently from bee stings to persons hypersensitive to the venom.

Insects as medical pests are much more important in an indirect way, as vectors of human pathogens. Here, some insects, ticks, and mites harbor nematodes, rickettsiae, viruses, protozoa, and bacteria. In some instances, part of the microorganism's life cycle, often including multiplication, occurs in the insect body. Among the most important diseases transmitted by insects bites are **malaria, yellow fever, filariasis**, and several types of **encephalitis**. The causes are malaria from a protozoan, filariasis from filarial worms, and encephalitides (plural of encephalitis) from viruses called **arboviruses** (an abbreviation of the term arthropod-borne viruses). Of these diseases, malaria is the most significant today, especially in the tropics, where it causes hundreds of thousands of deaths each year. Although diseases transmitted by insects are not presently of major importance in North America and Europe, arthropod-borne diseases remain some of the greatest medical problems in most of the developing nations.

Insects affect domesticated animals similarly to the way they affect humans. In most instances today, insect parasites of livestock and poultry cause monetary losses by weakening animals, rather than killing them. Animals in a weakened or agitated state eat less and, in turn, gain less weight. Some of the most important pests of livestock are the true flies, including the stable fly, *Stomoxys calcitrans*; horse fly, *Tabanus* species; horn fly, *Haematobia irritans*; and face fly, *Musca autumnalis*. Another fly, the screwworm, *Cochliomyia hominivorax*, found in the southern United States, Mexico, and Central and South America, is a serious livestock pest that is now largely under control in the United States and Mexico. Because of the spectacular success of a sterile-fly release program, this fly problem was eliminated from Florida, Texas, and other western states, where millions of dollars in losses are being averted. Chewing lice are probably the most important insect pests of poultry, with heavy infestations causing skin irritation that results in reduced weight gain and egg production.

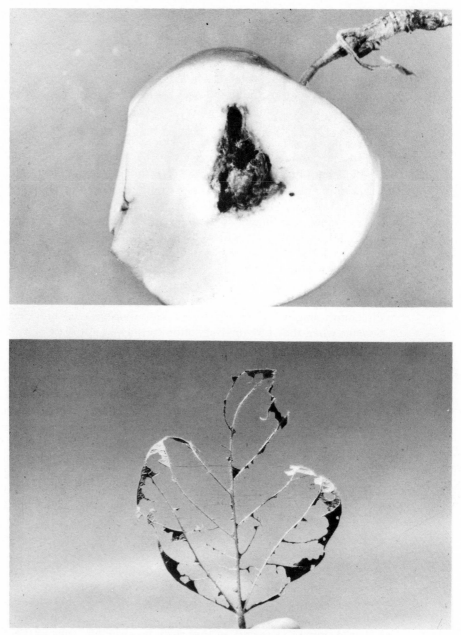

Figure 1.19 Direct injury, represented by codling moth, *Cydia pomonella*, larval feeding in apple (top) and indirect injury, represented by green cloverworm, *Plathypena scabra*, larval feeding on soybean leaflet. (Top photo by D. Lewis, Iowa State University Extension Service)

Figure 1.20 Leg Elephantiasis caused by infection with filarial nematodes. These nematodes are transmitted by biting mosquitoes, and enlargement of the legs is caused by blockage of lymph drainage. Arthropod-borne diseases remain some of the most important medical problems in many developing nations. (Courtesy W. Rowley, Iowa State University)

Figure 1.21 Gulf Coast tick, *Amblyomma maculatum*, on ear of cow. Such arthropod ectoparasites cause significant livestock and poultry losses by weakening and sometimes killing the host. (Courtesy W. Rowley, Iowa State University)

Destruction or value depreciation of stored products and possessions is a category whose importance is sometimes understated. However, losses can amount to millions of dollars annually even in countries like the United States, where storage facilities are good and control technology readily available. Indeed, in some instances losses to food in storage may exceed those inflicted by insects before crops are harvested.

Most of the important insect pests of stored grains and processed foods in North America are beetles (for example, granary weevil, *Sitophilus granarius*; and red flour beetle, *Tribolium castaneum*) and moths (for example, Angoumois grain moth, *Sitotroga cerealella*; and Indianmeal moth, *Plodia interpunctella*). Meat and cheese in storage may also be attacked by the larder beetle, *Dermestes lardarius*, and cheese skipper, *Piophila casei*.

In addition to stored food pests, other insects affect many household possessions. Clothes moths and carpet beetles eat anything containing animal fibers, including woolens, upholstery, and carpet. Other insects like cockroaches spot and taint food and living areas. Termites are another frequent household problem in most localities, consuming wood on foundations, in floors, and in walls that often require expensive repairs.

The balance sheet. When we weigh the good and the bad of it, would the world really be better for humans without insects? This question could arouse a debate whose outcome probably is less important than our having argued the question. The fact is, insects are part of our everyday lives and are here to stay. Perhaps having considered the question, our perception may have changed, if ever so slightly, toward a greater understanding of nature. More understanding can result in a greater appreciation and enjoyment of the things we like and greater tolerance of the things and events we dislike.

Figure 1.22 Granary weevil, *Sitophilus granarius*, infesting stored wheat. Beetles are some of the most important pests of stored grain in North America. (Courtesy USDA)

Regrettably, many people believe all insects are bad. Indeed, we might say that insects suffer from poor public relations. This view was expressed by J. W. Mertins in a review of insects and other arthropods that have been used as subject matter in moviemaking. In films, arthropods usually represent threatening, sinister figures or images of danger or death. They are also shown as distasteful or shocking images for the squeamish or images of silliness for ridicule by "rational" persons. In a similar vein, Mertins found that entomologists are usually shown as "detached from reality, as eccentric buffoons, often as psychotics, or, at best, ineffectual dupes." He found no instances where popular movies present the scientific study of insects in a positive light. Such movies may not only tend to reflect public opinion but also may help to form and direct it. Unfortunately, commercial films in countries like the United States have portrayed insects mostly in an inaccurate, unflattering way, and this same medium serves to educate masses of impressionable young people.

It is not only in the movies the young are taught an aversion to insects. It also occurs in the home environment when curious youngsters are scolded into throwing away a caterpillar tucked in the hand or are encouraged to step on a beetle running across the sidewalk. Unfounded admonitions such as, "It will bite you, get rid of it" and "Kill it before it gets away" serve to create attitudes of both **fear** and **confrontation** that are passed along from generation to generation.

Figure 1.23 In the film *Mothra*, (1962), Japan is subjected to the catastrophic backwash from a gigantic moth. Most commercial films have portrayed insects in an inaccurate, unflattering way. (From *A Pictorial History of Science Fiction Films* by Jeff Rovin, published by arrangement with Lyle Stuart)

The ultimate development of fear of insects can be seen in certain psychological disturbances found in a small number of people. An irrational and lasting fear of real insects, resulting in unconventional behavior, is known as **entomophobia**. Another somewhat different fear is expressed by persons suffering from **delusory parasitosis**. Here, sufferers have hallucinations of "bugs" on their bodies. In either instance, professional help is usually required to overcome the affliction.

A strong confrontational attitude and technological ability to effectively kill insects has given rise to the **only-good-bug-is-a-dead-bug** or **if-it-moves-kill-it syndrome**. Such feelings have created the fetish of an insect-free environment that exists not only around the home (for example, individuals spraying chemicals in the yard to eliminate mosquitoes and fleas), but also in agriculture and forest production. Although some improvements in attitude have been made, this fetish continues to be the source of certain environmental problems such as those with water quality.

Although maligned in general, a few insects have a relatively good image. These include the producers of useful materials like bees and silkworms and famous natural enemies of pests, particularly ladybird beetles. Anthropomorphic images of industry, for example, "busy as a bee," and cooperation, for example, "the farmer's friend," understandably are ascribed to these insects.

Other insects, however, have good images as objects of beauty. This image is true of colorful butterflies, which are found in collections among other beautiful objects. Today, so-called "butterfly gardens," plantings that attract a variety of butterflies, are becoming increasingly popular. But why do we find butterflies beautiful and sometimes virtuous—even symbolic of the human spirit (*psyche*, from Greek mythology, a young girl with the wings of a butterfly)? Is it their color? Other insects are also beautiful yet are considered repugnant by some

Figure 1.24 A ladybird beetle larva (Coleoptera: Coccinellidae) feeding on aphids (Homoptera: Aphididae). A good image is usually ascribed to such insects. (Courtesy USDA)

Figure 1.25 Most children are not inherently fearful of insects and other arthropods. They are taught fear and disgust by adults. (Photo by P. Fleury, with permission)

people. Is it the seemingly warm appearance of butterflies—a delusion created by their furlike covering of scales?

Clearly, our culture and environment could be improved with a general change in attitude about insects. If we viewed other insect species as we do butterflies, we could not believe the world would be better off without insects. With such an outlook, we might accept that the benefits of insects far outweigh the losses caused by them, and our fetish of creating insect-free environments would be abolished, along with significant quantities of environmentally harmful chemicals.

The key to this change in attitude is knowledge and understanding about insects and the important part insects play in the ecology of our planet. From past mistakes, it would seem that the most advantageous attitude should be to live in harmony as much as possible, with all elements of nature, including insects. Just how such an attitude can be promoted is the basis for insect pest management, the sensible approach to insect problems.

INSECT PEST MANAGEMENT

A great many approaches to alleviate insect pest problems have been attempted through the years. Recently, no approach has been more popular than **pest management**, also called **integrated pest management**. Pest management is a general approach to dealing with all kinds of pests including insects, mites and other arthropods, plant-parasitic nematodes, microbial and viral plant pathogens, weeds, and vertebrates. The primary objective of the approach is to reduce losses from pests in ways that are effective, economically sound, and ecologically compatible.

Pest management sometimes is equated with **plant protection**, an approach that

Figure 1.26   Pest management is a general approach to dealing with all kinds of pests, and many sources of information are available on this topic. (Photo by D. Lewis, Iowa State University Extention Service)

focuses more on the object of value rather than on the pest. However, the scope of pest management is broader, because in addition to crop plants, it addresses pest problems with livestock, urban dwellings, and landscape, and includes certain aspects of human health, for example, mosquitoes and other vectors of human disease.

Pest management also is characterized by the way several techniques are employed, simultaneously, to solve specific pest problems. Therefore, a single pest management program may involve uses of pesticides, host-plant resistance, tillage, sanitation, and biological control. To use these tools effectively, the approach relies on an intimate understanding of ecology, pest biology, and the appropriate integration of information. An important goal of pest management programs is to arrive at long-term solutions to problems, rather than to simply derive short-term protection.

Insect pest management, the primary topic of this book, is a division of integrated pest management that emphasizes insects and other arthropods. It seeks to reduce the status of pests by following principles of ecology and using the latest advancements in technology. Much of the foundation of insect pest management rests on determining whether an insect is truly a pest and, if so, just how serious a problem it is causing—in other words, determining its status.

The Concept of Pest

Pest species are those that interfere with human activities. According to Australian entomologist P. W. Geier and his colleagues, the quality of being a pest is **anthropocentric** (considering humans as the central fact or final aim and end of a system) and **circumstantial**. Termites feeding on dead wood in a forest serve an important ecological function, one of degradation in the process of returning nutrients to the soil. Clearly, termites are not pests in this context; they are beneficial to humankind. The same species, performing the same ecological function in the environment of a human home, is an important pest. An understanding of the often significant ecological roles played by "pest" species in unmanaged environments, as well as agricultural ones, frequently gives insights into how to deal with them. Moreover, this understanding can aid in developing more tolerant attitudes toward pest presence.

Probably, our greatest share of insect pest problems are encountered in agricultural and forest production systems. Here, problems arise because of **large numbers** of insects, not simply because of the presence of a species. Most insect pests of agriculture (including agronomy, animal science, horticulture, and forestry) are those species whose activities, enhanced by population numbers, cause economic losses.

Other pests include those whose mere **presence** is objectionable. We might refer to these as **aesthetic pests**. Economic losses based on aesthetics are difficult to determine but are nonetheless real. In instances where insects enter the household to spend the winter, for example, boxelder bugs (*Leptocoris trivittatus*), some persons may be motivated to spend money on insecticidal control and therefore suffer an economic loss. Others may not be as bothered by insect presence. Presence of insects or insect parts in food, however, is another matter, being regulated by law. Insect presence in food processing and storage requires costly sanitation measures and, often, the disposal of tainted products. Such losses can be measured objectively and are significant.

Another group of insects can be termed **medical pests**. Losses caused by medical pests also are very significant but, like some aesthetic pests, are extremely difficult to measure. Loss of work or work efficiency may be measurable in

Figure 1.27 Boxelder bug, *Leptocoris trivittatus*. Such insects are sometimes considered aesthetic pests when they enter homes to overwinter. They do no harm there, but their mere presence is objectionable to many people. (Courtesy USDA)

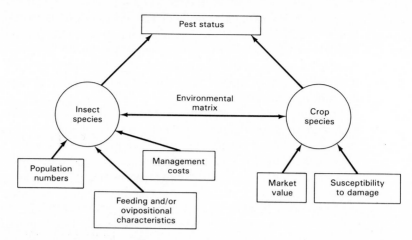

Figure 1.28 Diagrammatic representation of the major factors contributing to pest status.

economic terms, but how can economic loss be assigned to discomfort, pain, and even loss of life? Such problems make management of medical pests and some aesthetic pests quite difficult.

The Concept of Pest Status

Pest status concerns the ranking of a pest relative to the economics of dealing with the species. This aspect is variable for a pest and depends on many factors. With agricultural pests, status depends greatly on the crop involved and the environment in which the pest-crop interaction occurs.

The major factors contributing to pest status are shown in Figure 1.28. Among these, **market value** is one of the most variable. With all other factors held constant, a pest can assume higher or lower levels of importance because of changing economics. **Susceptibility of the crop to pest injury** is another important, but often little understood, factor. Variable weather factors (particularly moisture) and cultural procedures (such as fertilization) can profoundly affect crop vigor. Under optimal conditions, pest status can be lowered because of compensatory growth compared with adverse growing conditions.

But it is the environment, including the human social environment, that is the basis of change in the factors governing pest status. Factors of the environment mediate the importance of pests and determine the measure and constraints of insect pest management programs.

In the following chapters, we attempt to understand the approach of pest management to the solution of insect problems. To do this, we first need to learn about insect biology and life processes, then proceed to understand modern protection technology and the optimal application of this knowledge.

Selected References

ATKINS, M. D. 1978. *Insects in Perspective*. New York: Macmillan, chapters 18–22.

EISNER, T., and E. WILSON. 1977. General introduction: the conquerors of the land, pp. 2–15. In T. Eisner and E. Wilson, eds., *The Insects. Readings from Scientific American*. San Francisco: Freeman.

GEIER, P. W., L. R. CLARK, and D. T. BRIESE. 1983. Principles for the control of arthropod pests, I. Elements and functions involved in pest control. *Protection Ecology* 5:1–96.

GREENBURG, B. 1985. Forensic entomology: case studies. *Bulletin Entomological Society America* 31:25–28.

KAESTNER, A. 1968. *Invertebrate Zoology.* Translated by H. W. Levi and L. R. Levi, vol. 2. New York: Wiley.

MCNEILL, W. H. 1976. *Plagues and Peoples.* Garden City, N.Y. : Anchor Press/Doubleday, chapters 1–2.

MERTINS, J. W. 1986. Arthropods on the screen. *Bulletin Entomological Society America* 32:85–90.

PEDIGO, L. P. 1985. Integrated pest management. In *McGraw-Hill Yearbook of Science and Technology.* New York: McGraw-Hill, pp. 22–31.

SKELETON, T. E. 1976. Insects and human welfare. *American Biology Teacher* 38:208–210.

SOUTHWOOD, T. R. E. 1977. Entomology and mankind. *American Scientist* 65:30–39.

2 INSECT STRUCTURES AND LIFE PROCESSES

BY ANY MEASURE, the success of organisms is due to the particular biological features they possess. As conquerors of the land, insects have evolved an assemblage of special features that have insured success. These features have allowed the animals to overcome an array of inhospitable environments through effective solutions to the problems of body maintenance, reproduction, and dispersal. In large part, these important adaptations are derived from an amazing body comprising specialized structures that function in harmony with one another and with their external environment.

Consequently, our understanding of insects in nature and our explanation of them as pests depends on a knowledge of their biological features. As we shall see, pest management relies heavily on predicting pest behavior, and basic to predicting behavior is understanding insect structures and their functions.

THE INSECT BODY

General Organization

Tagmosis and the body wall. Insects have segmented bodies, as had their wormlike ancestors. Some authorities believe that as many as 20 segments existed in these ancestors. Today's insects show a reduction in this number, with certain segments fusing to form three usually well-defined regions: head, **thorax**, and abdomen. The grouping of segments into functional regions is known as **tagmosis**.

The head shows few signs of segmentation in the adult stage but is believed to have been formed from the fusion of a primitive head and several trunk segments. The thorax consists of three distinctly modified segments, each bearing a pair of segmented legs. Least modified of the body regions is the abdomen, which may have as many as eleven segments.

These insect structures are formed from the body wall, which also supplies a support system known as the **exoskeleton**. The exoskeleton provides a rigid foundation for the body and serves for the attachment of muscles, something like the internal skeleton of vertebrates. The flexible exoskeleton is similar to a spring, but it does not stretch (except at certain times during growth). In addition to its support function, the exoskeleton serves as a covering to protect internal organs, and it helps prevent desiccation.

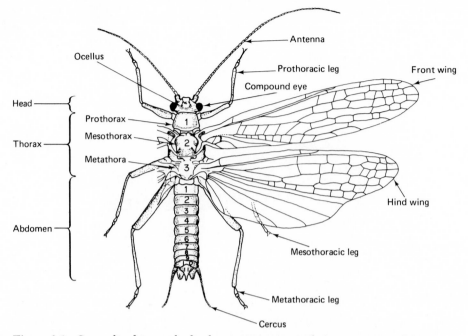

Figure 2.1 Generalized insect body showing tagmosis and major anatomical features. (Reprinted with permission of Macmillan Publishing Company from *Insects in Perspective* by Michael D. Atkins. Copyright ©1978 by Michael D. Atkins)

Certain parts of the body wall are hardened, or **sclerotized**, in the form of plates called **sclerites**. Between these plates, the body wall is membranous, or soft, which allows movement and expansion of the body during eating and egg development. Other types of membranous connections found in insect bodies and appendages include telescoping ring segments, simple appendage joints, and ball-and-socket articulated joints.

In addition to sclerites and membranes, the body wall is characterized by other external and internal features. Externally, these features include wrinkles, spurs, scales, spines, and hairs. Internally, inpushings of the body wall called **apodemes** are prominent in the head and thorax. These strengthen the exoskeleton and provide areas for muscle attachment.

Detail of the body wall. The body wall, or **integument**, of insects is composed of a single layer of cells called the **epidermis** that is bounded on the inside by the **basement membrane** and on the outside by the **cuticle**. Among the epidermal cells are scattered wax-secreting cells and other cells that secrete molting fluid involved in the growth process. The cuticle covers the entire outer body surface and also lines the insect's air tubes, salivary glands, and parts of the digestive tract. It is produced by the epidermis.

The cuticle is made up of three primary layers: a thin outer **epicuticle**, a thicker **exocuticle**, and a still thicker **endocuticle**. The epicuticle itself is divided into four layers: an innermost **homogeneous** layer that is overlain, in succession, by layers of **cuticulin** (a lipoprotein), **wax** (of exceptionally long hydrocarbon chains), and, in most instances, **cement**, a tanned protein substance that serves as a varnishlike covering. The cuticulin layer is of utmost importance in the growth process, for it is permeable to the important chemicals and nutrients associated with growth,

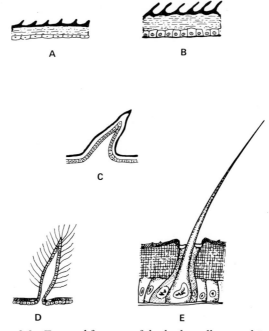

Figure 2.2 External features of the body wall. A. and B. Noncellular surface configurations include ridges and spurs. C., D., and E. Cellular structures include setae and multicellular spines. (Reprinted with permission of Macmillan Publishing Company from *Insects in Perspective* by Michael D. Atkins. Copyright ©1978 by Michael D. Atkins)

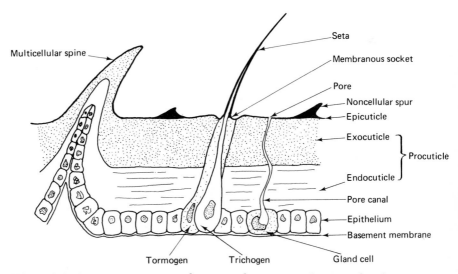

Figure 2.3 Insect integument and associated structures. (Reprinted with permission of Macmillan Publishing Company from *Insects in Perspective* by Michael D. Atkins. Copyright ©1978 by Michael D. Atkins)

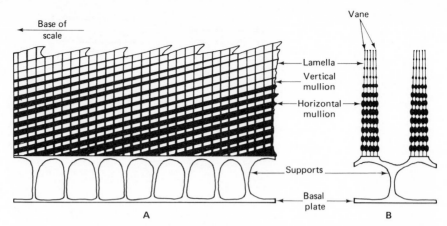

Figure 2.4 Representation of the multilayered structure of a *Morpho* butterfly (Lepidoptera) wing scale. Such structured scales produce beautiful iridescent colors and are examples of interference coloration. (Reprinted with permission of Macmillan Publishing Company from *Insects in Perspective* by Michael D. Atkins. Copyright ©1978 by Michael D. Atkins)

yet it is impermeable to enzymes that break down parts of the old cuticle before it is shed. There are numerous vertical channels through the cuticle for sensory nerve fibers and ducts for wax glands. Helical **pore canals** that transport materials to the body surface are also found in the cuticle of many insects.

The exocuticle gives the cuticle its characteristic strength and resilience. It is formed of **chitin**, a polymerized compound (nitrogenous polysaccharide linked to a protein) found not just in insects, but in other arthropods as well. The molecules of polymerized compounds are linked in long chains; they are common in nature as a base for materials like wood, hair, and horn.

Other constituents of the cuticle include **quinones**. Some quinones, derived mainly from the blood, are incorporated directly into the cuticle, and some are polymerized to form dark brown or black pigments known as **melanin**. Yet other quinones are utilized in forming **sclerotin**, which gives the exoskeleton additional hardness and strength, as found in beetles.

Although the source of color is in the cuticle of most black and dark brown insects, many brilliant colors arise from pigments in the epidermis or fat cells beneath a transparent cuticle. Yet other colors are produced by the diffraction of light waves, giving brilliance and iridescence. Diffraction gratings may be formed by rows of closely set grooves in the surface of the cuticle, and interference colors are produced with alternating layers of the cuticle with different refractive indices. The exceptionally beautiful iridescent blue wings of *Morpho* butterflies are excellent examples of interference colors.

The Head

The head of insects is a structure specialized for feeding and for sensing and integrating information about the environment. It is made up mostly of a **cranium**, which bears mouthparts, **antennae**, and eyes.

Cranium. The insect cranium is a hardened capsule with an opening leading to the mouth and to the thorax. It is attached to the thorax by way of a short neck or **cervix**. The orientation of the cranium varies among different insects, being positioned either vertically (mouthparts pointed down, hypognathus), horizontally

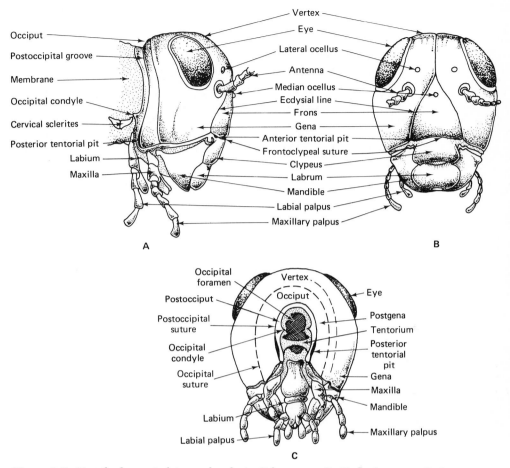

Figure 2.5 Detail of a typical insect head. A. Side aspect. B. Frontal aspect. C. Rear aspect. (Reprinted with permission of Macmillan Publishing Company from *Insects in Perspective* by Michael D. Atkins. Copyright ©1978 by Michael D. Atkins)

(mouthparts pointed forward, prognathous), or obliquely (mouthparts pointed backward, opisthognathus). Insects like ground beetles and grasshoppers with chewing mouthparts have heavy crania, adapted for muscles involved in capturing prey and biting off leaf tissue. Insects like leafhoppers and mosquitoes, which pierce and suck liquid food, often have much more delicate crania. Internally, the cranium is strengthened by a set of invaginations of the body wall fused to form the **tentorium**. Other pertinent details of the head are given in Figure 2.5.

Mouthparts. Mouthparts are some of the most distinctive features of insects, and their structure tells a great deal about the feeding habits of a species. Mouthparts are greatly varied, but in their primitive form, for example, in grasshoppers and beetles, they include a **labrum** (upper lip), a pair of chewing **mandibles** (jaws), a pair of **maxillae** (second jaws), and a **labium** (lower lip). These structures surround the mouth and form the pre-oral cavity. In addition, a central tongue-like **hypopharynx** drops from the membranous floor of the cranium, behind the mouth, and bears the opening of the salivary ducts. The interior, fleshy surface of the labrum, endowed with numerous sensory structures, is referred to as the **labrum-epipharynx**.

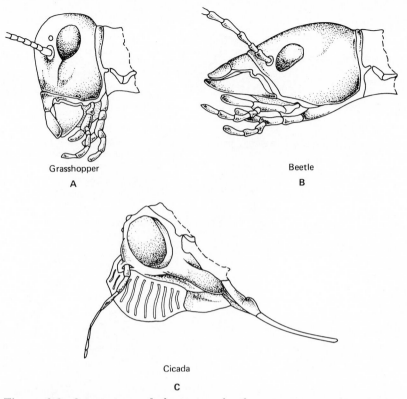

Grasshopper

A

Beetle

B

Cicada

C

Figure 2.6 Orientations of the insect head. A. Hypognathus (vertical), grasshopper (Orthoptera). B. Prognathus (horizontal), ground beetle (Coleoptera). C. Opisthognathus (oblique), cicada (Homoptera). (Reprinted with permission of Macmillan Publishing Company from *Insects in Perspective* by Michael D. Atkins. Copyright ©1978 by Michael D. Atkins)

This arrangement of mouthparts, called chewing mouthparts, enables the insect to bite off pieces of food with the cutting area of the mandible. Both mandibles and maxillae then orient and manipulate the food and, to a degree, the grinding area of the mandible masticates it. During this activity, saliva is applied, after which the food bolus is guided to the mouth opening with the aid of the maxillae, labium, and hypopharynx.

Several other types of mouthparts occur in insects whose modifications are believed to have evolved from the basic chewing type. Therefore, in addition to chewing, insects have types that include piercing-sucking, rasping-sucking, siphoning, sponging, cutting-sponging, and chewing-lapping.

Piercing-sucking mouthparts are very common among insects. They are modified to pierce the epidermis of plants or the skin of animals and to suck up sap or blood. In plant feeders like cicadas and aphids (Homoptera), the piercing-sucking needle is formed from four hairlike stylets fitted closely together. The outer stylets are derived from mandibles, and the inner stylets are derived from maxillae. The maxillae are double-grooved on the inner face and, when held together, form two channels; one is for the passage of saliva into the plant to facilitate food flow and digestion, and the other is for the uptake of plant juices. The labium forms a protective sheath for the stylets. Other plant feeders like leafhoppers, aphids,

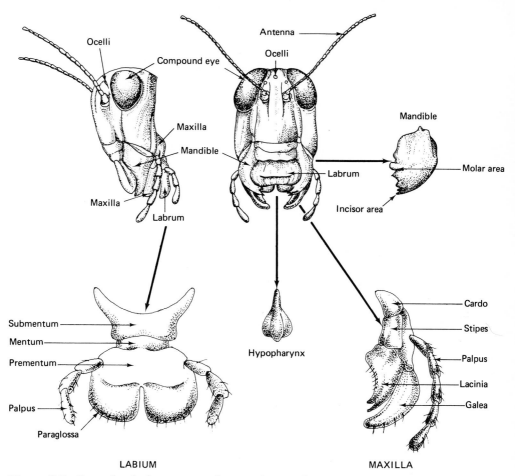

Figure 2.7 Generalized chewing mouthparts of a grasshopper (Orthoptera). (Reprinted with permission of Macmillan Publishing Company from *Insects in Perspective* by Michael D. Atkins. Copyright ©1978 by Michael D. Atkins)

and scales have the same basic mouthpart structure, and with many, disease-causing organisms like viruses are injected into the plant along with the saliva.

In blood-feeding insects like mosquitoes (Diptera: Culicidae), an elongate labium also forms a protective sheath for six, rather than four, stylets. Here, the mandible- and maxilla-formed stylets are accompanied by an additional pair, modified from the hypopharynx and labrum-epipharynx, which forms a food channel. The back of the food channel is closed by the hypopharynx with its salivary duct. This duct carries saliva with enzymes and anticoagulins that reduce blood clotting in the host and improve flow of blood into the mosquito. In feeding, the maxillary and mandibular stylets work together as a needle to penetrate the host's skin.

Rasping-sucking mouthparts are a primitive form of the piercing-sucking type and are found in thrips (Thysanoptera). Thrips mouthparts have a cone-shaped beak formed from the clypeus (a plate above the labrum in chewing insects), labrum, parts of the maxillae, and the labium. This beak contains the maxillae, hypopharynx, and the left mandible; together these structures form a stylet.

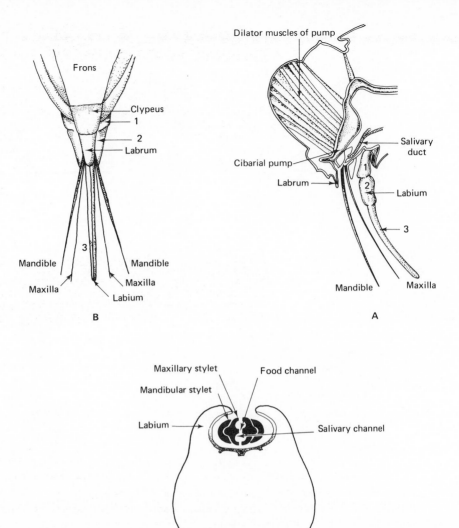

Figure 2.8 Piercing-sucking mouthparts of a cicada (Homoptera). A. Lateral cross sectional aspect. B. Frontal aspect. C. Section through mouthparts showing relative position of structures and channels. (Reprinted with permission of Macmillan Publishing Company from *Insects in Perspective* by Michael D. Atkins. Copyright ©1978 by Michael D. Atkins)

Thrips use the beak to rasp host tissues and take up liquid food through the stylet.

Siphoning mouthparts are a highly specialized type, modified for the uptake of flower nectar and other liquids by butterflies and moths (Lepidoptera). Here, a long **proboscis** composed of maxillary elements forms a tube through which food passes. The tube is held in a coiled-spring fashion when not in use.

Sponging mouthparts also are highly specialized structures in house flies (*Musca domestica*) and their close relatives. With this type, the mandibles and maxillae are nonfunctional, and the remaining parts form a proboscis with a fan-shaped sponge at the tip. Liquid food is "mopped up" by the capillary action of

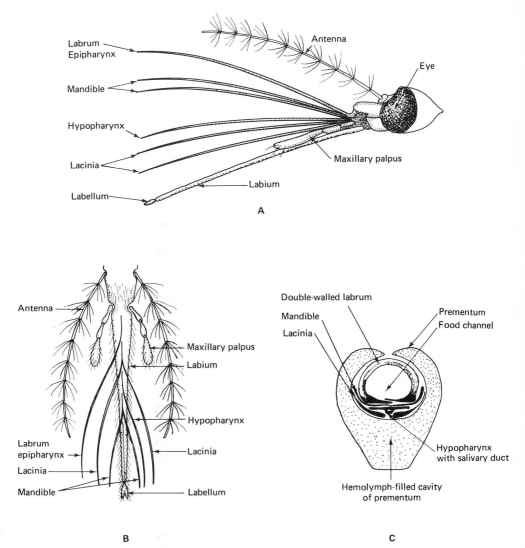

Figure 2.9 Piercing-sucking mouthparts of a mosquito (Diptera). A. Side aspect. B. Frontal aspect. C. Section through mouthparts showing relative position of structures and food channel. (Reprinted with permission of Macmillan Publishing Company from *Insects in Perspective* by Michael D. Atkins. Copyright ©1978 by Michael D. Atkins)

this sponge, and, if not liquid, salivary secretions through the mouthparts make it so. Cutting-sponging mouthparts in biting flies, for example, in horse flies (Tabanidae), are similar to the sponging type but have well-developed mandibles, forming sharp blades, and maxillae, forming probing stylets. Together, these structures cut and tear the skin of mammals, causing blood to flow. Subsequently, the blood is taken up by the sponge and conducted to the fly's gut.

Another mouthpart type used in eating liquid food is the chewing-lapping type in bees and wasps (Hymenoptera). With this type, the mandibles and labrum are similar to the chewing type and are used in molding wax, manipulating nest materials, and grasping prey. The maxillae and labium form a channeled proboscis through which saliva is discharged and nectar is drawn up.

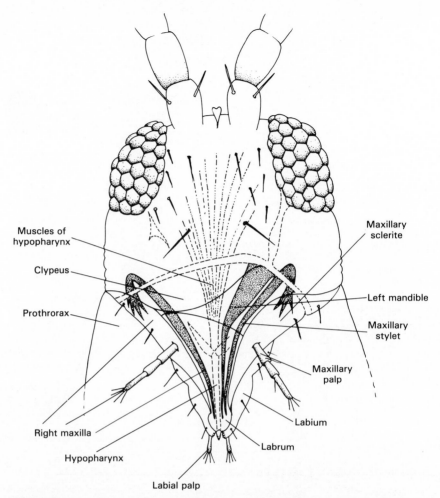

Figure 2.10 Rasping-sucking mouthparts of a thrips (Thysanoptera). (Redrawn from Metcalf, et al., 1962, *Destructive and Useful Insects*, McGraw-Hill Book Company)

Antennae. A pair of variously shaped appendages, called **antennae**, are found on the head of insects, below or between the eyes. These are movable and contain sensory structures that allow insects to detect odors, vibrations, and other environmental stimuli. Antennal shape is strikingly different among insect species and, often, between sexes of the same species. For this reason, antennae frequently are used in insect identification and sex determination. Anatomically, the antenna is divided into three parts, consisting of two basal segments, the **scape** and the **pedicel**, and a tip usually of several subsegments, the **flagellum**. Major antennal types include **filiform** (threadlike), **serrate** (sawlike), **moniliform** (beadlike), **clavate** (club-shaped), **capitate** (head-shaped), **lamellate** (platelike), **pectinate** (comb-shaped), and **plumose** (featherlike).

Eyes. Located on each side of the head of most adult insects are prominent **compound eyes** (see Figure 2.5), which consist of many hexagonal elements. These elements, known as **facets**, number from only a few to as many as 28,000 in dragonflies (Odonata). In addition to the compound eyes, most adult insects

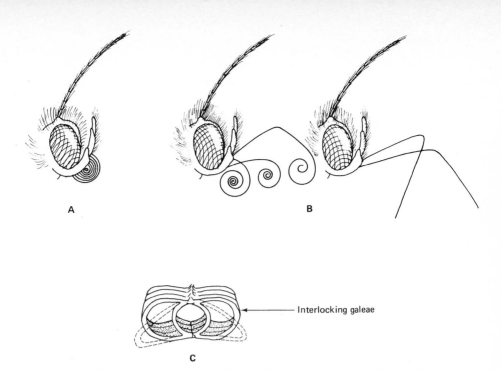

Figure 2.11 Siphoning mouthparts of a moth (Lepidoptera). A. The proboscis in its coiled position. B. Stages of uncoiling and bending the probiscis. C. Section through the proboscis showing interlocking maxillary elements, the galeae. Broken line shows changes in proboscis shape to produce uncoiling. (Reprinted with permission of Macmillan Publishing Company from *Insects in Perspective* by Michael D. Atkins. Copyright ©1978 by Michael D. Atkins)

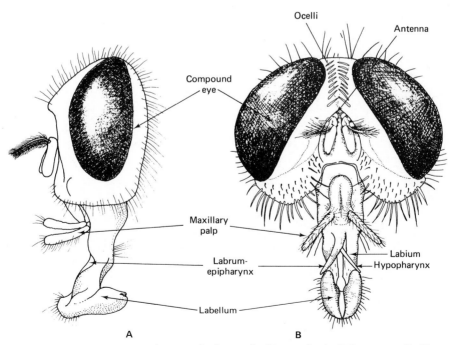

Figure 2.12 Sponging mouthparts of a house fly (Diptera). A. Side aspect. B. Frontal aspect. (Reprinted with permission of Macmillan Publishing Company from *Insects in Perspective* by Michael D. Atkins. Copyright ©1978 by Michael D. Atkins)

45

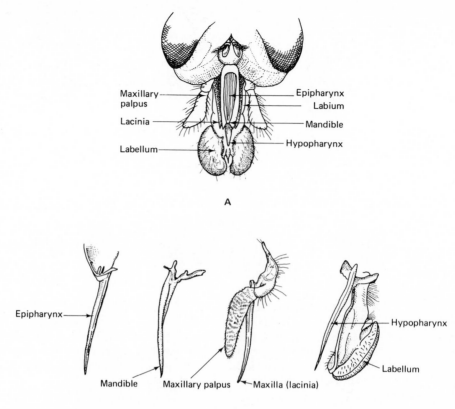

Figure 2.13 Cutting-sponging mouthparts of a horse fly (Diptera). A. Frontal aspect. B. Side aspect with structures separated from front (left) to back. (Reprinted with permission of Macmillan Publishing Company from *Insects in Perspective* by Michael D. Atkins. Copyright ©1978 by Michael D. Atkins)

also have several simple eyes, or **ocelli**. The ocelli are located between the compound eyes on the front of the head (see Figure 2.5). Most often, they are small and have a single lens. Many immature insects, for example, caterpillars (Lepidoptera), have only simple eyes, and other insects may have no eyes at all, for example, some springtails (Collembola).

The Thorax

The **thorax** is a rigid "box" of three distinct segments. From the head back, these segments are the **prothorax**, **mesothorax**, and **metathorax**. Each thoracic segment bears a pair of jointed legs, and in most adult insects, the mesothorax and metathorax each have a pair of wings.

Each thoracic segment is composed of hardened plates that give it rigidity. The upper plate is called the **notum**, and the lower plate is called the **sternum**. Side plates called **pleura** (sing. **pleuron**) also are present and are most well developed on the winged segments. Figure 2.17 shows the structural details of a mesothoracic segment.

Legs. Insects legs are articulated appendages comprising five segments. From the body outward, these segments are the **coxa**, **trochanter**, **femur**, **tibia**, and **tarsus**. The coxa fits into a cuplike depression in the body, allowing multidirec-

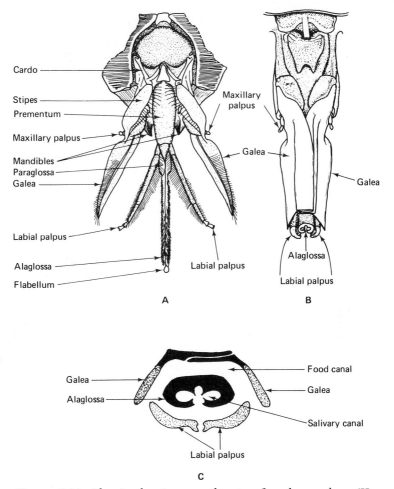

Figure 2.14 Chewing-lapping mouthparts of a honey bee (Hymenoptera). A. Rear aspect with structures spread apart. B. Frontal aspect with mouthparts folded together. C. Section through mouthparts showing relative position of structures and canals. (Reprinted with permission of Macmillan Publishing Company from *Insects in Perspective* by Michael D. Atkins. Copyright ©1978 by Michael D. Atkins)

tional movement. The trochanter connects the coxa with the femur, usually the largest leg segment. The tibia is often a long, slender structure with downward-pointing spines that aid in climbing. The tarsus, usually made up of several subsegments called **tarsomeres**, terminates in a pretarsus. The pretarsus usually includes a pair of claws. A pad called the **arolium** may be found between the claws for adhesion; if a pad is present at the base of each claw, the pads are called **pulvilli**.

Wings. Along with birds and bats, insects are the only other animals with wings. Insect wings show much diversity, varying in shape, texture, and coloration. Most insects have two pairs, except flies and a few other groups, which have only one pair. In flies each metathoracic wing has been modified to form a knobbed guidance organ called a **halter** (pl. halteres). Other wing modifications include the tough, protective cover in beetles (Coleoptera) formed from the front wings,

Figure 2.15 Major antennal types found in insects. (Reprinted with permission of Macmillan Publishing Company from *Insects in Perspective* by Michael D. Atkins. Copyright ©1978 by Michael D. Atkins)

the **elytra**. Yet other insects, such as thrips, have fringed wings, and the wings of butterflies, moths, and mosquitoes are covered with scales.

However, the most common wing is a membranous one with a system of **veins** running throughout. These veins give strength and rigidity to the wing and are used widely in classification and identification of insects. A system of naming veins, devised for a hypothetical wing, serves as a sort of road map in analyzing and identifying certain insect groups.

Some insects have no wings. A few, for example, silverfish (Thysanura), are believed to have had wingless ancestors (primitively wingless), whereas others, like fleas and lice, probably evolved from a winged ancestor (secondarily wingless). Aphids have nonwinged forms only at certain times of the year. Except for mayflies (Ephemeroptera), all immature insects lack functional wings.

Figure 2.16 Photomicrograph of a compound eye of a bark beetle showing hexagonal facets. (Courtesy Canadian Forestry Service)

The Abdomen

The insect abdomen usually consists of six to ten segments and terminates in the **paraproct**, where the anus opens. This paraproct may be further modified to form a lobelike **epiproct** above the anus and a pair of lateral paraprocts around it.

As with the thorax, each abdominal segment has a tergum (usually called a notum in the thorax) and a sternum, and these are connected on each side by a lateral membrane. The terminal segment in some insects bears a pair of sensory **cerci**. Although adult insects lack functional legs on the abdomen, many immatures such as caterpillars have fleshy prolegs that aid in locomotion. Also, some aquatic immatures such as mayflies have respiratory gills associated with the abdominal segments.

Paired reproductive structures, arising from the primitive eight and ninth abdominal segments, form an egg-laying device, the **ovipositor**, in most females. Most males have a copulatory organ, the **aedeagus**, developed from a pair of appendages of the primitive ninth segment and sometimes parts of the tenth. Because of their tremendous diversity and distinctiveness, copulatory organs are often used to identify insect species.

Figure 2.17 Structural details of a generalized mesothoracic segment. A. Top (dorsal) aspect. B. Side (lateral) aspect. C. Bottom (ventral) aspect. Major plates or sclerites are separated by membranous areas. Sclerites are subdivided into areas by lines and sutures, which frequently are used as identifying characteristics. (Reprinted with permission of Macmillan Publishing Company from *Insects in Perspective* by Michael D. Atkins. Copyright ©1978 by Michael D. Atkins)

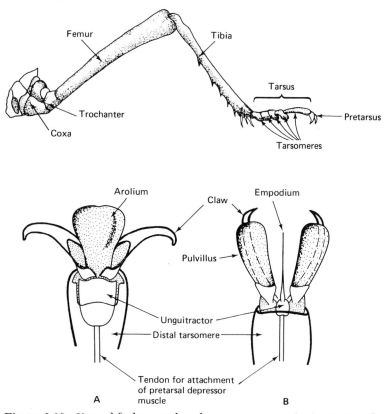

Figure 2.18 Unmodified insect leg showing segments (top). A. Details of a cockroach pretarsus. B. Details of a fly pretarsus. (Reprinted with permission of Macmillan Publishing Company from *Insects in Perspective* by Michael D. Atkins. Copyright ©1978 by Michael D. Atkins)

MAINTENANCE AND LOCOMOTION

As with other forms of life, the insect body is a cosmos of interdependent and interacting systems. Basic to the organism is the insect cell. Collections of cells make up tissues, which make up organs, and organs, in turn, form organ systems. For energy, insect cells require food, water, oxygen, and minerals, which are obtained from eating, drinking, and respiratory activities. Organic molecules and oxygen are then disseminated within the insect body, where metabolism occurs. The by-products of metabolism are excreted and wastes disposed of, and appropriate salt and water balances are maintained.

Individual organ systems, which function simultaneously to achieve the overall life process, include feeding and digestion, respiration, blood circulation, waste excretion, musculature, and nerves and sensory receptors. Although we discuss the details of these separately, we should not lose the perspective of these systems operating interactively and concurrently.

Feeding and Digestion

Feeding. As a class, insects feed on and digest a wide variety of foods. Part of their success as a group has depended on this activity. According to the type

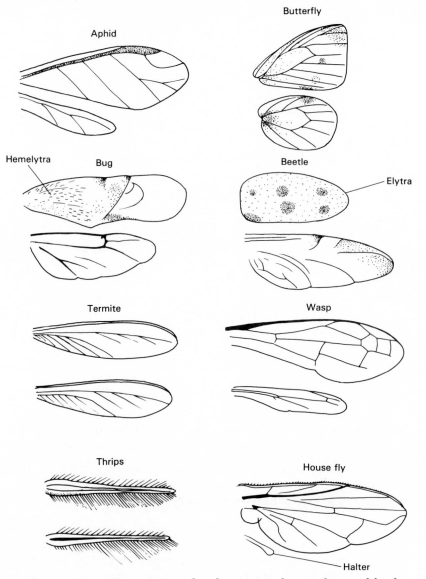

Figure 2.19 Major wing types found in insects showing front and hind wing. A. Aphid (Homoptera). B. Bug (Hemiptera). C. Termite (Isoptera). D. Thrips (Thysanoptera). E. Butterfly (Lepidoptera). F. Beetle (Coleoptera). G. Wasp (Hymenoptera). H. House fly (Diptera), showing halter. (Redrawn from Fenemore, 1982, *Plant Pests and Their Control*, Butterworths Publishing Company)

of food fed upon, insects can be described as **phytophagous**, **zoophagous**, or **saprophagous**.

Phytophagous insects feed on plants, and as mentioned earlier, most insect species fall into this category. They may feed on any part of the plant by removing tissues of leaves, stems, roots, and reproductive structures. Insects like butterflies that feed on nectar or other plant products are also considered phytophagous, as well as those that puncture the plant to feed on sap, for example, leafhoppers (Homoptera).

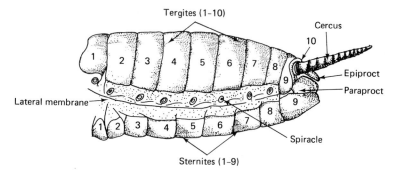

Figure 2.20 Side aspect of a generalized insect abdomen. (Reprinted with permission of Macmillan Publishing Company from *Insects in Perspective* by Michael D. Atkins. Copyright ©1978 by Michael D. Atkins)

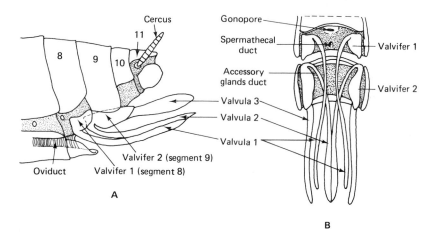

Figure 2.21 Diagrammatic representation of ovipositor structure. A. Side aspect. B. Bottom aspect. (Reprinted with permission of Macmillan Publishing Company from *Insects in Perspective* by Michael D. Atkins. Copyright ©1978 by Michael D. Atkins)

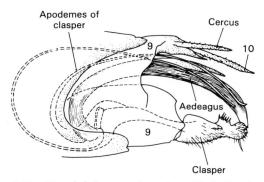

Figure 2.22 Tip of abdomen of a male caddisfly (Trichoptera) showing side aspect of the aedeagus. Numbers identify abdominal segments. (Courtesy Illinois Natural History Survey)

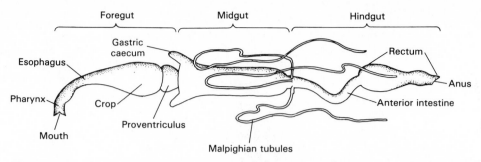

Figure 2.23 Diagrammatic representation of the gut or alimentary canal of insects (Reprinted by permission of John Wiley & Sons, Inc. from *A Textbook of Entomology* by H. H. Ross, C. A. Ross, and J. R. Ross. Copyright ©1987)

Zoophagous insects feed on other animals, including almost all vertebrates and invertebrates. Those feeding on vertebrates are mostly parasites like fleas (Siphonaptera) or mosquitoes (Diptera), but a few can be predators, such as predaceous diving beetles (Coleoptera), which may kill and eat small fish. Most zoophagous insects, however, kill and eat or parasitize other invertebrates, including insects.

Saprophagous insects feed on nonliving organic matter. As discussed previously, they play an important role in cycling nutrients in the environment. Here, we find general scavengers (cockroaches, Orthoptera), dung feeders (dung beetles, Coleoptera), dead plant feeders (termites, Isoptera), carrion feeders (carrion beetles, Coleoptera), and humus feeders (springtails, Collembola).

Although these categories are a reasonable representation of insect food habits, some insects may cross categories. For instance, some stink bugs (Hemiptera) that normally eat other insects may switch to feeding on plant tissue if prey is not found. Such insects are considered **omnivorous** in their feeding habits.

The digestive system. The insect gut, or alimentary canal, is the basic structure of the digestive system; essentially, it is a tube that extends from the mouth to the anus. The gut can be differentiated into three parts: foregut (**stomodaeum**), midgut (**mesenteron**), and hindgut (**proctodaeum**). These divisions are usually separated by valves, the **cardiac valve** in the front and the **pyloric valve** in the rear. Both the foregut and the hindgut are formed from inpushings of the

Figure 2.24 Portion of the midgut on an earwig (Dermaptera), showing the peritrophic membrane. (Redrawn from Wigglesworth, 1953, *The Principles of Insect Physiology*, Methuen and Company)

body wall during embryonic development (**ectoderm**), but the midgut arises from a separate embryonic cell layer, the **endoderm**.

The foregut is commonly described as having divisions, some poorly defined. Beginning at the mouth and moving toward the back, these divisions include the **pharynx** (throat), the **esophagus** (gullet), an expanded **crop**, and a valvelike **proventriculus**. In some insects such as cockroaches, the proventriculus may be muscular or gizzardlike and possess teeth for masticating and straining food.

The midgut has no divisions but often has associated with it **gastric caeca**, blind sacs that open into its forward end. Internally, the midgut often has a semipermeable, chitinous **peritrophic membrane** that forms a protective layer between food material and delicate epithelial cells. The peritrophic membrane is absent in many fluid-feeding insects.

The hindgut is greatly variable among insects but is normally divided into a tubular **intestine** and a short, expanded **rectum** connected to the anus.

Other structures associated with the gut include paired salivary glands, with ducts that open into the preoral cavity by way of the hypopharynx, and a number of slender **Malpighian tubules** that join the intestine just back of the pyloric valve. These tubules make up the main excretory organ.

Digestion. Limited digestion of food begins when it comes in contact with enzymes present in the saliva. Some of this digestion may occur externally before food is ingested and when saliva is injected into host tissues (for example, leafhoppers, Homoptera) or exuded over the food (for example, house flies, Diptera). In extreme cases, much of the digestion occurs externally, as with some diving beetles (Coleoptera), which regurgitate digestive juices from the gut and inject them with the mandibles into prey.

Usually, slightly digested food is passed along to the crop, where it can be stored or moved to the proventriculus. In insects with a well-developed proventriculus, it is ground or strained, thereby increasing surface area for action by digestive enzymes. When food reaches the midgut it comes into contact with the major digestive enzymes. Much enzyme activity occurs within and in the area of the gastric caeca.

Insects possess most of the major enzymes found in other animals but all may not be found in any one species. Therefore, we find an array of proteases, lipases, and carbohydrases, each breaking down a basic food group. Insects that feed on a variety of food groups have such enzymes as **amylase**, **maltase**, **lipase**, **invertase**, and **exo-** and **endopeptidases**. Blood-feeding species have primarily **proteolytic** enzymes. In wood-boring insects we may even find **cellulase** capable of breaking down wood tissue. Many of these insects, for example, termites (Isoptera), produce the cellulase by **symbiotic** microorganisms living in the gut.

After the food is digested, nutrients pass through the **peritrophic membrane** and are absorbed by the midgut epithelium. This membrane permits the passage of liquids and solutes but blocks the movement of larger fragments. It also protects the epithelial cells from abrasion. The peritrophic membrane passes on with the food to the hindgut and is replaced by special cells of the midgut. Some insect groups that feed only on liquid food, for example, blood-sucking flies and butterflies, lack the peritrophic membrane.

In the hindgut, undigested wastes are moved along and eliminated through the anus. Here, the rectum plays a particularly important role by resorbing water from fecal pellets and thereby helping to maintain the balance of body water and salts. Some aquatic insects like dragonfly naiads (Odonata) also have **tracheal gills** in the rectum that aid in respiration.

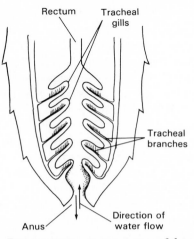

Figure 2.25 Diagrammatic representation of the tracheal gills in an immature dragonfly (Odonata). (Redrawn from Wigglesworth, 1953, *The Principles of Insect Physiology*, Methuen and Company)

Nutrition. Insects have similar nutritional requirements to other animals. **Carbohydrates** are necessary for energy and are usually supplied as glucose or sucrose. **Amino acids** are necessary for protein synthesis and the development of new tissue. Fats and oils are not generally considered essential in insect nutrition. The **vitamins** A, C, D, E, and those of the B complex are used in various metabolic functions but all are not required in all insects. Insects do not synthesize sterols; thus they require supplements of **cholesterol** in their diet. Water also is required along with a large number of minerals, including phosphorus, potassium, iron, zinc, calcium, copper, and cobalt.

Excretion

As nutrients are taken up and metabolism occurs, a number of waste products are produced that are toxic to insect cells and tissues. These are eliminated from the insect body through a process called excretion. Although there are others, the main waste materials usually considered in excretion are nitrogenous wastes, which accumulate from the degradation of proteins and nucleic acids.

The system that serves in the excretion of nitrogenous wastes also plays an important role in the maintenance and regulation of salts and water balance in body fluids. This regulation is crucial in allowing the insect to perform its life processes and yet survive in dry environments.

The excretory system. The most important organs of the excretory system are the Malpighian tubules and the rectum. The Malpighian tubules vary in number from two in some scale insects (Homoptera) to 200 or more in honey bees (Hymenoptera). In most instances these tubules are closed at their tips and float freely in the body cavity but sometimes are embedded in the outer part of the midgut wall.

Excretion. Excretion occurs when blood flows over the Malpighian tubules and substances diffuse or are actively transported into the tubules. Most of the nitrogen taken up by the tubules is in the form of uric acid salts, which are formed in

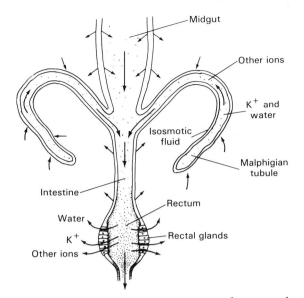

Figure 2.26 Diagrammatic representation of water and salt circulation in an insect excretory system. (Redrawn after Ross, et al., 1982, *A Textbook of Entomology*, Wiley & Sons)

the fat body and passed into the blood before removal. Other substances picked up by the tubules include amino acids, various ions, and water. A potassium-ion gradient in the tubule causes the urine to flow.

The urine then is discharged from the tubules into the hindgut, where it passes to the rectum. Here, most water resorption occurs, producing relatively dry feces in many insects. The lining of the rectum also serves to remove needed inorganic ions and amino acids from the urine and return them to the blood in controlled amounts, thereby maintaining an appropriate ionic balance.

Other excretory modes. In some insects, uric acid is not eliminated from the body but is inactivated and stored in special **urate cells** in the **fat body** (a group of loosely organized cells in the body cavity that also function in intermediate metabolism), for example, in immatures of parasitic wasps (Hymenoptera). In other insects, the uric acid is deposited in the cuticle as white crystals, producing patterns of white color.

Circulation of Blood

In many animals, such as vertebrates, the blood circulation is accomplished by movement through a series of vessels (veins, arteries, capillaries). This is termed a closed system.

Insects, on the other hand, have an open system of circulation. Their blood flows through the open body cavity, the **hemocoel**, thereby supplying organs and tissues.

The circulatory system. The **dorsal vessel** is the main organ of circulation in insects. This vessel lies at the top of the hemocoel and extends from the back of the abdomen to the head. It comprises two parts, a **heart** at the rear and an **aorta** in the front. The heart is a chambered organ, drawing blood in through openings in each chamber, called **ostia**, and contracting to pump it forward to the aorta.

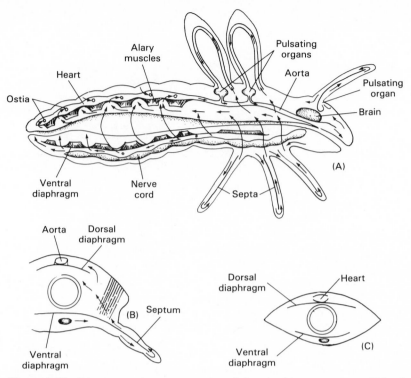

Figure 2.27 Diagrammatic representation of the circulatory system and blood circulation in insects. (Redrawn from Ross, et al., 1982, *A Textbook of Entomology*, Wiley & Sons)

The aorta is usually a simple tube that carries blood forward and dumps it into the head capsule.

The heart of most insects is bordered below by bands of wing-shaped muscles, known as **alary muscles**. These connect the heart to lateral portions of the terga. These muscles can form an almost complete partition between the heart and the body cavity. This partition is known as the **dorsal diaphragm** and the chamber occupied by the heart, the **dorsal sinus**.

In some insects, accessory pulsating organs exist in the thorax and lower body cavity. **Thoracic pulsating organs**, found in hawk moths (Lepidoptera), occur in the mesothorax and draw blood through the wings. In many insects there is also a **ventral diaphragm**, similar in form to the dorsal diaphragm, which guides blood flow back and to the sides.

Blood.   The blood of insects, called **hemolymph**, is a watery fluid that may be colored yellow or green. It comprises liquid plasma, and about 10 percent of its volume is made up of blood cells, or **hemocytes**.

Unlike vertebrates, most insect blood cells contain no hemoglobin, the transporter of oxygen. Therefore, insect blood does not play a role in respiration, except in certain groups such as bloodworms (immatures of *Chironomus* species).

The main function of blood in insects is to transport nutrients, wastes, and hormones. The blood absorbs nutrients from the digestive system and carries them through the body cavity, feeding tissues and organs. Wastes from metabolism at cell sites also are absorbed by the blood and carried to excretory organs, where they are removed. Hormones from glands are transported in the blood to sites of activity. Moreover, blood has a role in the immune system of insects, having specialized cells that dispose of harmful microorganisms; it also functions in the healing of wounds. Additionally, in intermediate metabolism it stores and converts substances (for example, trehalose stored and converted to glucose) and serves as a pressure source (hydraulic function) to expand body parts during hatching and growth.

Circulation.   Insect blood taken in by the heart circulates forward because of the peristaltic heart movements (Figure 2.27). It is carried to the head, flows over the brain, and percolates rearward where it is picked up and recycled by the heart. Longitudinal membranes, or **septa**, in the appendages of some insects aid in blood flow through these structures. In some insects, thoracic pulsating organs aid in circulating blood through the wing veins.

Respiration

Free energy is required to perform life functions in insects, and it is derived mostly from the oxidation of nutrients. The supply of oxygen for this purpose is obtained through respiration. Respiration involves the intake, transport, and utilization of oxygen by cells and tissues and the removal of carbon dioxide from the body. In insects, the primary mode of respiration is the diffusion of atmospheric gases directly across membranes into the cells.

The tracheal system.   The main system for the transport of gases in insects is the **tracheal system**. It is composed of a series of branching tubes called **tracheae**. Tracheae are formed into groups in each body segment and open to the outside by way of openings, **spiracles**, on each side of most body segments. In some insects, the spiracles have operable valves that close these openings for periods of time, thereby avoiding excessive water loss through them.

Spiracles open into **tracheal trunks**, large tracheae on each side that run the length of the body and connect the grouped tracheae of each segment. In fast-flying insects like the house fly (Diptera), tracheal trunks have enlargements, called **air sacs**, that serve as bellows to increase ventilation. Tracheal branches arise from the trunks in each segment and become increasingly fine and branched. Ultimately, the fine tips of the tracheae divide into minute (less than 1 micron in diameter) **tracheoles**. These fine capillary tubes subsequently branch around cells and tissues and penetrate deeply into muscle fibers. At the primary branch of several tracheoles is a weblike **tracheole cell** with thin protoplasmic extensions.

The tracheae are formed from inpushings of the body wall and, therefore, are lined with cuticle. As with other parts of the integument, this lining, the **intima**, is secreted by the epithelial cells but has different characteristics of permeability than the body wall. The inner surface of the intima contains helical thickenings, called **taenidia**, that support and shape the tracheae. Tracheoles also have taenidia but lack a regular layer of epithelium.

The system with spiracles that open to the outside of the body is called an open system. This open system is variously modified in different insects. A closed system occurs in some insects, where the spiracles become nonfunctional or are entirely absent. Many aquatic immatures have the closed system (for example, mayflies, Ephemeroptera, and stoneflies, Plecoptera), and instead of spiracles,

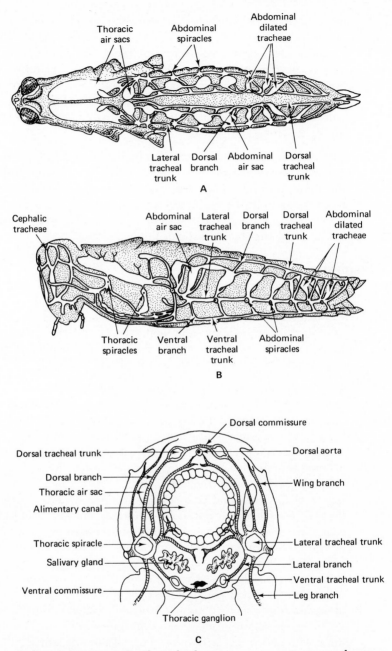

Figure 2.28 Generalized tracheal system as seen in a grasshopper (Orthoptera). A. Top aspect, B. Side aspect, C. Section through body. (Reprinted with permission of Macmillan Publishing Company from *Insects in Perspective* by Michael D. Atkins. Copyright ©1978 by Michael D. Atkins)

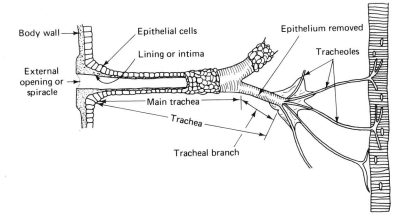

Figure 2.29 Detail of the insect tracheal system showing tracheoles. (Reprinted with permission of Macmillan Publishing Company from *Insects in Perspective* by Michael D. Atkins. Copyright © 1978 by Michael D. Atkins)

a fine network of tracheoles runs under the skin or into external gills. The unusual rectal gills of dragonfly (Odonata) naiads have already been mentioned. Other specialized forms of respiration found in aquatic insects include diving air stores (for example, in water scavenger beetles and predaceous diving beetles, Coleoptera), where a film or bubble of air is attached to some part of the body. In some instances, the air bubble does not dissipate and serves as a physical gill, called a **plastron**, across which gases diffuse. Other aquatic insects possess air tubes used like a snorkle to siphon air from the surface (for example, immature mosquitoes and water scorpions).

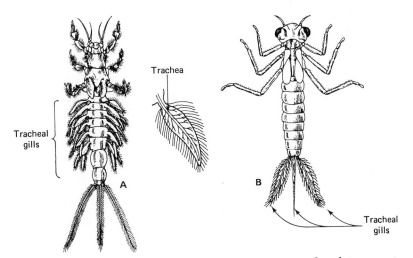

Figure 2.30 Examples of closed respiratory systems, found in aquatic insects. A. Immature mayfly (Ephemeroptera), showing side (lateral) gills. B. Immature dragonfly (Odonata), showing terminal gills. (Reprinted with permission of Macmillan Publishing Company from *Insects in Perspective* by Michael D. Atkins. Copyright ©1978 by Michael D. Atkins)

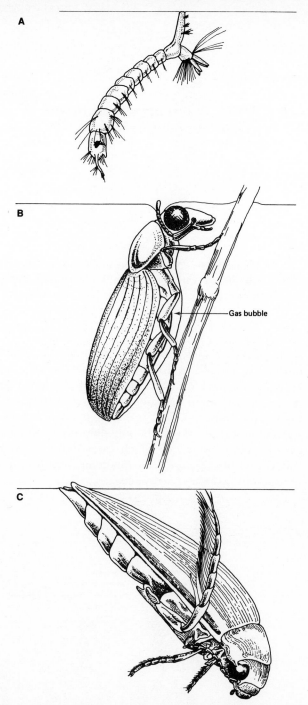

Figure 2.31 Examples of open respiratory systems in aquatic insects. A. Mosquito larva (Diptera) siphoning air. B. Water scavenger beetle (Coleoptera) replenishing gas bubble by breaking water-surface film with antennae. C. Predaceous diving beetle replenishing air store located beneath front wing covers. (Reprinted with permission of Macmillan Publishing Company from *Insects in Perspective* by Michael D. Atkins. Copyright ©1978 by Michael D. Atkins)

Respiratory process. It is believed that respiration occurs by the diffusion of oxygen and carbon dioxide through the tracheal system, aided by mechanical ventilation of abdominal tracheae and air sacs. Diffusion of oxygen into the system occurs because of a drop in oxygen pressure at the tip of the tracheoles. Similarly, carbon dioxide diffuses out through the tracheal system; and because insect tissues are more permeable to it, significant amounts of carbon dioxide also escape through the body wall.

Musculature and Locomotion

Movement of the insect body and its appendages is produced by an intricate system of musculature. The coordinated movement of muscles and appendages is critical to such important activities as feeding, walking, jumping, and flying.

Muscle system. The musculature of insects occurs as conspicuous layers and bands distributed differently in various parts of the body. For convenience, muscles can be grouped into three categories: **visceral**, **segmental**, and **appendage** muscles.

Visceral muscles occur in circular, longitudinal, or oblique bands around the digestive tract. Here, they produce peristaltic movements of the gut that are important in moving food and wastes along its length. Specialized muscles also are associated with valves of the spiracles and are responsible for their operation. Others cause peristalsis of the heart.

Segmental muscles occur as a series of bands that connect body segments. In the abdomen, tergites are connected to one another by longitudinal bands, as are the sternites below. The tergite of an abdominal segment is connected to the sternite of the same segment by oblique muscles on each side of the body. In the thorax, the predominant muscles are large cordlike bands that move the legs and wings. Other smaller muscles form complicated patterns of attachment.

Movable appendages have muscles either in them or attached to them. Usually, appendages like the legs, which are divided into segments, have attachments

Ventral
bands

Dorsal
bands

Figure 2.32 Segmental musculature in mesothorax and metathorax of a caterpillar (Lepidoptera). (Redrawn from Snodgrass, 1935, *Principles of Insect Morphology*, McGraw-Hill Book Company)

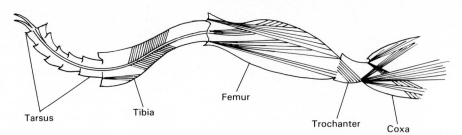

Figure 2.33 Diagrammatical representation of an insect leg and muscle attachments. (Redrawn from Ross, et al., 1982, *A Textbook of Entomology*, Wiley & Sons)

for muscles housed in the body as well as for muscles housed directly in the appendage. The body muscles generally move the whole appendage, and the segmental muscles move parts of the appendage. Muscles of nonsegmented appendages, as mandibles, have operating muscles housed only in the body proper.

Muscle function. All muscles in insects are composed of **striated fibers**. Each of the fibers is made up of many **myofibrils** situated in a matrix of multinucleated cytoplasm, **sarcoplasm**. Along with the myofibrils are numerous **mitochondria**. These organelles contain enzymes for the oxidation process that releases free energy for muscle reaction.

When stimulated by a nerve cell impulse, insect muscles contract similarly to those of other animals. The period of contraction is followed by a period of recovery. During recovery, the muscle assumes its original elongated shape.

Locomotion. The basic mode of locomotion of insects on land is walking. Adult insects, with three pairs of legs, take each step on a tripod; that is, two legs are on the ground on one side and one leg on the other side. The other three legs are raised as the grounded legs push backward. The alternation of pushing tripods results in a slightly zigzag path of forward motion. This gait may be modified to

Figure 2.34 Crawling locomotion used by inchworms or loopers (Lepidoptera), with *d* being distance traveled in each step. (Reprinted with permission of Macmillan Publishing Company from *Insects in Perspective* by Michael D. Atkins. Copyright ©1978 by Michael D. Atkins)

sequential, single-leg movements in very slow walking, and it varies in insects like praying mantids (Orthoptera), which do not use the front legs in walking (but for capturing and manipulating prey in this example).

Crawling is the basic mode of land locomotion commonly seen in worm-shaped immatures (larvae). Many of us have seen the rippling body of a caterpillar as it inches its way forward. This movement is accomplished when the longitudinal and dorsoventral muscles that oppose each other in each segment operate. Synchronized operation of these produces a wave of contraction down the body. Fleshy **prolegs** on the abdominal segments assist this motion by anchoring the abdomen, preventing it from slipping backward and causing the whole body to

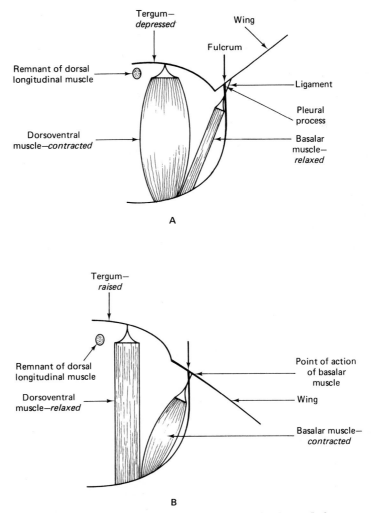

Figure 2.35 Diagrammatical representation of the direct-flight muscle system in insects. Contraction of the dorsoventral muscle depresses tergum, which pushes wing bases below the pleural wing process (fulcrum), thereby causing wing upstroke. Relaxation of the dorsoventral muscles and contraction of the basal muscle pulls wing down against the fulcrum for the power stroke. (Reprinted with permission of Macmillan Publishing Company from *Insects in Perspective* by Michael D. Atkins. Copyright ©1978 by Michael D. Atkins)

Figure 2.36 Diagrammatical representation of the indirect-flight muscle system in insects. A. Dorsoventral muscle contraction depresses the notum, raising the wings as in the indirect system. B. Dorsal longitudinal muscle contraction distorts the notum in such a way as to depress the wings for the power stroke. (Reprinted with permission of Macmillan Publishing Company from *Insects in Perspective* by Michael D. Atkins. Copyright ©1978 by Michael D. Atkins)

move forward. In other immatures, so-called inchworms or loopers (Lepidoptera), crawling is accomplished in a sort of looping motion, where the prolegs are brought forward and anchored, causing a "loop" in the abdomen, followed by a straightening out of the whole body.

One of the most studied modes of locomotion in insects is flying. Insects are the only invertebrates that can fly, and only adults can accomplish flight (with the exception of mayflies, Ephemeroptera, where weak flights may occur just before full sexual maturity). The flight of insects involves the large cordlike muscles of the thorax that attach to tergites and sternites, **indirect flight muscles**, and others that attach to the wing bases, **direct flight muscles**.

In the more primitive insects (for example, dragonflies, Odonata, and cockroaches, Orthoptera) and in beetles (Coleoptera), the up-and-down wing-beat cycle begins with contraction of the **dorsoventral muscle** (indirect flight muscle), which depresses the tergum, forcing the wing base down and the wing up. The thoracic pleurite serves as a fulcrum in this process. In the next cycle, contraction of the **basalar muscle** (direct flight muscle) and relaxation of the dorsoventral muscle forces the wing down, and the tergum is raised. In more advanced fliers like honey bees, up-and-down movement is produced only by indirect muscles (dorsoventral and **dorsal longitudinal**) that cause flexing of the thoracic segment to produce wing beat. In addition to these up-and-down movements, wings display a synchronized rotation that gives more surface on the downstroke than the upstroke and pushes air backward to cause forward movement. This rotation is accomplished by muscles that distort the tergum and others that apply pressure alternately to the front and back of the wing base.

Other forms of locomotion also are important to insects, including jumping, swimming, and water surface locomotion. Being somewhat more specialized activities, they will not be discussed here.

SENSING THE ENVIRONMENT AND INTEGRATING ACTIVITIES

For insects to exist, they must sense the realities of their environment and govern their activities accordingly. They must locate food, find mates, avoid enemies, make nests, and perform internal functions. This can be accomplished only by perceiving conditions and integrating this information to perform the appropriate behavior. Perception is achieved by a number of different sense organs, and behavior results from integration of information and stimulation by the nervous system.

Sense Organs

The basic sense organs of insects include those involved in sight, smell, taste, touch, and hearing. Entomologists usually group these organs by category, the most basic of which are **photoreceptors**, **chemoreceptors**, and **mechanoreceptors**.

Photoreceptors.　Photoreception is the sensing of light, and when images are produced, it is called sight. To a certain extent, all animal cells are light-sensitive, but certain ones in insects are specialized to sense light presence, day length, light intensity, color, and other aspects.

Probably the most complex photoreceptors in insects are those involved in forming images, the eyes. Compound eyes and simple eyes have already been mentioned as prominent organs on the head of many insects. Compound eyes are believed to have evolved along with flight and are not found in primitively wingless insects. They also are lacking in immatures of the most advanced insects (division Endopterygota). Only simple eyes are present in many of these species, and most insects with compound eyes also have simple eyes. Some insects have no eyes and sense light through the cuticle (**dermal photoreception**).

The compound eye is made up of a number of individual sensory units called **ommatidia**. Light is gathered in the ommatidium by the lens or **cornea** that focuses light through a **crystalline cone** to a light-receptor apparatus. This apparatus comprises six to eight **retinula cells** that combine to produce a central light sensor, the **rhabdom**. Both the crystalline cone and the light-sensing apparatus are surrounded by pigment cells that isolate, to various degrees, one ommatidium from another. The retinula cells turn light into electrical energy that is, in turn, carried by nerve fibers to the brain. Images produced by the collection of ommatidia are believed to result in an overall mosaic of the object, with each ommatidium supplying only a small piece of vision.

Figure 2.37 Diagrammatical representaion of a compound eye unit, the ommatidium. (Reprinted with permission of Macmillan Publishing Company from *Insects in Perspective* by Michael D. Atkins. Copyright ©1978 by Michael D. Atkins)

Simple eyes, or ocelli, are believed to perceive images, color, and movement, as do compound eyes, except a less complete mosaic is probably produced. However, some insects with only ocelli "scan" the environment by moving their head back and forth. The structure of ocelli is quite variable among insects, with some being similar to individual ommatidia (for example, caterpillars, Lepidoptera) and others having a single cornea that overlays several retinula cells and rhabdoms.

Chemoreceptors.   The senses of taste (**gustation**) and smell (**olfaction**) are based on the detection of certain molecules by receptor organs that subsequently produce a nerve impulse. The distinction between these anthropomorphic senses is trivial, being separated almost solely on the basis of distance from a source. In other words, food is tasted only when in the mouth, but when it is not in the mouth (farther away), it is smelled. In either instance, the mechanism of chemoreception is the same.

Chemoreceptors usually occur in the form of short pegs or hairs on various body parts. Taste receptors sense molecules from liquids and are often hairlike and have fine nerve endings exposed to the environment at the hair's tip. Smell receptors may be more peglike and have a greater number of nerve endings at the surface than taste receptors. As might be expected, taste receptors are abundant on the mouthparts, but they are also dense on the tarsi of many insects, thereby helping the insect to detect food. Smell receptors are most numerous on insect antennae but may also be abundant on appendages (**palpi**) of the mouthparts.

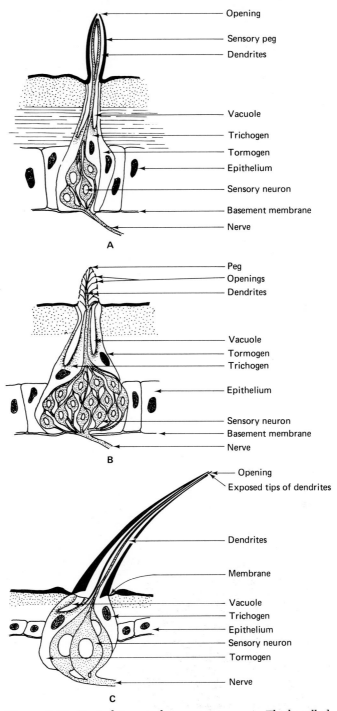

Figure 2.38 Typical insect chemoreceptors. A. Thick-walled peg that serves in general chemical perception. B. Thin-walled peg that often serves in smell. C. Setal-type receptor frequently involved in taste. (Reprinted with permission of Macmillan Publishing Company from *Insects in Perspective* by Michael D. Atkins. Copyright ©1978 by Michael D. Atkins)

Mechanoreceptors. Mechanoreceptors, called **sensilla** (sing. sensillum), are the most numerous sensory structures on insects, being found over much of the body surface. These sensilla may be hairlike, in which instance they are called **trichoid**, or they may be domelike (**campaniform**) or platelike (**placoid**).

Some of these receptors are sensitive to touch and respond to pressure by sending a flow of impulses to the nervous system. These sensilla have been termed **tonic**. Other receptors respond mostly to the rate of bending and are most sensitive to such things as air or water vibrations. These are often termed **phasic**. Some specialized sensilla also function in relaying information about the position of an insect in its environment, whereas others provide information on position of body parts relative to one another. These have been termed **proprioreceptors**.

A few mechanoreceptors have no external structure associated with them. These are the **chordotonal sensilla** that are composed of a bundle of bipolar nerve cells stretched between two surfaces of the integument. These detect pressure on the body wall and movements of the insect. Specialized groups of chordotonal sensilla make up **Johnston's organ**, a structure on the second antennal segment of adult insects that responds to movement of the antennae and may be involved in hearing. The main hearing organ of insects, however, is composed of a membranous **tympanum** whose vibrations are detected by a

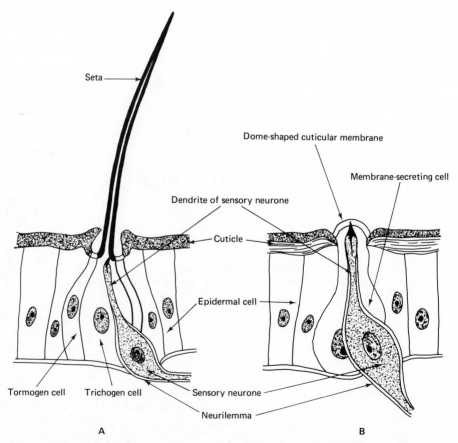

Figure 2.39 Details of two common mechanoreceptors. A. Hairlike sensillum. B. Campaniform sensillum. (Reprinted with permission of Macmillan Publishing Company from *Insects in Perspective* by Michael D. Atkins. Copyright ©1978 by Michael D. Atkins)

group of chordotonal sensilla. Such tympanal organs are found in grasshoppers (first abdominal segment), crickets (tibiae), moths (abdomen or metathorax), and many other insects.

Other receptors. In addition to the sense organs described above, insects also have receptors for perceiving ambient moisture and temperature, although little is known about these. Perceiving moisture in the air is called **hygroreception**; moisture levels are thought to be sensed, at least in some insects, by hairs or tufts of hairs that absorb moisture. Insects have a general sense of temperature and utilize this sense to locate suitable environments for their activities. As an example a wood-boring beetle, *Melanophila* species (Buprestidae), has sensory pits on the underside of the mesothorax that are sensitive to the heat of recently fire-damaged trees, which the beetle prefers to attack. **Geomagnetic receptors** that sense magnetic fields also occur in such insects as honey bees who use them in orientation and other activities. However, the actual sense organs used in geomagnetic reception have not been identified by electrophysiology.

Nervous System

The nervous system of insects functions to generate and transport electrical impulses, to integrate information received, and to stimulate muscles for movement. This system can be divided into a central nervous system and a visceral nervous system.

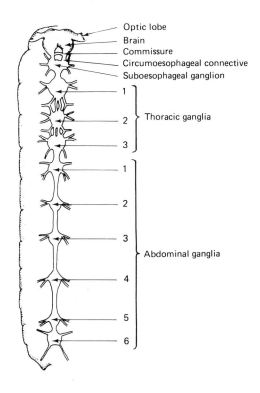

Figure 2.40 Organization of the central nervous system in generalized form. (Reprinted with permission of Macmillan Publishing Company from *Insects in Perspective* by Michael D. Atkins. Copyright ©1978 by Michael D. Atkins)

Central nervous system. Basically, the central nervous system is formed from a brain, located in the head, and a ventral nerve cord running from the brain through the abdomen along the base of the body cavity. The central nervous system supervises and coordinates activities of the insect body.

The brain is located in the upper part of the cranium above the esophagus and, therefore, is often referred to as the **supraesophageal ganglion**. The brain is structured as paired parts consisting of the **protocerebrum**, innervating the compound eyes and ocelli; the **deutocerebrum**, innervating the antennae; and the **tritocerebrum**, which connects to the visceral nervous system.

The brain connects to a large nerve **ganglion** (nerve cell bundle) located under the esophagous by way of a paired nerve cord that branches around the esophagous. This special ganglion, called the **subesophageal ganglion**, innervates the mouthparts and the salivary duct and attaches to the ventral nerve cord.

The ventral nerve cord is composed of a series of ganglia attached to one another by paired connectives, forming a kind of chain. Typically, there is a ganglion in each thoracic segment and, in many species, one in each abdominal segment. In specialized groups of insects, the thoracic ganglia fuse to form a larger unit, and abdominal ganglia become reduced in number or are lost almost completely. Nerves arise from each ganglion, branch, and innervate both sensory structures and muscles.

Visceral nervous system. The main component of the visceral nervous system is the so-called **stomodeal nervous system**. The stomodeal system controls activities of the anterior gut and the dorsal vessel. This system consists of a **frontal ganglion** connected to the brain and then to other small ganglia. These ganglia give rise to paired nerves that innervate mainly the digestive tract and neuroendocrine glands, the **corpora cardiaca** and the **corpora allata** (see Figure 2.41). Both glands are involved in insect growth, discussed in Chapter 4. Other parts of the visceral nervous system include the ventral sympathetic system, which innervates the spiracles of each segment in some insects, and a caudal sympathetic system, which is involved in activities of internal sexual organs.

Nerve-Impulse Transmission and Integration

Nerve-impulse transmission. The basic unit of the nervous system that functions in nerve-impulse transmission is the nerve cell, or **neuron**. A neuron is composed of a **cell body**, one or more receptor fibrils, and an **axon** that branches at the tip. According to function, there are three types of neurons: **sensory neurons**, **motor neurons**, and **interneurons**. Sensory neurons have receptor fibers arising directly from the cell body that are connected to sense organs. The axon of sensory neurons carries impulses to ganglia in the central nervous system. The cell body and receptor fibrils of motor neurons are located within the central nervous system and have an axon that branches into muscle tissue. The axon of motor neurons carries nerve impulses away from the central nervous system.

Interneurons are situated totally within the central nervous system and connect sensory and motor neurons.

Nerve impulses traveling along axons are electrical. These impulses arise from the flow of positive sodium ions through the cell membrane, causing depolarization. Depolarization creates an action potential that passes, as a wave, quickly down the axon. The action potential lasts only a few thousandths of a

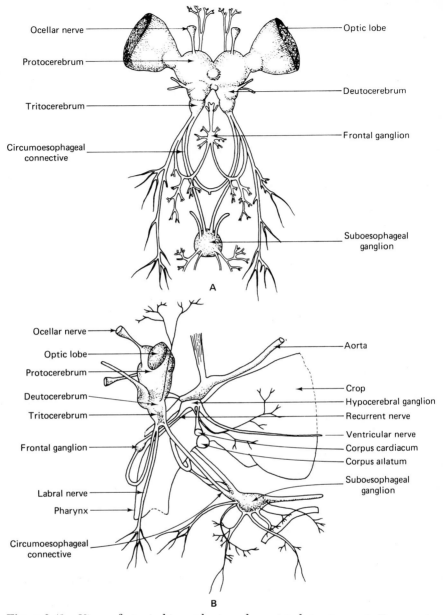

Figure 2.41   Views of a typical insect brain and associated structures. A. Top aspect. B. Side aspect. (Reprinted with permission of Macmillan Publishing Company from *Insects in Perspective* by Michael D. Atkins. Copyright ©1978 by Michael D. Atkins)

second, at which time sodium gates of the membrane close. When the sodium gates close, positive potassium ions flow out and restore the cell's resting potential (slightly negative). This is followed by an exchange of sodium inside with potassium outside the cell and a return of the original (resting) ion distribution. This type of nerve transmission is termed axonic transmission; it conveys an impulse from an arrival point along the axon to another neuron, muscle, or gland, or from a receptor cell.

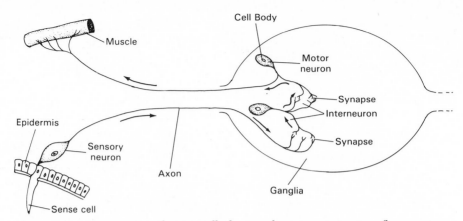

Figure 2.42 Arrangement of nerve cells for impulse transmission in reflex reactions. Arrows indicate direction of nerve impulse. (Redrawn from Wigglesworth, 1953, *The Principles of Insect Physiology*, Methuen and Company)

Another type of impulse transmission is synaptic transmission, which is mainly chemical. A **synapse** (Figure 2.42) is the junction between neurons and other cells. Among other places, a synapse occurs at the junction between an interneuron and a sensory neuron, and another occurs between the same interneuron and a motor neuron. A completed reaction from a sensory neuron through an interneuron and directly to a motor neuron, causing muscle contraction, is called a **reflex reaction**. An example of this reaction occurs in the human body when a finger is jerked back from a hot surface, with no thought involved. Here, the nerve impulse travels a short route from the finger to the spinal cord and directly to an arm muscle; integration of information by the brain is not involved.

When an impulse moving along the axon reaches a synapse, it dies out. As this happens, a chemical is secreted that crosses the synapse and induces and stimulates an impulse in an adjacent neuron or stimulates a muscle or gland. The most well known of these chemical transmitters is **acetylcholine**, although there are many others. Following transmission, the synapse is returned to the resting state by enzymes like **acetylcholinesterase**, which break down the chemical transmitter in the case of acetylcholine. As we will see in Chapter 11, many modern insecticides act by upsetting chemical reactions at nerve synapses.

Integration. Many life-and-death activities of insects involve reflex reactions. Large nerve fibers allow rapid passage of the impulse, allowing a galvanic twitch of legs for jumping or flipping of the body, to escape an enemy.

Usually, smaller nerves carry impulses that may cause a slow wave of muscle contraction, resulting in more delicate and precise movements like those involved in feeding, walking, and nest building.

The brain is a coordinating center. It receives information from antennae, compound and simple eyes, and labrum and other sensory structures and sends impulses for appropriate actions. Unlike the brain of vertebrates, the insect brain inhibits rather than initiates locomotion. If the brain is removed, the insect will walk steadily forward but cannot climb or go backwards.

The subesophageal ganglion is the coordinating center for feeding. It receives information and controls the movements of the mandibles, maxillae, and labium.

Generally, each thoracic and abdominal ganglion organizes and coordinates the

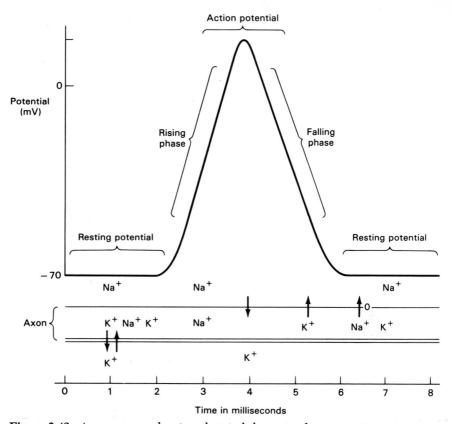

Figure 2.43 A nerve axon showing electrical dynamics during axonic transmission. At rest, the axon has a high potassium level and a low sodium level; the reverse is found outside. The action potential occurs (center) when the first sodium ions rush in (depolarization), followed by potassium ions rushing out (repolarization). After the action potential has occurred, sodium ions inside the axon are exchanged with potassium ions outside. (Reprinted with permission of Benjamin/Cummings Publishing Company from *Insect Biology* by Howard E. Evans. Copyright ©1984)

behavior of its own segment. A striking example of this segmental behavior can be seen in praying mantids (Orthoptera) when the male continues to copulate as the female eats first the male's head and then his thorax. Therefore, the insect can be seen behaving as a set of individually functioning segments, each coordinated by a nerve ganglion. The brain acts as a center for receiving stimuli from important sensory structures of the head and for governing the release of segmental and intersegmental activities.

INSECT REPRODUCTION

The power of insects to reproduce is one of their most important biological features, accounting for much of their evolutionary success. Most insects are **dioecious**, that is, there is a male and a female that mate to produce a **zygote** (fertilized egg). The reproductive organs of females and males are somewhat similar, usually occurring near the rear of the abdominal cavity.

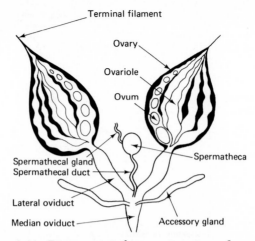

Figure 2.44 Diagrammatical representation of a typical female reproductive system. (Reprinted with permission of Macmillan Publishing Company from *Insects in Perspective* by Michael D. Atkins. Copyright ©1978 by Michael D. Atkins)

The Female System

The main organs of the female reproductive system are the paired **ovaries**. Each ovary is usually made up of a bundle of several tubular **ovarioles** in which eggs are formed. Each ovariole is attached with a thread called the **terminal filament**, and germ cells form in the ovariole just below this filament. The germ cells develop as they move along and arrive as fully formed eggs at the ovariole base, or **pedicel**.

Eggs pass through the pedicels (collectively, the calyx) into paired **lateral oviducts** and then into a common **oviduct**. From the oviduct, they move to the **vagina**, where they are fertilized and held for laying. The organ involved in fertilization is the **spermatheca**, which receives and stores sperm after copulation. A **spermathecal gland** attached to the spermatheca supplies nutrients for main-

Figure 2.45 Diagrammatical representation of a typical male reproductive system. (Reprinted with permission of Macmillan Publishing Company from *Insects in Perspective* by Michael D. Atkins. Copyright ©1978 by Michael D. Atkins)

taining the sperm before it is dispensed, and paired **accessory glands** secrete adhesive and a protective covering for the egg after it has been fertilized. Many modifications of this basic system occur among insect groups.

The Male System

The primary organs of the male reproductive system are paired **testes**, found in approximately the same position as the female ovaries. Each testis is made up of a number of sperm tubes enclosed in a sheath. Sperm are produced in the sperm tubes and move through narrow **vasa efferentia**, which empty into a common duct, the **vas deferens**. Sperm continue through the vas deferens and are held in a storage structure, the **seminal vesicle**, where they combine with secretions of paired accessory glands to form semen. In some insects, the sperm are enclosed in a capsule, the **spermatophore**. During copulation, semen from the seminal vesicle moves through the **ejaculatory duct** and out the **penis**. The aedeagus, mentioned earlier, is associated with the penis and aids in the copulatory process.

Selected References

ATKINS, M. D. 1978. *Insects in Perspective*. New York: Macmillan, chapters 5–10.

BARRINGTON, E. J. W. 1979. *Invertebrate Structure and Function*. 2nd ed. New York: Wiley.

BLUM, M. S. , ed. 1985. *Fundamentals of Insect Physiology*. New York: Wiley.

CHAPMAN, R. F. 1982. *The Insects, Structure and Function*. Cambridge, Mass. : Harvard University Press.

DETHIER, V. G. 1963. *The Physiology of Insect Senses*. New York: Wiley.

GILMOUR, D. 1970. General anatomy and physiology. In Commonwealth Scientific and Industrial Research Organization (CSIRO). *The Insects of Australia*. Carlton, Victoria, Australia: Melbourne University Press, chapter 2.

HAPP, G. M. 1984. Major Life Systems. In H. E. Evans, *Insect Biology*. Reading, Mass. : Addison-Wesley, chapter 3.

HORN, D. J. 1976. *Biology of Insects*. Philadelphia: Saunders, chapters 4–7.

MATSUDA, R. 1970. Morphology and evolution of the insect thorax. *Memoirs Entomological Society Canada*, no. 76.

MATSUDA, R. 1976. *Morphology and Evolution of the Insect Abdomen*. New York: Pergamon Press.

PRINGLE, J. W. S. 1965. Locomotion: flight. In M. Rockstein, ed. , *The Physiology of Insecta*, vol. 2. New York: Academic Press, pp. 283–329.

RICHARDS. O. W. , and R. G. DAVIES. 1977. *Imms' General Textbook of Entomology*, vol. 1, 10th ed. London: Chapman and Hall.

ROSS, H. , C. A. ROSS, and J. R. P. ROSS. 1982. *A Textbook of Entomology*, 4th ed. New York: Wiley, chapters 3–6.

SELMAN, B. 1979. Basic biology. In *Insects, An Illustrated Survey of the Most Successful Animals on Earth*. London: Hamlyn Pub. Group, pp 44–81.

SNODGRASS, R. E. 1935. *Principles of Insect Morphology*. New York: McGraw-Hill.

WILSON, D. M. 1966. Insect walking. *Annual Review of Entomology* 11:103–122.

3 INSECT CLASSIFICATION

THE NAMING AND ORDERING of objects into groups is probably the most fundamental step in the development of scientific principles. For instance, in chemistry the discovery and naming of elements and organizing them into the periodic table became the basis for the development of the science.

In biology, the naming of organisms is referred to as **nomenclature**, and ordering them into a hierarchy of categories is known as **classification**. A related science, **taxonomy**, involves the theoretical basis for classification and the study of classification schemes. Specialists working in these areas usually are referred to as **systematists**; their overall activity, **systematics**, is the study of the diversity and classification of organisms.

Because of the tremendous size of the class Insecta, the naming and classification of all insects would seem to be a difficult, if not impossible, achievement. However, hundreds of insect systematists around the world work daily on this task, and great strides have been made toward understanding insect diversity. Indeed, some experts estimate that nearly 7,000 species new to science are being discovered, named, and classified each year! But there is still much systematics work to be done before insects will be as well known as other animals, for example, birds and mammals.

OBJECTIVES OF CLASSIFICATION

Like the periodic table of chemistry, classification allows us to order what we know about insects and to compare and contrast characteristics. From these comparisons, we formulate predictions about relationships, including those with both evolutionary and ecological meaning. For instance, members of the same **species** are expected to behave similarly in their food habits, tolerances to environmental extremes, developmental patterns, and in other ways. A group of similar species, put together in a higher category called a **genus**, could also be predicted to share somewhat similar ecologies and also to have evolved from the same ancestor. Moving to higher and higher groupings in classification, we expect more and more diversity within the grouping.

A major application of classifications is in **identification** of insect specimens. Identifications of major groups like insect **orders** can usually be made at a glance; however, finer identifications often require the use of **keys**. Most keys comprise a sequence of paired statements and questions that allow the user to eliminate alternative options and eventually associate the unknown specimen with a name. Many keys exist for **orders** and **families** of insects. Some of the most useful are those written by D. J. Borrer, D. M. DeLong, and C. A. Triplehorn (*An Introduction to the Study of Insects*) for the insects of North America and by C. T. Brues, A. L. Melander, and F. M. Carpenter (*Classification of Insects*) for the insects of the entire world.

Correct identification is the first step, and probably the most important one, in dealing with a pest. It allows us to retrieve the information required for insect pest management. Without identification we have no basis for predicting injury and advising action.

Table 3.1. An example of an identification key as represented by one for the classes of Arthropoda. This key is used by starting with the couplet and following couplet routing according to the characteristics of the specimen until the class is identified.

Key to Classes of Arthropoda
This key includes all living classes of Arthropoda except Pentastomida (or Linguatulida) and Tardigrada. They are small groups of small to minute, seldom seen animals that lack antennae and are not likely to be taken for Arthropoda. In fact, some authorities do not consider them such.

---

1. Without antennae (in immature stages, or even in some adults of certain insects, antennae are much reduced or lacking; such forms will not key out well in this part of the key) --- 2
   With antennae-- 4
2. Usually with 7 pairs of appendages, 5 of them legs; abdomen rudimentary (sea spiders)-------------------------------- PYCNOGONIDA
   With 6 (rarely fewer) pairs of appendages, 4 (rarely 5) of them legs; abdomen usually well developed as a body region but sometimes fused with cephalothorax --- 3
3. Abdomen with booklike gills on underside of abdomen; large animals up to 50 cm long, with hard expanded shell and long spinelike tail (horseshoe crabs) -------------- class MEROSTOMATA, order or subclass XIPHOSURA
   Abdomen without booklike gills; smaller forms rarely over 7 cm long, body not as above (spiders, scorpions, mites, etc.)-----------------------ARACHNIDA
4. With 2 pairs of antennae (1 pair may be rudimentary in sowbugs); head merged with thorax to form cephalothorax; breathing by gills (crabs, lobsters, shrimps, sowbugs, etc.)------------------------------------- CRUSTACEA
   With 1 pair of antennae; head separate from thorax; breathing by tracheae 5
5. With 3 pairs of legs at some stage in life cycle; body in 3 divisions–head, thorax, and abdomen; true legs only on thorax, but abdomen sometimes with appendages resembling but differing from those on thorax; wings often present (all insects)-- INSECTA
   With 9 or more pairs of legs; most body segments behind head with legs; wings lacking; body more or less wormlike (myriapodan classes) --------------- 6
6. Most segments of body with double pairs of legs; slow-moving animals (millipedes) ---DIPLOPODA
   Legs not in double pairs; most segments of body with 1 pair -------------- 7
7. Body more or less flattened; not minute, with 15 or more pairs of legs; rapidly moving animals (centipedes)-------------------------------- CHILOPODA
   Body usually cylindrical; small to minute forms not over 8 mm long, with 9-12 pairs of legs --- 8
8. Antenna branched; 9 pairs of legs; minute animals 1-1.5 mm long, found in leaf litter, etc. (pauropods) --------------------------------- PAUROPODA
   Antenna not branched; 10-12 pairs of legs; white cylindrical, centipedelike animals 1-8 mm long (symphylans) ---------------------------- SYMPHYLA

SOURCE: Steyskal et al. (1986), Courtesy USDA.

ELEMENTS OF CLASSIFICATION

The classification of organisms is based on a hierarchy of categories, with the most inclusive occurring at the top and the least inclusive at the bottom. The major categories used in animal classification are **phylum, class, order, family, genus,** and **species**. But for added distinction in large, diverse groups, many other categories fall between these major ones. For example, a **subclass** category is commonly present below the class category and a **superfamily** category above the family category.

An example of the major categories for the European corn borer shows the following classification:

Phylum—Arthropoda
Class—Insecta
Order—Lepidoptera
Family—Pyralidae
Genus—*Ostrinia*
Species—*Ostrinia nubilalis*

Note that each category consists of only one word except the species category. The scientific name of a species is **binomial**; it is composed of two names, a genus name and a **specific name**, also called a **specific epithet**. Unlike all the higher categories, the specific name cannot stand alone; it must be used with the genus name. In zoology, it is conventional, but not required, for the species name to have the name of the person who first described the species as a suffix. Therefore, we might see the species name written as *Ostrinia nubilalis* (Hübner). The parentheses around the author name indicates that when Hübner originally described the species, he placed it in another genus. An author name *without* parentheses means that the species remains in the genus originally used by the describing author. Author names are used often in technical systematic literature; they are omitted in this book.

The system of **binomial nomenclature** we use today for classification was advanced by Swedish naturalist Carolus Linnaeus, who first used it consistently in 1758. Strict rules and conventions apply for name assignment, and for zoology, these are stated in the *International Code of Zoological Nomenclature*. Although they may be based in any language, scientific names are latinized and usually refer to some characteristic of the animal or group named. The binomial of a species is always printed in italics or, if hand-written, is underlined to indicate italics. The names of genera and higher categories begin with a capital letter, but the specific name of the species and subspecific names always begin with a lowercase letter, as in the subspecies *Diabrotica undecimpunctata undecimpunctata* Mannerheim. In this example the third name indicates a subspecies, a group of individuals of the same species that have differences in body form or color and geographical distribution.

A frequently used name for the species, the common name, is not covered by the formal rules of nomenclature. The name "European corn borer" is a common name, somewhat similar to a nickname. Such names are often used in insect pest management because they are easily pronounced compared with the Latin name. However, because no formal rules govern these names, we may find different names or versions used in different localities for the same species, and confusion may arise. Because there is only one scientific name, it is always the safest name to use to avoid problems of semantics. In the United States,

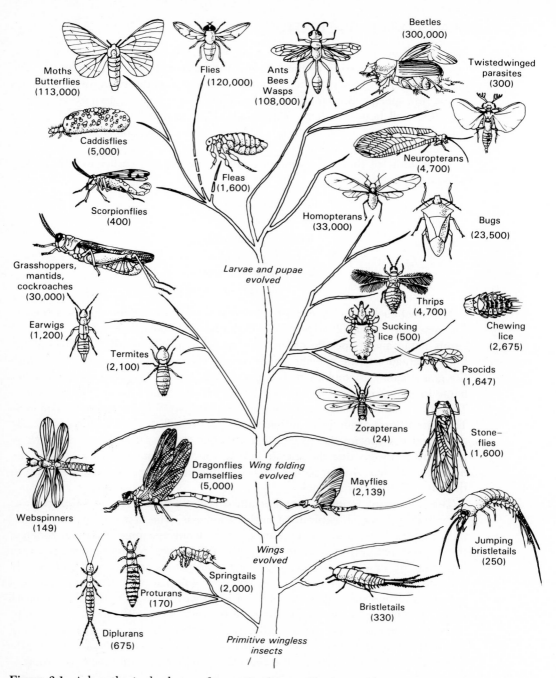

Figure 3.1 A hypothesized scheme of insect evolution. Groups at the base of the tree are the most primitive, and those at the top are the most highly evolved. Classifications of orders, in a general way, are based on such schemes. Numbers in parentheses are number of species. (Redrawn from Ross, 1962, *How to Collect and Preserve Insects*, Circular 39, Illinois Natural History Survey)

a list of common names approved by the Entomological Society of America is presented in *Common Names of Insects and Related Organisms*. But these names are merely recommended and are not necessarily followed by the international scientific community.

The species category plays the central role in systems of classification. It is the only category that is real; in other words, it is the only natural group. The species is a complex unit that is sometimes difficult to define, but a widely used definition applied to the species concept is as follows: a group of interbreeding individuals that are similar in body structure and that produce fertile offspring; moreover, these groups are reproductively isolated from other such groups. This is the biological species definition.

All other categories of the hierarchy are based on the species category but are abstractions. Even though they are derived from suspected pathways of insect evolution, these other categories are human contrivances used for convenience. Because this is so, we find many differences of opinion as to where to "lump" and where to "split" categories; therefore, many different schemes of insect classification result.

Consequently, there are no "correct" and "incorrect" systems. Rather, we have "successful" and "unsuccessful" systems based on their usefulness. In this book, we use the widely adopted scheme of D. J. Borrer and associates, mentioned earlier, with the exception of their class name.

GENERAL CLASSIFICATION OF INSECTS

Clearly, no one person can know all the kinds of insects in our environment. Not even trained entomologists instantly recognize every insect specimen presented them. However, a good working knowledge of the major insect groups is possible with a modicum of reading and laboratory study. The usual level of insect identification for communication with specialists is the order category. There are 28 orders in the scheme we will use, and discussions will focus on identifying characteristics and biological properties of each order. Seven of the 28 orders are the most diverse and account for the greatest share of our pest problems. Major families will be named and discussed in these orders. An overall scheme of insect ordinal classification follows, with major orders indicated by an asterisk (*). In general, this list proceeds from the most primitive insects to the most highly evolved.

Class Insecta

Subclass Apterygota—primitively wingless insects

    1. Protura—proturans
    2. Collembola—springtails
    3. Diplura—diplurans
    4. Thysanura—bristletails
    5. Microcoryphia—jumping bristletails

Subclass Pterygota—winged and secondarily wingless insects

    Division Exopterygota—simple body change during growth

        6. Ephemeroptera—mayflies
        7. Odonata—dragonflies and damselflies

*8. Orthoptera—grasshoppers, crickets, cockroaches, mantids, and walking-sticks
9. Dermaptera—earwigs
10. Isoptera—termites
11. Embioptera—webspinners
12. Plecoptera—stoneflies
13. Zoraptera—zorapterans
14. Psocoptera—psocids
15. Mallophaga—chewing lice
16. Anoplura—sucking lice
17. Thysanoptera—thrips
*18. Hemiptera—bugs
*19. Homoptera—aphids, scale insects, hoppers, cicadas, psyllids, and whiteflies

Division Endopterygota—complex body change during growth

20. Neuroptera—alderflies, antlions, dobsonflies, fishflies, lacewings, snakeflies, and owlflies
*21. Coleoptera—beetles
22. Strepsiptera—twisted-winged parasites
23. Mecoptera—scorpionflies
24. Trichoptera—caddisflies
*25. Lepidoptera—butterflies and moths
*26. Diptera—flies and mosquitoes
27. Siphonaptera—fleas
*28. Hymenoptera—ants, bees, wasps, and sawflies

A key for identification of specimens to order is given in Appendix 1.

Subclass Apterygota

The Apterygota are an assemblage of the most primitive insects, consisting of the orders Protura, Collembola, Diplura, Thysanura, and Microcoryphia. Insects in all these orders are said to be primitively wingless; they lack wings and their ancestors are believed to have been wingless. Moreover, internal structures that strengthen the thorax for flight in winged insects are absent in these orders. All members of this subclass grow with very little change in form during development, a pattern called **no metamorphosis**.

Entomologists differ considerably on the classification of this group. Many systematists feel that the orders Protura, Collembola, and Diplura are sufficiently different from the rest of the insects to be excluded from the class Insecta. In particular, insects in these orders have mouthparts housed within a cavity, and only the tips protrude when the insects are feeding. Such mouthparts are called **entognathous**. All other insects have exposed, or **ectognathous**, mouthparts. Whether or not considered insects, they are still commonly dealt with in the science of entomology.

ORDER PROTURA—proturans

Proturans are some of the most unusual insects. They are small and whitish (0.6–2.0 mm long), with styletlike mandibles and no eyes or **antennae**.

Figure 3.2 Order Protura, proturans. Insects in this order lack eyes and antennae. (Reprinted with permission of Macmillan Publishing Company from *Insects in Perspective* by Michael D. Atkins. Copyright © 1978 by Michael D. Atkins)

However, the front legs are held out in front of the head, possess many sensilla, and function somewhat like antennae. Another unusual feature of proturans, is that the young begin with nine segments and add three more during development, a phenomenon known as **anamorphosis**. The final segment in the adult abdomen is a primitive structure, the **telson**, hence the sometimes used name telsontails.

Proturans live in soil and decomposing plant material, where they feed on organic debris and fungal spores. They are not known to be pests.

ORDER COLLEMBOLA—springtails

The Collembola are an ancient group whose members possess a central extensible, tubelike structure attached to the underside of the first abdominal segment called the **collophore**, hence their ordinal name. The collophore, or "glue bar," is eversible and maintains secretions at the tip. It was believed that the structure was used as an aid in climbing on smooth surfaces, but more recent evidence suggests it may be used in water uptake or preening activity.

Collembolans are small, ranging from 0.2 to 10 mm in length, and may be gray, off-white, or more brightly colored green or purple. Body shape is either long and slender or rounded (globular). Mouthparts are usually the chewing type, but in some species, the maxillae and mandibles are modified into long, sharp stylets.

Figure 3.3 Order Collembola, springtails. Family Sminthuridae (top) and family Entomobryidae. Insects in this order have a ventral tubular collophore, and most have a ventral furcula near the tip of the abdomen.

Each eye occurs as a loose aggregation of single ommatidia, but eyes are absent in many true soil-dwelling and cave-dwelling forms.

The common name springtail comes from the **furcula** arising from the underside of the abdomen, near the tip. Most of the time, the furcula is "cocked" by a clasp, the **tenaculum**. When a springtail is disturbed, the furcula is released, propelling the insect for a distance several times its body length. The furcula is absent in species that inhabit deeper levels of soil.

Springtails are found most frequently in moist environments, including soil, decaying leaves and wood on forest floors, at the edges of ponds, on snow, and among fleshy fungi. Some species of the family Sminthuridae (globular forms) are found on green plants, where they are occasional pests. The most infamous in this regard is the Lucerneflea, *Sminthurus viridis*, which has been a notable pest of alfalfa in Australia, and the garden springtail, *Bourletiella hortensis*, a pest of garden vegetables. Other species, for example, *Hypogastrura armata*, also are pests of commercially grown mushrooms.

ORDER DIPLURA—diplurans

Diplurans are small, blind, whitish insects, less than 7 mm long, with chewing mouthparts and many-segmented antennae. Their name is derived from the two prominent **cerci** at the tip of the abdomen that occur as styli in some (for example, the family Campodeidae) and forceps in others (family Japygidae).

Because of their small size, low abundance, and secretive habits, diplurans are rarely seen. They occur most often in soil and soil surface debris. Very little is known about the ecology of these insects, and no species is known to be a pest.

ORDER THYSANURA—bristletails

Thysanura are a group of medium-sized insects, 7–19 mm long, that have a somewhat flattened body and ectognathous chewing mouthparts. The most distinctive features of these insects are the two long cerci and a median caudal filament that occur at the end of the abdomen, giving the appearance of three "tails." Scales,

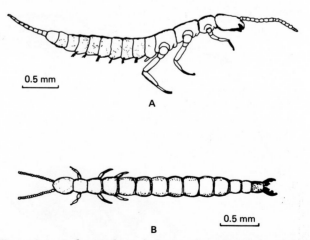

Figure 3.4 Order Diplura, diplurans. Families Campodeidae (top) and Japygidae. Insects in this order are blind and have two cerci at the tip of the abdomen. (Reprinted with permission of Macmillan Publishing Company from *Insects in Perspective* by Michael D. Atkins. Copyright ©1978 by Michael D. Atkins)

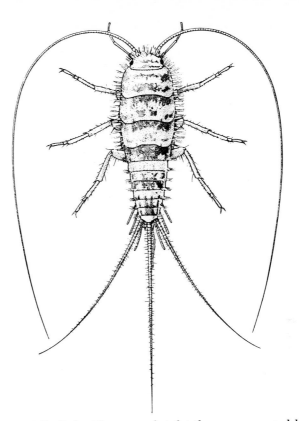

Figure 3.5 Order Thysanura, bristletails, as represented by the common household pest, the firebrat, *Thermobia domestica*. The most common features in this order are the two long cerci and the median filament at the tip of the abdomen. (After Steyskal, et al., 1986, courtesy USDA)

which cover the body and often give a silver sheen, occur in many species, as well as small compound eyes. Eyes or scales may be lacking in others.

Some bristletails are found in woodlands under decaying bark, but others may occur in caves, mammal burrows, or in ant and termite nests. The most well known in North America occur in households and other human dwellings. Here, the two most frequently encountered species are silverfish, *Lepisma saccharina*, and firebrats, *Thermobia domestica*; both are members of the family Lepismatidae. Temperature requirements may vary from cool, damp situations favored by silverfish to warm, dry areas like furnace rooms, inhabited by firebrats.

Although they are minor household pests, silverfish and firebrats do cause damage to possessions by feeding on starchy substances. Where books are stored, they feed on starch in bindings and on labels. In other instances they have been known to feed on starched clothing, curtains, and starch paste in wallpaper.

ORDER MICROCORYPHIA—jumping bristletails

Jumping bristletails are very much like the bristletails in size and appearance. They differ from bristletails in having a cylindrical body and several abdominal styli arising beneath the abdomen, in addition to the three "tails." All species have large, compound eyes and chewing mouthparts.

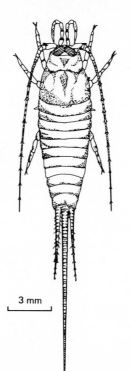

Figure 3.6 Order Microcoryphia, jumping bristletails. Although these insects look somewhat like bristletails, their body is more cylindrical, and their compound eyes are large and contiguous. (Reprinted with permission of Macmillan Publishing Company from *Insects in Perspective* by Michael D. Atkins. Copyright ©1978 by Michael D. Atkins)

The name of this order is derived from the ability of these insects to jump (sometimes distances of 25 to 30 cm) when disturbed. They occur most frequently in wooded habitats under leaves, rocks, and bark of decaying logs. Jumping bristletails are believed to feed mainly on algae, but they may also eat lichens, mosses, and other materials. No species of Microcoryphia is considered a pest.

Subclass Pterygota

The most important characteristic of the subclass Pterygota is the possession of wings in the adult stage. Although some adult Pterygota lack wings, for example, fleas, they are believed to have evolved this condition secondarily; their ancestors probably had wings.

Pterygota are variously divided by systematists into large groups of orders based on the presence or absence of wing-flexing mechanisms, internal or external growth of wings, or complexity of the developmental process. Orders of insects whose wings do not flex back over the body are usually placed in the infraclass **Paleoptera**; orders with folding wings are placed in the infraclass **Neoptera**. Neoptera comprise by far the largest infraclass. In other classification schemes, orders with individuals that have a simple form of body change during growth and external wing development are placed in the division **Exopterygota**; those with complex changes in body form and internal wing development are placed in the division **Endopterygota**.

ORDER EPHEMEROPTERA—mayflies

Mayflies are among the most primitive of the winged insects and are characterized by triangular, net-veined, membranous wings and two or three filamentous appendages at the tip of the abdomen. These insects are soft-bodied, 10–40 mm long, with short, bristlelike antennae and mouthparts that are reduced and nonfunctional in the adult. Frequently, adults are found near lakes, ponds, and streams and are attracted in large numbers to lights.

Mayflies lay eggs on the surface of water or on objects in the water. The young remain and develop in the water, feeding on algae and detritus. An immature looks like the adult in overall body shape but has well-developed chewing mouthparts, no functional wings, leaflike gills on the sides of the abdomen, and usually three "tails."

Adult mayflies emerge in huge swarms over bodies of water and mate; females lay eggs within a few minutes to a few hours after mating. The adults do not feed and die within a day or two, therefore the name Ephemeroptera (*ephemera*; for a day, short-lived). Upon death, mayfly bodies may pile up to a depth of over a meter along shorelines, and sometimes their presence makes roadways and bridges slick and dangerous for traffic.

The major importance of mayflies is food for freshwater fish and other aquatic animals. Indeed, many fishermen recognize the importance of mayflies as fish

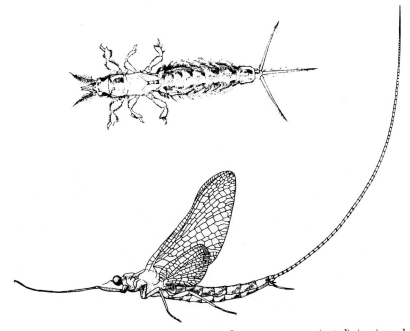

Figure 3.7 Order Ephemeroptera, mayflies. Immature (naiad) (top), and adult. The most common features of this order are leaflike abdominal gills and three "tails" on naiads. Adults have triangular net-veined wings and two or three filamentous abdominal appendages. (After Steyskal, et al., 1986, courtesy USDA)

food, designing fishing "flies" modeled after mayflies. Because of different habitat preferences among species, mayflies also are useful as indicators of aquatic conditions and concentrations of pollutants upstream.

ORDER ODONATA—dragonflies and damselflies

Members of the Odonata are large insects that are usually conspicuous because of their sometimes brightly colored bodies and banded wings. Dragonflies and damselflies have four net-veined, membranous wings and large heads, with large, protruding eyes and long, slender bodies. Antennae are short and inconspicuous, and mouthparts are the well-developed chewing type.

Adults feed on insects caught in flight, including mosquitoes and midges. Females lay eggs in water. Immatures are aquatic and also prey on other insects and small animals in ponds, lakes, and backwaters of streams. These immatures do not swim but walk along the bottom, among vegetation and debris, in search of prey. Development to the adult stage requires from one to four years.

Figure 3.8 Order Odonata, dragonfly (top) and damselfly. The most common characteristics of this order are four equal-sized, net-veined wings; large, protruding eyes; and well-developed chewing mouthparts. Dragonflies (suborder Anisoptera) are more heavy-bodied than damselflies (suborder Zygoptera) and hold their wings out when at rest. Damselflies hold their wing together above their backs. (After Steyskal et al., 1986, courtesy USDA)

Suborder Anisoptera—dragonflies. Adult dragonflies are heavy-bodied and have strong, well-controlled flight. Males often establish territories that they patrol in search of food and defend against other dragonfly males. The rustle of wings is often heard when two males clash.

Dragonfly immatures are robust-bodied and usually found in mud at the bottom of lakes and ponds. They lack external gills but have internal rectal gills, as mentioned in Chapter 2.

Suborder Zygoptera—damselflies. Damselflies are slender and more delicate than dragonflies. Adults display a fluttering type of flight, as opposed to the strong directional flight of dragonflies. When at rest, the wings of damselflies are held together about the back, almost parallel with the body.

Immature damselflies also are more slender than immature dragonflies. They possess three leaflike gills at the tip of the abdomen and are more often found near submerged vegetation than on the very bottom of lakes and ponds.

*ORDER ORTHOPTERA

This is the first of the major orders that contain important pest species. Although some are wingless, most have two pairs of wings, with parchmentlike front wings, the **tegmina**, and membranous, fanlike hind wings. These wings are folded over the back when not in use. Mouthparts are the typical chewing type, and a pair of short cerci is present at the top of the abdomen. There is little change in body form during growth and development (gradual metamorphosis). In temperate climates, Orthoptera often have only one generation each year, and the winter is usually passed in the egg stage.

Sound production is very common among the Orthoptera, particularly in grasshoppers and crickets. Sounds are made by rubbing their body parts together. This behavior, called **stridulation**, usually is involved in mate finding.

Family Acrididae—grasshoppers or locusts. Grasshoppers are some of the best known insects. They are characterized by threadlike antennae that are shorter than the body and by hind legs modified for jumping. The ovipositor of females is short, and the **tarsi** are three-segmented.

Grasshopper eggs are usually laid in soil and frequently in grass sod. In temperate regions, winters are spent in the egg stage, and hatching occurs in the

Figure 3.9 Order Orthoptera, family Acrididae, represented by the American grasshopper, *Schistocerca americana*. As with most other members of this order, the front wings are parchmentlike, and the hind wings are membranous. Grasshoppers or locusts have antennae that are shorter than the body and legs with three-segmented tarsi. (After Riley, courtesy USDA)

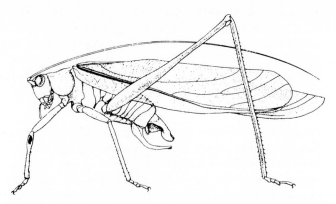

Figure 3.10 Order Orthoptera, family Tettigoniidae, represented by a katydid. Sometimes called "longhorn grasshoppers," members of this family have longer antennae than body and legs and four segmented tarsi. (After Steyskal, et al., 1986, courtesy USDA)

spring. Grasshoppers feed on a wide array of green plants and are some of the most serious agricultural pests in the world. Outbreaks occur most often in the western United States and in drier parts of the world, or following protracted droughts in wetter areas.

Family Tettigoniidae—longhorn grasshoppers and katydids. Longhorn grasshoppers have antennae longer than the body and some, such as the well-known Mormon cricket (*Anabrus simplex*), are wingless. All have four-segmented tarsi, and most females possess a saberlike ovipositor. Many longhorn grasshoppers are green, as with most katydids and false katydids, and several make elaborately patterned sounds. Eggs are laid singly or in groups, but without pods, on leaves or twigs of trees and shrubs, within plant tissues, or in the soil.

Very few species of the Tettigoniidae are known as pests. Occasionally katydids injure orange trees, but the most outstanding pest in this family is the Mormon cricket. This pest occurs in the Great Basin region of the Rocky Mountains in North America, where outbreaks have caused severe injury to rangeland and cultivated crops. The most famous outbreak occurred in 1848 when this pest threatened complete destruction of crops of the Mormon settlement. However, flocks of seagulls moved in to consume these insects, and disaster was averted. A monument to commemorate this event was erected in Utah.

Family Gryllidae—crickets. Crickets are known by almost everyone, as are their chirping sounds. These insects are usually recognized by their long antennae, a lance-shaped ovipositor, and three-segmented tarsi.

Field crickets, *Gryllus* species, are some of the most widely distributed gryllids in North America. Most are black or brown. They feed on a wide array of foods, ranging from plants to other insects, and sometimes they are cannibalistic. They may damage field crops or invade homes, where they may chew holes in clothing.

Other pests in the Gryllidae include the tree crickets, *Oecanthus* species, which insert eggs in bark and stems of trees and shrubs, sometimes causing young twigs to die.

Family Phasmatidae—walkingsticks. These are some of the most unusual-

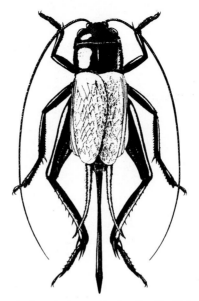

Figure 3.11 Order Orthoptera, family Gryllidae, a cricket. Members of this family have long antennae, a lanceshaped ovipositor, and three-segmented tarsi. (After Steyskal, et al., 1986, courtesy USDA)

looking American insects; they are usually wingless and bear a striking resemblance in shape and color to the tree twig or grass stem on which they are found. They move slowly through the vegetation, consuming leaf material. Females simply drop their eggs to the ground. There is usually only one generation produced per year.

The longest insect in the United States is a walkingstick, *Megaphasma denticrus*, reaching lengths of 15 to 18 cm. Tropical walkingsticks may exceed 30 cm in length!

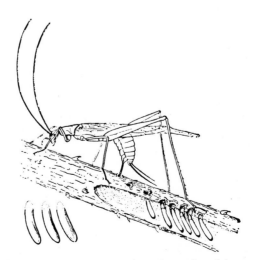

Figure 3.12 A snowy tree cricket, *Oecanthus fultoni*, inserting an egg into a raspberry cane. (Courtesy USDA)

10 mm

Figure 3.13 Order Orthoptera, family Phasmatidae, walkingsticks. These insects resemble tree twigs or grass stems in color and shape. (Reprinted with permission of Macmillan Publishing Company from *Insects in Perspective* by Michael D. Atkins. Copyright ©1978 by Michael D. Atkins)

Family Mantidae—mantids. The mantids, or praying mantids, also have a striking appearance because of their highly modified front legs and elongated prothoraces. The common name, praying mantis (soothsayer, holy or wise man, prophet), comes from the habit of holding the front legs up in a "praying" attitude while waiting for prey. They grasp other insects (including other mantids) with the spined front legs and hold them with these while they eat. Being wholly predaceous, mantids are the only group of the Orthoptera beneficial to humans outside of the laboratory and are sometimes collected for "pets."

Eggs are laid in a grayish egg case, which is glued to tree branches, fence posts, weeds, or other objects, and one generation occurs each year.

Figure 3.14 Order Orthoptera, family Mantidae, as represented by the Carolina mantid, *Stagmomantis carolina*.Members of this family have front legs modified for grasping prey and an elongated prothorax. (After Steyskal, et al., 1986, courtesy USDA)

Superfamily Blattoidea—cockroaches. Many people find cockroaches the most repulsive of all insects. This feeling arises because some species live in close association with people, inhabiting cupboards, sink cabinets, and other places where food is prepared or stored. However, not all cockroaches are found in human dwellings, and most species occur out of doors in tropical or subtropical climates. Cockroaches prefer warm, dark, humid habitats, where they breed year-round. Their common name is derived from the Spanish *cucaracha*.

Cockroaches have an oval shape and a somewhat flattened body and are usually brown or brownish-black. The head is oriented downward and is barely visible from above because an extended **pronotum** covers most of it. Very often, two pairs of fully developed wings are present, but occasionally, they are reduced or lacking. Cockroaches feed on many kinds of plant and animal products, including human food, sewage, garbage, binding and sizing of books, and wallpaper paste. Most species lay eggs in egg pods, or **oothecae**, and reproduction occurs year-round. Most urban species cannot survive out-of-doors in temperate climates.

Cockroaches are economically important because of the loss of revenue in managing them. Many home dwellers attempt pest control themselves using relatively safe household pesticides. Elimination of cockroaches in commercial establishments is usually achieved by contracting the services of structural pest control firms. Most pest species in the United States are members of the families Blattidae and Blattellidae.

ORDER DERMAPTERA—earwigs

Earwigs are medium-sized insects, about 13 mm long, that superficially resemble certain types of beetles. They are elongate insects with a pair of pincerlike cerci at the tip of the abdomen. These insects may have wings or be wingless. Those with wings have front wings that form short, leathery coverings for the large membranous hind wings. Earwigs have chewing mouthparts and three-segmented tarsi, and show little change in body form during growth (gradual metamorphosis). Some species have glands with openings on the abdomen that emit a foul-smelling liquid believed to serve as protection from enemies.

Figure 3.15 Order Orthoptera, superfamily Blattoidea, represented by the German cockroach. Insects in this superfamily have an oval, flattened shape and head concealed under the pronotum. (After Steyskal, et al., 1986, courtesy USDA)

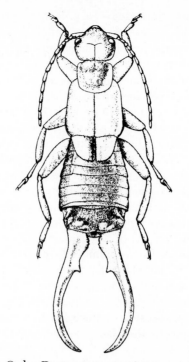

Figure 3.16 Order Dermaptera, earwigs, represented by the Eurpoean earwig, *Forficula auricularia*. Insects in this order have short front wings that form leathery coverings for the hind wings and a pair of pincerlike cerci. (After Steyskal, et al., 1986, courtesy USDA)

Earwigs are active at night and hide during the day. They feed mostly on decaying organic matter, but a few occasionally feed on plants, and others are predaceous. Eggs are laid in a cell in soil, and females watch over the eggs and young. In temperate climates, there is only one generation per year.

The name earwig is believed to have come from the erroneous belief that these insects crawl into people's ears. They do not bite but will pinch, sometimes painfully so, when handled. The greatest concern with earwigs is as a nuisance when they enter households.

ORDER ISOPTERA—termites

Termites are the most dreaded insect pests of human dwellings and other structures because of their ability to destroy wood and wood products. Termites are often called "white ants" because of their color and social habit. Although about the same size, their bodies differ from those of ants by having an abdomen broadly joined to the thorax versus one joined by a narrow waist. The front and hind wings of termites are the same shape and size, but in ants the front wings are the larger. Moreover, termites spend their lives in wood, soil, or earthen tunnels they build, but ants move openly above ground. Termite mouthparts are the chewing type, and there is little body change during growth (gradual metamorphosis). Antennal segments are beadlike.

Termites are social insects, living in communal nests and having a division of labor among individuals. The termite society is based on castes, with members

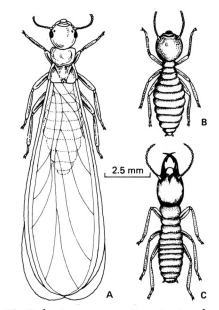

Figure 3.17 Order Isoptera, termites. A winged reproductive (left), a worker (top right), and a soldier. These insects superficially resemble ants but their abdomen is broadly joined to the thorax. (Reprinted with permission of Macmillan Publishing Company from *Insects in Perspective* by Michael D. Atkins. Copyright ©1978 by Michael D. Atkins)

of each caste differing from those of other castes in body structure and behavior. In most common species, these castes include primary reproductives (kings and queens), supplementary reproductives, workers, and soldiers.

New termite colonies are formed, usually in the spring or fall, by winged primary reproductives that swarm from the parent colony. After a short flight, the wings break off, and males and females form pairs. Each pair excavates a cell in soil or wood, and they mate. The original king and queen remain mated for life, and both are involved in rearing the first termite brood. Workers and soldiers are produced in the first brood, after which the queen's primary activity is egg laying.

Workers are immatures or undifferentiated individuals that are wingless and lack compound eyes. They do the work of the colony, including gathering food; feeding reproductives, soldiers, and the very young; constructing nests, associated tunnels, and gallaries; and other duties.

Soldiers have enlarged heads and/or mandibles, and they may or may not have compound eyes. When present in a colony, soldiers attack intruders or protect the structural integrity of the nest. In some species soldiers are called **nasuti**; these individuals use a glandular snout to spew a sticky material over intruders.

In colonies where they occur, supplementary reproductives replace the reproduction of the primary king and queen when the latter decline and die or are lost.

Most termites are saprophagous, feeding on dead wood and other nonliving plant materials. The cellulose of wood is a major termite food component. As mentioned in Chapter 2, cellulose is digested with the aid of symbiotic microorganisms that supply the needed enzymes for this purpose. By having the ability to break down woody tissues, termites have extraordinary ecological importance in the humification of soil.

Two groups of termites in the United States are of major economic importance: dry-wood termites and subterranean termites.

Family Kalotermitidae—drywood termites. These insects are found in the southwestern United States and attack sound, dry wood of buildings, stored lumber, and furniture. They do not require external moisture or contact with the ground, as do subterranean termites. Another group in this family, powderpost termites, occurs in the southeastern United States and also attacks dry wood, turning it into sawdust.

Family Rhinotermitidae—subterranean termites. These termites cause damage throughout most of the United States and account for approximately 95 percent of monetary losses from termites. Subterranean termite colonies must always have contact with the soil. They move above ground only by building earthen tubes that lead to wood not touching the soil surface. These tubes allow the insects to get food and also obtain moisture for survival.

Damage from subterranean termites is prevented by proper construction— keeping wood away from soil. Mechanical barriers composed of metal plates separating concrete foundations from wood framing are also useful. In addition, the soil of the building site can be treated with long-lasting insecticides as an added preventive. Drywood termite damage is usually prevented by insecticide fumigation of buildings. This fumigation involves enclosing the entire structure in a plastic envelope.

ORDER EMBIOPTERA—webspinners

Webspinners are small insects, 4–7 mm long, that are rarely encountered. Males may be winged or wingless, and females are always wingless. Mouthparts are

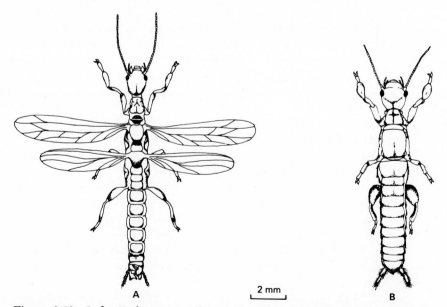

Figure 3.18 Order Embioptera, webspinners. Winged male (left) and wingless female. Members of this order have threadlike antennae, three-segmented tarsi, and short cerci. (Reprinted with permission of Macmillan Publishing Company from *Insects in Perspective* by Michael D. Atkins. Copyright © 1978 by Michael D. Atkins)

the chewing type, antennae are threadlike, and tarsi are three-segmented. There is a pair of short cerci at the tip of the abdomen. An unusual characteristic of webspinners is the presence of silk glands and spinning hairs on the first segment of each front tarsus. There is little change in body form from young to adult stages (gradual metamorphosis).

These insects are most common in tropical and subtropical areas. They are gregarious, and some build communal silken tunnels under stones, tree bark, logs, and in other environments. Small numbers of eggs are laid in these tunnels.

Webspinners feed on dead plant material, lichens, moss, and bark, and some prefer fungi. Eggs within the silk-lined gallaries are tended by the females, and there is probably one generation per year. No webspinners are known as pests.

ORDER PLECOPTERA—stoneflies

Like mayflies, stoneflies are associated with aquatic environments. They are drab gray, green, brown, or black, and adults have two pairs of membranous,

Figure 3.19 Order Plecoptera, stoneflies, represented by *Neoperla clymene*. Adults in this order have membranous, veined wings longer than the body; long, slender antennae; and three segmented tarsi. (After Steyskal, et al., 1986, courtesy USDA)

multiveined wings that exceed the length of the body. The wings fold flat over the abdomen at rest. Antennae are long and slender, and tarsi are three-segmented. There is a prominent pair of cerci at the tip of the abdomen, and mouthparts, although reduced in the adults of some species, are the chewing type.

Body form is similar in young and adults. Stonefly young resemble those of mayflies but have two tails rather than three and have branched gills on the thorax and leg bases rather than along the abdomen.

The name stonefly comes from the habits of the immatures, which crawl under stones on the bottoms of streams and along lake shores. Food of the young of some species is plants, but other species are predaceous or even omnivorous. As their name implies, winter stoneflies are active as adults and reproduce during the winter. Adults are weak fliers and, in some species, do not feed. The smaller stonefly species produce one generation per year, but those with larger individuals may require more than one year to complete development.

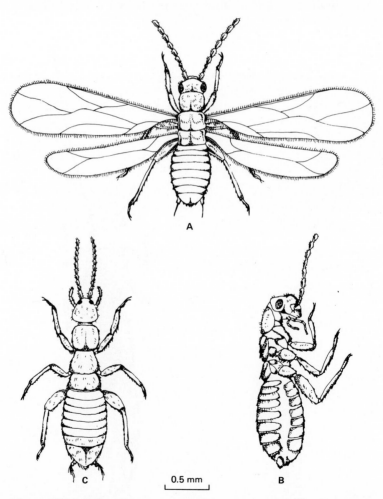

Figure 3.20 Order Zoraptera, zorapterans, represented by *Zoroptypus hubbardi.* Winged female (top), wingless female (bottom, left), and dewinged female. (Reprinted with permission of Macmillan Publishing Company from *Insects in Perspective* by Michael D. Atkins. Copyright ©1978 by Michael D. Atkins)

Stoneflies are largely beneficial insects, serving as food for fish and other aquatic animals. None is known to be a pest.

ORDER ZORAPTERA—zorapterans

Zorapterans are very rare, minute insects (fewer than 3 mm long) that superficially resemble termites. Adults of either sex may be winged or wingless. When present, both pairs of wings are membranous with few veins. Wingless forms lack eyes. Antennae are beadlike, and a pair of short cerci that each terminate in a bristle are found at the tip of the abdomen. The tarsi are two-segmented, and mouthparts are the chewing type. Immatures resemble adults (gradual metamorphosis).

Aggregations of zorapterans may be found in rotten logs, under bark, and in piles of sawdust. Food seems to be fungal spores and dead arthropod tissue. Zorapterans are not known as pests.

ORDER PSOCOPTERA—psocids

Psocids are small, delicate insects, fewer than 6 mm long. Adults usually have two pairs of membranous, sparsely veined wings that are held rooflike over the body when not in use and antennae that are long and threadlike. These insects have chewing mouthparts and two- or three-segmented tarsi.

Most psocids occur on tree bark, in the foliage of trees and shrubs, and under stones and dead leaves. These species are often called barklice. However, the psocids most often seen by the average person occur in buildings among books and papers; these are frequently called booklice. The indoor species are mostly pale and wingless.

Psocid eggs are laid singly or in clusters and may be covered by a silken web. There is little change in body form during growth to the adult stage (gradual metamorphosis). Some species are gregarious, and most feed on pollen, fungi,

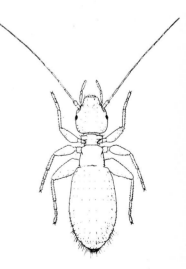

Figure 3.21 Order Psocoptera, psocids, represented by *Ectopsocus pumilis* (left) and *Liposcelis divinitorius*. Psocids may be winged or wingless and have long, threadlike antennae and two- or three-segmented tarsi. (After Steyskal, et al., 1986, courtesy USDA)

cereals, and nonliving organic material. Psocids cannot be considered pests because of damage. They may be considered so only because of their unwanted presence.

ORDER MALLOPHAGA—chewing lice

Chewing lice are parasites of both birds and mammals, but because many species are found on birds, they are often called bird lice. These small, flattened parasites are **wingless** and **have a large head that is wider than the thorax**. They have reduced eyes and short antennae. As the common name implies, they have **chewing mouthparts**. The young greatly resemble adults (gradual metamorphosis).

The tarsi of chewing lice have either one or two claws at their tips. Although a few exceptions exist, those with one claw usually parasitize mammals, and those with two claws parasitize birds. These claws form clasping organs adapted for clinging to hair and feathers.

The Mallophaga are ectoparasites; they live and feed on the outside of their hosts. Food of these insects consists mainly of feathers, hair, skin scales, blood clots, and other surface materials of the host. They are usually quite specific in choosing hosts, and the entire life is spent on the host. Eggs, commonly called "nits," most often are glued near the bases on feathers or hairs, and three to seven weeks may be required for development of adults.

The principal injury caused by these insects is irritation of the host. In domestic animals, such irritation can reduce the rate of weight gain to cause a subsequent loss in profit. However, deaths of young chickens are known as a result of louse infestations. Chewing lice do not transmit disease organisms and do not parasitize humans. Two of the more important pests of poultry are the chicken head louse, *Cuclotogaster heterographus*, and the large turkey louse, *Chelopistes*

Figure 3.22 Order Mallophaga, chewing lice, represented by *Meromenopon meropis*. Chewing lice are wingless ectoparasites with a head wider than the thorax. (After Steyskal, et al., 1986, courtesy USDA)

meleagridis. Both are members of the family Philopteridae, the largest family in the Mallophaga.

Infestations by these parasites are usually managed by dipping animals or dusting them with suitable insecticides.

ORDER ANOPLURA—sucking lice

Sucking lice are small, wingless ectoparasites that feed on the blood of mammals, including that of humans. Anoplura are distinguished by their narrow heads (narrower than the thorax), mouthparts formed for piercing-sucking, and flattened bodies. Eyes are reduced or absent, and legs bear a single claw that folds back against the **tibia**. This claw and tibia arrangement forms an efficient mechanism for grasping hair. There is little change in body form during growth (gradual metamorphosis), and the entire life is spent on the host.

These parasites are pests because of the irritation they cause, but some can transmit disease organisms as well. The most important lice that infest humans are two species: head and body lice and crab lice.

Family Pediculidae—head and body lice. Head and body lice belong to the same species, *Pediculus humanus*, but are divided into subtaxa, according to the region of the body they infect. The head louse, *Pediculus humanus capitis*, is found only on the head and is transmitted from person to person by the sharing of caps, hair brushes, combs, and other such items. Body lice, or "cooties," *Pediculus humanus humanus*, are found on the body and are transmitted by shared clothing and bedding or migration from one pile of clothes to another.

Figure 3.23 Order Anoplura, sucking lice, as represented by *Polyplax stephensi*. Sucking lice are wingless ectoparasites, as are the Mallophaga, but have a head narrower than the throax and mouthparts formed for piercing and sucking. (After Steyskal, et al., 1986, courtesy USDA)

Head lice attach their eggs to hair, and body lice lay them along seams of clothing. These lice feed at frequent intervals by piercing the host's skin to suck a blood meal. The irritation from their activities causes itching. The body louse transmits microorganisms that cause epidemic typhus in humans. Infection occurs when louse feces or the crushed body of the louse is scratched into the skin but not by the louse's bite. Until the development of DDT, typhus epidemics associated with war often killed more people than combat! Other diseases associated with body louse transmission are **relapsing fever** and trench fever, the latter important in World War I.

Today, body lice are not as common as in the past because of more frequent bathing and laundering of clothes. However, in many cities of the United States, head louse epidemics continue to be frequent among young school children.

Family Pthiridae—crab lice. Crab lice, *Pthirus pubis*, are shorter and more rounded than head and body lice. These parasites occur mainly in human pubic hair but may be present on other parts of the body, including eyelashes and facial hair. Eggs are glued to body hair. This species is irritating but does not transmit disease organisms.

Sucking lice of domestic animals. Several sucking lice are important pests of domestic animals. Some of the more significant are the hog louse, *Haematopinus suis*, on hogs and the shortnosed cattle louse, *Haematopinus eurysternus*, on cattle. Most lice on domestic animals are managed with appropriate insecticidal sprays and dips.

ORDER THYSANOPTERA—thrips

Thrips (the singular is also thrips) are very small insects, usually 2 to 3 mm long. The most striking features of thrips are the two pair of wings that are fringed with long hairs and the conelike rasping-sucking mouthparts. The tarsi are also quite unusual; each is one- or two-segmented, lacks claws at the tip, and has a bladderlike organ that protrudes when the tarsus touches an object. The body form of immatures is generally the same as that of adults, and a quiescent period occurs in full-grown immatures just before they reach adulthood (yet still considered a gradual metamorphosis). Thrips have relatively large compound eyes and six- to nine-segmented antennae.

Most thrips reproduce sexually; however, some need not mate to produce young. Thrips may or may not have an ovipositor; those that do lay eggs in plant tissues, and those that do not lay eggs in crevices or under loose bark. Several generations usually are produced each year.

Thrips often occur in great numbers in flowers and feed on a wide array of cultivated plants. They can be pests on onions, pears, cotton, soybean, citrus, and greenhouse plants. Some are predaceous. Plant injury is produced when these insects rasp the surface of leaves, stems, buds, and flowers, often causing necrotic strips (silvery appearance) or the dieback of stems. In some instances, injury is most severe in the seedling stage, when feeding is concentrated on a small surface. Some plant disease organisms are vectored by thrips, for example, the virus that causes tobacco ringspot disease in soybean.

Some of the important pest species are onion thrips, *Thrips tabaci*, attacking onions, melons, and other garden crops; flower thrips, *Frankliniella tritici*, attacking flowers, wheat, strawberry, and other crops; and citrus thrips, *Scirtothrips citri*, feeding on buds, new growth, and fruit of citrus trees.

Figure 3.24 Order Thysanoptera, thrips, represented by the flower thrips, *Frankliniella tritici*. Major features found in this order include wings fringed with long hairs; conelike, rasping-sucking mouthparts; and two-segmented tarsi without claws. (After Steyskal, et al., 1986, courtesy USDA)

*ORDER HEMIPTERA—bugs

Although many people refer to all insects as "bugs," the only true bugs are hemipterans. These are some of the most common insects, occurring in both terrestrial and aquatic habitats. Although many of our most important plant pests are bugs, several also are important natural enemies that help to destroy insect pests. Yet other bugs are ectoparasites of humans and other warm-blooded animals.

Bugs have a somewhat flattened body, and they have an abdomen that is broadly joined to the thorax. Most have two pairs of wings; the front wings, called **hemelytra**, are the most distinctive characteristic of the order. Hemelytra are thickened at the bases and membranous at the tips, and when at rest, they overlay the membranous hind wings. Between the bases of the wings, there is usually a triangular plate, called a **scutellum**, that may cover much of the body in some species.

The mouthparts of hemipterans are the piercing-sucking type, with which bugs pierce tissues of plants or animals and extract sap or blood. Many bugs have scent glands from which they secrete disagreeable odors, presumably for protection. Hemipteran antennae are moderately long, and compound eyes are usually well developed.

Eggs are laid most often in plant tissues, on the surface of plants, or in crevices. Growth and development is gradual, with immatures resembling adults but lacking fully developed wings and, sometimes, adult coloration.

The Hemiptera comprises 44 families that are often grouped into three suborders. Only a few of the prominent terrestrial families are mentioned here.

Family Pentatomidae—stink bugs. Stink bugs are relatively large, triangular bugs, with five-segmented antennae and a large scutellum. The common name is derived from the disagreeable odor they produce. Most overwinter as inactive adults in the northern United States, but they may be active year-round in the South. Eggs are usually barrel-shaped and are laid in clusters.

Stink bugs are largely plant feeders, sucking sap from leaves, stems, and reproductive parts. An imported species, the southern green stink bug, *Nezara viridula*, is a particularly important pest in the southern United States, where it injures soybean, peas, beans, cotton, tomato, pecan, citrus, and other crops. On the other hand, the spined soldier bug, *Podisus maculiventris*, is beneficial to humans by feeding on many plant pests and particularly on caterpillars. Other species, such as the green stink bug, *Acrosternum hilare*, are both phytophagous and predaceous.

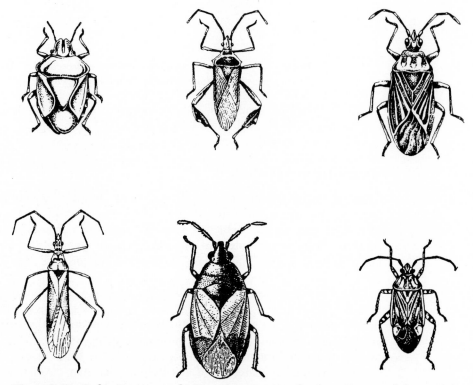

Figure 3.25 Order Hemiptera showing representative families. Top row (left to right): Family Pentatomidae (stink bug), family Coreidae (leaffooted bug), and family Lygaeidae (seed bug). Bottom row: Family Nabidae (damsel bug), family Anthocoridae (minute pirate bug), and family Miridae (plant bug). The major features of the Hemiptera include an abdomen broadly joined to the thorax, and front wings (hemelytra) that are thickened at the base and membranous at the tip. (Reprinted with permission of Macmillan Publishing Company from *Insects in Perspective* by Michael D. Atkins. Copyright ©1978 by Michael D. Atkins)

Family Coreidae—leaffooted bugs.   The leaffooted bugs are so named for the hind tibia, which is expanded and leaflike in some species. These insects have four-segmented beaks and antennae. Most are plant feeders, and the squash bug, *Anasa tristis*, which injures cucurbits, is probably the most important pest.

Family Lygaeidae—seed bugs.   This is a large family of small bugs much like the Coreidae, but they have only a few veins in the membrane of the hemelytra. The most important pest of this group is the chinch bug, *Blissus leucopterus*, which has been a devastating pest of wheat, corn, and other cereals. Most recently it has become an important pest of turf grasses. Another group of lygaeids, the Geocorinae (*Geocoris* species and others), is important as natural enemies of insect pests.

Family Nabidae—damsel bugs.   These bugs are slender and have a number of small cells formed by veins around the membranous margins of the hemelytra. Damsel bugs, *Nabis americoferus* and other species, are predaceous on many different insects, particularly aphids and caterpillars.

Family Anthocoridae—minute pirate bugs.   These are very small bugs (2–5 mm long) that often have black-and-white markings and a three-segmented beak. Many of the species are predaceous on other small insects and insect eggs. One of the most beneficial of these natural enemies is the insidious flower bug, *Orius insidiosus*, an important predator of the corn earworm, *Heliothis zea*.

Family Miridae—plant bugs.   This family contains more species than any other in the order. These insects are recognized by the presence of a **cuneus** (a special part of the hemelytron set off by a crease) and one or two closed cells in the membranous part of the wing. Although the name implies that these are all plant feeders, a few species are predaceous. One of the most important pests in this group is the tarnished plant bug, *Lygus lineolaris*, which causes damage to alfalfa, vegetables, and flowers in the eastern United States. A related species, *Lygus hesperus*, is an important pest of alfalfa and cotton in the western United States.

*ORDER HOMOPTERA

The Homoptera is a very large, important group of insects closely related to the Hemiptera. In fact, some authorities group the Homoptera as a suborder of the Hemiptera. They are so diverse that no single common name represents the whole order.

Homopteran mouthparts are piercing-sucking like those in the Hemiptera, but unlike the Hemiptera, wherein the beak arises from the front of the head, the homopteran beak joins to the back of the head. Homopterans usually have two pairs of wings; the front pair is membranous (or slightly thickened) and uniformly thick throughout, whereas the back pair is membranous. When at rest the wings are held rooflike over the body. Winged or wingless individuals may occur in one sex or the other in certain species; in other species, both conditions may be found in the same sex. The males of scale insects have only one pair of wings.

Antennae vary from short and bristlelike to long and threadlike. Compound eyes are usually present and well developed, but ocelli may be absent in some groups.

Most homopterans develop in a simple manner from immatures that resemble adults (gradual metamorphosis). However, whiteflies have a quiescent stage just before adulthood, as found in more complex forms of development. All homopterans are plant feeders, and many are important agricultural pests.

The order Homoptera can be divided into two suborders, the Auchenorrhyncha and the Sternorrhyncha. The Sternorrhyncha may be further subdivided into six superfamilies, the largest of which is the superfamily Coccoidea, the scale insects. There are 38 homopterous families, of which only a few representatives are mentioned here.

Family Cicadidae—cicadas. These insects are widely known by many people who recognize their calling sound coming from trees in summer. They are sometimes mistakenly called "locusts," but this name is appropriate only for some of the shorthorn grasshoppers.

The cicadas are the largest of the Homoptera, ranging from 25 to 50 mm in length. They can be recognized by their large size and two pairs of membranous wings. The most common types are the large dogday cicadas (several species) and the smaller periodical cicadas (*Magicicada* species). Periodical cicadas require 13 or 17 years to develop, depending on the particular brood. Mass emergence of these insects is a particularly noticeable event, with some broods being very numerous and loud.

Cicadas injure woody plants by inserting eggs into twigs. This injury may cause the terminal parts of twigs to die.

Family Cicadellidae—leafhoppers. These small insects usually are fewer than 13 mm long and resemble some of the other "hoppers" of the Homoptera. Some are brightly colored and patterned, but all are distinguished by the one or more rows of small spines on the hind tibiae.

Leafhoppers feed on all types of plants, from woody to herbaceous species. The feeding site is usually leaves, where the insect sucks plant juices. During the feeding process, chlorophyll content is reduced, and in certain instances, conductive tissues are blocked. Many plant disease organisms also are transmitted by leafhoppers, including those causing curly top in sugar beets, aster yellows, corn stunt, and phloem necrosis of elm. A migratory species, the potato leafhopper, *Empoasca fabae*, is one of the most ubiquitous leafhopper pests.

Superfamily Fulgoroidea, many families including Fulgoridae—planthoppers. These insects are usually small in the United States, 10–13 mm long, but in the tropics they may be as large as 50 mm. Planthoppers differ from leafhoppers and another closely related group, spittlebugs, in having only a few small spines on the hind tibiae and antennae that arise below the compound eyes. The heads of different planthoppers may look pointed or bulbous or show other bizarre shapes. Planthoppers feed on a vast number of plants, but few cause damage to crops.

Family Psyllidae—jumping plantlice. These very small insects have the shape of cicadas. They have well-developed legs for jumping and long antennae. Adult jumping plantlice have two pairs of membranous wings and three-segmented beaks. Immatures may look like fuzzy, white balls because of waxy secretions that cover the body. These insects feed on all kinds of plants; some, like the pear psylla, *Psylla pyricola*, are important pests of fruit trees.

Family Aleyrodidae—whiteflies. Whiteflies are tiny, white insects, 2–3 mm long, that are often seen flying around plants in the greenhouse. Both sexes are winged, and the white color comes from waxy powder that covers the wings.

Figure 3.26 Order Homoptera showing representative families. Top row (left to right): Family Cicadidae (cicada) and family Aleyrodidae (whitefly). Middle row: Family Cicadellidae (leafhopper), family Psyllidae (jumping plantlouse), and family Aphididae (aphid). Bottom row: Superfamily Coccoidea, (left and center) female scale insects and (right) a winged male scale insect. (Reprinted with permission of Macmillan Publishing Company from *Insects in Perspective* by Michael D. Atkins. Copyright ©1978 by Michael D. Atkins)

During most of their development, immatures are stationary and covered by a scalelike waxy secretion. They undergo a quiescent period just before adulthood. Whiteflies are important pests of citrus in the West Indies and Mexico and have caused problems in Texas and Florida. They are a difficult nuisance to manage in greenhouses.

Family Aphididae—aphids.   Aphids are some of the most important of all insect pests. They are highly adaptive to new situations and have tremendous powers of reproduction. Nearly every plant grower has problems with aphids at some time.

Aphids are small, delicate insects, usually fewer than 5 mm long. They can be recognized from their characteristic pearlike shape and the pair of tubelike **cornicles** that protrude from the back of the abdomen. The cornicles secrete a waxlike substance believed to function in protection. As do some other homopterans, aphids secrete **honeydew**, a sugary liquid exuded through the anus. When excreted, honeydew falls onto leaves, twigs, fruit, or other surfaces below, attracting ants that sometimes care for and defend the aphids from predators. Honeydew on plants also may produce an unsightly appearance when sooty mold grows on it.

The seasonal changes of aphid generations may be very complex, with the occurrence of various forms and various reproductive patterns.

Aphids well may be the most destructive pests in the Homoptera. In addition to direct injury, they are the most important vectors of plant viruses, most of which cause mosaic symptoms. A single species may transmit only a single virus, or it may, as in the case of the green peach aphid, *Myzus persicae*, transmit several different viruses.

Superfamily Coccoidea—scale insects.   Scale insects are a group of small, greatly modified Homoptera that may be divided into 16 separate families. Adult female scale insects are wingless and usually lack legs. They remain stationary through much of their lives and are covered with waxy or resinous secretions, thickened **integuments**, or waxy scales. Males have legs and usually one pair of wings on the mesothorax. The males resemble small gnats but lack mouthparts (they do not feed as adults) and have a styletlike protrusion at the end of their abdomens.

Scale insects have various modes of development, most of which are very complex. Newly emerged immatures, called **crawlers**, have legs and antennae and are active. As development proceeds, legs and antennae are lost in females, and their bodies are covered by secretions. Females complete development under these coverings and produce young there. Males develop much like females except for a quiescent period just before adulthood.

These insects are often divided into three informal groups, including mealybugs, soft scales, and armored scales. Scale insects attack a wide variety of plants; some of the most important pests injure orchard crops, particularly citrus, and ornamentals.

ORDER NEUROPTERA—nervewinged insects

These are medium to large insects, with two pairs of membranous wings held rooflike over the abdomen when not in use. The most distinctive feature of adults is the wing venation made up of numerous longitudinal and cross veins that give a netlike appearance (as a group of branching nerves).

The mouthparts of neuropterans are the chewing type, and antennae are usually long and many-segmented. The tarsi have five segments and there are no cerci at the tip of the abdomen.

Neuropterans are the first insects discussed whose members undergo a dramatic change from young to adult. This type of change is called complete metamorphosis. It is discussed in more detail in Chapter 4.

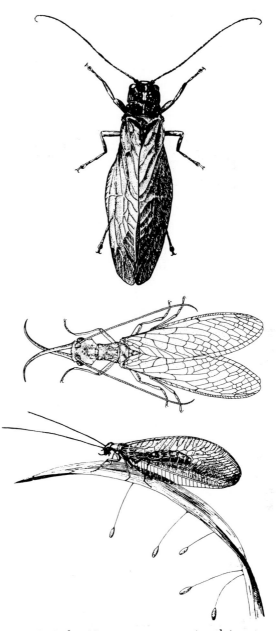

Figure 3.27 Order Neuroptera, nervewinged insects, represented by the families Sialidae (alderfly) (top), Corydalidae (male dobsonfly) (middle), and Chrysopidae (lacewing with eggs). The most distinguishing feature of this order is the membranous wings with numerous longitudinal and cross veins. These veins give the appearance of branching nerves. (After Steyskal, et al., 1986, courtesy USDA)

Nearly all neuropterans are predaceous in both the immature and adult stages. They are found in terrestrial and aquatic habitats. Some of the more important terrestrial groups include the lacewings (family Chrysopidae), which consume such pests as aphids, scale insects, and thrips; antlions (family Myrmeleontidae), whose larvae sometimes build pits for capturing ants; and mantisflies (family Mantispidae), which as adults superficially resemble praying mantids but as immatures are mainly parasitic in the egg sacs of wolf spiders.

The major aquatic neuropterans (suborder Megaloptera) are the dobsonflies, fishflies, and alderflies. The immatures of these groups live in lakes and streams and have gills along the sides of the abdomen. They feed on other aquatic insects and themselves serve as food for fish. Indeed, large larvae of dobsonflies, called **hellgrammites**, are a favorite fish bait.

*ORDER COLEOPTERA—beetles

Beetles are, by far, the largest order of the Insecta—with over one-quarter million described species! In North America, beetles range in length from less than 1 mm to approximately 75 mm.

Most beetles possess two pairs of wings; the front pair is specialized and the most unifying characteristic of the order. These wings are thickened to form leathery or hard coverings called **elytra**. When the insect is not flying, these elytra are folded to form a protective cover over the back. The hind wings are membranous and are used for flight. They fold under the elytra when not in use.

Most beetles are hard-bodied, with unusually tough, thick exoskeletons. Both immatures and adults have typical chewing mouthparts, although in some, the head is formed into a long snout, with mouthparts located at the tip. The growth of immatures follows the pattern of complete metamorphosis.

Beetles are found in almost any type of habitat—on plants, on the soil surface, in soil, in water, inside seeds and fruits, and in ant nests, to name a few. Most are phytophagous or predaceous, but a few are scavengers, fungus consumers, or parasites.

Beetle species are divided by authorities into over 100 families and four suborders. The two most common suborders are the Adephaga and the Polyphaga. The Adephaga are the smaller group, characterized by hind coxae that divide the first abdominal segment. The Polyphaga is, by far, the larger group and has hind coxae not dividing the first abdominal segment. Only one family, the Carabidae, are mentioned to represent the Adephaga. The remaining families discussed are a small sampling of the Polyphaga.

Family Carabidae—ground beetles. This large family of beetles is generally beneficial. Most species feed on other insects in both adult and immature stages. Many species are dark and shiny, but a few are bright green, blue, or other colors. These insects hide under rocks and surface debris during the day, and most search for prey on the ground surface at night. They run rapidly when disturbed.

Family Staphylinidae—rove beetles. These beetles are usually small, elongate insects with short elytra exposing several abdominal segments when viewed from above. Rove beetles run rapidly when disturbed and hold the abdomen up like a scorpion. Most species are predaceous and can be found under stones and other objects, near water, and in nests of birds, mammals, or other insects. Some are parasitic.

Figure 3.28 Order Coleoptera, beetles. Top row (left to right): Family Carabidae (ground beetle), family Dermestidae (dermestid), family Meloidae (blister beetle). Second row: Family Scolytidae (bark beetle), family Elateridae (click beetle), family Cerambycidae (longhorned beetle), family Buprestidae (metallic wood-boring beetle). Third row: Chrysomelidae (leaf beetle), family Coccinellidae (ladybird beetle), family Scarabaeidae (scarab). Bottom row: Family Staphylinidae (rove beetle), family Cleridae (checkered beetle), family Curculionidae (weevil). The most conspicuous feature of this order is the front wings (elytra), which form leathery or hard coverings of the hind wings. (Reprinted with permission of Macmillan Publishing Company from *Insects in Perspective* by Michael D. Atkins. Copyright ©1978 by Michael D. Atkins)

Family Scarabaeidae—scarab beetles. This family consists of a large number of diverse species. Generally, they are oval and bulky or somewhat elongate beetles with five-segmented tarsi and antennae that form platelike lamellae or compact clubs at the tips. As a group, scarabs feed on plants, dung, and decomposing plant and animal material, among other things. Plant-feeding pests injure foliage, roots, fruits, and flowers. Common pests like white grubs, *Phyllophaga* species and others, cause serious damage to lawns and golf greens. The Japanese beetle, *Popillia japonica*, an introduced species that causes widespread damage to horticultural and agronomic crops, is the most important American scarab pest.

Family Buprestidae—metallic wood borers. Insects in this family are mostly metallic green, blue, or black as adults and fly readily when approached. Buprestid grubs live in shallow burrows beneath tree bark and have a broad, flat thorax. Therefore, they are also known as flatheaded wood borers. Adults are usually attracted to unhealthy and dying trees, where they lay eggs in bark crevices. Usually, one generation is produced each year. The flatheaded appletree borer, *Chrysobothris femorata*, is an important pest of fruit, shade, and forest trees. The bronze birch borer, *Agrilus anxius*, is another important buprestid that especially affects white or paper birch trees in regions of marginal tree growth.

Family Elateridae—click beetles. Adult click beetles are often brownish insects that are somewhat tapered at both ends. They derive their common name from the noticeable sound they make. A click beetle on its back will flip into the air with the use of a special spine and notch on its underside. As it flips, an audible click is produced, and the beetle usually lands upright on its tarsi. Click beetle immatures are tan or yellowish and cylindrical. Because of their appearance, they are referred to as **wireworms**. Wireworms usually develop in the soil and are injurious in this stage. They attack many cultivated crops and are particularly

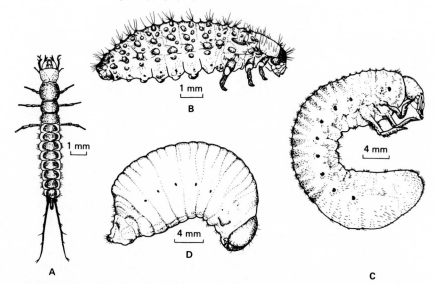

Figure 3.29 Major forms of beetle larvae. A. Carabaeiform. B. Campodeiform. C. Scarabaeiform. D. Apodus. (Reprinted with permission of Macmillan Publishing Company from *Insects in Perspective* by Michael D. Atkins. Copyright ©1978 by Michael D. Atkins)

destructive of newly planted seed. In the midwestern United States, corn planted in recently plowed sod may be completely destroyed by these pests, requiring replanting.

Family Dermestidae—dermestids. Dermestids are small, plump, oval beetles. Many are covered with minute scales that may be brown, red, or white; some are uniform brown or black and hairy. Immature dermestids are bristly or hairy grubs and are important pests. Although most adults feed on plant pollen, grubs are scavengers of dried organic materials, primarily of animal origin, including leather, furs, skins, wool, silk, and stored food. One important pest, the larder beetle, *Dermestes lardarius*, is particularly a problem in kitchens. Here, it infests nearly any stored food, including meats and cheeses. Grubs of some carpet beetles, *Anthrenus* species, feed on and ruin household possessions made of animal fibers, including such items as carpet, clothing, upholstery, and curtains. They may even ruin specimens in insect collections.

Family Coccinellidae—ladybird beetles. Probably no insect is more accepted by the general public than a ladybird beetle. Ladybird beetles, or simply lady beetles, are often colorful little insects well known for their benefits to the farmer and home gardener. Most adults are oval-shaped and convex. They are often tan, black, or red, and spotted with contrasting colors of yellow, white, black, or red. Immatures are elongate, have a tapered abdomen, and are covered with spines. They are often dark with bright orange or red markings.

With few exceptions, both adults and immatures are beneficial, feeding voraciously on any arthropod they can handle, but particularly on aphids, scale insects, and spider mites. Several species, including the famous vedalia beetle, *Rodolia cardinalis*, which saved the California citrus industry from disaster, have been imported and released as a pest management tactic. The native convergent ladybird beetle, *Hippodamia convergens*, is the most common coccinellid species in North America.

An exception to the beneficial nature of ladybird beetles is the Mexican bean beetle, *Epilachna varivestis*. This coccinellid is an important defoliator of snap beans, lima beans, and soybean in the Ohio River valley region of the United States.

Family Meloidae—blister beetles. Blister beetles are medium-sized cylindrical insects with leathery elytra. The thorax is narrower than the elytra and they have a complex mode of development referred to as hypermetamorphosis (discussed in Chapter 4). Blister beetles are often black, gray, or tan with black stripes. Their common name comes from the effect of their body fluids on human skin. They contain **cantharidin**, a substance that sometimes causes blisters. As a **keratolytic**, cantharidin preparations are prescribed by physicians for the removal of benign epithelial growth (wart removal).

Some immature blister beetles are predators of grasshopper eggs, and others eat eggs of solitary bees along with associated food stores. Adults are pests that often feed gregariously, causing patchy areas of defoliation in plantings. All types of vegetable and agronomic crops may be injured. A recent concern comes from the presence of blister beetle bodies in baled hay. Horses feeding on this hay can die from the ingestion of cantharidin from only a few dead beetles. Such occurrences have resulted in substantial economic losses of thoroughbred race horses.

Family Cerambycidae—longhorned beetles. These usually large, cylindrial beetles are sometimes strikingly marked or patterned. The antennae are long, at least the length of the combined head and thorax and often longer than the entire body. Like the metallic wood borers, cerambycid grubs bore into wood, but they differ in that their bodies are straight and cylindrical, with swollen thoraces. Many of these insects attack unhealthy trees, but some, like the locust borer, *Megacyllene robiniae*, inhabit healthy ones.

Family Chrysomelidae—leaf beetles. Leaf beetles are some of the most important agricultural pests in North America. They are rather small insects, usually fewer than 12 mm long, that are closely related to the cerambycids but are smaller and have much shorter antennae. Many common species are yellow or tan with black markings. In many species, like the well-known potato pest, the Colorado potato beetle, *Leptinotarsa decemlineata*, both immatures and adults feed on plant foliage and flowers. In others, like the northern corn rootworm, *Diabrotica barberi*, and the bean leaf beetle, *Cerotoma trifurcata*, immatures feed in the soil on roots and associated structures. In a few species, the immatures feed by mining leaves.

Family Curculionidae—weevils. Weevils are an economically important group of Coleoptera that, along with other families in the superfamily Curculionoidea, have long snouts with chewing mandibles at the tips. This is the largest insect family. The antennae of curculioinids are attached to the snout about halfway along its length. These weevils use the snout to feed internally on plant tissues and notch out egg-laying sites. Immature weevils are light-colored, without legs, and most feed within plant tissues. Adults drill holes and feed in seeds, fruits, and other reproductive parts; some feed on leaves. Weevils rank near the top as pests of certain commodities. Some of the most notable include the boll weevil, *Anthonomus grandis*, on cotton, the alfalfa weevil, *Hypera postica*, on alfalfa, and the rice weevil, *Sitophilus oryzae*, in stored grain.

Family Scolytidae—bark beetles. Scolytids are small cylindrical beetles fewer than 9 mm long. The antennae are elbowed and clubbed at the ends, and the end of the abdomen appears squared off. Immatures are small, white, legless grubs found in tunnels beneath tree bark or burrowing into the heartwood. These insects spend most of their lives in the tunnels, and adults usually leave for only a short time to find new trees to colonize.

Members of this family are often divided into two groups: bark beetles, which feed internally on tree tissues, and ambrosia beetles, which occupy sapwood and feed on ambrosia (fungus) that they cultivate in their burrows. In North America, these beetles cause greater destruction of timber than any other insect. The most important pests occur in the genera *Dendroctonus*, *Ips*, and *Scolytus*. Elm bark beetles, *Scolytus multistriatus* and *Hylurgopinus rufipes*, are principally important in spreading the fungus *Ceratocystis ulmi*. This organism is the cause of Dutch elm disease, a fatal disease of certain elms in the United States.

ORDER STREPSIPTERA—twistedwinged parasites

The Strepsiptera are minute insects, 3 mm long and smaller, that usually are internal parasites of certain Diptera, Hemiptera, Homoptera, Hymenoptera, Orthoptera, and Thysanura.

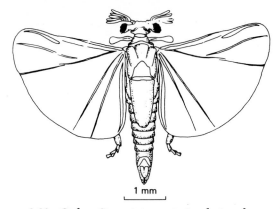

1 mm

Figure 3.30 Order Strepsiptera, twistedwinged parasites. The most notable feature of this order is the paddlelike front wings and larger triangular hind wings of males. Females (not shown) are wingless and often have no legs. (Reprinted with permission of Macmillan Publishing Company from *Insects in Perspective* by Michael D. Atkins. Copyright ©1978 by Michael D. Atkins)

Male and female strepsipterans differ greatly in their appearance. Males have two pairs of wings, but the front wings are reduced to a pair of paddlelike vestiges. The hind wings are broad and triangular. Males have large, protruding eyes and antennae with lobes extending from some of the segments. They are free-living.

Females are wingless and often without legs; many lack eyes and antennae and have heads fused with thoraces. Most remain within hosts.

All strepsipterans have complete metamorphosis. Because their development resembles that of some beetles, a few authorities group strepsipterans as a family of the Coleoptera. They are of little importance in insect pest management.

ORDER MECOPTERA—scorpionflies

The Mecoptera are medium-sized insects, 18–25 mm long, and most have an elongated head that forms a beak with chewing mouthparts at the tip. Usually mecopterans have two pairs of multiveined membranous wings and antennae that are long and slender. The name scorpionfly comes from the pincerlike copulatory organ at the tip of the male abdomen that resemble the stinging organ of a scorpion.

Immature scorpionflies look a little like caterpillars but have more abdominal legs. Development is by complete metamorphosis. Both adults and immatures live in moist places with rank vegetation. Most are scavengers, feeding on dead insects and other organic debris, but adult hangingflies (family Bittacidae) are predators. No scorpionflies are pests.

ORDER TRICHOPTERA—caddisflies

Caddisflies are mothlike insects with two pairs of membranous, usually hairy wings. In the resting position the wings are held rooflike over the abdomen. Antennae are long and threadlike, and mouthparts are the chewing type; how-

Figure 3.31 Order Mecoptera, scorpionflies, represented by *Panorpa rufescens*. The most notable features of this order are the elongated head that forms a beak and the pincerlike organs at the tip of the male abdomen. (After Steyskal, et al., 1986, courtesy USDA)

Figure 3.32 Order Trichoptera, caddisflies, represented by *Phryganea vestita*. Adults of this order have two pairs of membranous wings that are usually hairy. (After Steyskal, et al., 1986, courtesy USDA)

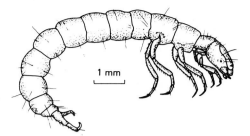

Figure 3.33 Order Trichoptera, a caddisfly larva. Some larvae construct nets for capturing food, and others build portable cases (not shown) in which they live. (Reprinted with permission of Macmillan Publishing Company from *Insects in Perspective* by Michael D. Atkins. Copyright ©1978 by Michael D. Atkins)

ever, mandibles are greatly reduced, and palpi are well developed. Adults are found near water, where they lay eggs.

Immatures develop in water and show complete metamorphosis. They look somewhat like caterpillars but have only one pair of abdominal legs and usually a series of abdominal gills. Mouthparts are the well-developed chewing type, and food is animal and vegetable debris found in the water. A few species are predaceous, feeding on other aquatic insects.

Some immature caddisflies build portable cases, and others construct nets for capturing food. The case-building immatures build their cases with silk from modified salivary glands. In some species, characteristic materials like sand pebbles, shells, or debris are incorporated into the case, but in others, only silk is used. Case builders live inside these protective coverings, coming partly out to feed and retracting into the cases at other times. Net spinners live in fast-flowing water or along lakeshores, where they build funnellike nets of silk. The nets are tied to rocks or other objects and held open by water flowing through them. The immatures hide near these nets and feed on materials caught in them. Predaceous caddisfly immatures do not build cases or nets.

As a group caddisflies may be considered beneficial in that they are food for fish. Although adults may be a nuisance when attracted to lights, none is considered a pest.

*ORDER LEPIDOPTERA—butterflies and moths

This very large order includes some of the most beautiful species and some of the most important pests in the Insecta. Butterflies and moths are distinguished by their large, membranous wings covered with minute overlapping scales. These scales give wings rigidity and color. Only occasionally are wings absent or nearly without scales.

Adult mouthparts are usually of the siphoning type and are coiled under the head when not in use. These mouthparts are adapted for feeding on flower nectar and plant juices, and flower pollination often accompanies the feeding process. Some lepidopterans have spines at the tip of the siphoning tube that allow the insects to break the skin of fruit and suck the juices. In a few instances,

Figure **3.34** Examples of caterpillars, order Lepidoptera. Top row (left to right): Family Danaidae (monarch butterfly) and family Noctuidae (corn earworm). Bottom row: Family Sphingidae (tomato hornworm) and family Lymantriidae (whitemarked tussock moth). (After Steyskal, et al., 1986, courtesy USDA)

mouthparts may be reduced, and the adult does not feed. In another rare instance (Micropterigidae) the mouthparts are developed for chewing.

Immature butterflies and moths are usually called caterpillars, and development in the order is through complete metamorphosis. Caterpillars have a wormlike appearance and three pairs of jointed legs on the thorax; usually two to five pairs of fleshy **prolegs** are present on the abdomen. Hooks or claws, known as **crochets**, at the tips of the prolegs aid in locomotion. Caterpillars have short antennae, simple eyes, and chewing mouthparts.

Almost all immatures have well-developed silk glands formed from modified salivary glands. Silk may be used in tying plant parts together to form a sheltered environment for feeding, to make nests for aggregations of caterpillars, to build cocoons, or in other ways. Cocoons are shelters built by full-grown moth caterpillars for transformation to the adult stage. Most butterfly caterpillars do not build cocoons. Instead, the transformation stage, often called a **chrysalis**, is naked and held in place on a twig or other object by a silk pad and a spiny **cremaster** at the tip of the abdomen. Some chrysalids also employ silken threads around their midsections that hold them upright.

With the exception of clothes moths (family Tineidae) and a few predators and scavengers, caterpillars feed on plants. Almost all plant species have at least one caterpillar pest feeding on them, making this order of great concern to insect pest management.

Many schemes of classification exist for the Lepidoptera. Some divide the order into five suborders. A more informal method divides the order into the Microlepidoptera and the Macrolepidoptera according to average body size of the included species.

Another informal scheme divides the Lepidoptera into moths, butterflies, and skippers. Usually moths are distinguished from butterflies by having wings folded rooflike or wrapped around the body when at rest and filamentous or featherlike antennae. Butterflies hold their wings vertically when at rest and their antennae are clubbed at the tips. Another group, the skippers, have antennae with hooks

Figure 3.35 Order Lepidoptera, Family Tineidae, clothes moths. A case-making clothes moth, *Tinea pellionella* (left), and a webbing clothes moth, *Tineola bisselliella*. (Courtesy USDA)

at the tips. Behaviorally, moths are usually active at night and butterflies and skippers are usually active during the day.

The suborder Ditrysia is by far the most important of the five formal suborders. Only a few of the most economically important families in this suborder are mentioned.

Family Tineidae—clothes moths and others. Tineids are small moths (wingspan about 10 mm), usually with large maxillary palps. The most important family member is the webbing clothes moth, *Tineola bisselliella*, whose caterpillars feed on wool, furs, feathers, and other organic products, causing considerable damage to fabrics. These caterpillars produce a weblike mass over fabric as they feed, whereas larvae of another clothes moth, the casemaking clothes moth, *Tinea pellionella*, spin a portable case in which they reside during development.

Family Psychidae—bagworm moths. Bagworm moths are so named because of the bag they construct and inhabit. Males have wings with very few scales. Females are wingless and lay eggs that overwinter in silken bags. Immatures leave the mother's bag in the spring and build their own bags that become larger and larger as caterpillars grow. The bags are characterized by bits of leaves,

Figure 3.36 Order Lepidoptera, family Psychidae, bagworm, represented by *Thyridopteryx ephemeraeformis*. Adult (left), case containing eggs (middle), and larva in case. (Courtesy USDA)

twigs, and other plant material incorporated into them. Immatures carry their bags with them as they move and extend their heads and legs out of their bags to feed. Adult males leave their bags, but females remain in theirs and lay eggs there. Mating occurs through the lower opening of the bag. These insects are particularly important pests of woody ornamentals.

Family Tortricidae—tortricid moths.   The tortricids are small moths. Tortricid caterpillars sometimes roll leaves and feed, inside out, on the rolled leaf. Others feed on fruits, buds, or stems. Adults are characterized by having broad front wings that are squared off at the tips. Many pest species in this family are destructive of fruit tree and forest crops. One of the most important of the fruit tree group is the codling moth, *Cydia pomonella*, whose caterpillars enter the blossom end of apples and pears and feed internally on the fruit. Two of the most destructive forest pests are the spruce budworm, *Choristoneura fumiferana*, in the East, and the western spruce budworm, *Choristoneura occidentalis*, in the West. These species feed on buds and new foliage, severely defoliating spruces and firs.

Family Pyralidae—snout and grass moths.   The name "snout moth" comes from the well-developed labial palps found in some members of this group. Most pyralids are small, fragile moths that have a scaled **proboscis**. One significant pyralid pest is the European corn borer, *Ostrinia nubilalis*, a species introduced into North America from Hungary or Italy in shipments of broom corn about 1908 or 1909. It ranks as one of the most destructive pests of corn, but it also damages peppers and other garden crops. In corn, early-season larval tunneling disrupts normal plant physiological processes, and late-season tunneling causes structural weakening of the plant and eardrop.

Family Papilionidae—swallowtail butterflies.   Members of this group are medium to large butterflies. The swallowtails are particularly well known as very colorful insects with long, taillike extensions on the hind wings. Some of the largest butterflies in the world belong to this group, having wingspans as much as 255 mm! One of the most common in North America is the black and yellow tiger swallowtail, *Papilio glaucus*.

Family   Nymphalidae—brushfooted   butterflies.   Members   of   this   large

Figure 3.37   Order Lepidoptera, family Tortricidae, tortricid moths, represented by the fruittree leafroller, *Archips argyrospilus*. Larva (left), pupae (middle,top) egg mass (middle, bottom), and adult. (Courtesy Oregon Agricultural Experiment Station)

Figure 3.38 Order Lepidoptera, family Papilionidae, swallow-tails, represented by the tiger swallowtail, *Papilio glaucus*. (After Steyskal, et al., 1986, courtesy USDA)

family of many common butterflies, along with some other families, have reduced front legs without claws. Only the middle and hind legs are used in walking. The painted lady butterfly, *Vanessa cardui*, a migratory butterfly found throughout the world, is one of the most common representatives of this family in North America. Occasionally it is a pest.

Family Lasiocampidae—tent caterpillars and lappet moths. These medium-sized moths have heavy bodies and feathery antennae. The caterpillars are also covered with long hairs and feed on the foliage of a number of deciduous trees, including fruit trees. Tent caterpillars, for example, the eastern tent caterpillar, *Malacosoma americanum*, obtain their name from building silken nests in the crotches of cherry, apple, peach, plum, and other trees. Caterpillars move back and forth from the tents to feeding areas during development, using the tents for shelter. Caterpillars of a related species, the forest tent caterpillar, *Malacosoma disstria*, do not make tents.

Figure 3.39 Order Lepidoptera, family Nymphalidae, brush-footed butterflies, represented by an angle wing butterfly. (Reprinted with permission of Macmillan Publishing Company from *Insects in Perspective* by Michael D. Atkins. Copyright ©1978 by Michael D. Atkins)

Figure 3.40 Order Lepidoptera, family Lasiocampidae, tent caterpillars. Shown are tents of the eastern tent caterpillar, *Malacosoma americanum*. (Courtesy Ohio Agricultural Research and Development Center)

Family Sphingidae—sphinx moths. The front wings of sphingids are narrow and pointed at the tips; with their spindle-shaped body, the family is readily recognizable. These moths often have large, stout bodies and hover in front of flowers, like hummingbirds, as they feed. The caterpillars are usually called hornworms because of the large curved horn arising on the upper surface and at the end of the abdomen in most species. One of the common hornworms in this family is the tomato hornworm, *Manduca quinquemaculata*, a large green caterpillar that is one of the most important insect pests of tomatoes and potatoes.

Family Noctuidae—noctuid moths. This is the largest family of the Lepidoptera. Probably most of the moths seen around street and yard lights in summer are noctuids. Most adults are somewhat triangle-shaped when the wings are at rest, and they are often dull gray, brown, or black (with the notable exception of underwing moths, which have colorfully striped hind wings). Most noctuid caterpillars are foliage feeders, but some bore into stems and fruits. Many of our most devastating agricultural pests are found in this family, including several species of armyworms, cutworms, and loopers.

Perhaps the most destructive pest in this group, or in the order for that matter, is the corn earworm, *Heliothis zea*. Caterpillars of this noctuid are most destructive when they feed on the fruits of plants and on flower buds. However, they also feed on foliage and young stems. Important crops attacked include corn, cotton, and tomatoes.

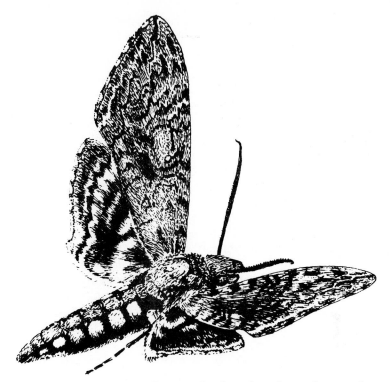

Figure 3.41 Order Lepidoptera, family Sphingidae, sphinx moths. Shown is an adult tomato hornworm, *Manduca quinquemaculata*. (After Steyskal, et al., 1986, courtesy USDA)

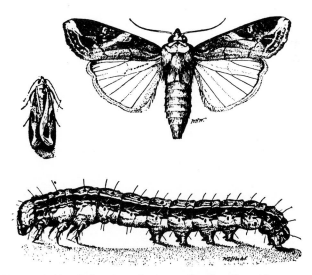

Figure 3.42 Order Lepidoptera, family Noctuidae, noctuid moths, represented by the fall armyworm, *Spodoptera frugiperda*. Adults (top) and larva. (Courtesy USDA)

Other important noctuid pests worthy of mention are armyworms and cutworms. Like the corn earworm, species in these groups attack a wide variety of crop plants.

*ORDER DIPTERA—flies

Many kinds of insects have the word "fly" as part of their common names. Some of the ones we have discussed include dragonflies, mayflies, and butterflies. However, none of these is a "true" fly, and entomologists conventionally designate this fact by writing the word "fly" in combination with a descriptor to form one word. True flies occur only in the order Diptera, and in this instance, "fly" is conventionally written as a separate word from the descriptor, as "house fly" and "horse fly."

Diptera are very diverse and very successful insects. They are highly evolved, and many are significant pests that directly and indirectly affect the health of humans, domestic animals, and plants. Others are important natural enemies of insects, and some are considered beneficial as plant pollinators. They occur in every conceivable terrestrial and freshwater habitat; they even occur in puddles, as well as in brackish and alkaline water, and in pools of crude oil.

True flies are easily distinguished from the rest of the Insecta on the basis of their wings. They have only one pair of membranous wings arising from the **mesothorax**, and the hind pair has been reduced to form balancing organs called **halteres**. Wings are absent in some Diptera, but halteres are usually present.

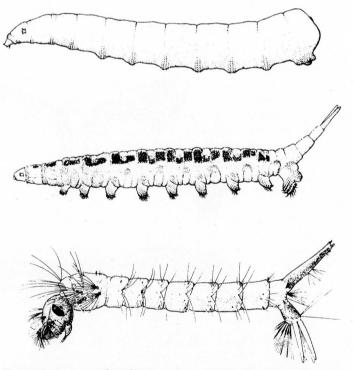

Figure 3.43 Examples of Diptera larvae. Muscidae (house fly) (top), Ephydridae (shore fly) (middle), and Culicidae (northern house mosquito). (After Steyskal, et al., courtesy USDA)

Only male scale insects have one pair of wings and metathoracic balancing structures, but these insects have a stylus at the end of the abdomen.

The mouthparts of the Diptera are some of the most diverse in the Insecta. As previously mentioned, mosquitoes have piercing-sucking mouthparts; house flies and related insects have sponging mouthparts; and biting flies have cutting-sponging mouthparts. Mouthparts are not functional in a few adult Diptera. Development is by complete metamorphosis, and the immatures take on a variety of shapes and sizes.

The Diptera are often divided into three suborders: Nematocera, Brachycera, and Cyclorrhapha. Adult nematocerans like mosquitoes and gnats have antennae with six or more segments, and immatures have a well-developed head and mandibles. In the Brachycera (for example, horse flies) and Cyclorrhapha (for example, house flies) the antennae are usually three-segmented. In the Brachycera, the immatures have sclerotized (hardened) heads, whereas only the mouthparts are sclerotized in the heads of Cyclorrhapha. Only a few important families representative of the Diptea are discussed here.

Family Culicidae—mosquitoes. As mentioned in Chapter 1, mosquitoes have been some of the most significant pests in the history of humankind. Their importance as medical pests continues today, particularly through vectoring organisms that cause malaria.

Adult mosquitoes are distinguished by their long proboscis and scales along the wing veins. Adult females feed on blood and are responsible for transmitting disease organisms. Adult males (and occasionally females) feed on nectar and other plant exudates. The feeding immatures (larvae or wrigglers) are found in water, where most consume algae and organic debris.

Members of the genus *Anopheles* are known as the malaria mosquitoes and lay their eggs singly on the water surface. Those in the genus *Culex* lay eggs in clusters as rafts on the water surface. Members of another important genus, *Aedes* (floodwater mosquitoes), lay eggs in depressions, along high water lines, and in almost any container of rain water. These eggs lie dormant until flooded with water, after which they hatch, and wrigglers develop in the flooded area. As a vector of the yellow fever pathogen, *Aedes aegypti* is one of the most important floodwater mosquitoes.

Family Cecidomyiidae—gall midges. Gall midges are small, fragile nematocerans with long antennae and reduced wing venation. Almost all gall midges feed on plants as immatures, and many cause abnormal swellings in plant tissues called **galls**. A given species forms galls only on certain plant parts, and the gall is usually very distinctive. One of the most important pests in this family is the Hessian fly, *Mayetiola destructor*, a species introduced to North America from Europe that attacks wheat. Winter wheat is injured in the spring when maggots feed between the leaf sheath and the stem, weakening plants and sometimes causing them to die.

Family Tabanidae—horse flies and deer flies. These representatives of the suborder Brachycera are medium to large flies with cutting-sponging mouthparts. As do mosquitoes, adult females feed on blood, and adult males consume pollen and nectar of flowers. Tabanid larvae feed on small invertebrates in water or in moist soil.

The two most common genera are *Tabanus*, the horse flies, and *Chrysops*,

Figure 3.44 Order Diptera showing representative families. Top row (left to right): Family Cecidomyiidae (gall midge) and family Culicidae (mosquito). Middle row: Family Syrphidae (flower fly) and Muscidae (house fly). Bottom row: Family Tabanidae (horse fly), Family Asilidae (robber fly), and Tachinidae (parasitic tachinid). Adult dipterans are distinguished by their one pair of membranous wings and presence of balancing organs (halteres) in place of the hind wings. (Reprinted with permission of Macmillan Publishing Company *Insects in Perspective* by Michael D. Atkins. Copyright ©1978 by Michael D. Atkins)

the deer flies. Horse flies are known to transmit pathogens causing anthrax and tularemia, but most harm is done through their vicious bite and extraction of blood. Deer flies are usually smaller than horse flies and will buzz around the back of people's heads and attempt to get into their hair. The bites of deer flies can be painful.

Family Syrphidae—flower flies. Flower flies are the first mentioned here of the large suborder Cyclorrhapha. Adult flower flies are often mistaken for bees because of their bright yellow bands and their habit of hovering about over flowers. Flower fly larvae are of special importance because many feed on aphids and other pests of cultivated crops. Others feed in polluted water and have long breathing tubes; these are called rattailed maggots.

Family Tephritidae—fruit flies. Tephritid fruit flies (not to be confused with flies in the family Drosophilidae) are small- to medium-sized flies that usually have banded or patterned wings. Adults occur on vegetation, and females often lay eggs in fruit. Developing maggots feed inside the fruit, often causing serious losses to apples, cherries, and citrus. An important pest of apples in the northern United States is the apple maggot, *Rhagoletis pomonella*. Another species, the Mediterranean fruit fly, *Ceratitis capitata*, is a devastating pest of fruit in tropical and subtropical parts of the world. It has been introduced into Florida and California on several occasions, and eradication programs successfully eliminated it each time in Florida. Persistent sightings in California have required continued surveillance and eradication activities.

Family Muscidae—muscid flies. This large group of common flies are very important as pests of humans and domestic animals. The family includes the house fly, *Musca domestica*, whose maggots feed on all kinds of decaying matter. The fly is known to transmit organisms that cause typhoid fever, yaws, anthrax, dysentery, and forms of conjunctivitis. Another pest similar to the house fly is the face fly, *Musca autumnalis*, a species that clusters on the faces of cattle. Neither the house fly nor the face fly bites.

Some other species in this family are biting flies, including stable flies, *Stomoxys calcitrans*, and horn flies, *Haematobia irritans*, both important pests of cattle. The infamous tsetse flies, *Glossina* species, of Africa also are biting muscid flies.

Family Tachinidae—tachinid flies. This is a very large family of Diptera. Many tachinids look similar to house flies and other muscids, but others are larger and covered with strong bristles. This is a most important group because they are

Figure 3.45 Order Diptera, family Tephritidae, fruit flies, represented by an adult Mexican fruit fly, *Anastrepha ludens*. (Courtesy USDA)

parasitic on other insects, particularly immature Coleoptera, Lepidoptera, and Hymenoptera. In most instances, females lay their eggs on the body of the host, and after egg hatch, the maggots tunnel inward through the body wall. These maggots feed internally, eventually killing their host, and when fully grown, they emerge to develop into adults. In some species eggs are laid on plant foliage, and larvae attach to appropriate hosts as these hosts pass by. There are many other life cycle variations as well.

ORDER SIPHONAPTERA—Fleas

Fleas are very small, secondarily wingless insects that, in the adult stage, feed on the blood of mammals and birds. Most adult fleas are characterized by a body that is flattened from side to side and covered with rows of bristles that point backwards. Flea antennae are short and lie in grooves on the sides of the head. The mouthparts, modified for piercing and sucking blood, are composed of three stylets. Eyes may be present or absent.

The legs of adult fleas are long with long coxae, which allow a flea to jump distances many times its own body length. Such a capability permits fleas to effectively move from host to host.

Fleas develop by complete metamorphosis, and the eggs are usually laid on the ground or in the nest of the host. When laid on the host, they fall off into the nest or onto the ground. Egg hatch produces white, legless immatures. These are worm-shaped, with a well-developed head, no eyes, chewing mouthparts, and small hooks at the tip of the abdomen. These immatures feed mostly on organic debris and feces of adult fleas. When fully grown, they spin cocoons in which transformation to the adult occurs.

Although some have common names referring to hosts, most fleas are not very specific feeders. For instance, two common species, the cat flea, *Ctenocephalides felis*, and the dog flea, *Ctenocephalides canis*, both are parasitic on cats, dogs, humans, and other hosts. Similarly, the human flea, *Pulex irritans*, attacks humans, dogs, cats, rats, horses, and other animals.

Figure 3.46 Order Siphonaptera, fleas, represented by a cat flea, *Ctenocephalides felis*. Adult fleas are characterized by a body flattened from side to side and covered with rows of bristles that point backwards. (After Steyskal, et al., 1986, courtesy USDA)

Fleas are probably most important as vectors of disease-causing organisms in humans. The most important disease resulting from transmission is **plague**, or black death. Plague transmitted by fleas is called bubonic plague, caused by the bacterium *Yersinia pestis*. Through the ages, it has resulted in tremendous loss of life. The reservoir for the disease is rats, and fleas like the oriental rat flea, *Xenopsylla cheopis*, move off of rats dying of the disease and onto humans. Here, they transmit the bacterium, mostly through regurgitation at the time of biting.

*ORDER HYMENOPTERA

The Hymenoptera is another of the major orders and is considered by many people to be the most beneficial. This view is based on activities of the Hymenoptera in plant pollination, in producing honey and wax, and in killing pest insects. However, certain sawflies, ants, and other forms are pests and can be quite destructive.

A great diversity of form occurs in the Hymenoptera, but most species have two pairs of membranous wings, with the hind pair smaller than the front. The front and hind wings are held together by rows of tiny hooks on the hind wings called **hamuli**. Wing venation is often reduced or almost lacking in some groups. Mouthparts are basically of the chewing type but are modified for sucking or lapping in highly evolved forms, such as bees. Most hymenopteran females have a well-developed ovipositor, and in some groups, it has been modified to form an effective stinging organ used for protection.

Development in the Hymenoptera occurs by complete metamorphosis, and the immatures (larvae) have well-developed head capsules and chewing mouthparts. On attaining full growth, the immatures sometimes build silken or parchmentlike cocoons.

The order Hymenoptera is usually divided into two suborders, the Symphyta and the Apocrita. The Symphyta include sawflies and horntails, which have abdomens broadly joined to their thoraces. The ovipositor is usually well developed and sawlike in form. Larvae of the Symphyta look much like the caterpillars of moths but have more than five pairs of abdominal prolegs and no crochets. Most Symphyta feed externally on plants.

The Apocrita are characterized by a "thin waist," with the abdomen attached to the thorax by a stalklike **petiole**. This suborder includes ants, wasps, bees, and many forms of parasitic wasps. Most larvae of this suborder are legless and maggotlike or grublike. Food habits of this large suborder are very diverse. As larvae, most species are parasitic on other insects, but others are predators, feed on plant materials, or eat other products. Adults often consume nectar, pollen, sap, and other plant materials. Females of some parasitic species may ingest body fluids of hosts they sting when laying eggs.

Figure 3.47 Examples of sawfly larvae. Diprionidae (left) and Tenthredinidae. (Courtesy USDA)

Figure 3.48 Order Hymenoptera, family Tenthredinidae, common sawflies, represented by the cherry fruit sawfly, *Hoplocampa cookei*. Egg and egg position on blossom (upper left), larva (upper right), adult (lower middle) and infested cherries. (Courtesy USDA)

Family Tenthredinidae—tenthredinid sawflies. These sawflies compose a very large family of medium-sized (less than 21 mm long) and sometimes brightly colored species. Adults are wasplike, and many larvae curl over the edge of a leaf when feeding. Most species feed on trees and shrubs, and some can be quite destructive. The cherry fruit sawfly, *Hoplocampa cookei*, is an example of an important pest in this family, attacking fruit of cherries and plum in the western United States.

Family Cephidae—stem sawflies. These elongate sawflies, about 9 mm long, are flattened from side to side. Cephid immatures burrow in the stems of grasses,

Figure 3.49 Order Hymenoptera, family Cephidae, stem sawflies, represented by the wheat stem sawfly, *Cephus cinctus*. Adult (left) and larva. (Courtesy USDA)

Figure 3.50 Order Hymenoptera, family Braconidae, braconids, represented by *Cotesia thompsoni,* a parasite of the European corn borer (Lepidoptera: Pyralidae). (Courtesy USDA)

shrubs, and other plants, in many instances causing them to break over. One of the most important pests in this family is the wheat stem sawfly, *Cephus cinctus,* occurring in the Great Plains area of the United States.

Family Braconidae—braconids. Braconids are a family of small wasps, usually fewer than 15 mm long, that parasitize and kill many different kinds of insects. Depending on the species, one to more than a hundred braconid larvae can develop in a single host. Some braconid larvae leave the host when full grown and build silken cocoons, where they transform into adults. Others complete adult development inside the host. Braconids are important in the natural regulation of many insect populations and have been reared and released in programs of pest management. One of the most important genera in this family for pest management is *Cotesia.*

Family Ichneumonidae—ichneumons. This is a very large family whose smaller members resemble braconids. They mainly differ in the venation of the forewing. Ichneumons vary greatly in size, with some attaining lengths of 40 mm or more. The ovipositors of many ichneumons are long, sometimes longer than their bodies. All ichneumons are parasitic, and most insect groups have species with at least one ichneumon parasite. Certain ichneumons are also parasites of spiders.

Family Trichogrammatidae—minute egg parasites. Members of this family are tiny wasps, 0.3–1.0 mm long, that parasitize insect eggs. They are characterized by three-segmented tarsi and wings with rows of microscopic hairs. One of the

Figure 3.51 Order Hymenoptera, family Ichneumonidae, ichneumons. A female ichneumon. (Courtesy USDA)

most important genera in this family is *Trichogramma*. Possibly more effort has gone into the production and release of various *Trichogramma* species worldwide for pest management than any other parasite. These parasites are readily available from several commercial insectaries in the United States.

Family Formicidae—ants. Ants are widespread and recognized even by the novice. They are believed by some to be the most successful group of insects. All ants are social insects, living together in colonies. Ants may be winged or wingless and are recognized by an upright lobe that occurs on the **pedicel** of the abdomen. The antennae of ants is usually elbowed, with the first segment longer than the others.

Ant colonies usually contain three castes: queens, males, and workers. Mating occurs in swarms of flying queens and males. After mating, males die and females establish new colonies. Usually, the young queen's wings break off, and she makes a cavity in the soil under a stone or bark. After a month or more in the sealed cavity, the queen lays eggs and feeds developing larvae her saliva until they mature. The first adults produced are all small workers (sterile females), which on becoming adults, take over the duties of the colony. Worker activities include nest construction, food gathering, and care of the young. As the colony expands, more and more chambers are added to the nest. After several years, winged males and females (queens) are produced to leave the nest and establish new colonies of their own. Queens are usually long-lived; some may live 15 years or more. Colonies may exist for as long as a human generation. In addition to soil, ants may nest in plant parts (for example, stems and nuts), or damp, decaying wood, which is sometimes excavated for galleries.

Several species of ants are household pests but mainly as a nuisance. Others, like the Texas leafcutting ant, *Atta texana*, may cause serious damage to trees, shrubs, and cultivated crops by stripping their foliage. The foliage is cut into pieces and carried inside the nest to certain gallaries where it is used as a medium for growing fungus. This cultivated fungus then is eaten by the ants.

Family Vespidae—vespid wasps. Vespid wasps are a small but common group of the Hymenoptera that includes paper wasps, yellowjackets, and hornets. Many of these insects are familiar black or reddish brown wasps with yellow or off-white markings. Like ants, these social insects have queens, males, and workers.

Figure 3.52 Order Hymenoptera, family Formicidae, ants, represented by *Camponotus castaneus*. (Courtesy USDA)

Figure 3.53 Order Hymenoptera, family Vespidae, vespids, represented by the paper wasp, *Polistes fuscatus pallipes*, on a nest. (Courtesy USDA)

However, unlike ants new colonies are established annually. Queens and workers have well-developed stinging organs and can inflict painful stings. These wasps construct nests of papery material that they produce by chewing wood fibers with saliva. Overwintered queens begin a colony in the spring by starting a new nest or using one of a previous year. The first brood produced is composed of workers (sterile females), which take over nest duties when they mature. The immatures are fed with insects (often pests) and other invertebrates. Paper wasp nests are commonly attached to the eaves of houses, and many people believe that wasp activities outside the home are a nuisance.

Family Apidae—bumble bees, honey bees, and others.   These mostly common bees are some of the most beneficial insects. Most are social and important as plant pollinators.

Bumble bees (genera *Bombus* and *Psithyrus*) are relatively large, hairy bees, 20 mm or longer, that are usually colored black and yellow. They build nests in ground depressions, deserted rodent nests, hollow logs, and other places. Like vespids, colonies in temperate regions are established on an annual basis by overwintered, fertilized queens. Workers produced in the first brood take over nest duties during the production of subsequent broods. The proboscis of bumble bees is very long, making them effective pollinators of plants with long corollas, such as red clover.

The honey bee, *Apis mellifera*, is a species introduced into North America by early English and Spanish settlers for use in beekeeping. Beekeeping, the culturing of bees for their honey and wax, is one of the oldest agricultural pursuits and is an important industry everywhere. However, the pollination service provided by honey bees far outweighs the value of their material products.

Unlike bumble bees, honey bees maintain colonies on a perennial basis. A colony consists of three castes: a single queen, males (called **drones**), and workers (sterile females). Nests in nature are built in logs, hollow trees, caves, and other such places. Inside the nests, workers construct and maintain vertical **combs** of wax, with hexagonal cells for the storage of honey, pollen, and immature bees. The primary function of the queen in this colony is to lay eggs, and she sometimes produces 1,500–2,000 daily! Fertilization of queens occurs during mating flights, and a queen may live for several years.

New queens are reared in special cells, called queen cells, and are fed exclusively on **royal jelly**, a glandular secretion of workers. Queens are produced from fertilized eggs. Drones are reared in drone cells and develop from unfertilized eggs. After the mating season, drones are driven out of the hive to die, thereby helping to conserve the store of honey for the winter. Workers develop from

Figure 3.54 Order Hymenoptera, family Apidae, represented by the honey bee, *Apis mellifera*, on a comb. (Courtesy USDA)

fertilized eggs in other cells of the comb; they are fed royal jelly initially but complete development on pollen and nectar. New workers feed and care for the queen and brood, secrete wax, build combs, clean the hive, convert nectar to honey, act as guards, and perform other duties. After about three weeks, workers became field bees, mainly gathering nectar and pollen and bringing water to the hive. New colonies are formed when new queens are produced, and the old queen moves out with a group of workers to **swarm** and build a new nest.

Selected References

ATKINS, M. D. 1978. *Insects in Perspective*. New York: Macmillan.

BORRER, D. J. , D. M. DeLong, and C. A. Triplehorn. 1981. *An Introduction to the Study of Insects*. 5th ed. New York: Saunders.

BRUES, C. T. , A. L. Melander, and F. M. Carpenter. 1954. *Classification of Insects*. *Bulletin of the Museum of Comparative Zoology*, Harvard University, vol. 108.

DAVIDSON, R. H. , and W. F. Lyon. 1987. *Insect Pests of Farm Garden, and Orchard*, 8th ed. New York: Wiley.

LITTLE, V. A. 1972. *General and Applied Entomology*, 3rd ed. New York: Harper & Row.

PFADT, R. E. , ed. 1985. *Fundamentals of Applied Entomology*, 4th ed. New York: Macmillan.

ROSS, H. H. , C. A. Ross, and J. R. P. Ross. 1982. *A Textbook of Entomology*, 4th ed. New York: Wiley.

STEYSKAL, G. C. W. L. Murphy, and E. M. Hoover, eds. 1986. *Insects and Mites: Techniques for Collection and Preservation*. United States Department of Agriculture, Agricultural Research Service, Miscellaneous Publication 1443.

WENER, F. G. , ed. 1982. *Common Names of Insects and Related Organisms*. College Park, Md.: Entomological Society of America.

4 THE INSECT LIFE CYCLE

INSECTS PERFORM AN AMAZING FEAT—survival and persistence in the face of seemingly insurmountable odds. The bitterness of winter, the harshness of drought, and the punishment of wind and rain would seem to argue against the success of these small, cold-blooded animals. But survive they do and, at certain times of the year, become a major force to be reckoned with in our daily lives.

The story of insect survival and persistence really begins with innate patterns of living—biological templates for species characterized by reproduction, growth, and development of individuals in the population. We refer to these patterns or templates as **life cycles**.

More formally, a life cycle is the chain or sequence of biological events that occurs during the lifetime of an individual insect. The cycle is usually considered to begin with deposition of the egg and to end with egg laying by the adult female. A life cycle lies in the boundaries of a single generation.

For convenience, we will consider the life cycle as having three major divisions and discuss it from this aspect. These divisions include (1) reproduction and embryonic development, (2) postembryonic growth and development, and (3) maturity.

Related to the life cycle is the insect **seasonal cycle**. The seasonal cycle is the sequence of life cycles of a species that occurs over a one-year period. By addressing the topic of seasonal cycles, we will be able to more fully understand some of the ways insects can survive adverse times of the year.

An understanding of both life cycles and seasonal cycles for a pest species is required for proper management. Prediction of injurious stages, times when the insect is vulnerable, and probable habitats all depend on this knowledge. Without it, development of effective, long-term solutions to insect pest problems would be virtually impossible.

REPRODUCTION AND EMBRYONIC DEVELOPMENT

Types of Reproduction

By far, the most frequently observed form of reproduction in insects is **sexual** reproduction. In most species, both females and males are prevalent and mating occurs, with eggs being fertilized in the female oviduct as they pass the duct of the sperm storage sac (**spermatheca**). Infrequently, fertilization may occur while eggs are still in the ovarioles or in the hemocoele.

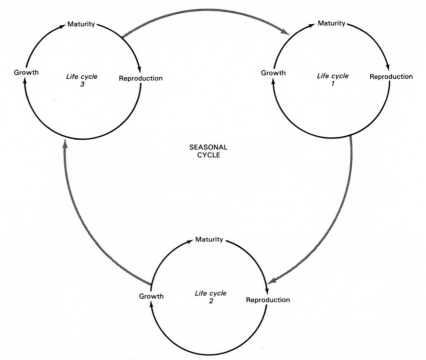

Figure 4.1 Diagram showing the relationship of a life cycle to the insect seasonal cycle. Most insect species undergo one or more life cycles during a seasonal cycle.

Reproduction in the absence of male gametes is not as common. This form, asexual reproduction, or **parthenogenesis** is a derived and specialized characteristic in some insects. Males are not found in species with obligate parthenogenesis, and unfertilized eggs are females. Examples of this type of reproduction include the pear sawfly, *Caliroa cerasi*, several black flies (Diptera: Simuliidae),

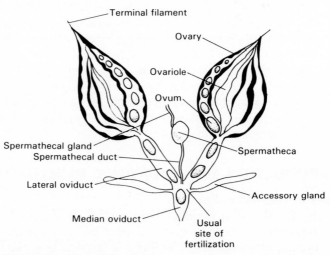

Figure 4.2 Typical female insect reproductive system showing usual site of fertilization. (Redrawn from Atkins, 1978, *Insects in Perspective*, Macmillan Publishing Company)

a number of weevils (Coleoptera: Curculionidae), and many aphids (Homoptera: Aphididae).

A combination of sexual and asexual modes, **haplodiploidy**, occurs in a few species of four insect orders, Coleoptera, Hymenoptera, Thysanoptera, and Hemiptera. The honey bee is one of the best known examples of this mode, which is sometimes called facultative parthenogenesis. Here, unfertilized eggs (haploids) are males who become drones, and fertilized eggs (diploids) are females, most of whom become workers. The queen bee, or reproductive female, has the ability to control the sex of her offspring, and this characteristic was an important factor in the evolution of social life in bees, wasps, and ants.

Another unusual form of reproduction, occurring in one beetle family (Coleoptera: Micromalthidae), several genera of gall and fungus gnats (Diptera:

BOX 4.1 **HONEY BEE**

SPECIES: *Apis mellifera* Linnaeus (Hymenoptera: Apidae)

DISTRIBUTION: The honey bee is an introduced species to the United States, with most colonies residing in human made hives. Escaped swarms, however, survive in the wild and usually live in hollow trees. The species is completely domesticated (as opposed to the African honey bee, *Apis mellifera adansoni*) and has a cosmopolitan distribution.

IMPORTANCE: Honey bees play a a vital and indispensable role in agriculture and horticulture as pollinators. Although wild flowers are often sufficiently pollinated by wild bees, large-scale agricultural plantings often require domesticated bees to be placed in the proximity of the crop for adequate pollination and subsequent grain/fruit production. In addition, raising bees (apiculture) for honey and beeswax is a multimillion dollar industry in the United States. Honey, as a commodity, is a popular food source and is often used as a sugar substitute. Beeswax is another important by-product of apiculture and is frequently used in shaving creams (to prevent drying), cold creams, cosmetics, crayons, candles, and other similar items.

APPEARANCE: Adult queens are 16–20 mm long; the workers (sterile females) are 11–13 mm long; and the drones (males) are 15–16 mm long. The bodies of workers are generally dark brown with a reddish brown abdomen and black apical segments (banded). Adults are lightly covered with hair.

LIFE CYCLE: Workers are the most abundant caste, providing the bulk of colony maintenance. Upon eclosion, workers care for the queen within the hive by feeding her and taking her eggs to cells where they develop as larvae. The workers service developing larvae by providing food throughout the complete metamorphosis. Toward the end of their life, workers leave the hive and begin to forage for nectar and pollen. In this activity, nectar is picked up from several flowers by the adult worker and transferred back to the hive, where it is converted to honey for larvae and queen. Workers also defend the hive by attacking and stinging hostile invaders. Once they expend their venom, however, they soon die. The single queen has sole responsibility for egg production for the hive and is frequently fertilized by drones. New colonies are formed when a queen, accompanied by a portion of the workers, swarms from the old hive to establish a new location.

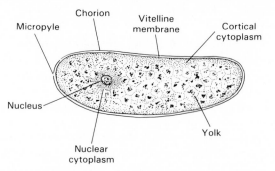

Figure 4.3 Section of a typical insect egg showing important components. (Redrawn from Snodgrass, 1935, *Principles of Insect Morphology*, McGraw-Hill Publishing Company)

Cecidomyiidae), and a common flower fly (Diptera: Syrphidae), is **paedogenesis**, which refers to reproduction by juveniles. In paedogenic flower flies, the ovaries become functional in females with immature (larval) body form, and eggs develop parthenogenetically. Eventually, the mother larva becomes filled with developing daughter larvae, and these eventually emerge from the mother. Under certain environmental conditions, winged males and females are produced that, in turn, reproduce sexually.

Fertilization

As mentioned, fertilization of eggs usually occurs in the female oviduct. Here, the sperm enters the egg through an opening in the egg covering or **chorion** called the **micropyle**. One to several sperm enter, with one reaching the nucleus to form a zygote. Subsequently, cleavage of the egg nucleus begins.

Development of the Embryo

Unlike vertebrates and some other animals, most insects (except Collembola) do not undergo complete cell division during cleavage. Instead, the nucleus divides

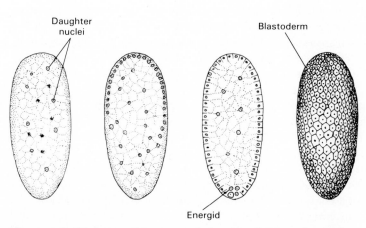

Figure 4.4 Diagrams of insect eggs showing division of a nucleus into daughter nuclei (far left), migration of energids to egg periphery (middle), and final formation of multicellular blastoderm. (Redrawn from Snodgrass, 1935, *Principles of Insect Morphology*, McGraw-Hill Publishing Company)

Germ band

Figure 4.5 Formation of the germ band in the blastoderm. The germ band later gives rise to the insect embryo. Transverse section (left) and longitudinal section. (Redrawn from Snodgrass, 1935, *Principles of Insect Morphology*, McGraw-Hill Publishing Company)

into a number of daughter nuclei. These gather a mass of cytoplasm, further divide, and migrate to the periphery of the egg. On reaching the periphery, a membrane forms around each nucleus and cytoplasm mass (**energid**) to form a cell. These cells create a wall around the interior of the egg, one cell layer thick, forming a stage called the **blastula**. A crowding of cells then begins in a ventral area of the egg, forming a thickened area known as the **germ band**, or, if small and platelike, the germ disk. It is from the germ band or germ disk that the new embryo develops.

Development in most species proceeds when the germ band forms an infolding that eventually produces a three-layered embryo (**gastrula**). The inner layer (**mesoderm**) gives rise to muscles, body fat, heart, blood cells, and reproductive organs. The outer cell layer (**ectoderm**) produces the body wall, fore- and hindgut, nervous system, respiratory system, and many glands. Rudiments at either or both ends of the body give rise to a third embryonic tissue (**endoderm**) that forms the midgut.

Segmentation (transverse creases) of the body begins at an early stage of embryonic development, and segments fuse later to form the head. Appendages develop soon after segmentation, some of which become legs, mouthparts, and posterior abdominal structures.

In advanced stages of embryonic development, an observer can often see eye spots through the sometimes semitransparent chorion. In some species like the European corn borer, *Ostrinia nubilalis*, this is an indication that the egg will soon hatch.

In most insect species, only one embryo forms in a single egg. In some parasitic

Mesoderm

Ectoderm

Figure 4.6 Cross section of a blastoderm (top left), followed by successive development toward the two-layered embryo (gastrula). In this instance, mesoderm is formed by invagination; however, other processes of mesoderm formation also are found in insects. (Redrawn from Snodgrass, 1935, *Principles of Insect Morphology*, McGraw-Hill Publishing Company)

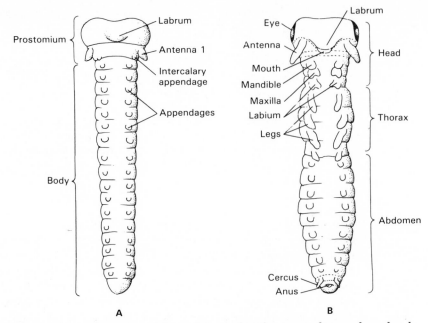

Figure 4.7 Segmentation of the embryo and early stages of appendage development. A. Very young embryo with rudimentary appendages on all but the last segment. B. Older embryo showing development of body regions, as well as appendages. (Reprinted by permission of John Wiley & Sons, Inc. from *A Textbook of Entomology* by H. H. Ross, C. A. Ross, and J. R. Ross. Copyright ©1982 John Wiley & Sons, Inc.)

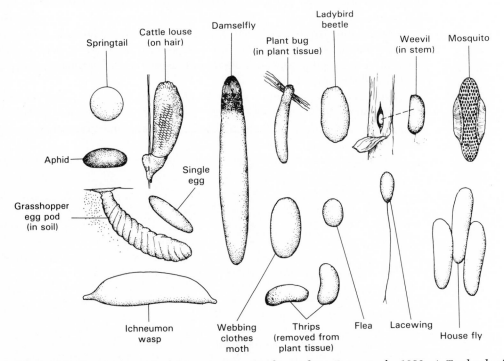

Figure 4.8 Various examples of insect eggs. (Redrawn from Ross, et al., 1982, *A Textbook of Entomology*, Wiley & Sons, after Essig and others)

Hymenoptera, however, more than one embryo may be formed through a process of asexual division. The numbers of embryos produced in this process of **polyembryony** usually vary from two to a few, but in some species, from 100 to 3,000 embryos are produced from one egg. In polyembryony, each embryo develops into an active larva. Such a phenomenon offers a great advantage to small parasites, which can quickly insert only one egg and yet produce a large number of offspring in a single host.

In most insects, embryonic development proceeds within the egg after it has been laid or deposited. In this instance, the egg is surrounded by a protective chorion, which gives it a characteristic shape and surface appearance. Insects giving birth in this manner are called **oviparous**. In other words, birth, the act of young leaving the female, is achieved by egg laying.

BOX 4.2 **TSETSE FLIES**

SPECIES: numerous *Glossina* species (Diptera: Muscidae)

DISTRIBUTION: Tsetse fly distribution is somewhat limited, with populations currently restricted to Africa south of the Sahara Desert. However, these pests were more widely distributed in earlier geologic times. Fossils of four *Glossina* species have been described from shale samples collected in Colorado. Most vector species are "riverine," meaning they can be found near shores and banks of rivers or lakes near forested areas.

IMPORTANCE: Of the 22 species of *Glossina* currently described, 6 species are of specific medical or veterinary importance. Specifically, *Glossina* flies are the primary vectors of pathogens that cause sleeping sickness in Africa (also known as African trypanosomiasis). The disease, natively known as Nagana, also affects domestic cattle. These pathogens are flagellate protozoa, which are transmitted with the bite of a tsetse fly. Sleeping sickness is a disabling and usually fatal disease. It is a primary factor limiting the development of Africa south of the Sahara Desert and is responsible for about 7,000 human deaths annually.

APPEARANCE: Tsetse flies range from 7.5 to 14 mm long, depending on species. The body is somewhat wasplike and brown. The wings extend beyond the tip of the abdomen when at rest and are crossed in a scissorslike fashion. The wings have a very distinct cleaver or hatchet cell in the venation that makes *Glossina* easy to distinguish from other genera of Muscidae. Larvae are pale yellow to creamy white, but puparia are blackish brown and barrel-shaped. Both immature forms are found just under the surface in sandy areas.

LIFE CYCLE: Tseste flies are very atypical in their development because larvae are reared internally within the adult female (viviparous development). After development through two stadia, the third instar is deposited on sand, where it burrows a few centimeters and completes its development. Larvae are susceptible to predation or parasitization in this stage but form puparia within a few hours and are relatively safe from natural enemies. Fourth instars occur within the puparia prior to true pupation, which lasts 30 days or more. Only 1 larva is carried by the female at a time, with 8 to 25 days needed for development. Adult flies emerge and continue the cycle. Breeding is continuous in the warm latitudes of tropical Africa.

In a few species of many insect orders, eggs with a chorion are held within the female until embryonic development is nearly complete. Hatching usually occurs when these eggs are still in the female body or in egg-laying organs. These insects, for example, many flesh flies (Diptera: Sarcophagidae), are referred to as **ovoviviparous**. Such an adaptation is particularly advantageous to insects whose food supply deteriorates quickly—in other words, before the eggs, if laid, would hatch.

A very specialized form of birth occurs in **viviparous** species of insects. Members of the fly genus, *Glossina*, which includes the tsetse fly, exhibit viviparity. Here, one larva develops at a time in a specialized "uterus." Each larva grows by receiving nutriments from special uterine ("milk") glands. Tsetse fly larvae are fully developed when "born" and begin adult development soon after birth. Most aphid species also are viviparous but only at certain times of the year.

POSTEMBRYONIC GROWTH AND DEVELOPMENT

Egg Hatching

When embryonic development is complete in oviparous insects, eggs are predisposed to **hatch**. The hatching process may begin immediately on completion of

Figure 4.9 Unhatched egg of a painted lady butterfly (top) and empty chorion of the egg after hatch. (Photos by T. Meyers)

embryonic development or may be delayed for a time, thereby taking advantage of more favorable environmental conditions.

At this point in the life cycle, the embryo within the egg must break out of its enclosure. The hatching process often begins when the embryo swallows fluids or air in the egg. This action gives the embryo more bulk and turgidity.

Subsequently, the embryo must rupture the egg covering to escape. Ruptures may be caused when the insect produces rhythmic muscular activity and presses or strikes against the covering with its head. The eggshell rupture may occur irregularly along the egg surface, as in grasshoppers (Orthoptera: Acrididae) or along preformed lines of weakness. The green stink bug (Hemiptera: Pentatomidae) represents the latter type; its eggs have easily ruptured "caps" that the insects open like a lid. In some insect groups, for example, dragonflies (Odonata), a T- or Y-shaped central "tooth" (egg burster) forces the rupture, and in

BOX 4.3 GREEN STINK BUG

SPECIES: *Acrosternum hilare* (Say) (Hemiptera: Pentatomidae)

DISTRIBUTION: The green stink bug is found in both North and Central America but is more injurious in the southern United States. Individuals of this species can be found in field borders before fruit set and move into fields as fruits and pods become available.

IMPORTANCE: The green stink bug attacks many wild and cultivated plants, including alfalfa, lima bean, peach, cotton, and soybean. With cotton, the piercing-sucking mouthparts are inserted into the bolls, and sap is removed from immature seeds. The punctures cause shedding of small bolls, whereas larger ones have stained lint. In soybeans, nymphs and adults pierce pods and feed directly on the seeds. This feeding results in smaller seeds and shriveled seed coats. Such injury results in a downgrading of seed quality at sale, reducing the market value of the entire harvest.

APPEARANCE: Adults are shield-shaped, typical pentatomids, with a large triangular scutellum. This species has a distinctive long spine between the thoracic legs, which dis-

tinguishes it from a similar species, the southern green stink bug, *Nezara viridula* (Linnaeus). Nymphs resemble adults, except they are smaller and wingless. In addition, nymphs are dark with white lines on the abdomen (southern green stink bugs have white dots on the abdomen). Stink bug eggs are characteristically barrel-shaped and are typically found in compact clusters of regular rows. It is not uncommon to find parasitized stink bug eggs. When parasitized, eggs are black and become inviable.

LIFE CYCLE: Adults overwinter in sheltered areas such as fence rows where plant refuse is abundant. In spring they become active and females oviposit 300 to 500 eggs, in clusters of about 30, on foliage during a 30-day period. Hatching occurs about seven days after oviposition, and five stadia follow, showing gradual metamorphosis. The adult age is reached in approximately six weeks. Multiple generations occur annually with overlap typical. Later generations, however, frequently are the most damaging to crops because of the presence of fruiting structures.

others spines or teeth on the head or body are used to "saw" through the eggshell. Many butterflies and moths (Lepidoptera) simply chew their way out of the eggshell.

Upon causing the rupture, the insect works its way out of its enclosure. Insects that possess a special embryonic cuticle or skin shed it as they leave the egg or shortly thereafter. The act of the larva leaving the egg is called **eclosion**.

Growth of Immatures

After eclosion, the new free-living insect may be similar to its parents or have an entirely different form. In either instance, the immature stage is usually characterized by feeding behavior. The immature is a virtual "feeding machine" at this point in the life cycle. Frequently, injury by pest species is most severe at this time and, in many instances, occurs exclusively during this stage (for example, caterpillar pests).

This feeding behavior results in substantial growth in size of the insect. Growth, as measured by weight gain, usually progresses in a stepwise fashion, with each step being delineated by shedding of the immature's old skin (**exuviae**). This shedding process, called **molting**, results from the inability of the insect's skin or cuticle to expand. Therefore, the insect grows to a size where its outer covering becomes limiting. It sheds that covering, develops a new larger one, grows to fill up the new skin, sheds it, and so on, until growth is complete.

The molting process usually begins with the cessation of feeding and the clearing of gut contents. The process is initiated and controlled by hormones that circulate in the blood. The major hormones involved are **brain hormone** (AH, activation hormone, or PTTH, prothoracicotropic hormone, as it is called by some authors) and **ecdysone**. Brain hormone is produced by neurosecretory cells in the brain and dispensed into the blood cavity through brain accessory structures. Once in the blood, brain hormone circulates to an activity site in the insect's prothorax. A small gland in the prothorax, the **prothoracic gland**, is stimulated to secrete ecdysone (molting hormone, also called prothoracic gland hormone), which initiates the growth and molting activities of cells.

Upon stimulation by ecdysone, cells in the epidermis (the cellular part of the exoskeleton) divide and become closely packed. As a result, the cuticle (noncellular part of the exoskeleton) becomes separated from the epidermis, a process called **apolysis**. As the cuticle separates from the epidermis, **molting fluid** is produced. Molting fluid contains proteinase and chitinase enzymes that digest up to 90 percent of the old cuticle. This digested material is absorbed and metabolized. While the old cuticle is being digested, a new cuticle is started. It grows by forming different layers, the outer layer being resistant to enzymes of the molting fluid, thus protecting new layers from the digestive process.

Once the new cuticle is complete and digested material is absorbed, **ecdysis** occurs. **Ecdysis** is the process of shedding the remainder of the old cuticle, which is split along weakened lines by movements of the insect inside. The initiation of these movements in many insects is believed to be caused by another hormone called **eclosion hormone**. Eclosion hormone is secreted by specific brain cells at a certain time of day. In many instances, the insect swallows air or water and thereby increases blood pressure. Muscle action pumps blood to a certain part of the body, usually the thorax, to expand it. This expansion causes the old cuticle

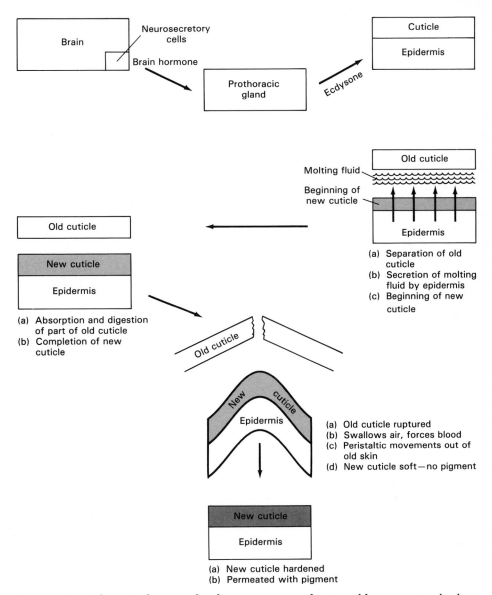

Figure 4.10 Schematic diagram of molting in insects and principal hormones involved.

to rupture. The insect then exits through this rupture, usually head and thorax first, followed by abdomen and appendages.

Immediately after emergence the new cuticle is soft and pliable. The insect expands this new covering before it hardens. This process may again involve swallowing air or water and using blood pressure to expand the body, one part at a time. Once expanded, the new cuticle hardens and is permeated with pigment in a process called **sclerotization**. Recently, it has been shown that yet another hormone from the nervous system, **bursicon**, plays a role in the control of these postecdysal processes.

The rate of insect growth is greatly influenced by the physical environment, particularly temperature. Within certain limits, the higher the temperature, the faster the rate of growth. By understanding this relationship, it is possible to make certain predictions about the stage and size of insects during a season. As we shall see later, such predictions are very important in insect pest management.

Metamorphosis

As insects grow, various changes, some of which are quite dramatic, occur during the life cycle. The developmental process that takes place from eclosion of the free-living insect until adulthood is called **metamorphosis**, "change in form." Metamorphosis runs a gamut involving hardly noticeable to quite drastic changes in body form. In the former instance, immature insects look similar to their parents, differing only in body proportion and lacking reproductive appendages. In the latter instance, the immatures are very different from their parents, often changing from a wormlike, crawling form to an ovoid, inactive form and, finally, to a winged, flying form.

Metamorphosis represents orderly, genetically programmed changes in insect form during completion of the life cycle. Insects with radical changes in body form, for example, butterflies, undergo internal processes that cause rearrangements within cells, tissues, and organs. These processes can be ordered into two phases, **histolysis** and **histogenesis**. Histolysis is the breakdown of body tissues, which often begins when the insect is at the end of its primary growth stage; it continues into the transformation stage. During histolysis, stored nutrients (carbohydrates, fats, glycogen) supply energy for numerous biochemical reactions, and enzymes with the aid of special blood cells (**phagocytes**) convert the insect fat body and muscle tissue into a nutritive mix. This mix is transported in the blood to growing tissues. Histogenesis is the formation of new body tissues; it occurs at the same time as histolysis. In particular, complete exchanges of immature muscle tissues for adult muscle tissues occur during this time, as well as exclusive adult organs like wings and reproductive structures. The result, in many instances, is an adult that is quite different in appearance from the immature form. Specific metamorphic patterns are discussed later in this chapter.

Metamorphosis is primarily under the control of another insect hormone called **juvenile hormone** (JH). JH is produced by glands accessory to the brain, the **corpora allata**. Production is believed to be mediated by brain hormone. Secretion of JH in the blood functions to suppress adult characteristics, causing juvenile structures to be retained. An insect having a high titer of JH in the blood, therefore, molts but retains the same form in the next stage. At a critical stage of growth, the titer of JH is reduced, and after the next molt the insect takes another genetically programmed form. Finally, JH production is suspended or declines to a very low level, and after the next molt, the insect assumes an adult form. Therefore, it can be seen that growth and development are interrelated, with control of the entire process assumed by three very important enzymes: brain hormone, ecdysone, and juvenile hormone.

An understanding of the hormones governing growth and development is of great importance to insect pest management. As we shall see, some of our newest insecticides (third-generation insecticides, biorationals, insect-growth regulators) mimic the action of these hormones and, when applied properly, kill the targeted pests.

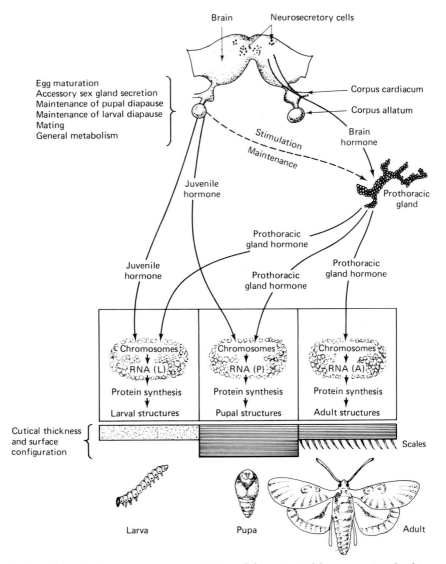

Figure 4.11 Diagrammatic representation of the principal hormones involved in insect growth and metamorphosis and their action. (Reprinted with permission of Macmillan Publishing Company from *Insects in Perspective* by Michael D. Atkins. Copyright ©1978 by Michael D. Atkins)

Some Terminology

Entomologists have used various terms to describe the developing insect during its life cycle. Some of the common ones are mentioned here to acquaint readers with them so as to form a basis for further discussion of life cycles and seasonal cycles.

One of the most widely used terms is **generation**. A generation is a cohort of offspring from a parent population moving through the life cycle together. The term **brood** is often used interchangeably with generation, as for instance second-brood European corn borers. This is an incorrect usage, however, because brood is more appropriately used when a parent or parent population produces several cohorts of offspring at different times or in different places (more than one nest), as do many social insects.

Other terms that are used (and sometimes misused) frequently are stage, stadium, and instar. **Stage** refers to the insect's developmental status. Therefore, if we want to emphasize development attained at a given point in the life cycle we speak of the egg stage, larval stage, or adult stage. We can also use stage to show developmental status after specific molts, for instance, a third-stage larva. This insect would be the immature form after the second molt. **Stadium** is the time period between molts, and **instar** is the actual insect between molts. As an example to put these terms in order, we might say the following: The sixth-stage larva displays the greatest rate of consumption during the life cycle, and its stadium is three days. We have counted ten sixth-instars in the sample, and they are green with white stripes. Never use, for example, sixth-instar larvae, because elements of the term are redundant.

MATURITY

Emergence of the Adult

Adults emerge from the last-immature exuviae in a way similar to immatures exiting their old skins. After splitting the old cuticle, the new adult pulls itself out and, if winged, expands its newly formed wings by pumping blood through them. As with many newly molted immatures, the new adult at first is soft and may lack much of its final pigmentation. At this time, it is very vulnerable to natural enemies and other environmental adversities. Within a short time, however, the cuticle hardens and becomes pigmented, and the wings are fully expanded and functional.

The degree of sexual maturation at the time of adult emergence varies between sexes and among species. Most male insects emerge with a store of sperm and some are fully capable of mating within a short time. However, in some species the female may require a longer time to become sexually mature. During female reproductive maturation, eggs form, the reproductive tract completes development, and other glands associated with reproduction develop. In certain short-lived species, for example, mayflies (Ephemeroptera), both sexes are mature within a day or so, and mating and egg laying occur shortly thereafter.

Mating Behavior

Mating behavior consists of all the events involved in the transfer of sperm from the male to the female. The two most important events in this behavior are finding mates and copulation.

Mate finding is an important behavior in all but a very few insects. For example, some Collembola transfer sperm indirectly; the male simply places a sperm packet, called a **spermatophore**, on a substrate, and, subsequently it is picked up by a female, presumably by chance. In most instances, however, locating a mate is necessary, and this act may involve complex processes, among which are certain cues to locate and recognize an appropriate mate. Such cues may be visual, olfactory, auditory, tactile, or combinations of these.

Olfactory cues are probably the most commonly used type in mate finding. Many insect species secrete volatile chemical attractants that are highly specific and can be used by one sex to attract its opposite over considerable distances.

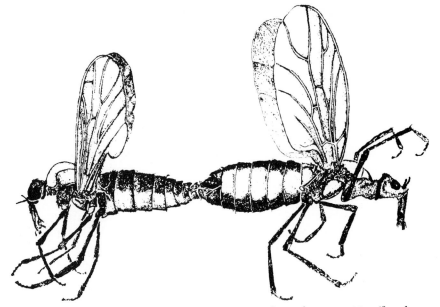

Figure 4.12 Lovebugs, *Plecia nearctica*, in typical copulatory position (female on right). (Reprinted by permission of John Wiley & Sons, Inc. from *Insect Behavior* by R. W. Matthews and J. R. Matthews. Copyright ©1978 John Wiley & Sons, Inc.)

Such volatiles are called **sex pheromones**. Male-attracting sex pheromones are the common type found in many insect orders, particularly Coleoptera and Lepidoptera. In many moths, females begin "calling" (everting the pheromone-releasing glands and releasing the volatile) about one to two hours before dawn. Males detect differences in concentration of the volatile and tend to follow the gradient of increasing concentration until they locate the female. When the male gets close to a female, he hovers and lands, then moves about until he collides with her. In other insects, like the boll weevil, *Anthonomus grandis*, the male produces a sex pheromone, and the female is attracted.

Copulation may follow immediately after mate finding or, in some species, after specific forms of courtship behavior. Depending on the species, a female may copulate several times during her lifetime or only once. Following the act of copulation (joining of male and female genital structures), **insemination** (introducing of sperm into the female reproductive tract) occurs.

As we will see in later chapters, an understanding of mating behavior has important implications for insect pest management. By taking advantage of the compelling nature of sex pheromones, entomologists have developed more effective insect sampling programs, and they have attempted to utilize pheromones and synthetic mimics of them to disrupt the mating process.

Oviposition

Once mating has occurred and eggs are matured and fertilized, egg deposition is the ultimate event in the life cycle of an insect. The act of laying eggs by the adult female is called **oviposition**.

Insects differ greatly in the way they lay their eggs, or oviposit. Eggs may be laid singly, (for example, lacewings), in clumps (stink bugs), or in a contiguous mass (European corn borer). Some insects, such as cockroaches, produce an **ootheca**, which is a protective pod that contains several eggs "glued" together by a secretion.

Insect eggs simply may be dropped on the ground, as with walkingsticks (Orthoptera: Phasmatidae), or they may be placed on or in host tissues, where newly eclosed immatures have ready access to food. In the latter instance these eggs may be glued to plant leaves or stems, as with moths and butterflies, or inserted inside stems and leaf petioles, as with many leafhoppers, for example, the potato leafhopper, *Empoasca fabae*. Another important placement pattern for many insects is oviposition in soil; for example, grasshoppers lay masses of eggs in special chambers in grass sod. Oviposition in plant and animal tissues and soil provides a more or less protected environment for egg development.

Female insects lay their eggs with a specialized structure called an **ovipositor**. Ovipositor structure varies from a simple extensile tube through which eggs pass (for example, in Lepidoptera and Diptera) to a well-developed "saw" found in sawflies (Hymenoptera: Cephidae). This saw-toothed ovipositor cuts into plant tissues for egg insertion. Highly developed ovipositors with dual functions are found in other Hymenoptera. They are used to "sting" a host, usually another insect, and place eggs inside (for example, parasitic wasps) or sting another insect (or other animal) for defensive or paralyzing purposes (for example, bees, ants, and wasps). In the latter instance, a venom is injected when the "stinger" is inserted. With hunting wasps, the stung prey (for example, caterpillars,

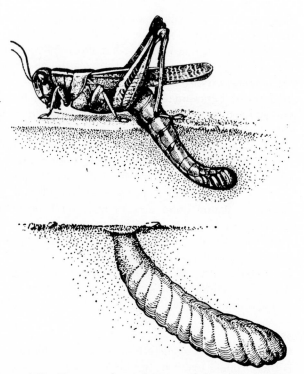

Figure 4.13 Twostriped grasshopper, *Melanoplus, bivittatus*, ovipositing in soil (top) and egg pod with eggs after oviposition. (Courtesy USDA)

grasshoppers, crickets) becomes stupefied and is carried back to the wasp nest; subsequently, eggs are laid on it. After hatching, newly emerged immatures begin feeding on the stunned but still living prey. When bees, wasps, and ants sting people, pain, discomfort, or even death may result from the venom.

GENERAL MODELS OF THE LIFE CYCLE

A study of insects shows that each life cycle is species-specific; in other words each species has its own peculiar mode of living, and it is difficult to generalize from one species to the next. But we do see patterns in metamorphosis among insect species that relate to their activities and modes of living in the environment, and we can use these patterns as general models of the life cycle.

As insects have evolved, there has been an overall trend toward more complex life cycles. In advanced forms, this complexity has resulted from greater differences in body structure and behavior between immature and adult insects. Along the evolutionary scale, the immature insect has become increasingly specialized in feeding and growth, whereas the adult has become specialized in reproduction and dispersal.

By emphasizing degree of complexity and metamorphic pattern, we can recognize four insect life cycle models: (1) no metamorphosis, (2) gradual metamorphosis, (3) incomplete metamorphosis, and (4) complete metamorphosis. These categories are given for convenience in discussing life cycles, not for physiological classification of metamorphosis. For such classification, most authorities would group models (2) and (3) into a single category, gradual metamorphosis.

The No-metamorphosis Model

The no-metamorphosis model is found in primitively wingless insects (subclass Apterygota), for example, springtails, silverfish, and firebrats, as well as in other noninsect arthropods. In these groups, the life cycle progresses from egg to **juvenile** to adult. The transition from the first juvenile to the adult is gradual, with several molts taking place in the juvenile stage. During the life cycle, the juvenile appears very similar to the adult, differing mainly in body size and proportion and lacking functional genitalia. In these insects, all stages can be found in the

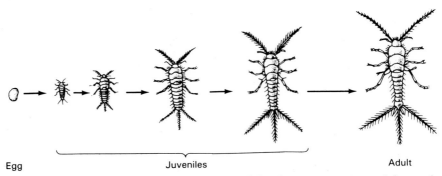

Egg   Juveniles   Adult

Figure 4.14 The no-metamorphosis model of development represented by a silverfish, *Lepisma sacharina*. (Redrawn from Little, 1972, *General and Applied Entomology*, Harper and Row Publishing Company)

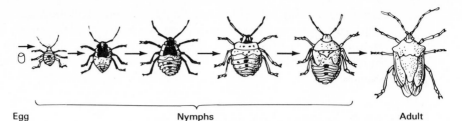

Egg   Nymphs   Adult

Figure 4.15 The gradual-metamorphosis model of development represented by a stink bug (Hemiptera: Pentatomidae). (Redrawn from Little, 1972, *General and Applied Entomology,* Harper and Row Publishing Company)

same habitat, and juveniles feed similarly to adults. Unlike with other insects, molting continues in the adult stage, and because females lose the lining of the spermatheca with each molt, they are inseminated several times during the life cycle.

The Gradual-metamorphosis Model

The life cycle of insects with gradual metamorphosis has three major stages: egg, **nymph**, and adult. Common examples of this model are grasshoppers, bugs, cockroaches, and crickets, among others. Here, the nymph resembles the adult but lacks fully formed wings and external genitalia. Nymphs may be colored quite differently from adults (for example, some stink bugs), and they show the beginnings of wings, called **wing pads**, in later developmental stages. In most instances, eggs, nymphs, and adults with this life cycle are found in the same habitat, and nymphs and adults feed similarly.

The Incomplete-metamorphosis Model

The incomplete-metamorphosis model is one in which immatures may or may not resemble adults, have external wing pads, and may have special "tracheal gills" that allow aquatic respiration. There are many variations in this type of life cycle, but in general, eggs are found in or near the water and immatures, called **naiads**, feed and develop in water. Adults of these insects may be found flying above or near the water or, with some dragonflies, quite some distance from it.

Egg   Naiads   Adult

Figure 4.16 The incomplete-metamorphosis model of development represented by a mayfly (Ephemeroptera). (Redrawn from Little, 1972, *General and Applied Entomology,* Harper and Row Publishing Company)

Unlike the gradual-metamorphosis model, immatures and adults of species with incomplete metamorphosis spend most of their time in very different habitats and feed quite differently. Although they do not represent a cohesive group in an evolutionary sense, mayflies (Ephemeroptera), dragonflies and damselflies (Odonata), and stoneflies (Plecoptera) all follow this life cycle model.

The Complete-metamorphosis Model

This form of the life cycle is found in the most highly evolved insects. Here, there are four distinct life stages in the cycle: egg, **larva**, **pupa**, and adult. The presence of the larval and pupal stages are the most notable aspects of this life cycle model. With few exceptions, larvae are very different from adults. They have a different form, lack compound eyes, have reduced antennae, and lack external evidence of wing formation. In most, wing development begins in the early larval stages but occurs internally from rudiments known as **imaginal disks**. The pupal stage is represented by an insect usually in a quiescent state (a few insect species have active pupae). Most quiescent pupae are found in hidden, protected habitats. Some, like the moths, are present in **cocoons**, a covering of silk or silk-bound debris constructed by the last larval instars (**prepupae**). Pupae of the more advanced Diptera are found within the last larval skin, which hardens to form a protective **puparium** (Figure 4.17). Wing pads appear on the pupa, and the active physiological processes of histolysis and histogenesis take place internally during this stage.

The evolution of a pupal stage has allowed the development of specialized larval structures that are not continued in the adult stage. With this arrangement, larvae and adults often have evolved in very different directions, with larvae specializing in food gathering and adults developing advanced means of reproduction and dispersal. In addition to larvae consuming much more food than most adults, they also may consume different food, thereby eliminating much of the competition between stages. For instance, moth larvae (caterpillars) feed on green plant tissue for growth, but adult moths do not feed or feed on nectar for maintenance. When both larvae and adults are specialized for food gathering, their food source may differ. For example, mosquito larvae are found in water and feed on microorganisms there, but adult females are specialized to find and extract blood from birds and mammals.

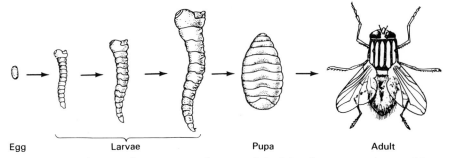

Egg   Larvae   Pupa   Adult

Figure 4.17 The complete-metamorphosis model of development represented by a house fly, *Musca domestica*. The pupa of this insect is found within the hardened skin of the oldest larva, the puparium. (Redrawn from Little, 1972, *General and Applied Entomology*, Harper and Row Publishing Company)

In the majority of life cycles with complete metamorphosis, larvae in all stages behave similarly; they live in the same habitat and consume the same food. However, in some insects like blister beetles (Coleoptera: Meloidae) or many parasitic Hymenoptera, body form, feeding behavior, and sometimes habitat change as larvae age. This phenomenon is termed **hypermetamorphosis**. For example, in the black blister beetle, *Epicauta pennsylvanica*, the first-stage larva,

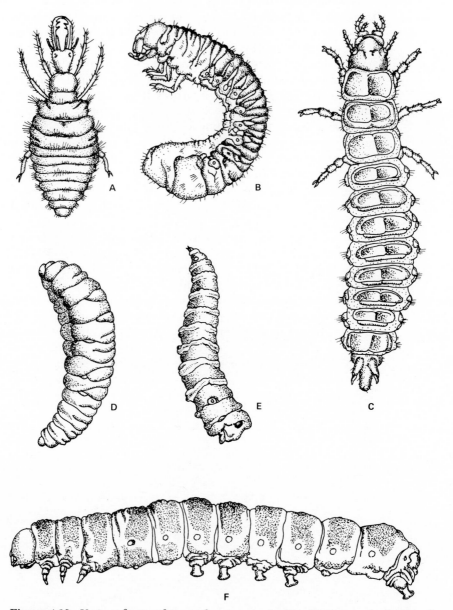

Figure 4.18 Various forms of insect larvae. A. Antlion (Neuroptera). B. Scarab beetle (Coleoptera). C. Ground beetle (Coleoptera). D. Weevil (Coleoptera). E. Fly maggot (Diptera). F. Caterpillar (Lepidoptera). (Reprinted with permission of Macmillan Publishing Company from *Insects in Perspective* by Michael D. Atkins. Copyright ©1978 by Michael D. Atkins)

Figure 4.19 Stages of growth of imaginal disks. (Redrawn from Ross, et al., 1982, *A Textbook of Entomology*, Wiley & Sons)

BOX 4.4 **BLACK BLISTER BEETLE**

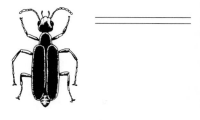

SPECIES: *Epicauta pennsylvanica* (De Geer) (Coleoptera: Meloidae)

DISTRIBUTION: Blister beetles are found in all parts of the United States and Canada. The relative abundance within this range is highly variable and regulated by abundance of grasshopper eggs (food source) in the region. In dry years, grasshopper egg survival is very high, and there will usually be an associated abundance of blister beetles.

IMPORTANCE: Adult blister beetles feed on foliage of several host plants, including alfalfa, soybean, sugar beet, and a wide variety of garden vegetables and ornamental species. Behaviorally, adults feed gregariously, so that damage may be localized within a large area. In addition to plant damage, blister beetles are a serious threat to livestock, especially horses. The danger arises because blister beetles contain high levels of cantharidin, which is a severe irritant of the gastrointestinal and urinary tracts of many animals. Poisoning of livestock following accidental consumption of alfalfa hay containing dead blister beetles is frequently reported. Ironically, larvae are considered to be beneficial because they feed almost exclusively on grasshopper eggs.

APPEARANCE: Adults have a distinct elongate and slender appearance. Black blister beetles are 7 to 13 mm long and relatively soft-bodied and have a head wider than the pronotum. Adults are commonly found on goldenrod flowers. Larvae develop by hyper-metamorphosis (each larval instar contains two or more distinct body types). The first instar is an active, long-legged form called a triungulin. The second instar is similar, but the legs are shorter. The third, fourth, and fifth instars are robust and somewhat grublike. The sixth instar lacks functional appendages, is harder and darker, and is frequently called a pseudopupa. The seventh instar is small, white, and active (still legless) but does not feed and develops into the pupa.

LIFE CYCLE: Pseudopupae overwinter in soil and complete development in spring. Adults emerge in June or July and feed on plant hosts in localized groups. Females oviposit eggs in clusters of 100 to 200 in holes prepared in soil. In about 12 days the eggs hatch and triungulin larvae search for a grasshopper egg mass, where they gnaw into the egg pod and feed on the eggs throughout larval development. Typically, there is one generation per year, with the survival of larval populations dependent on the availability of grasshopper egg pods.

Figure 4.20 Various forms of pupae. A., B., C. Lepidoptera. D. Hymenoptera. E. Neuroptera (dobsonfly). F. Diptera (mosquito). E. and F. are active pupae. (Reprinted with permission of Macmillan Publishing Company from *Insects in Perspective* by Michael D. Atkins. Copyright ©1978 by Michael D. Atkins)

the **triungulin,** actively seeks out grasshopper eggs, punctures an egg, and feeds on its contents. The second, third, fourth, and fifth instars are inactive and continue to feed on the egg contents. When fully grown, the fifth instar leaves the food mass, burrows farther into the soil, and builds an earthen cell. There, it molts to become a sixth instar, a hardened oval form (**coarctate form**) with exceptionally small legs that does not feed and overwinters. This form is particularly resistant

Figure 4.21 Hypermetamorphosis as seen in a black blister beetle, *Epicauta pennsylvanica*. A. Unfed first instar. B. Fully fed first instar. C., D., E. Second, third, and fourth instars. F. Newly molted fifth instar. G. Gorged fifth instar. H. Sixth instar. I. Seventh instar. J. Pupa. (Reprinted by permission of John Wiley & Sons, Inc. from *A Textbook of Entomology* by H. H. Ross, C. A. Ross, and J. R. Ross. Copyright ©1982 John Wiley & Sons, Inc.)

to drought, and if the following summer is abnormally dry, it will not progress to the seventh stage for another year or even two years. When appropriate conditions occur, a seventh instar appears, and this larva molts to become the pupa. Adult blister beetles then may feed on and become important pests of potatoes, tomatoes, squash, some legumes, and other crops.

SEASONAL CYCLES

Thus far, we have focused attention on insects within a single generation and emphasized the major events taking place in a life cycle. A logical extension of our focus would be to examine the cycles of insects from an annual perspective, in other words, look at the cycles as they pass through different seasons of the year. This progression of one or more life cycles occurring during a one-year period (for example, winter to winter) is termed the **seasonal cycle**.

Seasonal cycles are very important to the survival of insects, because most environments of which insects are a part undergo annual cycles. Therefore, to achieve success, insects have evolved ways to time their activities to take advantage of environmental resources and avoid unfavorable extremes. Knowledge of patterns in seasonal cycles and the timing of important biological events (for example, dormancy, migration, development, and reproduction) relative to environmental cycles is crucial when attempting to reduce pest damage.

Seasonal cycles can be grouped according to the number of generations that occur in a year; this number is referred to as a population's **voltinity**. Voltinity types are univoltine, multivoltine, and delayed voltine.

Univoltine Cycles

The **univoltine** cycle refers to a single generation each year. This type is common among most grasshoppers, corn rootworms (*Diabrotica* species), and many other insects. In this type, the generations are easily distinguished in nature, with generation overlap occurring when adults of one generation are laying eggs of the next. Therefore, when observed in the field at any point in time (except during egg laying), most of the population would be in the same growth stage. Because of this simplicity, these populations are often easiest to study and understand.

Multivoltine Cycles

Multivoltine cycles have more than one generation per year. Numbers of generations per year may range from two to four or more, depending on developmental time requirements and environmental conditions. Examples of multivoltine insects include house flies (*Musca domestica*), thrips (Thysanoptera), aphids, and the European corn borer. The European corn borer is somewhat unusual in having both multivoltine (for example, bivoltine in Iowa) and univoltine (for example, in Nova Scotia, Canada) types. As numbers of generations per year increases, overlap between these becomes greater with seasonal progression. This overlap may be great enough toward late season so that all stages of the insect are present, and generations may become indistinguishable. Such a phenomenon usually occurs with thrips and mites; this complexity makes study of their populations very difficult.

Multivoltine cycles may be subdivided into types, depending on their pattern.

Figure 4.22 Schematic drawing of the relationship between a seasonal cycle (SC) and a life cycle (LC) in univoltine populations.

Figure 4.23 Schematic drawing of the relationship between a seasonal cycle (SC) and life cycles (LC) in multivoltine populations.

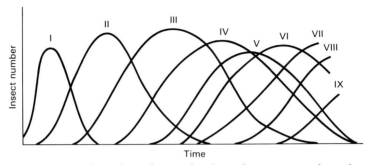

Figure 4.24 A hypothetical example of population curves of a multivoltine insect population. Roman numerals indicate generations during a season. Note the increasing degree of overlap as the season progresses.

The most common, simplest pattern occurs with species having repetitious generations. With these, each successive life cycle is essentially the same. Therefore, each generation of cockroach has similar characteristics: the same morphology, eating habits, and reproductive behavior. Periods of dormancy that may occur with some insects in certain life cycles are not considered a major alteration in the fundamental life cycle pattern. Consequently, a multivoltine species whose overwintering generation differs from others only by virtue of its dormancy is still considered to have repetitious generations.

Multivoltine species in which succeeding generations are different in mode of reproduction or morphology are said to have alternating generations. One of the best examples of alternating generations is aphids. In a relatively simple aphid life cycle, winter is spent in the egg stage. Eggs hatch in early spring, and all new nymphs are wingless females. When mature, these so-called **stem mothers** reproduce parthenogenetically and viviparously, giving rise to other generations of parthenogenic, viviparous females. By midsummer, generations of both winged (**alate**) and wingless (**apterous**) females are produced, with the winged forms dispersing to begin new populations elsewhere. In late summer and fall, winged males and wingless females are produced. This phenomenon is accomplished through a loss of the X chromosome in a portion of the ova that produces males (XO = male, XX = female) and is believed to be under environmental control. Subsequently, mating occurs in this "sexual" generation, and eggs are produced for overwintering. Such an arrangement gives aphids the advantage of gene mixing for hybrid vigor and tremendous fecundity for population growth during the growing season. It is one reason why aphids are some of our most serious insect pests.

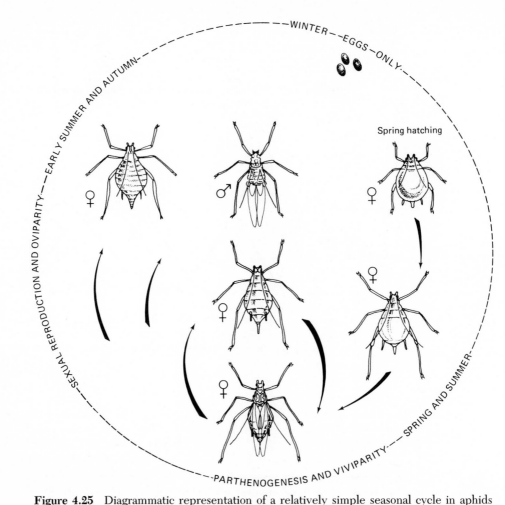

Figure 4.25 Diagrammatic representation of a relatively simple seasonal cycle in aphids (Homoptera: Aphididae). (Reprinted with permission of Macmillan Publishing Company from *Insects in Perspective* by Michael D. Atkins. Copyright ©1978 by Michael D. Atkins)

Delayed Voltine Cycles

In a few insect species, the life cycle requires more than one year for completion. In this instance, the insect must pass through more than one seasonal cycle to complete a single generation. Examples of delayed voltinism can be found in wireworms (Coleoptera: Elateridae) and many June beetles (Coleoptera: Scarabaeidae), with larvae requiring two to three years to mature. An extreme instance is the 17-year cicada (Homoptera: Cicadidae), which requires 17 years to develop—

Figure 4.26 Schematic drawing of the relationship between seasonal cycles (SC) and a life cycle (LC) in delayed-voltine populations.

the longest of any insect! In most instances of delayed voltinism, generations overlap so that adults are found every year in nature, and the activities of combined generations are often viewed in describing their seasonal cycles.

Seasonal Adaptations

During the seasonal cycle, insects adapt to many types of adverse environmental conditions. These conditions include irregular and unpredictable extremes in temperature, moisture, food, crowding, and other factors. They are usually localized and temporary. Insect response to these stresses tends to be immediate and may involve becoming **quiescent** (torpid) or moving out of the area. Other types of adverse conditions occur regularly with the seasons and therefore are predictable, for example, freezing temperatures in temperate climates, rains in the tropics, and droughts in deserts. Most insects avoid these predictable adversities in their environment by undergoing physiological and behavioral alterations that condition them for the approaching season. These alterations help to determine the particular pattern of an insect's seasonal cycle. The major forms of this conditioning may include dormancy, changes in body form and color, and migration.

Dormancy. Dormancy is particularly prevalent among insects as a means for escaping regular seasonal adversities. Dormancy can be defined as a seasonally recurring period in the life cycle when growth, development, and reproduction are suppressed. It may occur during summer, fall, winter, or spring, and during these periods dormancy is termed **aestivation**, autumnal dormancy, **hibernation**, and vernal dormancy, respectively. Some insects may be dormant through more than one season. In this instance, the dormancy terms are combined; for example, aestivo-hibernation is used for larvae of the seedcorn maggot, *Delia platura*, which aestivate most of summer and go directly into winter hibernation.

In temperate climates, hibernation is one of the most common characteristics of the insect seasonal cycle, and cold hardiness usually accompanies it. Cold hardiness is achieved in insects through the physiological processes of **supercooling** and **freezing tolerance**. Supercooling is a resistance to freezing by lowering the temperature at which freezing of body fluids begins, whereas freezing tolerance is survival despite freezing of body fluid.

Dormancy differs from quiescence in that during dormancy growth and reproduction may be suppressed even during periods when conditions are temporarily favorable for these functions. During quiescence, growth and reproduction resume immediately upon return of favorable conditions.

Diapause. Dormancy, as well as seasonal migration and morphological change, are controlled by a physiological condition called **diapause**. Diapause is usually characterized by low metabolism, little or no development, increased resistance to environmental extremes, and altered (often greatly reduced) behavioral activity. The mechanism found to underlie and orchestrate diapause is hormone activity, with some of the following involved: (1) larval and pupal diapause, lack of brain hormone, and consequently, ecdysone; (2) adult diapause, lack of brain hormone and juvenile hormone; and (3) egg diapause, release of a neurosecretory hormone from the subesophageal ganglia of the parent.

Diapause is induced by specific environmental cues called **token stimuli**. They are given this term because these environmental factors in themselves are neither favorable nor unfavorable but signal forthcoming changes in the environment.

BOX 4.5 SEEDCORN MAGGOT

SPECIES: *Delia platura* (Meigen) (Diptera: Anthomyiidae)

DISTRIBUTION: The seedcorn maggot is widely distributed in the temperate regions of the world and was first reported in North America in 1855. Surveys have shown that this species is found on all continents except Antarctica.

IMPORTANCE: The larvae (maggots) develop in the soil and feed on decaying organic matter. In the process of feeding, larvae readily feed on newly planted seeds, either prior to emergence or soon after. Feeding before emergence may damage the embryo and cause seed inviability, which translates into reduced plant density. Feeding after the seed has germinated may result in different plant responses. If the feeding is minor, the plant may be delayed in development, or it may be exposed to pathogenic soil organisms. If the feeding is moderate, the growing tip may be injured or destroyed, and the plant loses apical dominance, which results in a "Y"-plant (two main stems). Under severe feeding, the seedling may simply die from tissue loss. The host range is vast, including bean, pea, corn, cabbage, cauliflower, spinach, turnip, radish, and onion. Actually, any seed is susceptible if it is located in the spatial and temporal location of the larvae.

APPEARANCE: Adults are small, gray flies with black legs. They are about 5 mm long and have fine hairs on the undersurface of the scutellum. Larvae are pale and yellowish white. They are 6 to 7 mm long, tapering from posterior to anterior, legless, and tough-skinned. Pupae are dark brown, about 6 mm long, and oval.

LIFE CYCLE: In northern regions of the United States, the pupae overwinter. Adults emerge in spring and are common on flowing heads of grasses. After mating, female flies oviposit eggs singly or in small clusters just beneath the surface of the soil near larval food. Preferred sites are in moist soil containing decaying organic matter, such as newly cultivated fields. Over approximately six weeks, females produce an average of 270 eggs each. Eggs develop and hatch within nine days. Larvae develop rapidly when food is abundant and temperatures are appropriate. Three larval stages develop in 7 to 21 days, after which pupation occurs in the soil. Puparia develop initially and last about two days, and are followed by true pupae, which last from 7 to 26 days. The number of generations per year varies with region, depending on resident temperatures. In some areas, populations aestivate in mid to late summer before going into winter hibernation.

The most important token stimulus for insects is day length, or **photoperiod**. Temperature, moisture, and biotic factors also may act as token stimuli, either alone or in combination. The onset of diapause occurs well before most environmental conditions change, whereas dormancy is a concurrent response or reaction to changing conditions. Once diapause is induced, it continues, even though favorable environmental conditions prevail.

Usually diapause occurs at a specific stage of the insect's life cycle, with this stage being genetically fixed. All life stages have been known to diapause. For example, most grasshoppers and corn rootworms diapause as eggs; European corn borers and many bark beetles (Coleoptera: Scolytidae) diapause as larvae; cecropia moths, *Hyalophora cecropia*, diapause as pupae; and boll weevils, *Anthonomus grandis*, diapause as adults. The stage sensitive to token stimuli may be the diapausing stage (some crickets), certain periods in the parental generation (silkworms), or even periods in the grandparent generation (some aphids).

Once diapause is initiated, it continues for several weeks in most insects but may last much longer in some. Commonly, diapause is maintained by specific patterns in day length, temperature, or both, and these patterns regulate its intensity. Once ended, the insect enters a postdiapause transitional period of dormancy (postdiapause quiescence) that is not sensitive to token stimuli. Subsequently, the return of favorable conditions signals the end of the diapause syndrome, and development and reproduction are resumed.

The diapause phenomenon mediates the progression of insect generations and, in doing so, enhances survival in unfavorable environments. In univoltine species, every generation has a diapausing stage, and interruptions in development and reproduction occur in every life cycle. In multivoltine species, the phenomenon occurs only in a single stage of one of the generations, although any generation is capable of diapause, given the appropriate token stimulus. Clearly, an understanding of these and other aspects of the diapause syndrome are necessary before we can efficiently rear insects for experimentation and be able to predict the seasonal cycle with the precision required for insect pest management.

Selected References

ATKINS, M. D. 1978. *Insects in Perspective*. New York: Macmillan, chapter 9.

GILLOTT, C. 1980. *Entomology*. New York: Plenum Press, chapters 19–21.

HAPP, G. M. 1984. Development and reproduction. In *Insect Biology*, H. E. EVANS, ed. Reading, Mass.: Addison-Wesley, chapter 4.

LITTLE, V. A. 1972. *General and Applied Entomology*; 3rd ed. New York: Harper & Row, chapter 3.

MATTHEWS, R. W., and J. R. MATTHEWS. 1978. *Insect Behavior*. New York: Wiley.

ROSS, H. H., C. A. ROSS, and J. R. P. ROSS. 1982. *A Textbook of Entomology*; 4th ed. New York: Wiley, chapter 7.

TAUBER, M. J., C. A. TAUBER, and S. MASAKI. 1986. *Seasonal Adaptations of Insects*. New York: Oxford University Press.

5 INSECT ECOLOGY

THERE IS LITTLE QUESTION that insects are one of the most prevalent features of our environment. No one can deny their presence; no one can avoid dealing with them. Insects' lives are inextricably intertwined with our own.

Although some of us are fascinated and intrigued by them, others see insects only as a hindrance to human activities and desires. In either instance, it is usually insect numbers, not simply the presence of a solitary individual, that attracts our attention. We are either appalled or in awe at seeing insects virtually explode in numbers as the growing season develops. We question why they can be so abundant in one area and not in another or why they are numerous one year and not the next.

To understand insect numbers is to understand their inherited traits and the particular environment of their life cycles. The interaction between these two aspects, inherited traits and environment, results in given numbers of individuals, and because these aspects are dynamic (constantly changing), insect numbers also are dynamic.

It is a major objective of insect ecology to explain the dynamics of insect numbers in time and space. For this explanation, insect ecology relies on an understanding of physiology and behavior of insects as affected by their environment. The focal point of study can be the individual, collections of individuals of the same species, or interactions between or among species.

In addition to the dynamics of insect numbers, environment also affects timing of biological events in insects (**phenology**) and the diversity of species in an area. These additional topics addressed by insect ecology are no less important than the question of abundance.

By providing explanations of how environment (including the agricultural environment) affects abundance, timing, and diversity of insects, ecology forms the very basis of pest management. Because pest problems usually develop from too many insects occurring at inappropriate times, dealing with them necessitates a knowledge of ecological processes. With this knowledge, the occurrence and severity of potential problems can be predicted and appropriate management activities formulated.

In this chapter, we consider the ecological role of pests in agricultural environments and survey the major determinants of insect abundance. Then, we learn how environmental factors affect timing of seasonal events and function to regulate insect numbers.

THE ECOLOGICAL ROLE OF INSECT PESTS

The Idea of Populations

Insect populations are groups of individuals set in a frame that is limited in time and space. Often, the boundaries of time and space for a population are somewhat vague and are fixed, for convenience, by the ecologist. Therefore, it is not uncommon (or inappropriate) to speak of, for example, the black cutworm population in Boone County, Iowa, in 1987. The aspects of time and space are of utmost importance in studying the population and necessarily need to be defined for any consideration of population dynamics.

Every insect population has unique characteristics; to understand population dynamics, it is necessary to define and quantify these. The characteristics of a

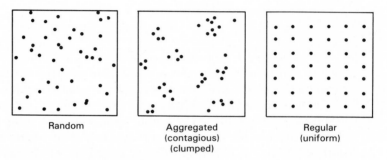

Random   Aggregated (contagious) (clumped)   Regular (uniform)

Figure 5.1 Diagrams showing types of spatial dispersions among insects and other organisms.

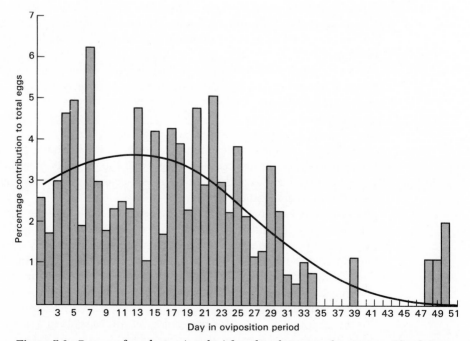

Figure 5.2 Pattern of egg laying (natality) found in the green cloverworm, *Playthypena scabra*. The curve in the graph is a fit to the data points, represented by bars. Such natality patterns are typical of many insect populations. (Redrawn from Wellik and Pedigo, 1978, *Environmental Entomology* 7:171-177)

BOX 5.1 REDLEGGED GRASSHOPPER

SPECIES: *Melanoplus femurrubrum* (De Geer)
(Orthoptera: Acrididae)

DISTRIBUTION: The redlegged grasshopper is distributed from central Mexico through all the United States and into southern Canada. Individuals are found in roadside ditches, along field margins, and in meadows. The southern redlegged grasshopper, *Melanoplus femurrubrum propinquus* Scudder, a sub-species, is the most common grasshopper in the southeastern United States.

IMPORTANCE: Under climatic conditions where rainfall is less than 30 inches per year, redlegged grasshoppers are very destructive pests of clover, alfalfa, and soybean. These grasshoppers defoliate legumes and feed on pods of soybean, exposing the seeds to pathogens and reducing yield and quality. This pest also feeds on small grains and corn, and when populations are high, it attacks almost any plant.

APPEARANCE: Adults are generally fewer than 25 mm long and are brownish red with the hind tibbiae pinkish red. Nymphs are similar to adults, differing in size and lack of wings. Twenty-five to 30 eggs are deposited in pods in the soil.

LIFE CYCLE: Eggs overwinter, and hatching occurs in early spring after the ground has warmed and received adequate rainfall. Nymphs begin feeding near egg-laying sites and eventually move to other areas where food is more plentiful. After molting five to six times (40 to 60 days), the adults emerge and mate, and females begin ovipositing. Adults continue to feed until they are killed by cold temperatures in fall. There is typically one generation per year in the temperate regions of the United States.

population are group attributes not possessed by single individuals in the group. These population attributes include: **density, dispersion, natality, mortality, age distribution,** and **growth form.**

Density and dispersion are complementary attributes. Density is the number of individuals per some unit of measure (for example, number of redlegged grasshoppers per square meter), whereas dispersion is the spatial arrangement of those numbers. Most insect populations are said to have clumped or aggregated dispersions, although in relatively uniform habitats (for example, grain fields with even terrain) dispersion may seem random. Randomness means that an insect is as likely to be found in one place as another.

Natality and **mortality** are rate processes. Natality is birth rate, often measured as the total number of eggs or eggs per female laid per unit time. Mortality is the death rate, or numbers dying per unit of time. These processes act in concert, natality adding numbers and mortality subtracting them, in determining population density and dispersion.

Age distribution is the particular proportions of individuals in different age groups at a given time. For instance, in early season the age distribution of an insect population may be adults, 75 percent; eggs, 20 percent; first-stage larvae, 5 percent. This distribution would change to mostly eggs and larvae as time progressed, and so on.

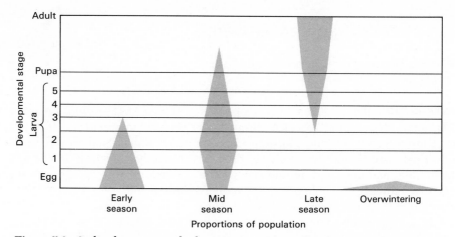

Figure 5.3 Stylized age pyramids showing seasonal age distribution of a hypothetical insect population.

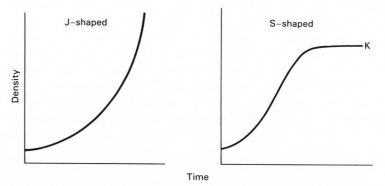

Figure 5.4 Simplified growth curves, J-shaped and S-shaped, found in animal populations. K in the S-shaped curve is the upper asymptote or limit of growth.

Growth form of a population refers to the particular shape of the density curves during a season or over a longer time period. Growth curves of many insect populations in temperate climates are J-shaped; the population grows exponentially, then falls off abruptly with the advent of winter. The particular shape of growth curves determines in part the amount of overlap between numbers of different life stages within a generation and between generations.

Ecosystems and Agroecosystems

When studying insect populations in a locale, one very quickly discovers that no population exists as an isolated entity. Individuals not only interact among themselves for mating, feeding, and other purposes, but also interact with individuals of other animal and plant populations. At the very least, individuals in a population feed upon and, in turn, are fed upon by individuals of other populations. This interacting "web" of populations in an area is called a **community**.

The interaction of populations in a community are strongly influenced by the physical environment. These assemblages of elements, communities, and

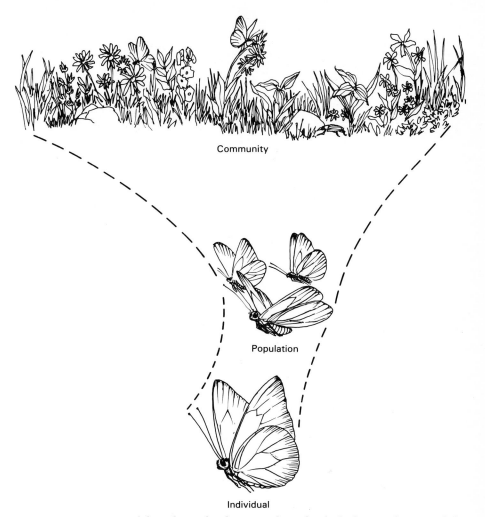

Community

Population

Individual

Figure 5.5 Diagram of the relationship between the individual, the population, and the community.

physical environments are usually referred to as **ecosystems**. Common examples of natural ecosystems are ponds, lakes, forests, and prairies. Ecosystems are usually thought of as the ultimate unit for study in ecology, because they are composed of both organisms and the nonliving environment.

But because of the great complexity of an entire ecosystem, it may be more convenient or desirable to consider subdivisions of it. Such subdivisions have been referred to as **life systems**. A life system is a subject species and its **effective environment**. The effective environment is composed of those elements in an ecosystem that have a direct influence on reproduction, survival, and movements of the subject species. Therefore, we can study the life system of an insect within an ecosystem and later combine our knowledge of several life systems to better understand that ecosystem.

Just as ponds and forests are systems of interacting elements, so are cropping areas. Consequently, we can think of the interacting biotic and abiotic elements

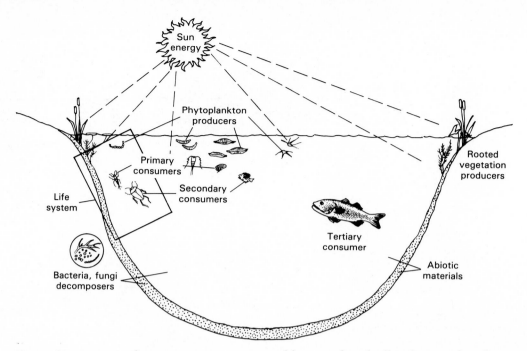

Figure 5.6 Diagram of an ecosystem as represented by a pond, with all its living and nonliving elements. A subcomponent of the ecosystem is the life system as represented by the boxed portion. (Redrawn from Odum, 1959, *Fundamentals of Ecology*, 2nd ed., W.B. Saunders)

of an orchard or grain field as an ecosystem. Any ecosystem largely created and maintained to satisfy a human want or need is called an **agroecosystem**. Just as ecosystems are the basic unit of study for ecology, agroecosystems are the basic unit of study for pest management, a branch of applied ecology. Much of a typical agroecosystem is composed of the more or less uniform crop-plant population, weed communities, animal communities (including insects), microbiotic communities, and the physical environment in which they interact. Much of the land mass in major agricultural areas is occupied by agroecosystems.

Although the theory of ecosystems extends to agroecosystems, several features make plant-based agroecosystems unique:

1. Agroecosystems often lack temporal continuity. Their existence may be of limited duration and may undergo immense, abrupt changes in microclimate because of cutting, plowing, disking, burning, chemical application, and other cultural practices.
2. Agroecosystems are dominated by plants selected by humans, many consisting of imported genetic material. Other crop plants not from imported material have been in protective cultivation so long that they hardly resemble the parent stock from which they were derived.
3. Most agroecosystems have little species diversity, and the crop species has little intraspecific diversity; in other words, the crop tends to be genetically uniform. Usually a single species dominates an agroecosystem, and the elimination of weed species further simplifies the environment.
4. With crop plants of similar type and age in the system, the vegetative structure is uniform, and a given phenological event (for example, flowering or podding) occurs in almost all the plants at the same time.
5. Nutrients are usually added to agroecosystems, which result in crop plants with uniformly succulent, nutrient-rich tissues.

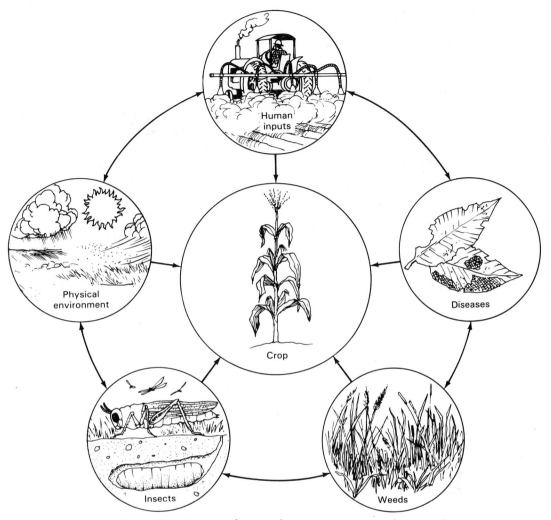

Figure 5.7   Diagram showing the major interacting elements of an agroecosystem.

6. Agroecosystems often have frequently occurring insect, weed, and disease outbreaks.

The last feature, pest outbreaks, largely results from the preceding five features. With insects, only a small fraction of the entire community finds the crop environment optimal, but those species that do are able to reproduce and survive very well, creating populations that can cause devastating losses.

The Ecological Role of Insect Outbreaks

We often ask, why do we have insects that cause damage to our crops, or what is the real role of damaging insects in the great plan of nature? Of course, an answer to these questions is teleological. However, a consideration of the ecological mechanisms operating in natural ecosystems may give some insights.

Some ecologists believe that in unmanaged ecosystems, a state of balance exists or will be reached; that is, species interact with each other and with their physical environment in such a way that, on average, individuals are

Figure 5.8 Severe insect injury as represented by caterpillar defoliation of broccoli. (Courtesy Iowa State University Extension Service)

able only to replace themselves. Subsequently, each species in the community achieves a certain status that becomes fixed for a period of time and is resistant to change; this is the familiar idea of the "balance of nature." When humans begin to manage the system or create a new system in the area— establish an agroecosystem where a natural ecosystem existed—the balance is altered. Humans strive to create a system in which a single species, the crop, can achieve its full reproductive potential. With such a stress caused by the modification of that ecosystem, exceptionally strong forces react in opposition to our imposed change—toward a return to the original system. An insect outbreak is one of those forces. Destruction by insects counteracts the disturbing factor ("overpopulation" by the crop species), promotes community diversity, and therefore enhances the tendency toward restoration of the original, unmanaged community. Accordingly, we see that insect pests are not ecological aberrations. They function as do other animal populations. Their only distinction in this regard is that their activities counter the wants and needs of human populations.

DYNAMICS OF INSECT LIFE SYSTEMS

Determinants of Insect Abundance

The determinants of insect abundance are found within a species' life system. They comprise the inherited properties of individuals in the species and attributes of the effective environment. These factors operate either to reduce or to promote insect numbers. Together, they explain differences in abundance among habitats and numerical change over time in the same habitat.

Inherited properties involve characteristics transferred genetically from individuals of one generation to those of the next. The inherited properties of indi-

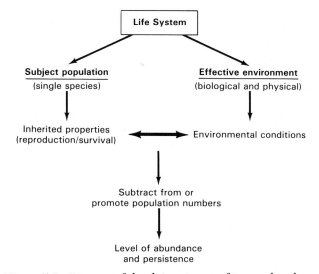

Figure 5.9 Diagram of the determinants of insect abundance in a life system.

viduals in a population combine to give the population its unique characteristics and mainly affect their ability to reproduce and survive.

Reproductive ability is often expressed as **potential natality**. Potential natality is the reproductive rate of individuals in an optimal environment. Potential natality varies greatly among insects. For example, house flies, *Musca domestica*, have a potential natality of approximately 500 eggs/female, whereas in sheep keds, *Melophagus ovinus*, potential natality is only about 15 eggs/female.

Survival rate also varies widely among insect species and depends much on feeding habits and protection of young. Insects that feed on a wide variety of food sources and protect their young to a degree have greater survival potential. In the latter instance, viviparous and ovoviviparous birth types generally have less mortality of newborns than oviparous birth types. This is because the offspring are in a protected environment, inside the females, much longer than are those of oviparous species.

Reproductive rate and survival rate are usually related in that species with high potential natalities tend to have relatively low survival rates and vice versa. Insect pests with high reproductive rates and low survival rates are called **r strategists**, named after the statistical parameter **r**, the symbol for the growth rate coefficient. Such pests may overwhelm the environment with new individuals, and although their losses are great, they can succeed because of sheer numbers. These pests, for example, aphids (Homoptera: Aphididae), increase astronomically with favorable environment, and can rebound quickly after environmental catastrophes. In contrast, **K strategists**, named after the symbol for the asymptote or flattened portion of a population growth curve (see Figure 5.4), reproduce slowly. However, they compete effectively for environmental resources, and therefore their survival rate is high. Pests like the codling moth, *Cydia pomonella*, are sometimes considered K strategists. Such species are often hard to deal with because of their resiliency to induced and natural hazards in their environment.

The effective environment is a force of equal importance to inherited properties in determining population numbers. Factors like weather, food quality

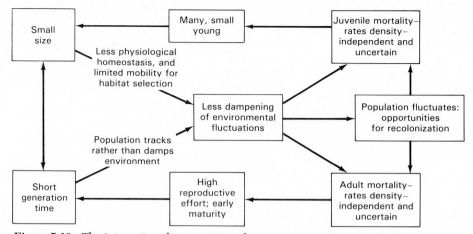

Figure 5.10 The interacting characteristics of an r strategist. (Reprinted with permission of Blackwell Scientific Publications, Ltd. from *Population Ecology*, 2nd ed. by M. Begon and M. Mortimer. Copyright ©1986)

and quantity, and living space may either support or inhibit population growth. Conversely, natural enemies, a prominent part of the effective environment, are strictly a subtractive force.

Overall, the actual numbers of a population are sometimes thought of as a sort of algebraic sum of the innate ability of the population to reproduce and survive

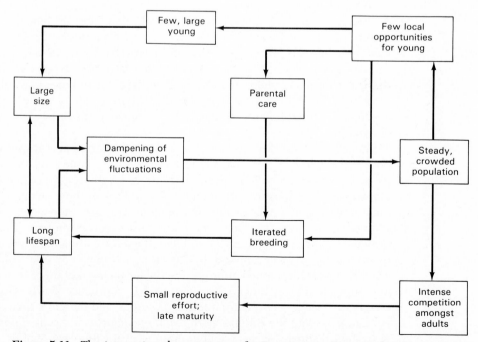

Figure 5.11 The interacting characteristics of a K strategist. (Reprinted with permission of Blackwell Scientific Publications, Ltd. from *Population Ecology*, 2nd ed. by M. Begon and M. Mortimer. Copyright ©1986)

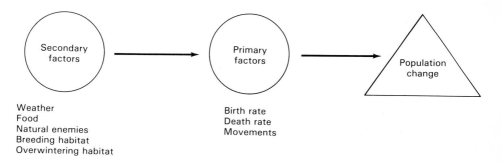

Figure 5.12 Diagram of secondary and primary factors affecting population change.

and the degree of restraint imposed by the environment. R. N. Chapman, an early ecologist, referred to this innate ability as **biotic potential** and the restraint as **environmental resistance**. He reasoned that populations grow at a rate determined by the dynamic power of a species pitted against the subtractive forces of the environment. His idea can be shown as

$$\text{Actual Abundance} = \text{Biotic Potential} - \text{Environmental Resistance}$$

Population Change

Change in numbers from one time to the next and from one place to the next is one of the most prevalent properties of any population. A widely used expression with the primary factors of this change is as follows

$$N_t = N_0 e^{(b-d)t} - E_t + I_t$$

where

N_t = number at the end of a short time period
N_0 = number at the beginning of the time period
e = base of natural logarithms = 2.7183
b = birth rate
d = death rate
t = time period
E = emigration = movement out of an area
I = immigration = movement into an area

The expression is a general model of change in any population and shows the mathematical relationship among the primary factors of this change: births, deaths, and movements.

In attempting to explain population numbers, we must account quantitatively for these primary factors. To attempt prediction of change, however, we must understand how environment modifies these primary factors.

The effective environment of a population can be thought of as having an array of secondary factors that determine the magnitude, duration, and frequency of the primary factors. Some secondary factors are weather, natural enemies, breeding habitat, and overwintering space. These affect the primary factors by altering the supply of necessary resources or directly affecting the existence of individuals.

Birth Rate

Birth rate is a major process for adding new individuals to a population. It is often expressed as numbers born per female or per 1,000 females in the population during a specific time period.

The major factors determining birth rate are **fecundity**, **fertility**, and **sex ratio**. Fecundity is the rate at which females produce ova, whereas fertility is the rate at which they produce new individuals, for example, fertilized eggs. With fecundity, the focus is strictly on female capability to produce reproductive units. With fertility, the focus is on the mating and fertilization process, the production of zygotes. Sex ratios in most populations are 1:1, male to female. Exceptions to this ratio are most common in parasitic Hymenoptera and, of course, parthenogenetic species. Overall, changes in birth rate may be caused by changes in the average production of eggs per female, changes in mating success, changes in the proportion of female individuals in the population, or a combination of these.

Both fecundity and fertility may vary greatly for an insect population. Secondary factors like temperature, moisture, and food may strongly influence the

BOX 5.2 SPRUCE BUDWORM

SPECIES: *Choristoneura fumiferana* (Clemens)
(Lepidoptera: Tortricidae)

DISTRIBUTION: The spruce budworm is found throughout the coniferous forests of southeastern Canada and the northeastern United States. This species was once thought to inhabit all the coniferous areas in North America, but recent studies show that several different, though related, species occur in western Canada and the western United States.

IMPORTANCE: Spruce budworm larvae primarily are pests of Balsam fir but also feed on larch, pine, hemlock, and white, red, and blue spruce. Where populations of the caterpillars are large, all new needles produced by the tree may be consumed. If the outbreak populations remain for three to five years, top-killing and tree mortality occur. Overwintering larvae emerge in early spring and attack new buds as well as bore into older needles. Half-grown larvae form distinctive nests of twigs tied together with silk.

APPEARANCE: Adult moths, which have a gray mottled appearance, have forewings with yellow or reddish brown blotches and hindwings that are dark brown with white fringe. Mature larvae are 25 to 30 mm long and have a brown head capsule and two rows of white spots along the back. These spots contain many short spinules. Eggs are pale green and oviposited in overlapping masses on the undersides of needles. Pupae are concealed in loose silken cocoons.

LIFE HISTORY: Second instars overwinter in silken hibernaculae on the twigs and branches of trees and emerge in early spring to feed on or in needles and new buds. As the larvae grow, they form twig and silk nests, and pupation occurs in late June. After approximately 10 days, adults emerge and are active from late June to early August. During this time, female moths oviposit shinglelike masses containing about 60 eggs. The eggs hatch in about 8 to 10 days, and the larvae feed until they become second instars. Subsequently, they prepare for overwintering by spinning the silken hibernaculae.

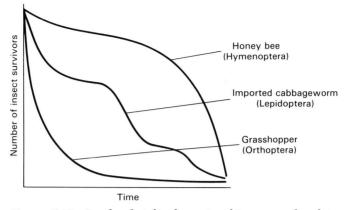

Figure 5.13 Graph of stylized survivorship curves found in insect populations. (Based on data from Ito, 1961, *Bulletin of Natural and Agricultural Science*, Series C, 13: 57–88)

number of eggs produced by a female. For instance, in controlled experiments with the European corn borer, *Ostrinia nubilalis*, 708 fertile eggs were produced per female at a temperature of 21°C, but only 533 were produced at 32°C. Likewise, female egg production can be greatly influenced by nutrition during the immature stage. For example, egg production by the Colorado potato beetle, *Leptinotarsa decemlineata*, depends on the variety of potato on which the larvae feed. In the spruce budworm, *Choristoneura fumiferana*, egg production increases to about 170 eggs/female when larvae are fed succulent new foliage versus about 100 eggs/female when larvae consume late-season foliage. Also, habitat and the physical environment are especially important during mating activities. It is well established that conditions essential to reproduction have a much narrower range than for other life functions, for example, movement and feeding.

Variability in birth rate of a population is often overlooked as a cause of population change. Although usually difficult to study, attempts to document and explain this variability are important, because the factors responsible may be the essence of a population outbreak.

Death Rate

Probably more has been written about death rates and causes of death than any other aspect of insect ecology. Numbers of insects dying over a period of time are often represented in **survivorship curves** or in **life tables**, a tabular form for accounting for deaths. In many insect populations, high death rates (mortalities) are the rule, with sometimes fewer than 1 percent of the individuals of a generation reaching adulthood.

Mortality factors may be biotic or abiotic and may operate in a single life stage or over many stages. For example, harsh weather like intense rain causes mortality of all stages of aphids, but the parasitoid *Rogas* (a small wasp) parasitizes and kills only middle-aged green cloverworm, *Plathypena scabra*, larvae.

Like birth rates, death rates can vary greatly from one time and place to the next. Documenting change in birth rate and causes of death have been a major preoccupation among entomologists in attempting to explain insect numbers.

Table 5.1. Life Table for the Spruce Budworm at Green River, New Brunswick, Canada.[a]

Stage	Number Living Beginning of Stage	Number Dying by End of Stage	Cause of Death	Percent Reduction During Stage
Egg	200	10.0	Parasites	
		20.0	Other	15
Early larva	170	136.0	Dispersal	80
Late larva	34	13.6	Parasites	
		6.8	Disease	
		10.2	Other	90
Pupa	3.4	0.3	Parasites	
		0.5	Other	25
Adult	2.5	0.5	Miscellaneous	20

SOURCE: Modified from Morris, 1957, The interpretation of mortality data in studies of population dynamics, *Canadian Entomologist* 89: 49–69.
[a]Numbers are insects per 10 ft^2 of conifer branch surface area.

Indeed, in many instances, relaxation of an important mortality factor may be the primary cause of a population outbreak.

L.R. Clark and his colleagues conveniently grouped the causes of mortality of insects into seven major categories.

1. Aging. Aging and death from old age is sometimes referred to as "physiological death." In most insect populations, only a small percentage of individuals live to achieve their longevity potential—a "ripe old age." Exceptions are many of the social insects (for example, Hymenoptera), where there is elaborate care of the young, a highly protected nest, and adults can effectively defend themselves. As a rule, physiological death is probably not highly variable and is most important in populations with relatively low reproductive rates.

2. Low vitality. As mentioned, part of an insect's characteristics is the ability to survive, in other words, inherent traits to resist adverse environmental factors. Individuals of a population vary in this ability, and sometimes the average vitality of a population may change because of the proportions of certain types of individuals. Therefore, unusually heavy mortality may be caused by an environmental factor, for example, rainfall, acting on a population with an unusually high proportion of susceptible individuals. Rather than being credited to the rainfall factor, this unusual mortality may be credited to low vitality, a genetic trait.

3. Accidents. Accidents encompass abnormal events in the insect life cycle resulting in death. Such events range from the physiological (for example, inability to completely shed the old larval skin or failure to expand wings after molting) to the purely ecological (for example, leaf-feeding insects being eaten along with leaves by a browsing deer). The proportion of individuals dying from accidents may be fairly constant in a population.

4. Physicochemical conditions. This category involves the physical and chemical conditions of air, water, and substrates in or on which the insect population lives. A particular feature with which we are concerned for most pest insects is weather. Components of weather, mainly temperature and moisture, can either promote population growth or cause population decline. Directly, severe weather increases death rates and may also retard or arrest physiological and behavioral processes

BOX 5.3 FOREST TENT CATERPILLAR

SPECIES: *Malacosoma disstria* Hübner (Lepidoptera: Lasiocampidae)

DISTRIBUTION: The forest tent caterpillar is distributed throughout Canada and the United States from Minnesota eastward, and is found on deciduous forest and shade trees. Older larvae may even be found feeding on field and vegetable crops.

IMPORTANCE: The young larvae (caterpillars) attack expanding buds, and older larvae consume the leaves of many deciduous trees. When outbreaks occur, larvae defoliate trees and migrate in huge "swarms" to new areas, often invading buildings and moving into field and vegetable crops. Outbreaks frequently continue for more than three years, resulting in widespread defoliation of deciduous trees in forests and urban areas.

APPEARANCE: Adults are large, hairy moths with a wingspan of 25 to 35 mm. The wings are orange to tan with two parallel brown lines across the forewing. Larvae, about 25 mm long when full grown, are light blue and sparsely covered with white setae. Each abdominal segment has a yellow dorsal key-hole-shaped marking and pale yellow horizontal stripes on the sides of the body. The 100 to 350 eggs are cemented together in masses on twigs and branches.

LIFE HISTORY: Eggs overwinter and hatch in mid-May. Larvae feed gregariously and form silken mats on branches and trunks in which they are protected after feeding. About six weeks after hatching, pupation occurs in cocoons, and adults emerge in 10 to 12 days. Swarms of adults can be seen in mid-July, and eggs are laid through August. There is only one generation per year throughout the pest's range.

---

so as to reduce birth rates. As a rule of thumb, cool, wet extremes are the most deleterious to insects because they promote disease, slow growth rates, and other adversities. But caution must be exercised when applying this rule, because some species, for example, mosquitoes (Diptera: Culicidae), the seedcorn maggot, *Delia platura*, and the forest tent caterpillar, *Malacosoma disstria*, survive best under such conditions.

In temperate climates, variable winter temperatures may be a major reason for changes in death rates of overwintering species. Indeed, outbreaks following a "mild" winter are often believed to be the result of lower death rates. Again, however, caution should be exercised when correlating these events, because some "mild" winters may cause more mortality than "harsh" winters. This condition can occur because during the "mild" winter abnormally warm days may cause some species to become active early. If this early activity is followed by days of normally cold temperatures, unusually high mortality may result because the insects are no longer protected by their previous state of dormancy.

The long-term influence of weather on insect mortality ultimately results in a particular geographical distribution of a species. However, this distribution may

Figure 5.14 Rove beetles, *Aleochara bimaculata*, attacking a cabbage maggot (Diptera: Anthomyiidae). Such insect predators are important natural enemies. (Courtesy USDA)

change during the course of a year, with some species occupying regions during the growing season that will not allow survival during the winter. Well-known examples of pest species that cannot survive winters of the upper midwestern United States but invade this area during the growing season include the potato leafhopper (*Empoasca fabae*), corn earworm (*Heliothis zea*), and black cutworm (*Agrotis ipsilon*). It is probable that many more pests exhibit this behavior than are known to exist today.

Because of its omnipotence, weather also affects death rate indirectly. For instance, rainfall can affect plant phenology and nutrient quality for insects. Moreover, asynchrony between a flowering or podding event of a plant and occurrence of a particular stage of the insect also can result in unusually heavy mortality. Asynchrony occasionally occurs with the apple blossom weevil, *Anthonomus pomorum*. This insect lays its eggs in flower buds and suffers high mortalities when blossoms open early in relation to egg hatch. In this instance, new larvae fall out of the blossom before they have time to tie the petals together.

Another indirect effect of weather is on natural enemies. Here, high humidities are known to enhance the growth and virulence of disease-causing organisms like fungi, and these in turn form widespread epidemics in insect populations. Conversely, abnormally low mortalities from parasitism may result in insect host populations (for example, some Diptera) because of unusually low summer temperatures; in other words, low temperature has a more deleterious effect on the parasite species than on the host species.

5. Natural enemies. Organisms that prey upon insects or parasitize them are called natural enemies. Natural enemies can be grouped, for convenience, as predators, parasites, and pathogenic microorganisms. These are discussed in greater detail in Chapter 9, but a brief mention here will allow their orientation among the other mortality factors.

Natural enemies are a very important component in the population dynamics of insect species. They have a strictly subtractive influence on total insect numbers,

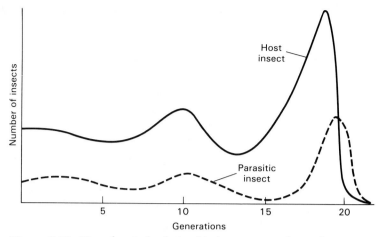

Figure 5.15 Hypothectical relationship between numbers of a parasite and those of its host, showing density-dependent action.

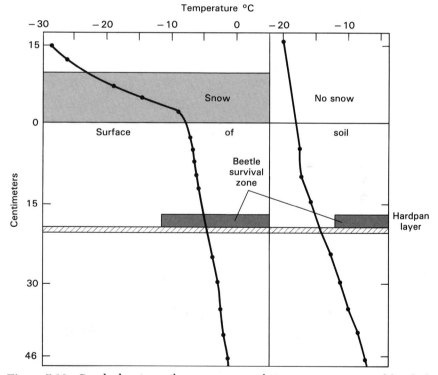

Figure 5.16 Graph showing soil temperatures relative to snow cover and level of a "hardpan." Overwintering Colorado potato beetles, *Leptinotarsa decemlineata*, cannot burrow through the hardpan, and survival in these soils is possible only in years and locations with adequate snow cover. (Redrawn from Mail and Salt, 1933, *Journal of Economic Entomology* 26:1068-1075)

and sometimes their impact is related to the density of their prey or host; their action is **density-dependent**. Therefore, in some life systems they become an all-important part of insuring the stability of insect numbers. In these instances, when numbers become large, death rate from natural enemies is great, and when numbers are small, death rate is low.

Because all pest species have natural enemies, understanding mortality caused by these enemies is critical to managing the pest. Indeed, the primary principle of all pest management systems is to conserve natural enemies in the system. We return to this principle several times throughout the book.

6. Food shortage. Food for insects largely depends on prevailing conditions in the ecosystem. In agroecosystems, the major food source, at least for plant-feeding insects, is the crop species, and to a major extent, supply of this food is determined by the grower. Except when weather seriously limits crop production, lack of food is not a major cause of insect pest mortality in agroecosystems. This is not true, however, for many insects that prey on or parasitize other insects.

These insect natural enemies must search through the habitat for prey or hosts of select species and, often, select age. Particularly when physicochemical conditions limit numbers of prey or hosts or the natural enemy decimates its own food source, starvation can be widespread. The latter phenomenon (growth of the natural enemy population when its host is abundant, subsequent decline of the host, followed by a decline of the natural enemy) is the very basis for classical biological control. Here, competition for food and starvation (or reduced reproduction) in natural enemy populations are the mechanisms by which regulation of some host populations is achieved.

7. Lack of shelter. Lack of shelter influences death rate indirectly. Protective habitats allow insects to avoid exposure to extremes in weather and natural enemies; the absence of such may produce considerable mortality in the population. For example, Colorado potato beetles succumb to winter temperatures below −12°C. In sandy soils, beetles burrow to depths of 36 to 91 cm, thereby escaping the lethal temperatures. In some soils, an impervious layer of clay, a hardpan, occurs at 20 cm, preventing beetles from reaching required shelter. Unless adequate snow cover insulates the soil, thus moderating temperatures, widespread beetle mortality occurs.

Movements

Knowledge of movements into an area (**immigration**) and out of an area (**emigration**) is vital to understanding population dynamics. Indeed, movements, rather than births or deaths, may be the main cause of rapid population change within a season; that is, predictable movement can be a major population characteristic. Information obtained from studying insect movements may allow such activities as pinpointing sources of plant-disease vectors or predicting the seasonal arrival of an important pest that does not overwinter locally.

Movement of all kinds is a rule during the life cycle of most insects. It may involve persistent crawling, walking, or hopping along terrestrial surfaces; swimming on or in the water; or flying or being carried passively (as with wingless insects) in the air. Most movements can be categorized as **trivial** (nonmigratory) or **migratory**.

Trivial movement involves displacements of insects within or close to the breeding habitat that are frequently interrupted when the insect encounters food,

BOX 5.4 COLORADO POTATO BEETLE

SPECIES: *Leptinotarsa decemlineata* (Say) (Coleoptera: Chrysomelidae)

DISTRIBUTION: The Colorado potato beetle is native to the semiarid eastern slopes of the Rocky Mountains, where it originally fed on sandbur. After the potato was introduced into Colorado, some beetles abandoned their weed hosts for the cultivated potato. At present, populations are found throughout the United States and Canada. The pest has been introduced into Europe, where it also is a serious pest of potato.

IMPORTANCE: This species is the most destructive potato pest in North America. In the past, the beetles have destroyed entire potato crops in several states. Injury is caused when adults and larvae feed on foliage and stems of potato plants, resulting in poor yields and/or death of the plants. Adults can also vector plant diseases such as spindle tuber, bacterial wilt, and root rot. The need to control these beetles resulted in the first use of Paris green, an insect stomach poison, in 1865.

Presently, however, the beetle is resistant to nearly all classes of insecticides and remains a serious pest.

APPEARANCE: Adults are approximately 9.5 mm long and about 6.4 mm wide, having 10 black stripes running lengthwise on yellow to orange elytra. Larvae are red, having black spots in rows along both sides of the hump-backed body. Pupae are yellow. Eggs are orange-yellow and are oviposited in groups of about 12 on the undersides of leaves.

LIFE HISTORY: Adults overwinter in the soil, becoming active about the time that adult potato plants emerge. Adults will feed and mate, and the females oviposit up to 300 eggs over a five-week period. The eggs hatch in four to nine days, and larvae begin feeding, molting four times in two to three weeks. Subsequently, larvae burrow into the soil and pupate, and adults emerge in about three days. There are typically one to three generations per year, depending on climate.

a mate, or an egg-laying site. Such movements usually occur throughout the life of individuals while carrying out their life functions, except during egg and pupal stages and during dormancy. An example of trivial movement is when a butterfly flits from flower to flower, sometimes moving among several habitats, feeding on nectar. Such movements are termed "trivial" because of the distances involved, not because they are trivial to the ecology and survival of the insect.

Conversely, migratory movements may involve great distances, perhaps hundreds of kilometers, during which time locomotion is not inhibited by food, mates, or oviposition sites. Migration is usually accomplished by flight. Migratory flights are composed of individuals which are predisposed to fly and are undistracted from this behavior for a lengthy period. This peculiar type of flight is an adaptation to periodically transport insects beyond the boundaries of their old reproductive sites and into new ones.

During migratory flights, great mortality may occur, because many individuals are deposited in areas where they cannot survive, for example, open seas, lakes, glaciers, and snow fields. However, a portion of the migrants (perhaps a minute

Figure 5.17 Painted lady butterfly, *Vanessa cardui*, an important species that migrates in the insect boundary layer. (Photo by T. Meyers)

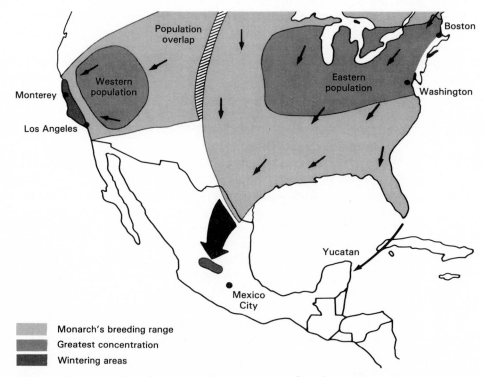

Figure 5.18 Map of North America showing summer breeding ranges and return migration of the monarch butterfly, *Danaus plexippus*. (Reprinted by permission of John Wiley & Sons, Inc. from *Insect Behavior* by R. W. Matthews and J. R. Matthews. Copyright ©1978 John Wiley & Sons, Inc.)

fraction) locate suitable habitats, often fortuitously, where they can reproduce. According to C. G. Johnson, the surface of the earth is scanned very effectively by millions of insects, flying on air currents, who continuously encounter suitable and unsuitable situations. They stop temporarily in suitable habitats and reproduce, and offspring migrate in directions determined either by the wind or by themselves.

The height of migration flights above the surface can vary from only a few meters (for example, some butterflies) to hundreds of meters (some leafhoppers, aphids, and moths). Migrating butterflies, flower flies, and dragonflies flying at

BOX 5.5 MONARCH BUTTERFLY

SPECIES: *Danaus plexippus* (Linnaeus) (Lepidoptera: Danaidae)

DISTRIBUTION: The monarch butterfly is distributed throughout nearly all North America, except for Alaska and the Pacific Northwest coastal region. This species can also be found throughout South America, as well as Australia and the Hawaiian Islands.

IMPORTANCE: The monarch butterfly is one of the most recognizable insects in the United States. This species is famous for its annual migrations both north and south along predictable routes. In the midwestern and eastern United States, the butterflies group and fly south to central Mexico, where they overwinter in dense concentrations in the Sierra Madre Mountains. Western monarchs fly to California, where they spend the winter in areas near Pacific Grove. The following spring, the individuals fly north, reproducing along the return route. Remarkably, the adults that overwintered in the south do not return to their original location. Instead, their offspring return and reproduce, and the cycle is repeated. Currently, conservation of monarch overwintering grounds in both Mexico and California have received great attention. This species is also known for its ability to ingest toxic chemicals (cardioglycosides) produced by its milkweed host plants and then store these toxins for its own protection. Both larvae and adults are toxic to potential predators and display this toxicity through bright warning coloration. The Viceroy butterfly, *Basilarchia archippus* (Cramer), which is palatable to most predators, mimics monarch coloration and thereby avoids some predation.

APPEARANCE: Adult monarchs are large, with a wingspan of 76 to 100 mm, and have rust-orange wings with black veins and a black marginal band, offset by many pairs of white spots. A pair of black spots appears on the hindwings of the male, visibly differentiating it from the female. Larvae, about 51 mm long, are white with alternating bands of black and yellow. Two filamentous appendages are projected from the front and the rear of the body. The chrysalides (pupal cases), about 28 mm long, are pale green and display gold spots. Eggs are pale green, ribbed, and about 1.2 mm long by 0.9 mm wide.

LIFE HISTORY: Females fly north in early spring, each ovipositing 400 to 500 eggs on milkweed plants. The developing larvae feed on milkweed leaves and, when full grown, larvae prepare chrysalides, where metamorphosis occurs. There are typically two to three generations in the northern part of the monarch's range and up to five generations in the southern part of its range.

lower levels are readily seen by the casual observer and appear to maintain a steady course. These insects are said to migrate within their **boundary layer**. The insect boundary layer (not to be confused with meteorologists' meaning of boundary layer) is the layer of air that extends from ground level upward through increasing wind speeds to a height where wind speed and insect flight speed are equal. Within this level insects determine their own flight track. Insects migrating at higher altitudes, above their boundary layer, may not be seen. The flight track of these insects is determined by the flight orientation of the insect and wind direction; in other words, they are largely transported by the wind.

One of the most famous insect migrations in North America is that of the monarch butterfly, *Danaus plexippus*. This migration occurs each spring when butterflies fly northward from overwintering sites in Mexico, California, and other southern locations. Destinations in the northern United States and Canada are

---

BOX 5.6 DESERT LOCUST

SPECIES: *Schistocerca gregaria* (Forskgål) (Orthoptera: Acrididae)

DISTRIBUTION: The desert locust is distributed throughout Africa, the Middle East, and India. This pest can be found on almost any plant, but it has a preference for cereal crops.

IMPORTANCE: The desert locust is an infamous pest that has plagued humans for thousands of years. Under the appropriate conditions, these insects form gigantic swarms, covering over 100 square miles, with 300 milion individuals per square mile. The swarming locusts strip the land of all vegetation, leaving a burntlike landscape. Indeed, locust, a Latin word, translates as "burnt place." Centuries ago, after the locusts devoured all the vegetation, famine and pestilence followed. Numerous references to this pest can be found in the Bible and other ancient writings. Under normal environmental conditions, the desert locust is a serious yet sporadic pest throughout its range. Outbreaks and swarming occur only when flooding provides ample food and is followed by a period of dry conditions. The locusts will increase in number and suffer little mortality. Crowding of the individuals changes color, body form, and

behavior. These pests begin to move as gregarious units, which results in swarming. The two phases of behavior, solitary and gregarious, are so completely different that it was once thought the two forms of desert locust were different species.

APPEARANCE: Adults are large insects, about 40 to 60 mm long and are pale yellow or brown. The forewings (tegmina) are greenish yellow with brown spots. First instars are vermiform (wormlike). All other nymphs resemble the adult, except that they have undeveloped wings. Eggs are oval, about 1.3 mm long by 0.8 mm wide, and are contained in pods.

LIFE HISTORY: Female locusts lay four to five pods in the soil, with each pod containing from 70 to 100 eggs. The vermiform first instars crawl to the surface and quickly molt into the more recognizable locust form. Nymphs undergo three more molts before the adult stage is reached. Depending on environmental conditions, the desert locust either enters a solitary or gregarious phase.

Figure 5.19 Ground view of migrating desert locusts, *Schistocerca gregaria*. (Courtesy FAO and Anti-Locust Research Centre)

reached in May and June, with the butterflies repopulating over 1 million square miles in two months. Reproduction occurs, and one generation (sometimes from more than one brood) usually develops by September. This northern generation then migrates back, often in large aggregations, to overwintering sites. By marking, releasing, and recapturing butterflies in the southerly migration, researchers have determined that between 18 and 109 days are required to make the return flight, which may cover as many as 3,465 kilometers (2,153 miles). The difference in time is probably determined by the distance that individuals travel to their destination and weather along the migration route. During the southern migration, butterflies aggregate (roost) in trees at night and during poor weather.

Another famous example of migration is that of the desert locust, *Schistocerca gregaria*. This insect occurs across north and central Africa to the Middle East, Arabia, and India, causing plagues that have been reported since biblical times. Vast swarms may occur across hundreds of kilometers and may be composed of as many as 10 billion individuals, with a weight of 15,000 tons! Such a swarm is estimated to have the same daily consumption as 1.5 million people! When they alight, the locusts consume all above-ground vegetation. The insect has both solitary and migratory phases, with phase being determined by degree of crowding of nymphs and their mothers. The migratory swarm appears as a mass flying purposively in a fixed direction. However, it has been shown that orientation of individuals in the swarm is random, and members at the

Figure 5.20 Distribution of desert locust (*Schistocerca gregaria*) swarms (black dots) relative to the Intertropical Convergence Zone. Winds from the north and south converge in this zone, causing rising air currents and rain (lower right). Rain results in vegetation growth required for reproduction of locusts. (Redrawn from Rainey, 1969, *Quarterly Journal of the Royal Meteorological Society* 95: 424-434)

swarm's edge move toward the center of the swarm. This flight behavior is adaptive, because the whole swarm is displaced downwind, toward zones of wind convergence. One such area is the Intertropical Convergence Zone, where winds of both sides of the equator meet. Borne on these winds, locusts accumulate in the convergence zone, and swarm displacement ends there. The environment in the convergence zone is favorable for the locusts because rising air here causes rainfall. The rainfall, in turn, allows the growth of vegetation, and locusts in these areas find environments appropriate for successful reproduction. The Intertropical Convergence Zone moves back and forth across the equator once each year, and because of individual flying behavior and meteorological conditions, locust swarms follow and take advantage of the resulting favorable environments.

The desert locust example shows that crowding and subsequent shortage of food can be an important stimulus for insects to migrate. Other important stimuli may include photoperiod (usually long day length) and weather factors.

In particular, weather factors may be important in determining the exodus flight. Convection caused by warming of the earth's surface provides thermal updrafts that transport some migrants to altitudes required for horizontal transport by the wind; the opportunity for night-flying insects to take advantage of such updrafts is not as great, and these insects often must gain altitude by their own flight activity. Rapid changes in atmospheric pressure may also stimulate migratory flight and allow the insect to take advantage of certain wind movements; for example, many aphids and flies are known to increase their takeoff activity with decreasing barometric pressure.

Once aloft, migrants seemingly respond to weather forces like wind speed

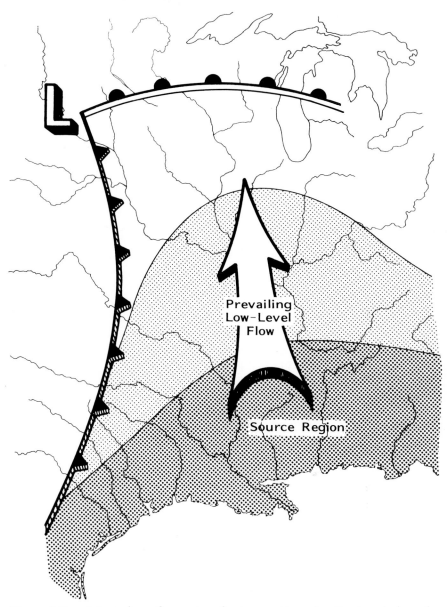

Figure 5.21 A typical weather system for transporting migrating insects from the southern United States into the upper Midwest. The letter L represents a low-pressure center. A trailing cold front is indicated by the spiked line and the upper warm front by the domed line. On the map, insects are advancing (lightly stippled area) from the source region (heavily stippled area). (Courtesy R. Smelser, Iowa State University)

and direction, temperature, barometric pressure, and dew point. These factors change with altitude, and the migrant insect may actively find an altitude where conditions are favorable. Finding the proper altitude, insects may continue to fly or assume a flight attitude and be carried for hundreds of kilometers with the wind.

BOX 5.7 **VELVETBEAN CATERPILLAR**

SPECIES: *Anticarsia gemmatalis* (Hübner) (Lepidoptera: Noctuidae)

DISTRIBUTION: The velvetbean caterpillar is a tropical species found year-round south of 28° North latitude. In spring and summer, the species migrates northward and has been found as far north as the New England states.

IMPORTANCE: This species is one of the most significant soybean pests in the western hemisphere. In the Gulf Coast region of the United States, the larvae are key defoliators of soybean, velvetbean, peanut, alfalfa, and cowpea. Larvae attack newly produced leaves first, then move to and feed on older leaves. All tissues on the leaves are eaten, except the veins. Older larvae often attack stems of host plants. Where outbreaks of the pest occur, entire fields can be defoliated in five to seven days. Populations are the most damaging in early fall.

APPEARANCE: Adults moths are light brown to black, with diagonal black lines running across the wings. Larvae, which have a yellow or orange head, are light to dark green and have dark lines bordered by lighter ones, which run lengthwise along the body. The eggs are very small and white.

LIFE HISTORY: Moths migrate north and appear at the Gulf states in June or July. Females oviposit, and the eggs hatch in about four days. Larvae feed for approximatley 21 days, then pupate just below the soil surface. Adults emerge about 10 days later and the cycle is repeated. There are typically three to four generations per year in the southern United States.

Transport by the wind continues for hours or days, until the migrant becomes grounded. Deposition may be gradual or abrupt. Gradual deposition, grounding of the migrants over a period of time and over a considerable area, often is caused by factors intrinsic to the population. These include changes in neurophysiological state, inefficient flight in cold air, exhaustion, or flight inhibition triggered by light intensity. Sudden deposition of migrants probably occurs as a result of atmospheric conditions. They may include downdrafts in cold fronts or cooling of the insect to the point that it cannot support itself in the air. Of all possible weather conditions, cold fronts seem to be the most important in deposition.

We thus see that migration is an important adaptive characteristic of insects, one can strongly influence their population dynamics. As the study of insect migration continues, it seems that many more insect species than heretofore believed use wind flow to relocate and colonize favorable habitats. Some insect pests whose migration patterns are well known are the potato leafhopper, black cutworm, velvetbean caterpillar (*Anticarsia gemmatalis*), fall armyworm (*Spodoptera frugiperda*), and spruce budworm. It is particularly important that we recognize this potential when predicting pest outbreaks, because rapid population increase may well be caused by migration alone.

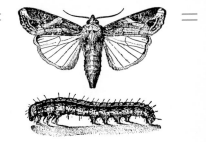

BOX 5.8 FALL ARMYWORM

SPECIES: *Spodoptera frugiperda* (Smith) (Lepidoptera: Noctuidae)

DISTRIBUTION: The fall armyworm in indigenous to Central America, South America, and the southeastern United States. However, this species migrates north each year and can travel as far as Canada.

IMPORTANCE: The caterpillars attack the leaves and tender stems of many cultivated crops, including cotton, tobacco, peanut, alfalfa, and garden vegetables. Larvae also feed on ears of corn, causing injury similar to that of corn earworms. Fall armyworms are most damaging in growing seasons preceded by a cold, wet spring. Late in the season, populations of the pest build up, and larvae move in large numbers seeking food. When outbreaks occur, entire areas may be defoliated.

APPEARANCE: Adult moths are gray with mottled black and white blotches on the wings. Each forewing is marked with a distinctive white tip. Larvae vary from brown to green to black, and the caterpillars have stripes running lengthwise along the body. A white inverted Y-shaped marking on the front of the larval head distinguishes the species, which reaches 40 mm long when full grown.

LIFE HISTORY: Adults migrate northward in the spring and summer, and females oviposit about 1,000 eggs in dome-shaped masses of 150 eggs on the host plant. In 2 to 10 days, the eggs hatch, and the larvae feed gregariously. After six molts (20 days), larvae pupate under about one inch of soil. Adults emerge in about 14 days. Female moths fly many miles before ovipositing. Typically, there are one to two generations in the north and as many six generations in the south.

EFFECTS OF ENVIRONMENT ON INSECT DEVELOPMENT

Like other organisms, a given insect species is capable of survival only within certain environmental limits, and when possible, individuals actively seek out preferred temperatures, humidities, and light intensities. Within this "favorable" range, these environmental factors usually influence rate responses of activities such as feeding, dispersal, egg laying, and development. In recent years, an understanding of insect developmental rates has played a particularly important role in insect pest management. This is because timing of management activities is crucial to implementation of pest management tactics. Knowing or predicting that an insect population is in the egg stage or larval stage, for example, may be one of the cues to begin insect sampling, initiate spraying, destroy crop residues, or plant a crop.

Of the environmental factors, temperature probably has the greatest affect on insect developmental rates. Primarily, this is because insects are **poikilothermic**, or cold-blooded. Within certain limits the higher the temperature, the faster the development of the insect. Developmental rate increases with temperature because chemical reactions occur more frequently and proceed more rapidly

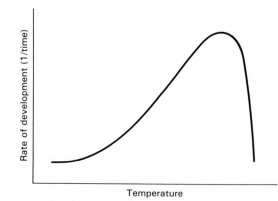

Figure 5.22 Graph of the relationship between rate of development (1/time) in insects and temperature.

at higher temperatures. As temperature increases, diffusion rates for substrates and enzymes also increase, resulting in greater formation of enzyme-substrate complexes. In addition, higher temperatures provide more thermal energy for the requirements of biochemical reactions.

Predicting Biological Events: The Degree-day Method

Because temperature is crucial to biochemical reactions, it can be used to predict rates of insect development. The idea of using temperature and time to describe development of cold-blooded animals is more than 250 years old, and several methods have been advanced to utilize the relationship for prediction. Of these, the most prevalent methods of measuring and estimating this so-called physiological time is the **degree-day** method.

Degree days represent the accumulation of heat units above some minimum temperature for a 24-hour period. Below this minimum, no development takes place, but above it, heat units are accumulated toward development. For degree-day accumulation, if the **developmental minimum** or **threshold** of an insect is 15°C and the average temperature for the day is 27°C, then 12 degree days, (symbolized as DD or °D) would have accumulated on that day.

To predict the stage of development from degree days, the **thermal constant** for an event must have been established. This thermal constant is the number of degree days required for an event to occur. By accumulating degree days each day and relating these to the thermal constant for an event, it can be estimated whether or not the event occurred on the day in question. The most common events for which insect thermal constants are established are hatching, each larval or nymphal molt, pupation, and adult emergence.

Both thermal constants and developmental thresholds vary among different species of insects. For instance, the thermal constant for egg to adult development of the painted lady butterfly, *Vanessa cardui*, is 440 Celsius degree days, but for the seedcorn maggot, it is 376 Celsius degree days. Developmental thresholds for these insects are 12°C and 4°C, respectively. Because of significant differences, threshold values must be determined for each species of interest.

The following formula has been devised to calculate degree days (DD) for a specific date:

BOX 5.9 PAINTED LADY BUTTERFLY

SPECIES: *Vanessa cardui* (Linneaus) (Lepidoptera: Nymphalidae)

DISTRIBUTION: The painted lady butterfly is found worldwide, except for South America and Australia. In North America, the species cannot survive cold temperatures; instead, it migrates northward each spring from the southwestern United States and Mexico and returns in the fall.

IMPORTANCE: Although the painted lady cannot overwinter in temperate areas, the species is among the most common butterflies of the world. In early spring, individuals begin migrating northward and by late spring the species suceeds in recolonizing the areas where it could not successfully overwinter. The caterpillars attack over 100 species of plants, including Canada thistle, soybean, sunflower, garden vegetables, and ornamentals. Occasionally, when large migrations occur, the species can achieve pest status, causing widespread defoliation of host plants. However, the painted lady, also known as the thistle caterpillar, is also considered a beneficial species because of its preference for Canada thistle, a common weed.

APPEARANCE: Adults butterflies are brown with red and orange mottling and have black and white spots. The hindwings have four eyespots in a row. Adults are approximately 25 mm long with wingspans from 51 to 57 mm. Larvae are about 32 mm long when full grown and are brown to black with a yellow stripe along each side of the body, which is covered with small, spiny hairs. The eggs are pale green and have several longitudinal ridges. Pupae (chrysalides) are about 22 mm long and are gold-colored.

LIFE HISTORY: Females oviposit eggs singly on host plants, and in about seven days the eggs hatch. After two to six weeks of feeding, the larvae pupate. In 7 to 17 days, adults emerge and the cycle is repeated. There are typically two generations per year in the northern part of the species range, but generations are continuous in the tropical and desert regions.

$$DD = \text{average daily temperature} - \text{developmental threshold}$$

The main difference among workers in making degree-day calculation is in computing average daily temperature. Although four basic procedures have been developed to estimate average daily temperature (rectangle method, sine-wave method, cosine curve, and triangle method), the most widely used is the rectangle, or simple-average, method. Although somewhat crude, this method probably offers enough precision for most practical programs of pest management. With simple averaging, the degree-day formula becomes

$$DD = \frac{\text{maximum temp.} + \text{minimum temp.}}{2} - \text{developmental threshold}$$

By simply adding degree days, a straight-line relationship between average daily temperature and development is assumed. In reality, however, the relationship is a straight line with curved ends (Figure 5.23). Because much of the

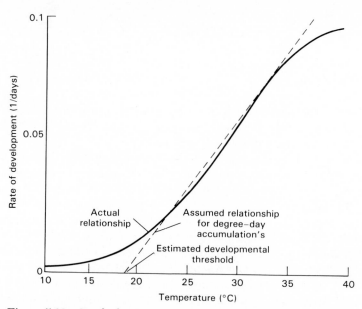

Figure 5.23 Graph showing the actual relationship of insect development and temperature (solid line) and the assumed relationship used with degree-day accumulations (broken line). Although some error is introduced in its use, the straight-line relationship is most practical for predicting insect development. Slope and intercept of the straight line are calculated from data of the curved line.

relationship is straight line, we can set limits on the upper curve and ignore minor deviations caused by the lower curve. In so doing, we gain simplicity but introduce error.

Several conventions or rules are used in making degree-day calculations, mostly for the purpose of reducing this error. These rules can be listed as follows:

1. If the maximum temperature did not exceed the developmental threshold, no degree days are accumulated. For example,

$$\text{developmental threshold} = 10°C$$

$$\text{maximum temperature} = 9°C$$

$$\text{minimum temperature} = 4°C$$

$$DD = 0$$

2. If the maximum temperature was above the developmental threshold, but the minimum temperature is below it, the minimum is set equal to the developmental threshold in calculating the average. For example,

$$\text{developmental threshold} = 10°C$$

$$\text{maximum temperature} = 20°C$$

$$\text{minimum temperature} = 8°C$$

$$\text{transformed minimum} = 10°C$$

$$DD = 5$$

3. If the maximum temperature exceeded that of the developmental optimum (the temperature at which developmental rate is highest), it is set equal to the optimum. For example,

$$\text{developmental threshold} = 10°C$$

$$\text{developmental optimum} = 27°C$$

$$\text{maximum temperature} = 29°C$$

$$\text{transformed maximum} = 27°C$$

$$\text{minimum temperature} = 21°C$$

$$DD = 14$$

Rule number 1 is always applied in calculating degree days, and rules number 2 and 3 are applied when so instructed by the particular degree-day program.

Degree-day programs for insect species are usually developed by research entomologists who record rate of development of insects placed in laboratory growth chambers, each set at a different temperature. Time required for development of the egg stage, each larval or nymphal stage, and pupal stage are recorded for each temperature. The reciprocal of time (1/days) for an event to occur is plotted for each temperature, and a straight line is fitted to these points (a statistical procedure called linear regression analysis is used to do this). The point at which this line crosses the horizontal axis is the developmental threshold. As can be seen, this point is a statistical estimation of development and not a true biological point; in other words, some development can occur below the estimated developmental threshold, but usually it is trivial for the purpose of estimation. Once a developmental threshold is determined, the experimental data can be used further to establish thermal constants for important events in the insect life cycle (Table 5.2).

In addition to manual calculations for degree-day accumulations, several programs have been written for computer calculation. One such program, DEGDAY, is functional on personal computers and derives degree-day accumulations by rectangle, sine-wave, and triangle methods (see Higley et al., 1986).

Degree-Day Programs in Insect Pest Management

As mentioned previously, degree-day programs are a basic component of many insect pest management programs. In most programs, degree-day accumulations are made from the beginning of a growing season and are continuously compared with thermal constants that indicate time of potential crop injury. Thereafter, samples are taken, density estimates are compared with economic threshold levels, and decisions are made as to whether suppression is necessary.

Although the degree-day method has been useful in the management of many insect pests, it is not always applicable. For some pests like corn rootworms, *Diabrotica* species, developmental time can be predicted best by date, without considering temperature. In such instances, some factor other than temperature may be more important, or temperatures in the microhabitat are relatively uni-

Table 5.2. Degree-day Requirements (Thermal Constants) for Certain Activities of the European Corn Borer.[a,b]

Stage	Activity	Degree Days
	First Generation	
Egg	Peak egg hatch	100
1st–2nd instars	Leaf feeding	200+
3rd instar	Stalk boring	350+
4th instar	Stalk boring	400+
5th instar	Stalk boring	550+
Pupa	Peak pupation	900+
Adult	Egg laying	1150–1700
	Second Generation	
Egg	Peak egg hatch	1260
1st instar	Sheath feeding	1350+
2nd instar	Sheath feeding	1500+
3rd instar	Sheath feeding	1600+
3rd instar	Initial stalk boring	1650

SOURCE: From Showers, et al., 1983, *Management of the European Corn Borer*, North Central Regional Publication No. 22.

[a]Developmental threshold = 50°F.

[b]Begin degree-day accumulation at first significant moth flight when extended corn leaf height is 26 inches or greater.

form, making them unimportant in prediction. Additionally, research information on degree-day requirements is not yet available for many insect species.

Other limitations of the degree-day method relate to accuracy. Accuracy of degree-day accumulations depends on temperature measurements used in the calculations. Consequently, degree days should be calculated with temperatures that represent environments where the species is present. Also, temperatures at one site give only a rough estimate of insect development at another site several miles away.

REGULATION OF INSECT POPULATIONS

Anyone who has seen hundreds of aphids crowded on a single plant leaf or witnessed the swarms of pomace flies hovering over ripened fruit will not doubt the reproductive capabilities of insects. Indeed, it would seem that population growth might cause numbers to overrun their food supply and thus cause their complete destruction from starvation. Likewise, at certain times of the year or at certain locations, the numbers of a species may seem so small that extinction would be imminent.

Neither of these catastrophes usually occurs to a species, at least in the duration of a human lifetime. Rather, population numbers fluctuate within certain bounds for long periods of time, neither increasing (overrunning resources) nor wasting away to extinction. The natural phenomenon of numbers being arrested short of extinction has been called **regulation**. Therefore, it might be said that all existing populations are regulated in one way or another. This regulation may occur at levels not bothersome to humans or at levels that are inconvenient to them, as with pests.

How regulation occurs or, indeed, if it occurs at all has been the subject of much biological debate. Some popular theories to explain how animal regulation occurs or, in some instances, simply why numbers change have been advanced by W. R. Thompson, A. J. Nicholson, H. G. Andrewartha and L. C. Birch, D. Chitty, V. C. Wynne-Edwards, D. Pimentel, and A. Milne.

In particular, Milne's theory provides a useful way of looking at natural regulation in insect populations, and data from many studies seem to support his tenets. As Milne's theory goes, there are basically three types of natural factors acting to regulate population numbers: (1) **perfectly density-dependent factors**, (2) **imperfectly density-dependent factors**, and (3) **density-independent factors**.

According to Milne, a perfectly density-dependent factor would be one that never fails to control the increase in population numbers. Such a factor acts strongly to subtract numbers from a population when density is high and increasing, and the factor reduces its pressure when density is low and decreasing. The only natural factor influenced solely by numbers in the subject population is **intraspecific competition**.

Intraspecific competition is a phenomenon whereby individuals in a population vie for a resource in limited supply like food. As population growth occurs, more and more of the resource is eliminated, causing greater and greater intraspecific competition. As individuals compete, some are eliminated, in this case from starvation, and density is reduced. With falling numbers, competition is relaxed, thus reducing the rate of decrease.

An imperfectly density-dependent factor functions similarly to a perfectly density-dependent factor, except that the imperfect factor may sometimes fail to limit numerical increase. Agents in this category include predators, parasites, and pathogenic microorganisms that cause insect death. Because the lives of these organisms are influenced by many environmental factors, one or more may limit their influence on a prey or host population. For example, a population of parasitic wasps may be very effective in limiting the population size of a caterpillar population, and the action of the parasite acts in a density-dependent manner; it causes a greater proportion of deaths when the host population is large than when that population is small. Occasionally, however, wet, cool weather may have a severe effect on the parasite and not on its host. Therefore, the density-dependent action of the parasite fails, and the host population increases.

Density-independent factors do not vary in their impact on a population according to its density. The primary example of this kind of factor is weather. Weather factors, including rainfall, temperature, and humidity, may have a strong effect on insect survival, but the size of the population has no significant influence on the degree of impact from these factors.

Milne visualizes animal numbers as fluctuating within three zones (Figure 5.24). Zone I is a level of very low numbers, and populations seldom fall here from Zone II. When they do, it is because of an unusual occurrence (duration and/or intensity) of unfavorable density-independent factors. Extinction from being in this zone usually does not occur (in a human lifetime) because density-independent factors, acting in random fashion, are not sufficiently deleterious long enough for this catastrophe to occur. On the return of favorable density-independent factors, the population moves back into Zone II.

Zone II is the level of usual numbers. Population density fluctuates within this zone for long periods of time. Here numerical change is a result of the combined action of density-independent and imperfectly density-dependent factors. Should the density-independent environment be very favorable and the imperfectly density-dependent factors fail, the population could move up into Zone III.

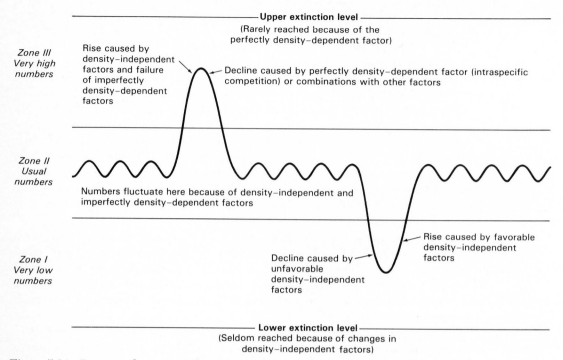

Figure 5.24 Diagram of insect population regulation based on the theory of Milne. (Redrawn from Milne, 1959, *Canadian Entomologist*, volume 89: 193-213)

Zone III is a level of unusually high numbers. Here the population is in danger of overrunning its environmental resources. It does not do so because competition acts perfectly to limit numbers and forces the population back down into Zone II. The reduction of numbers to lower levels may also be aided by the return of unfavorable density-independent factors and imperfectly density-dependent factors.

In this way, Milne believes that numbers of animals stay in existence through geological time periods (thousands of years). He believes that intraspecific competition is the ultimate controlling influence on population increase but that competition seldom occurs because numbers are held at lower levels by such things as weather and natural enemies.

As can be seen, Milne's population theory gives a general explanation of the mechanism of population regulation. As such, it plays an important role in pest management by orienting researchers and practitioners to the study of individual problems. As a primary benefit, detailed analyses of specific insect life systems based on such theory allow prediction of outbreaks. Ultimately, knowing the specifics of how numbers fluctuate gives clues to how insect life systems can be manipulated for the benefit of humankind.

Selected References

CHIANG, H. C. 1985. Insects and their environment. In R. E. Pfadt, ed., *Fundamentals of Applied Entomology*, 4th ed. New York: Macmillan, chapter 5.

CLARK, L. R. , P. W. GEIER, R. D. HUGHES, and R. F. MORRIS. 1967. *The Ecology of Insect Populations in Theory and Practice*. London: Methuen, chapters 1–4.

HIGLEY, L. G. , L. P. PEDIGO, and K. R. OSTLIE. 1986. DEGDAY: a program for calculating degree-days, and assumptions behind the degree-day approach. *Environmental Entomology* 15: 999–1016.

HIGLEY, L. G. , and W. WINTERSTEEN. 1987. *Using Degree Days in an Integrated Pest Management Program.* Iowa State University Cooperative Extension Service Pamphlet Pm-1296.

HUFFAKER, C. B. , and R. L. RABB. 1984. *Ecological Entomology.* New York: Wiley, chapters 1, 2, 11, 12, 13, and 17.

JOHNSON, C. G. 1969. *Migration and Dispersal of Insects by Flight.* London: Methuen, chapters 1, 2, 10, 22, and 24.

MILNE, A. 1959. The natural control of insect populations. *Canadian Entomologist* 89: 193–213.

PRICE, P. W. 1984. *Insect Ecology,* 2nd ed. New York: Wiley, chapters 10 and 12.

TURNBULL, A. L. 1969. The ecological role of pest populations. *Proceedings Tall Timbers Conference on Ecological Animal Control by Habitat Management* 1:219–232.

6 SURVEILLANCE AND SAMPLING

THE IMPORTANCE OF UNDERSTANDING insect activity to make decisions in insect pest management cannot be overstated. Indeed, this knowledge is indispensable to the paramount principle of pest management: to take no action against a pest unless that pest is known to be present and is an actual or potential threat.

Such insect activities as crop invasion, long-range migration, local movement, feeding, and reproduction are detected and documented through pest surveillance. Pest surveillance is the watch kept on a pest for the purpose of decision making. Depending on the kind of pest (native, newly introduced, potential invader), surveillance programs attempt to determine if a pest species is present, to estimate the numbers in a population and their distribution, and to assess how these factors change over time. More succinctly stated, the major objectives of surveillance are detection of species presence and determination of population density, dispersion, and dynamics.

Aside from private individuals conducting surveillance programs, for several governmental and commercial agencies pest surveillance is a major activity. The U.S. Department of Agriculture conducts programs of the widest scope in the United States, particularly through the Animal and Plant Health Inspection Service (APHIS). An administrative subdivision of APHIS, Plant Protection and Quarantine Programs (PPQ), is charged with detection of new pests entering the United States, determination of the spread of introduced pests into new areas, and determination of levels of infestation of established agricultural pests. Some of the major surveillance programs of PPQ are those for gypsy moth (*Lymantria dispar*), browntail moth (*Euproctis chrysorrhoea*), imported fire ant (*Solenopsis saevissima*), Japanese beetle (*Popillia japonica*), Mediterranean fruit fly (*Ceratitis capitata*), Mexican fruit fly (*Anastrepha ludens*), pink bollworm (*Pectinophora gossypiella*), whitefringed beetles (*Graphognathus* species), beet leafhopper (*Circulifer tenellus*), khapra beetle (*Trogoderma granarium*), and cereal leaf beetle (*Oulema melanopus*). PPQ also has responsibility in coordinating its surveillance activities with those of other agencies, for example, the Agricultural Research Service, the United States Forest Service, and state and private establishments.

Many states employ a state entomologist whose primary responsibility is surveillance of established and potential pest species and inspection of certain

BOX 6.1 IMPORTED FIRE ANTS

SPECIES: black imported fire ant, *Solenopsis richteri* Forel, red imported fire ant, *Solenopsis invicta* Buren (Hymenoptera: Formicidae)

DISTRIBUTION: The black imported fire ant was introduced into the United States (Mobile, Alabama) from South America in 1918. In the 1940s, the red imported fire ant was imported into the same region. Since the initial introductions, imported fire ants have spread into the southeastern United States as far north as North Carolina, and into the southwestern United States as far west as New Mexico. Presently, both species continue to expand their range.

IMPORTANCE: Both species are a nuisance to humans, livestock, and wildlife, because they viciously bite and sting when disturbed or threatened. Fire ants get their name from the effects of the characteristic sting, which is painful, concentrated, and firelike. The ants first bite into the skin and then proceed to sting the victim. During the stinging, a necrotizing poison is injected into the skin, resulting in pus formation in the wound. Wounds often result in scarring of the skin. Occasionally, victims are stung extensively by many fire ants, causing nausea and dizziness. Humans allergic to fire ant venom have died in situations where medical treatment was not available. The species also attack, kill, and eat young poultry and unhatched eggs. Fire ants attack germinating seedlings like soybean and corn and the distinctive hard-crusted mounds damage farm machinery in the field. Ironically though, fire ants are also considered beneficial insects because they prey on bollworms and boll weevils in cotton fields.

APPEARANCE: Adults are polymorphic, consisting of majors (6 mm long) and minors (3 mm long). Black imported fire ants are light brown to dark brown, whereas red imported fire ants are light reddish brown to dark brown. The abdomen of both species has a two-segmented petiole, and the antennae are slightly clubbed at the end. Larvae are white and grublike. Fire ant mounds are very hard and reach from 38 to 92 cm high.

LIFE HISTORY: Both species are active year-round. Newly mated queens dig brood chambers and oviposit from 75 to 100 eggs, which hatch in 8 to 10 days. Larvae are fed for 6 to 12 days, then pupate. After 9 to 16 days, adult workers emerge. Queens rear the first brood, but all subsequent broods are reared by workers. There are many broods per year.

commodities moving into and out of the state. Most states also have a land-grant university, where entomologists of the Cooperative Extension Service and those in research conduct regular surveillance programs. Additionally, a few states have county-based programs and mosquito abatement districts that conduct surveillance of specific pests.

Private firms can be another important source of pest surveillance information. Most of these are agricultural consulting firms that contract with producers to monitor pest levels and recommend where, when, and how to deal with problems. This type of consulting is rapidly growing, as indicated by the number of new firms established in the last decade.

The most successful surveillance programs carried out by the agencies mentioned are achieved through regular survey activities. An insect pest survey is a

BOX 6.2 **GYPSY MOTH**

SPECIES: *Lymantria dispar* (Linnaeus) (Lepidoptera: Lymantriidae)

DISTRIBUTION: The gypsy moth was introduced into North America (Massachusetts) from Europe in 1869 by a scientist interested in experimenting with silk production. The moths were accidentally released and soon spread to the New England states. Presently, the species covers most of the northeastern United States and southeastern Canada. In 1976, gypsy moths were discovered as far west as San Jose, California.

IMPORTANCE: Gypsy moth larvae are serious defoliators of trees in forests and urban areas. More than 500 species of trees are attacked by this pest, causing widespread defoliations often severe enough to kill trees. Female moths cannot fly; thus humans have been instrumental in their long-distance movements. Shipment of nursery stock and transport by automobile have aided the spread of the species. Young larvae that produce silken threads are also subject to transport, because they may be carried long distances by the wind.

APPEARANCE: Adult females are white with yellow hairs on the abdomen and have black wavy bands across the forewings. Adult males are smaller than females and are grayish brown with black wavy bands across the forewings. Larvae are very hairy and about 60 mm long when full-grown. They have distinctive yellow markings on the head and six pairs of paired red spots, followed by five pairs of paired blue spots on the back. The eggs, about 1 mm in diameter, are oviposited in hair-covered masses of 100 to 1,000 on tree trunks and limbs. Pupae are reddish brown and lightly covered with red hairs.

LIFE HISTORY: The eggs overwinter, and hatching begins in early May. Developing larvae devour leaves of their hosts for about 40 days and become full-grown by mid-July. Larvae then spin loose, silken cocoons, and adults emerge 10 to 14 days later. The adults mate, and the females oviposit on host trees during their 6- to 10-day existence. This species produces only one generation per year.

---

detailed collection of insect population information at a particular time in a given area. The survey program may be carried on for an entire growing season or at certain critical periods in the insect life cycle. The area involved in a survey may be as small as a field, a pond or a woodlot, or it may be as extensive as a state or region of the country.

Surveys may be classified as **qualitative** or **quantitative**. Qualitative surveys are the least complex and are generally aimed at pest detection. Usually, qualitative surveys yield lists of pest species discovered, along with a subjective reference to density, for example, abundant, common, rare. Such surveys are typical at international ports of entry, state boundaries, and similar areas where agricultural commodities are inspected. Qualitative surveys are also employed with newly introduced pests; in such instances attempts are made to understand the extent of the infestation. Often, qualitative surveys precede quantitative ones.

Quantitative surveys are the most common type employed in insect pest management. The quantitative survey attempts to define numerically the abundance of an insect population in time and space. Such information is used to predict future population trends and to assess damage potentials.

To collect information in the quantitative survey, a count of insects or a measure of their presence is required. Because of the great number and/or secretive nature of many insects, it is not feasible or even desirable to take a census: to count every individual in the population. Instead, it is usually more efficient to estimate population density by sampling.

Sampling requires that we take a representative part of the total population and base our estimate on that part. Therefore, to conduct a survey and make estimates, a sampling technique and a sampling program are required. A **sampling technique** is the method used to collect information for a single sample. Examples of sampling technique include counting all the rootworm beetles present on the ear of a corn plant by direct observation or counting all the black cutworm moths caught in a blacklight trap on one night. A **sampling program** is the procedure for employing the sampling technique in time and space. Sampling programs describe when sampling is to begin, location of samples, number of samples, and how often samples should be taken. Development of efficient, reliable sampling programs for insect pest management requires much biological information and considerable research by entomologists.

COMMON SAMPLING TECHNIQUES IN INSECT PEST MANAGEMENT

A vast array of different techniques is used for sampling insects. Each has advantages and disadvantages for the insect species being studied, and more than one technique may be appropriate for development of a sampling program. Only some of the most common techniques are discussed in this chapter.

In Situ Counts

In situ (Latin for "in place") counts, also known as direct counts and direct observation, often require no special equipment but rely on a good eye. With this method large and mostly conspicuous insects are viewed in the habitat, and counts are recorded. Usually the habitat viewed with this method is the plant canopy, a specific plant part, or a certain region on an animal (for example, the face of a cow). If numbers of insects are relatively low and the plant is small and isolated from adjacent plants (as with seedlings), all insects on the plant may be counted. Therefore, this procedure is frequently used with early-season pests of agronomic crops or in later season with crops where herbage is removed and regrowth is inspected, for example, alfalfa grown for hay. When the plant canopy is continuous and individual plant sampling is impractical, all the plants in an area, for example, one square meter or 30 cm of row, may be inspected. In this case, however, a piece of equipment like a square metal frame tossed over the area is used (e.g., counting grasshoppers in one square meter of pasture). In some instances, plants in an area are removed (cut off at ground level) and bagged, and the insects are counted later. Specialized equipment like an A-frame or clam trap is sometimes used for this purpose.

When plants are large, such as mature trees, shrubs, and late-season herbaceous vegetation, only certain numbers of leaves, stems, flowers, buds, or pods

Figure 6.1 *In situ* counts of insects on plants using a magnifying lens. White poster board placed on either side of plants provides a good background for observations.

Figure 6.2 An A-frame trap with detachable plastic bag (at right). The frame is placed over a plant canopy and closed. The plants then are cut off at the soil surface, and the contents are shaken down into the bag. Sample bags are returned to the laboratory for insect sorting and counting. (Photo by E. Bechinski)

Figure 6.3 Counting the number of corn rootworm adults *Diabrotica* species, in the ear zone of a corn plant. Subsequently, this information may be related to total number of insects per plant. (Photo by L. Higley)

may be counted. Orchard and forest pests are frequently sampled by counting numbers per shoot or branch. Similarly, agronomic crop pests may be counted as numbers per so many leaves, stems, or reproductive structures. For example, rootworm beetles may be counted in the "ear zone" of a corn plant and this information related to total numbers on the plant. When only part of the plant is inspected, it is important to know if insect numbers may be stratified in certain parts of the canopy and to account for any stratification when making estimates.

For any sampling technique, it is also important to assess the state of the plant and density of the sample units present. This assessment is relatively simple if the whole plant is the sample unit; count the number of plants per unit row or area. An estimate of the numbers of leaves or shoots available for insect habitation becomes much more complex and time-consuming, and estimation errors become greater. Still, it is important that these estimates be made because plant growth causes changes in the amount of available living space. Such changes do not allow the direct comparison of insect numbers from one time to the next. For instance, if numbers of aphids per leaf of a potato plant are falling during a period of

Figure 6.4 Face flies, *Musca autumnalis,* on the face of a Hereford. *In situ* counts are sometimes used as a sampling technique for this pest. (Courtesy Iowa State University Extension Service)

sampling dates, is this due to decreasing size of the aphid population, more leaf area on the potato plant, or both? With only the aphid per leaf information, we have a measure of population intensity, perhaps all that is needed if only a plant damage assessment is desired. If the actual trend in the aphid population is needed for predictive purposes, then a measure of insect population density is necessary. The population density estimate could be determined by multiplying the average number of aphids per leaf times the average number of habitable leaves per square meter.

The *in situ* count method is the most widely used with plants; for obvious reasons, use of the technique for damaging stages of poultry and livestock pests has been limited. Movements by animals being inspected and feathers or hair make direct counting difficult, at best. The method probably can be used only with large parasites, for example, ticks (Acari), or those that are exposed or very active, for example, sheep keds, *Melophagus ovinus.* Other parasites that can be counted directly on livestock include face flies on the faces of cattle and horn flies on the shoulders, backs, bellies, or horns of cattle or other livestock. Because of the many difficulties involved, in situ counting is most valuable for detection and qualitative surveys of these parasites on their hosts and for quantitative estimates in the parasites' free-living stages.

Knockdown

Knockdown is closely related to *in situ* counting, but in this instance, the insects are removed from the habitat by jarring, chemicals, or heating and then counted. Jarring is probably the most common method of knockdown from plants. It has been used particularly when sampling insects on the lower branches of trees and

Figure 6.5 Using a stick to beat an apple tree branch. Insects are jarred loose and fall to the white drop cloth on the ground. Counts are taken of those on the drop cloth. (Photo by L. Higley)

shrubs. Here, a cloth, tray, or other receptacle is placed on the ground, a branch is pulled down over the receptacle, and the branch is struck a prescribed number of times with a stick. Insects knocked off fall into the receptacle and are counted. Versions of this jarring approach also are used with some agronomic crops where the plant canopy can be bent over a ground cloth and shaken vigorously for a given time. In soybeans, for instance, this is the standard technique for surveying defoliating insects throughout most of the season. In alfalfa, alfalfa weevil, *Hypera postica*, samples are taken by picking stems and beating them against the side of a container to dislodge larvae for counting. The jarring approach is particularly useful with insects like adult weevils (Coleoptera: Curculionidae) and many beetles that "play dead" when disturbed.

Certain chemicals may be used in conjunction with the jarring technique to facilitate knockdown. This approach has found most use with plant parts placed in a container with the chemical. Here, the container is often shaken, after which insects drop off and are counted. Examples of this include aphids (Homoptera: Aphididae) exposed to vapors of methyl isobutyl ketone and thrips (Thysanoptera)

BOX 6.3 FACE FLY

SPECIES: *Musca autumnalis* De Geer (Diptera: Muscidae)

DISTRIBUTION: The face fly is native to Europe, Africa, and Asia, and was subsequently introduced into Nova Scotia and Canada from Europe about 1952. The species is currently distributed throughout Canada and the United States.

IMPORTANCE: The face fly is primarily a cattle pest but also attacks other domesticated animals. Adult flies feed on secretions produced from the eyes, nose, and lips of cattle and also on blood from open wounds or sores. The cattle become irritated and graze less, decreasing their weight gain and milk production. The species also vectors pinkeye and certain eyeworms, such as *Thelazia*. During outbreaks, there are often more than 100 face flies on each animal.

APPEARANCE: Adults are slightly larger than the house fly. Females have a black abdomen, whereas males have an orange one. Larvae (maggots) are yellow and from 8 to 19 mm long when full-grown. Eggs are pale yellow, and each egg has a respiratory stalk. Puparia are oval and grayish white.

LIFE HISTORY: Adults overwinter in masses in buildings where they are also known as cluster flies. In early spring, females oviposit 5 to 8 eggs in fresh dung. Females can oviposit up to 230 eggs in total. Usually after only one day, eggs hatch, and larvae feed in the dung, undergoing two molts in two to four days. Purpation occurs in puparia in the soil, and adults emerge in about 10 days. There are several generations per year.

Figure 6.6 Soybean plants are shaken over a white ground cloth (left) to dislodge insects. Counts are recorded from insects on the cloth. (Photo by R. Smelser)

Figure 6.7 Polyethylene enclosures are sometimes used with insecticides to kill or knock down insects for sampling purposes. After insecticidal action, the lid is removed, and dead insects are shaken into the receptacle at the base of the plants. Counts are made from insects in the receptacle.

exposed to turpentine vapors. When whole plants or several plants in an area are sampled, they may be enclosed in a polyethylene envelope and treated with a quick knockdown insecticide such as pyrethrum. After exposure, the plants may be shaken to facilitate knocking dead or stunned insects off the plant and onto a ground cloth or other receptacle.

Heating is most often used when plants, plant parts, or plant debris are removed and taken to a facility for insect counting. Here, plant samples may be placed in a special device, a Berlese funnel, that heats the sample, causing it to dry out. Finding the plant habitat increasingly inhospitable, insects move out of the plant material, fall through a funnel, and are retrieved from a receptacle for counting. This approach has the advantage of saving time and effort usually devoted to sorting plant material to locate insects. It has the disadvantage that some insects may die during the extraction process and not be counted. A closed modification of the Berlese funnel has been used commonly to extract alfalfa weevil larvae from alfalfa samples.

Although jarring, chemical knockdown, and heating are some of the most common techniques for extracting insects from plant material, other approaches have been used. Some of these include removing mites from leaves using a special brushing machine and washing insects from plants and plant parts with soap, alcohol, or other solutions.

Netting

Netting insects is one of the most widely used techniques in population sampling. Netting can usually be accomplished rather inexpensively and is adaptable to sampling a vast array of agricultural pests and their natural enemies.

Figure 6.8 A bank of modified Berlese funnels. Soil cores containing arthropods are placed in a screened tray at the opening of the funnel. A light bulb (or other heating element) in the shroud above the sample tray heats and dries the soil core, driving out the arthropods. These extracted arthropods then fall through the funnel into a vial of alcohol attached to the tip of the funnel.

Figure 6.9 Soil-core sampler used in obtaining cores for the Berlese funnel. The sampler is constructed in two pieces, which are held together with a clamp. After obtaining a core from the soil, the clamp is removed and the corer taken apart. This provides a soil section that is not compressed, allowing arthropods easy escape in the Berlese funnel.

Figure 6.10 Sweep-net sampling in alfalfa. The net is swung through the plant canopy, and insects jarred loose fall to the bottom of the net. (Photo by L. Higley)

When used to sample insects present in a plant canopy, the technique is closely related to knockdown by jarring. In this instance a muslin "net" is swung into the plant canopy, jarring the plants and causing insects on them to fall off and into the net. After a prescribed number of such swings, called sweeps, plant debris is removed and the insects are counted. Such a procedure constitutes a single sample. Compared with many other techniques, the sweep net sample is inexpensive, and adequate insect numbers often can be obtained for analysis of small populations by increasing the number of sweeps per sample. A frequent disadvantage of the technique is the great sampling variability among different surveyors from differences in swing force, net angle, and other factors. The technique has also been criticized for producing variable efficiency with different species, habitats, time of day, and weather, but most techniques can be criticized on these points.

Insect samples also can be taken from plants by a more sophisticated net, the vacuum net. This technique, often referred to by a trade name, D-Vac®, consists of a removable net attached inside a plastic cone, which in turn is connected to an engine-powered fan via a flexible hose. With the engine running, a strong vacuum is created in the cone; by placing the cone over or into the plant canopy, insects are sucked into the net. A sample may result from holding the net at one place for a prescribed period, for example, 15 seconds, placing the net in several locations, or using it to "sweep" through vegetation. Both backpack and hand-held models of this net are available. The vacuum net has been used most widely with light-bodied insects like leafhoppers in agronomic crops. Equipment efficiency declines rapidly as insect body size and clinging ability increases. Because of limited efficiency, cost, and the cumbersome nature of the equipment, vacuum nets are used only in special circumstances and mainly in research.

Other netting of insects is performed with "true," mesh-bag nets that allow capturing insects in flight. The most common aerial net is a lightweight, hand-held net that is swept through the air to capture an insect. This is really more

Figure 6.11 Contents of an inverted sweep-net bag after taking several sweeps through a plant canopy. Both insects and plant debris can be seen. Counts may be made directly from this sample, or the sample may be bagged and returned to the laboratory for processing later. (Photo by L. Higley)

of a collector's tool than a sampling technique; none the less, it is useful in surveillance for detection. Aerial nets used for sampling more commonly include the rotary net and the tow net. The rotary net, used especially with forest pests, consists of one or two nets at the ends of beams that are rotated by a motor-driven shaft. The insects collected in the net(s) are counted after a given period of time. By accumulating the volume of air encountered by the net(s) per hour, aerial density can be calculated. The tow net is a less widely used technique that usually involves the attachment of nets to moving aircraft, automobiles, or ships. Such nets are most often used for detection.

Trapping

Trapping involves some of the most important sampling techniques for insect survey. All trapping procedures have two basic requirements—the insects must move and the trap must hold captured insects. Most traps are set, left unattended

Figure 6.12 Sampling insects with a vacuum insect net. Here, a back-pack model has been mounted on a cart for ease of handling. With the device, insects are sucked into a cone that supports an internal net. After a sample is taken, the net is removed for insect counting.

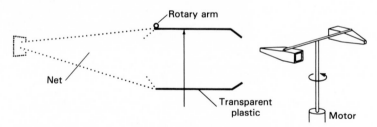

Figure 6.13 A rotary aerial net. (Redrawn from Southwood, 1978, *Ecological Methods*, Chapman and Hall Publishing Company)

for a period of time, then are visited to pick up the "catch." They are often operated continuously, integrating time, insect density, and insect activity to produce the numbers collected.

Traps can be either attractive or passive in their mode of collection. Attractive traps rely on a physical or chemical stimulus to lure insects into them; passive traps collect insects incidentally. It is difficult to categorize most trapping equipment as attractive or passive because, obviously, a passive trap can be made into an attractive trap by adding an attractant. As commonly used, visual traps and bait traps are the attractive types, and pitfall, window, Malaise, some sticky, water pan, and suction traps often lack attractants.

The light trap is the most widely used visual trap, having been particularly important in surveillance programs for moths (Lepidoptera) and mosquitoes (Diptera: Culicidae). Because many moths (particularly species of Noctuidae) and

BOX 6.4 HORSE FLIES AND DEER FLIES

SPECIES: many species, especially in the genera *Tabanus* and *Chrysops* (Diptera: Tabanidae)

DISTRIBUTION: Horse flies and deer flies are found throughout the world, with different species dominating specific regions.

IMPORTANCE: Both horse flies and deer flies are significant livestock pests in all areas of the world. Females inflict deep wounds in animals, causing substantial blood loss that the flies sponge up with their specialized mouthparts. Under heavy infestations, an average of 100 ml of blood may be lost per animal per day. Males feed only on nectar and honeydew. Fly attacks result in both reduced weight gain and milk production. The flies also vector many pathogens, such as bacteria, viruses, helminths, and protozoans. Members of the genus *Chrysops*, commonly called deer flies, often annoy and attack humans.

APPEARANCE: Many species differ greatly in appearance, but typically, horse flies are larger and less colorful than deer flies. Adult horse flies are from 10 to 25 mm long and are dark brown to black, with large bright-green or black eyes. Adult deer flies are yellow to black, with stripes on the abdomen and wings mottled with dark patches. Larvae (maggots) are tapered at both ends and are white to tan. Many species have black bands around each segment of the body, and some species are 50 mm long when full-grown. Eggs are cylindrical, from 1 to 2.5 mm long and are found in layers of up to 1,000 on objects near water.

LIFE HISTORY: Typical tabanids overwinter as larvae in mud at the bottom or edge of streams, ponds, and lakes. In spring, larvae begin feeding on vegetable matter or on small animals, depending on the species. When fully grown, larvae move to dry soil to pupate. After 5 to 35 days (depending on the species) adults emerge and mate, and females begin taking blood meals. Females then oviposit on objects near water, and in five to seven days, eggs hatch. Newly hatched larvae burow into the mud and begin feeding. Generation times vary greatly among species. Some species produce two generations per year, whereas others may produce only one every two to three years.

---

other insects are attracted by short wavelengths of the light spectrum, so-called blacklight lamps emitting ultraviolet light are widely used in trap design. A commonly used trap for collection of mosquitoes, the New Jersey trap, combines a blacklight and dry ice (CO_2) for attraction and a fan for sucking the insects into the trap when they fly near. Once inside the trap, insects may be taken live for counting or other uses or killed directly in the receptacle. Dichlorvos-impregnated vinyl strips (Vapona®) are frequently used for the latter purpose. Other visual traps rely on insect perception of shapes for attraction. These include the Manitoba trap utilizing a black or red sphere suspended under a transparent cone for collection of horse flies and the Missouri cutworm trap, which depends in part on a vertical screen for attraction. The latter trap also combines wheat seedlings and wheat bran bait for the attraction of black cutworm, *Agrotis ipsilon*, larvae in corn.

Figure 6.14 A standard blacklight trap with collecting funnel and receptacle (top) and close up of the receptacle showing catch from two nights. (Photo by L. Higley)

Bait traps rely on an insect's olfaction, or sense of smell, for attraction. A common attractant is food, for example, a mixture of yeast and molasses in a cone trap to sample seedcorn maggot, *Delia platura*, adults. Scents from food that attract insects or cause other behavioral responses are referred to as **kairomones**; study of these is an active field of entomological research. Other intensively researched attractants are **sex pheromones**. Traps containing pheromones or synthetic mimics of them, along with sticky surfaces for holding trapped insects, are sold commercially and have been useful in forest and orchard pest management programs. For example, traps baited with the synthetic sex pheromone **codlemone** are used to indicate potential damage by codling moth in apple, pear, plum, and walnut orchards. Pheromone-baited traps are also used widely in detection programs for exotic pests like the Mediterranean fruit fly, oriental fruit fly (*Dacus dorsalis*), and melon fly (*Dacus cucurbitae*) in California and Florida.

Most traps commonly used without attractants collect flying insects. Chief among these are the Malaise trap and the suction trap. The **Malaise trap** is basically an open-fronted tent made of cotton or nylon mesh that intercepts flying insects. The roof of the tent slopes upward to a peak where a receptacle is located. Insects tend to move upward and into the receptacle, where they are held for counting. The suction trap is more "active" in collecting, relying on suction to capture flying insects. Most suction traps are basically a wire-gauze funnel leading

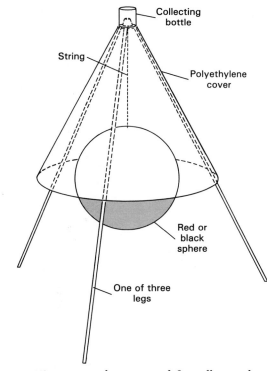

Figure 6.15 A Manitoba trap used for collecting horse flies (Diptera: Tabanidae). Flies are visually attracted to the suspended sphere, fly up into the polyethylene canopy, and are captured in the collecting bottle. (Courtesy USDA)

to a collecting jar, with a motor-driven fan situated below the funnel to create the suction. These traps have been particularly useful for sampling alate (winged) aphids and leafhoppers (Homoptera: Cicadellidae).

Window, water-pan, and sticky traps are examples of interception traps. The window trap consists of a large sheet of glass that sits in a collecting trough and is supported by wooden legs. Insects fly into the glass, are knocked down into a trough with soapy water, and are counted. The water-pan trap consists of an open pan, often a crankcase drainage pan, mounted on a wooden post and filled with soapy water. An omnidirectional baffle is usually set in the pan for the interception of insects. In some instances, as with aphids, the baffle and interior of the pan are painted (yellow for aphids) to add physical attraction to the trap. Sticky traps are of several types and may also be painted to attract certain insects. Such traps may be cylindrical (omnidirectional) or flat (unidirectional) and are usually mounted on a wooden stake at various heights above a plant canopy. Adhesives applied to the traps' surfaces include such materials as Tanglefoot®, Boltac®, and Tack Trap®. Pheromone traps (Figure 6.17) are really a kind of baited sticky trap.

Whereas the traps previously mentioned sample mostly flying and, therefore, mature insects, **pitfall traps** capture ground-moving insects. The pitfall trap basically consists of a collecting jar or bottle buried in the ground with a funnel at the soil surface that empties into the bottle. The landing to the funnel opening is

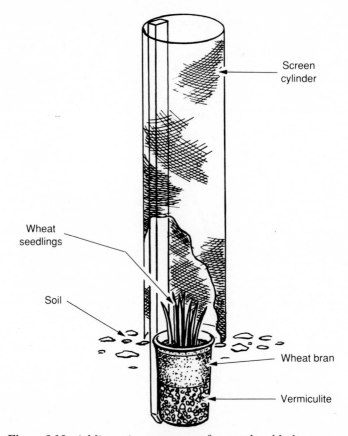

Figure 6.16 A Missouri cutworm trap for sampling black cutworms, *Agrotis ipsilon*. The vertical screen is a visual attractant, and larvae crawl underneath and into the container. Wheat seedlings and bran in the container also aid in attraction. (Courtesy W. Showers and USDA)

smoothed so that walking and crawling invertebrates are not impeded and fall into the funnel. Consequently, a combination of both mature and immature insects is collected. If collected insects are to be killed for counting, a preservative, usually ethanol, is placed in the receptacle. A problem with pitfall trapping is runoff from rainfall that may cause overflow of the receptacle's content and loss of collected specimens. This problem can be overcome somewhat with a more elaborate trap having rain deflectors.

Extraction from Soil

Soil is one of the most difficult environments for sampling insects because it is dense, and insects cannot usually be seen or easily extracted from it. Therefore, soil sampling techniques tend to be more complex and sampling more expensive than in other environments. Yet sampling here is very important because over 90 percent of insect species spend at least one stage in the soil or at the soil surface. Probably the most often used techniques for extraction of insects from soil samples are **Berlese funnels**, sieving, washing, and flotation.

In all these techniques, soil samples must first be taken, then delivered to a facility for extraction. Soil samples may be taken with conventional soil corers,

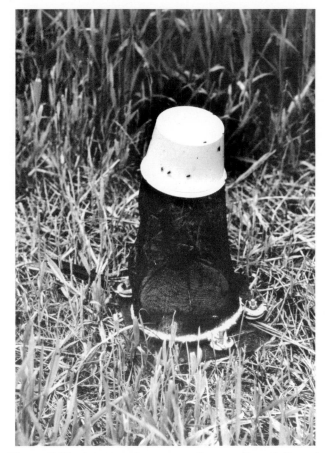

Figure 6.17 A screened-cone trap for sampling adult seed-corn maggots, *Delia platura*. The trap is baited with a container of yeast and molasses. Flies attracted to the bait fly up and are captured in sticky material applied to the inner surface of the plastic container (top of trap). (Photo by J. Funderburk)

golf-hole borers, bulb cutters, shovels, trowels, metal frames, or other equipment. Even commercial trenching machines have been used to collect soil for sampling rootworm eggs! A major consideration for selecting a soil sampling technique is whether or not the extracting equipment requires live insects, for example, the Berlese funnel. In this instance compacting the soil core when removing it from the corer may kill insects or close passages required for escape and collection. Here, sampling efficiency may be improved by using a split corer (Figure 6.9).

The Berlese funnel described earlier (Figure 6.8) is one of the most common devices for extracting insects from soil cores. This technique, originally developed by Italian entomologist A. Berlese in the early 1900s, utilized a water jacket surrounding a funnel to heat and dry the soil sample contained inside. A modification by A. Tullgren, a Swede, replaced this heat source with an incandescent lamp suspended above the sample. Other modifications for efficiency of extracting certain insect groups can also be found, but all such funnels operate

Figure 6.18 Typical trap baited with synthetic pheromones for monitoring orchard pests. Insects attracted inside the trap fly up and are held by sticky material applied to the inside surface.

Figure 6.19 A Malaise trap. Insects fly into the screened baffles and are directed upward. They are captured in the collecting jar at the top.

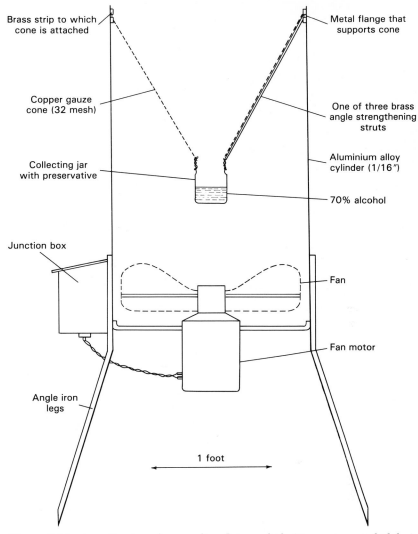

Figure 6.20 A suction trap for sampling flying aphids (Homoptera: Aphididae) and leafhoppers (Homoptera: Cicadellidae). (Redrawn from Southwood, 1978, *Ecological Methods*, Chapman and Hall Publishing Company)

on the same principles—making the soil environment so adverse that the insects inside are forced to move out or providing a stimulus to invoke withdrawal. With this technique, care must be taken to prevent undue mortality before insects leave the core. Cores are usually inverted in the funnel so insects can leave by the same passages used in entering the soil, and the core is heated gradually so they are not killed from desiccation. Berlese funnels have been used most frequently to sample small, abundant arthropods in the soil, for example, mites (Acari), springtails (Collembola), and beetles (Coleoptera), with efficiency varying greatly among species and physical characteristics of the soil. Obviously, this technique is useless for immobile insect stages in the soil like eggs, most pupae, and dormant stages.

Figure 6.21 A window trap for a sampling flying insects. Insects run into the window and fall into soapy water in the collecting trough. Soap is used as a wetting agent, and insects drown. (Redrawn from Southwood, 1978, *Ecological Methods*, Chapman and Hall Publishing Company)

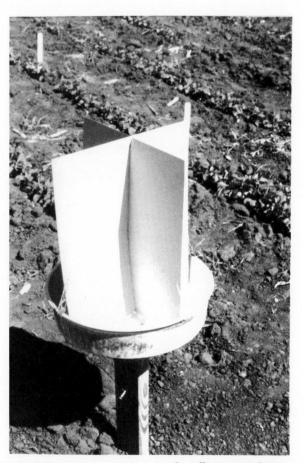

Figure 6.22 A water-pan trap with yellow, omnidirectional baffle for sampling flying aphids (Homoptera: Aphididae). (Photo by R. Hammond)

Figure 6.23 A unidirectional sticky trap showing insects adhering to the sticky surface. The sticky surface of this trap is removed from the holder and taken to the laboratory for insect counts.

Figure 6.24 A typical pitfall trap located in a plant row. Here, a metal can is buried in soil and supplied with a jar of alcohol. An aluminum funnel is inserted into the opening of the metal can, and the funnel tip extends into the collecting jar.

Figure 6.25 Wet sieving a soil sample (in malt cup) with rootworm (*Diabrotica* species) eggs (left). Sieved material is placed in a separatory funnel with a magnesium sulfate solution, and eggs float to the surface (right). Heavy debris is drawn off the bottom by opening the funnel stopcock. Water is added to the supernatant, and eggs sink to the bottom of the funnel. Subsequently, this sediment with eggs is drawn off into a petri dish for counting purposes. (Courtesy J. Tollefson and Iowa State University)

Sieving techniques are mechanical and, unlike the Berlese funnel, do not depend on insect movement for extraction. Both dry and wet sieving techniques have been used to sample insects. Dry sieving employs one or more sieves of varying coarseness, beginning with a coarse mesh, for example, number 20, and ending with a fine mesh, for example, number 80 or 100. First sieves are usually employed to screen out larger soil particles and other debris and to allow insects to pass through. Subsequent screens then are used to collect specific groups of insects, depending on body size. Standardized soil sieves may be purchased and used to sample insects or may be fabricated by constructing a wooden frame and attaching common hardware cloth. Frequently, seedcorn maggot pupae are sampled with "homemade" hardware cloth sieves. Wet sieving follows the same principles as dry sieving, except that water is used to facilitate the movement of particles, including insects, through the sieves. With this technique, the soil sample and sieves may be placed in a container of water, with the water being drained after a given time. Another method is to use a forced stream of water to wash particles through the sieves.

The flotation technique is another widely used extraction procedure that may be used alone or in conjunction with sieving and other techniques. Flotation operates on the principle that particles of lower specific gravity than their environmental medium float on the surface of that medium (like ice cubes float in a glass of tea). In certain instances, for example, extraction of rootworm (*Diabrotica* species) larvae, it is sufficient to use water as the flotation medium. In other situations, however, the efficiency of flotation is increased by adding a salt such as magnesium sulfate to increase the medium's specific gravity. Compromises are usually made in the specific gravity chosen to give reasonable extraction of insects without an inordinate amount of extraneous organic debris. Usually, different salts and concentrations are tried before deciding on an appropriate one.

Indirect Techniques

In all sampling techniques discussed previously, data for estimating population size were expressed as numbers of insects. Estimates can also be made from data on insect effects or products. Such estimates are often called **population indices**.

Measuring the effects of insects is probably the most common method of indexing an insect population. Root rating schemes indicating feeding of corn rootworms, percentage defoliation of soybeans, numbers of plants cut by cutworms, and number of plants with tillers or "deadheart" caused by borers are all examples of index data. Such data, of course, are also paramount in determining the economic impact of insects on the crop. The sampling technique involved with most of these measures is direct observation of results of insect injury and, sometimes,

Figure 6.26 Washing corn roots (top) so they can be rated. A rating scale of one to six is often used (bottom). Such ratings can be used as an index of insect population size. (Courtesy J. Tollefson and Iowa State University)

skilled ranking of the observations. A few techniques like corn rootworm injury ratings require special procedures: in this case pulling up corn plants, washing roots, then rating them according to amount of feeding.

Data on insect products include measures of larval and pupal skins, frass, and nests. With some of the larger insects where skins are conspicuous, as with cicada (Homoptera: Cicadidae) exuviae on tree branches, directly counting these may be an efficient method of sampling. Measuring the amount of frass, or insect excrement, may also be an efficient means of indexing an insect population. The measurement of frass drop in a collecting tray has been used to index population size of several moth (Lepidoptera) and sawfly (Hymenoptera) pests in forests. Nests of colonial insects, like webworms (Lepidoptera: Pyralidae) and tent caterpillars (Lepidoptera: Lasiocampidae) also may be easily counted, with these counts serving to index the population in a locality.

Auxiliary Survey Equipment

In addition to the equipment used to collect insect samples, surveyors or scouts need several other items. Although it is difficult to develop a complete list of these for all sampling programs, the following outline of equipment may be helpful.

A. Clothing—field clothes, comfortable boots, rain suit, rubber boots
B. Containers—knapsack, small plastic bags, plastic vials with alcohol, ice cream cartons, ice chest with ice (for storing samples)
C. Measuring equipment—hand counter, hand lens, tape measure, thermometer
D. Recording materials—clipboard, columned paper or scouting forms
E. Tools—trowel, utility knife
F. Other—mosquito repellent

Of course, not all these materials can be carried by the surveyor on the survey route. However, they should be kept easily accessible, for example, at the edge of the field or in the car.

THE SAMPLING PROGRAM

R. F. Morris, a noted Canadian entomologist, once expressed the opinion that the sampling program is a mixture of art, science, and drudgery and that the art and science are in the sampling design. Design involves selecting a sampling technique and detailing all the steps required to obtain an estimate. Usually, several designs are devised for an insect species or group, and these are tested to determine the most appropriate ones for a particular pest management program. The time and effort expended in obtaining good sampling designs is more than paid back later when the pest management program is implemented.

Kinds of Estimates

Selection of an appropriate sampling technique depends initially on the kind of estimate desired. Generally, estimates can be divided into two broad categories—absolute and relative.

Absolute estimates. Absolute estimates measure actual numbers in the insect population. Such estimates are expressed in numbers per ground surface area, for example, number per acre, number per hectare, number per square meter. Often, these are the most difficult estimates to make and frequently are the most costly. Absolute estimates are of utmost importance in research in insect population dynamics, but because of their cost, they are not used widely in pest management practice. Some examples of techniques used to make absolute estimates include (1) suction traps where the volume of air passing through the trap is known, (2) Berlese funnels where efficiency of extraction is known, (3) vegetation removal and laboratory processing of the sample, and (4) fumigation cage sampling. In all instances when a habitat unit is sampled, numbers of habitat units per ground surface area must be known to make the final absolute estimate. For example, after finding the number of cabbage loopers per head of cabbage, the number of cabbage plants per acre is used to arrive at number of loopers per acre.

If, in the previous example, the first type of estimate meets our needs by providing number of loopers per cabbage head, we would be using another type of estimate—the **population intensity estimate**. Population intensity is the number of insects per habitat unit, and it is an absolute-type estimate. The population intensity estimate is used frequently in pest management programs because it relates closely to crop injury. Such estimates have been particularly useful in establishing economic injury levels, and subsequently, these are expressed in population-intensity terms, for example, number of aphids per leaf, number of borers per stem, number of hornworm flies per side of steer or cow. *In situ* counts and other techniques used for true absolute estimates are used also for population intensity estimates.

Another estimate related to the absolute estimate is the **basic population estimate**. This estimate is intermediate between the absolute estimate and the population intensity estimate. It combines the habitat unit with a unit of measure. In forest entomology number of insects per 10 square feet of branch surface has been used. A common measure in row crops has been number of insects per row foot.

Relative estimates. These estimates differ greatly from absolute and related estimates because they do not translate directly into numbers per ground surface area. Instead, relative estimates are based on the kind of sampling technique used. For example, the sweep-net technique is used to make relative estimates, which are expressed as number of insects per sweep or insects per some fixed number of sweeps. These estimates do not directly answer the question, how many insects are present per acre, per plant, or per foot of row. Rather, they are used to compare population sizes in time and space according to a specific sampling technique. Thus, we can compare potato leafhopper, *Empoasca fabae*, populations per sweep of alfalfa in Baker county in 1983 and 1984, and we can compare the same between Baker and Jones counties in 1984, but we cannot determine, directly, number of leafhoppers per acre for the comparison. Why not? Because the estimate is relative; it depends on the technique used to gather the data. In some instances, relative estimates can be converted to absolute estimates, but considerable research and analysis is required first. Many techniques have been used to make relative estimates including several kinds of netting, all types of trapping, some in situ methods, and knockdown approaches. Relative estimates have the great advantage of being inexpensive and are often the most appropriate for insect pest management.

Population indices, as discussed earlier, are usually employed similarly to relative estimates for population comparisons. They have added value when used as an indicator of existing plant damage and as a basis for management decisions. They are expressed in numerous ways, depending on the insect effect measured, for example, cast skins, webbing, root feeding, leaf feeding. Percentage defoliation of a plant canopy is one of the most popular population indices in insect pest management.

Descriptive Statistics

Whether population estimates are expressed in absolute or relative terms, the proper use of statistics in summarizing data and designing sampling programs is extremely important. Statistics, of course, is the discipline of collecting, analyzing, and interpreting numerical data. Descriptive statistics used in insect sampling involve analysis of data to obtain a comprehensive, quantitative summary of insect population characteristics. Probably the most widely used statistics in describing insect population density are the **mean**, the **standard deviation**, and the **standard error**.

The mean is simply the arithmetic average of the sample numbers. For a set of N samples, where individual sample numbers are represented by $X_1, X_2, X_3, X_4...X_N$, the sample mean $\bar{x}$ is calculated as:

$$\bar{x} = \frac{X_1 + X_2 + X_3 + X_4... + X_N}{N}$$

For example, if six samples are taken and insect counts are 10, 13, 8, 12, 9, and 11, a mean of 10.5 would be obtained. If several of these means are obtained from a number of different sample sets and we calculate a grand mean, as above, from these sample means, we have m, our best estimate of the population size. By computing such values, we then can compare insect populations at different times and in different places.

Although the mean gives us one of the most important characteristics of an insect population—central tendency—it does not tell us anything about variation in the samples, in other words, numerical differences among different samples. This characteristic, referred to as distribution of the data, is most often expressed as the range or as the standard deviation. The range is the simpler of the two expressions, being the difference between the smallest and the largest sample number. Although the range gives some idea as to the spread of data about the mean, it is dependent only on extremes, some of which may be accidental and/or very rare.

The standard deviation of a set of sample numbers eliminates the problem of sole dependence on extremes by averaging deviations of sample counts from the mean value. The standard deviation s of a set of insect samples can be readily calculated on hand-held calculators, many of which are programmed for this statistic. If manual calculation is necessary, the raw data formula for s is most practical. The following example gives the formula and how it is used.

The value 7.40 is one standard deviation and gives us an idea of the magnitude of variation in our sample. If our data were distributed normally (described by a bell-shaped curve), then it would be expected that 68 percent of our sample numbers would fall within one standard deviation on either side of the mean; in other words, 17.8 minus and plus s gives a range from 10.4 to 25.2, and the majority of samples should fall within this range.

Table 6.1. Computation of the Standard Deviation Directly from Raw Data.

Sample Number	Sample Count X	X^2
1	27	729
2	15	225
3	23	529
4	18	324
5	9	81
6	12	144
7	10	100
8	18	324
9	30	900
10	16	256

SUM X = 178   SUM X^2 = 3,612

$$\bar{x} = 178/10 = 17.8$$

$$s = (1/N - 1)\sqrt{N \times (\text{SUM } X^2) - (\text{SUM } X)^2}$$

$$= (1/9)\sqrt{10(3,612) - (178)^2}$$

$$= 7.40$$

Although the standard deviation gives us a measure of variation in a set of samples, it is not our best estimate of the standard deviation of the insect population. For this estimate, we must consider the number of samples taken. This estimate is called the standard error of the mean or simply standard error. It is calculated as

$$s_{\bar{x}} = \frac{s}{\sqrt{N}}$$

where

$s_{\bar{x}}$ = standard error, sometimes designated SE
s = standard deviation of the sample numbers
N = number of samples taken

As can be seen, an important characteristic of the standard error is that its magnitude decreases as sample number increases. This statistic is particularly important in insect sampling programs because it can be used as a measure of precision, and we are striving for the greatest precision practicable in our estimates. Because the lower the standard error, the more precise the sampling program, it is possible to increase sample number and, thereby, increase precision. Greater detail on increasing precision will be discussed later.

Although the statistics discussed tell us something about population density and the variation in our samples, they do not describe the dispersion of insects in a population. Dispersion (not dispersal) is the pattern or arrangement of insects in space, or how they are spread out in an area. An understanding of dispersion is important because it gives us information about population dynamics and may influence the way we sample the population in an area.

As discussed earlier, the most common ways insects are dispersed in an environment are "random" and "clumped" (Figure 5.1). A random dispersion means that an insect has as good a chance of being present in one place as in another, and individuals do not affect each other's presence in a place. Random does not mean that the population has a uniform dispersion over an area. Random dispersions in insects probably occur most often in relatively uniform environments, for example, a field of soybeans on flat terrain. Clumped (contagious) dispersions probably are the most common type in insect populations. Clumping of individuals in a population means that if one insect of the species is found, chances are good that there are others in the same vicinity. Clumping may be caused by either behavioral (for example, mating and feeding) or environmental (for example, uneven habitat) factors or both. Statistically, random and clumped dispersions can be described by mathematical models. The most common models used in entomology are the **Poisson model** to describe a random dispersion and the **negative binomial model** to describe a clumped dispersion. Methods of analyzing dispersion pattern are beyond the scope of this discussion but may be obtained from textbooks that discuss ecological methods (for example, Southwood 1978, Andrewartha 1961).

Criteria of Estimates

Estimates obtained from newly designed sampling programs must be evaluated for their usefulness before final recommendations are made. Usually, several preliminary sampling plans are devised and used, with the data from these trials employed in the evaluation.

The major criteria for these evaluations are (1) fidelity, (2) precision, and (3) cost. **Fidelity** is the accuracy with which the estimates follow actual numbers in the insect population. Fidelity is usually determined by comparing a program of unknown fidelity with one known to be accurate, for example, numbers known through releases in an area or habitat removal sampling.

Precision measures the degree of error in making estimates and is usually expressed as percentage standard error of the mean. This value, known as relative variation or RV, is calculated as follows:

$$RV = (SE/\bar{x})\,100$$

where SE is the standard error of the mean and $\bar{x}$ is the mean.

With RV as a measure of precision, a good criterion for practical pest management programs is to obtain values near 25. For population research, greater precision is desirable and values near 10 are sought. Because SE decreases with increasing numbers of samples, precision can be improved by taking more samples.

The remaining, often overriding criterion is cost. Population estimates made with great fidelity and precision are useless if they are too expensive. Therefore great attention is given in accounting for costs of tentative sampling programs, and final design recommendations are those that yield the greatest fidelity and precision for the lowest cost.

Program Dimensions

After determining the kinds of estimates needed for a pest management program and sampling criteria, several programs are usually designed and tried. The design of a program involves at least four basic dimensions: (1) insect stage to sample, (2) number of samples to take, (3) time to sample, and (4) pattern of sampling.

Insect stage. In pest management, the insect stage sampled depends on the particular insect involved and on the management strategy for that insect. In making management decisions, the stage most often sampled is the damaging stage. This is especially true when little lead time is needed to suppress the population. When substantial lead time is needed, a stage prior to the damaging stage may be selected so that early predictions can be made, for example, sampling the nonfeeding adult stage to predict the subsequent damaging larval population.

Number of samples. The number of samples taken in a program is usually the result of a compromise between precision on the one hand and cost on the other. In many instances, the most affordable number is the prime determinant. When greatest affordable number of samples yields too little precision, the program should be modified or abandoned in favor of other designs. Considerable research may be required to find a program where number of samples needed is economically feasible. As a rule of thumb, many pest management sampling programs call for taking at least five samples in an average-size cropping area (about 35 acres). When greater precision is desired, sample number should be increased. A useful formula to calculate number of samples required to achieve a given degree of precision is as follows:

$$N = \left[(t \times s)/(D \times \bar{x}) \right]^2$$

where

N = number of samples required
$\bar{x}$ = mean density
D = required precision expressed as a decimal
      (for example, 0.25);
s = standard deviation
t = student's t value = a value obtained from
      statistical tables (usually the value of t for
      0.05 probability and sample number used to
      calculate the mean).

The mean and standard deviation values in the formula are obtained by using data from preliminary sampling.

Time to sample. The timing of sampling depends on the occurrence of the insect stage being sampled. For management decisions, it is most important for the sampling to coincide with the time of peak numbers of the stage. Such timing may be predicted using degree-day or other phenological models or by trial sampling. To trial sample, a conservative date of appearance is chosen, and samples are taken regularly (once or twice per week) until a peak is obvious from the cumulative data, in other words, successive dates of increase followed by successive dates of decrease. The greatest problem with the latter approach is that unacceptable damage may occur before a peak is detected. This problem does not occur if sampling is conducted early and frequently, and management decisions are based on threshold values before a potential peak occurs. It has been suggested that outdoor cages containing a control population be set up for observations and that timing of sampling be gauged from these observations. Time of day for sampling may also be a problem with some insects. If the sampling technique depends on a certain insect activity (for example, flight) and that activity occurs mostly at a certain time of day (for example, dusk to dawn) then samples should be taken during that activity period.

Pattern of sampling. The spatial sampling pattern at a location can vary widely, depending on the uniformity of the habitat and population dispersion. In the strictest sense, unbiased estimates of population density should be taken "at random"; that is, every sampling spot in the area has an equal chance of being selected for sampling. For this sampling, two random numbers (measurements of distance) along two coordinates in the area determine the sample site, for example, coordinate 1 = row 12 and coordinate 2 = 10 meters down the row.

Often, such sampling is too laborious (= expensive) for pest management surveillance, and subjective selections (= less expensive) of sampling sites are usually made. Usually, pest management scouting calls for walking a prescribed route through a growing area and taking samples somewhat regularly along this route. Such sampling is known as systematic sampling. Samples from such pattern are usually treated as if they were random in later analyses. This approach is more efficient when sampling many growing areas in a single day and is satisfactory unless cumulative errors arise because of the sampling pattern, for example, continually sampling sites of very low or very high density because of the pattern.

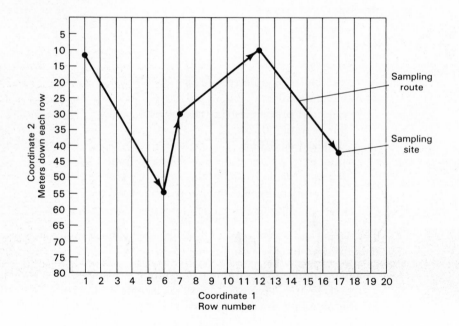

Random sampling program

Sample	Coordinate 1	Coordinate 2
1	12	10
2	7	30
3	1	12
4	6	55
5	17	42

Figure 6.27 An example of a random sample pattern using two coordinates. The coordinates are selected from a table of random numbers or generated by computer.

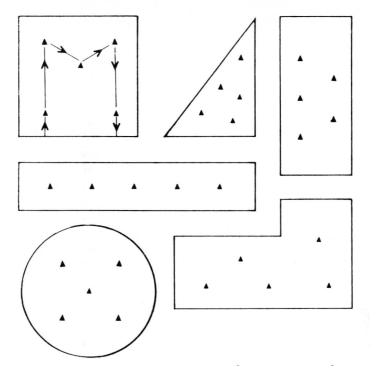

Figure 6.28 Some common patterns of systematic sampling. (After *Corn Pest Management for the Midwest*, North Central Publication 98)

In early stages of sampling-program development, it is good to test for cumulative errors by comparing strict random patterns with systematic patterns. As a rule of thumb, spread samples out over an area, making sure ridges and low areas are included. The edges of a growing area (first five meters or so) are usually avoided in making estimates unless the pest species normally invades and/or is more important at the edge, for example, armyworms (Lepidoptera: Noctuidae), grasshoppers (Orthoptera: Acrididae), and stalkborers (Lepidoptera: Noctuidae). Some common patterns of sampling are shown in Figure 6.28.

Pest Management Scouts and Scouting Records

Samples for making pest management decisions in agriculture may be taken by entomologists or other specialists, and reports are made to grower or grower agencies. Also growers may do their own sampling or supplement the sampling of specialists. Persons who sample growing areas regularly for the purpose of making pest management decisions are called pest management scouts, and their activity is called scouting. Therefore, we often hear of such descriptions as a "rootworm scouting program" or such activities as "scouting an apple orchard." Many pest management scouts are employed by private consulting firms that contract with growers to sample plantings for pests and provide timely advice.

To provide this advice, scouting records must be kept, and data must be analyzed to demonstrate pest population trends and make predictions. These records include not only the insect species present and its numbers but also may include information like field location, insect growth stage, current weather conditions, degree of damage to date, crop sampled, variety grown, crop condition and growth stage, and other pests present. An historical account of these factors helps in understanding both insect population dynamics and potential impact on crop yield. The records also help in interpreting the quality of the sampling estimates themselves because many of the factors recorded affect sampling efficiency. A typical scouting form is shown in Figure 6.29.

Alfalfa Weevil Sampling: An Example

Alfalfa weevils, particularly larvae, are early-season defoliators of alfalfa in the eastern half of the United States, causing significant damage to the first-cut crop of the season. Management of this insect involves scouting and, when appropriate,

BOX 6.5 ALFALFA WEEVIL

SPECIES: *Hypera postica* (Gyllenhal) (Coleoptera : Curculionidae)

DISTRIBUTION: The alfalfa weevil was originally imported into North America from southern Europe around 1900. In 1904, the first alfalfa weevils were discovered in Utah. The species was found in Maryland in 1951, as a result of a second introduction. Currently the pest is found throughout the United States and regions of Canada and Mexico. A similar species, the Egyptian alfalfa weevil, *Hypera brunneipennis* (Boheman), occurs in the western United States.

IMPORTANCE: The alfalfa weevil is one of the most serious alfalfa pests in North America. The species primarily damages first-growth alfalfa, but the second growth can also be damaged. The most significant injury is caused by larvae that damage the top growth of the plant by entering and feeding on the developing buds, and by feeding on the leaves. Injured plants appear frosted, owing to dried, damaged leaves. Adult weevils feed on older, lower leaves of alfalfa. Growth is stunted in infested alfalfa, resulting in both reduced quality and yield. Also, injured alfalfa is unsuitable for seed production.

APPEARANCE: Adults are about 5 mm long and are light brown, with a dark brown stripe running down the center of the back. As adults age, their color varies from brown to gray. Larvae are pale green with a black head and are 10 mm long when full-grown. They have a distinctive white stripe running down the back of the body. The eggs, laid in clusters of about 10, are initially yellow but turn black before hatching.

LIFE HISTORY: Adults overwinter in plant debris around the edges of fields; however, some individuals overwinter under leaf litter in wooded areas. In early spring, adults become active and begin feeding on alfalfa leaves. Females oviposit 400 to 1,000 eggs from April to May. Initially, egg clusters are deposited in dead alfalfa stems and field litter, but later they are laid in holes chewed into the live stems by the females. Eggs hatch in 7 to 21 days, depending on temperature. After three molts (25 days), larvae spin a cocoon that is either attached to alfalfa leaves or attached to plant debris on the ground. In 7 to 14 days, adults emerge and feed on alfalfa for a short period of time. Then the adults aestivate for the duration of the summer, becoming active in the fall only to prepare for winter hibernation. Typically, there is one generation per year, but some experts believe two generations occur in specific regions.

CORN INSECT SURVEY (Example)

Grower _____ County _____ Date _____

Township _____ Section _____ Field No. _____ Acres _____

Crop stage _____ Extended leaf height _____

PLANT POPULATION: (count number of plants in 20 feet at 5 locations)

Set 1	Set 2	Set 3	Set 4	Set 5	Total		
___ +	___ +	___ +	___ +	___ =	___	X A =	Row width _____
							PLANTS PER ACRE _____

Row width "A" factors: 30" = 174.24; 36" = 145.2; 38" = 137.561; 40" = 130.681

GENERAL FORM_____(Black cutworms, corn borer; for miscellaneous insects encountered, use this section with a sample size of 20 plants.)

	Set 1	Set 2	Set 3	Set 4	Set 5	Total	
Damaged plants/set	___ +	___ +	___ +	___ +	___ =	___	÷ C = % infestation _____
No. of insects	___ +	___ +	___ +	___ +	___ =	___	÷ D = No. of insects _____
Size	___	___	___	___	___		Size of insects _____

BLACK CUTWORM: Use General Form above with 50 plants per set, count worms around all cut plants per set, size worms in inches, C and D = 2.5.

Threshold: When 3% or more of the plants are being cut and worms are ½" - 1" in length, treatment may be necessary.

EUROPEAN CORN BORER: Egg masses/25 plants_____X 4 = Egg masses per 100 plants_____

Also use General Form above with 20 plants per set; count the number of insects on 2 infested plants per set; size is in instar (IV); C = 1; D = 10.

1st Brood   Tassel_____ ÷ Plant_____ = Tassel_____

            height_____ ÷ height_____ = ratio _____   TR avg. _____

Threshold **1st Brood:** If 50% or more of the plants have whorl feeding and live worms are present, treatment may be justified. **2nd Brood:** treatment may be justified if there are 1 or more egg masses or larvae per plant.

CORN LEAF APHIDS: Each set consists of 10 plants. Note beneficial insects found.

	Set 1	Set 2	Set 3	Set 4	Set 5	Total	
None present-few	___ +	___ +	___ +	___ +	___ =	___	× 2 = % low_____
▶50/Plant	___ +	___ +	___ +	___ +	___ =	___	× 2 = % moderate-high_____
tassel covered	___ +	___ +	___ +	___ +	___ =	___	× 2 = % tassel covered _____

Threshold: If 50% or more of the plants have 50 or more aphids per plant, treatment may be justified; also if 3% of the plants have the tassels covered and corn is under moisture stress.

CORN ROOTWORM BEETLE POPULATIONS: Each set consists of 5 plants for stalks (_____), and 5 different plants for ear zone_____

	Set 1	Set 2	Set 3	Set 4	Set 5	Total	Ear zone	Stalk
NCR	( ) +	( ) +	( ) +	( ) +	( ) =	___	× .04 = NCR's/_____	(_____)
WCR	( ) +	( ) +	( ) +	( ) +	( ) =	___	× .04 = WCR's/_____	(_____)

All ear zones + stalks combined = Beetles/plant _____

Threshold: If corn rootworms beetle populations exceed 5 per plant and silks have not begun to turn brown, pollination may be affected.

NOTES:

Figure 6.29 Portion of a typical scouting form for insect pest management using that for corn as an example. (After *Corn Pest Management for the Midwest*, North Central Publication 98)

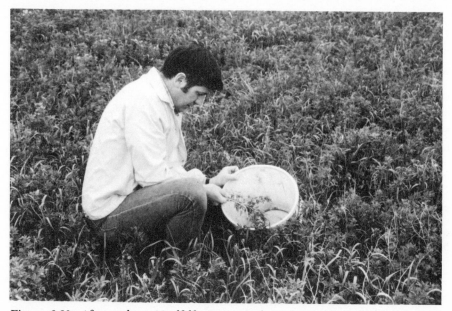

Figure 6.30 After picking 30 alfalfa stems, each is shaken vigorously over a pail to dislodge alfalfa weevil larvae. Number of larvae in the pail are then counted. (Courtesy Iowa State University Extension Service)

either early cutting or application of insecticides. In fact, scouting is the crucial element in this pest management program.

The typical scouting program in northern regions (areas where most weevil eggs are laid in the spring) involves keeping a record of accumulated degree days (48°F, minimum threshold temperature) from January 1. When 300 degree days have accumulated, the sampling program is begun, with samples taken approximately weekly. To scout a field, 30 stems are picked at equal intervals across a field. A U-shaped pattern is walked to pick the stems, and each stem is shaken vigorously over a pail to dislodge larvae. After removing larvae, the length of every third stem is measured. After all samples have been taken, larvae are transferred to a shallow pan and counted. The total larval count is divided by 30 to obtain an average, and plant height is also averaged. Using average larvae per stem and average plant height, tables (Table 6.2) are consulted to obtain pest management recommendations. Information in the table will advise the grower to spray, to harvest the crop, or to wait and resample the field in a specified number of days.

This good example of effective sampling design has been widely adopted for use in alfalfa weevil management. Such programs are the backbone of insect pest management practice.

Sequential Sampling

An increasingly important type of sampling program in insect pest management is the sequential sampling or sequential decision program. Such programs are based on insect dispersion patterns and economic decision levels. Sequential sampling was first developed during World War II for the inspection of war goods; it allows the placement of a given population into one of two or more categories. For

Table 6.2 Tables for making pest management decisions from a 30-stem sample. (After Iowa State University Cooperative Extension Service Publication IC-428, 1986.)

Table A. Alfalfa weevil pest management recommendations—Number of larvae collected from a 30-stem sample.[a]

Total degree days (dd)	Alfalfa height (inches)																
	2	3	4	5	6	7	8	9	10	11	12	13	14	15	16	17	18 or more
290–310																	
Spray		25	37	52	67	75	83	94	105	105	105						
Resample in 50 dd		0-24	0-36	0-51	0-66	0-74	0-82	0-93	0-104	0-104	0-104						
340–360																	
Spray					82	82	82	82	82	82	82	82	82	82	82		
Resample in 50 dd					14-81	14-81	14-81	14-81	14-81	14-81	17-81	17-81	17-81	17-81	17-81		
Resample in 100 dd					0-13	0-13	0-13	0-13	0-13	0-16	0-16	0-16	0-16	0-16	0-16		
390–510																	
Spray										52	52	58	64	68	72	76	80
Resample in 50 dd										8-51	8-51	8-57	14-63	14-67	14-71	18-75	18-79
Resample in 100 dd[b]										0-7	0-7	0-7	0-13	0-13	0-15	0-17	0-17
540 to harvest (see table B)																	
100 after harvest																	
Spray	23	33	43	48	53	58	63										
Resample in 50 dd	17-22	17-32	17-42	20-47	23-52	23-57	23-62										
Resample in 100 dd[c]	0-16	0-16	0-16	0-19	0-22	0-22	0-22										
150 or more after harvest (see table B)																	

Table B. Alfalfa weevil pest management recommendations[a]

Total degree days (dd)	Change in number of larvae since last sample		
	Decreased 10 or less	within 10	Increased 10 or more
540 to harvest			
Spray or harvest	73	63	53
Resample in 50 dd	23-72	18-62	13-52
Resample in 100 dd[b]	0-22	0-17	0-12
150 or more after harvest			
Spray	78	58	48
Resample in 50 dd	28-77	18-57	0-47
Quit sampling	0-27	0-17	

SOURCE: From University of Illinois Circular 1136.
[a]If this field was sprayed more than 7 days ago, you can wait 200 degree days to resample.
[b]If last preharvest sample had less than 30 larvae, the weevil season is over and you can quit sampling.

example, they efficiently tell the pest management scout if the population is an outbreak type requiring management or is endemic type requiring no action. Any good sampling program can do this if the economic decision level is known, but sequential sampling usually allows the decision to be made with fewer samples. Therefore, sequential sampling saves money in the pest management program.

With sequential sampling, the total number of samples to be taken in a field is variable and is not known when the sampling is begun. Using a table (or less often, a graph), one can start sampling and continue until the population can be classified. The number of samples required is relatively few if the insect population is either very large or very small. More samples are required when the population has a density near the dividing line between endemic and outbreak levels. This means we can be confident of classifying a population with only a few samples when it is at its high or low extremes, but we have to take more samples to be sure of our classification when it is at an intermediate level. Sequential sampling tells us how many samples to take before we can be confident about making a pest management decision.

An example of a sequential sampling table is shown using the green clover-worm, *Plathypena scabra*, program for soybeans. This insect defoliator is sampled by shaking 60 cm of soybean row (30 cm on each side of the scout) over a shake cloth and counting larvae in each sample. Sample numbers are accumulated and a table is consulted with the cumulative number (Figure 6.32). If the cumulative number is less than the lower critical value, sampling is discontinued and no action is recommended. If the cumulative number is greater than the higher critical value, sampling also is discontinued, and the population is treated with an insecticide. If the cumulative number is between the upper and lower critical values, another sample is taken. Sampling is continued in this manner until a decision is made, or 10 samples are taken in the field. If a decision cannot be made after 10 samples are taken, the scout stops sampling and returns to the field in two or three days to classify the population.

Sequential sampling plans have been developed for many insect pests and development continues for many others. Although intermediate population levels may still require considerable sampling, average savings of 50 percent can be expected of successful sequential sampling programs.

Figure 6.31 Graph of a sequential sampling program for a hypothetical pest. In the graph, samples were taken as long as the cumulative numbers were within the indecision zone. As soon as a number fell outside this zone, sampling was discontinued, and appropriate action was recommended. In field A, the pest population was classified as "economic," and, for example, insecticide application would be necessary. In field B, the population was classified "noneconomic," and no action would be advised.

Sequential decision plans for GCW larval management in soybeans

n^b	Cumulative total no. of GCW larvae[a]														
	Soybean stage V7–R1[c]					Soybean stage R2–R3[d]					Soybean stage R4–R5[e]				
		<		>			<		>			<		>	
1		—		—			—		—			—		—	
2		—		—			—		—			—		—	
3		37		71			67		113			52		92	
4	Noneconomic population	53	Continue sampling	91	Treatment required	Noneconomic population	94	Continue sampling	146	Treatment required	Noneconomic population	73	Continue sampling	119	Treatment required
5		68		112			121		179			94		146	
6		84		132			148		212			116		172	
7		100		152			175		245			138		198	
8		117		171			203		277			160		224	
9		133		191			231		309			182		250	
10		150		210			259		341			204		276	

[a]Cumulative number of first- to sixth-stage GCW larvae.
[b]n = Number of 60-cm plant shake samples.
[c]V7–R1 = Late vegetative to early flower bloom.
[d]R2–R3 = Late flower bloom to early pod development.
[e]R4–R5 = Late pod development to early bean development.

Figure 6.32 A table for use in sequential sampling of the green cloverworm (GCW), *Plathypena scabra*.

Selected References

ANDREWARTHA, H. G. 1961. *Introduction to the Study of Animal Populations*. Chicago: University of Chicago Press.

BECHINSKI, E. J. , G. D. BUNTIN, L. P. PEDIGO, and H. G. THORVILSON. 1983. Sequential count and decision plans for sampling green cloverworm (Lepidoptera: Noctuidae) larvae in soybean. *Journal of Economic Entomology* 76:806-812.

KOGAN, M. , and D. C. HERZOG, eds., 1980. *Sampling Methods in Soybean Entomology*. New York: Springer-Verlag.

MORRIS, R. F. 1955. The development of sampling techniques for forest insect defoliators, with particular reference to the spruce budworm. *Canadian Journal of Zoology* 33:225-294.

RUESINK, W. G. , and M. KOGAN. 1982. The quantitative basis of pest management: sampling and measuring. In R. L. Metcalf and W. H. Luckmann, eds., *Introduction to Insect Pest Management*. New York: Wiley, pp. 315–352.

SOUTHWOOD, T. R. E. 1978. *Ecological Methods with Special Reference to the Study of Insect Populations*, 2nd ed. New York: Chapman and Hall.

7 ECONOMIC DECISION LEVELS FOR PEST POPULATIONS

WITHOUT A DOUBT, economic decision levels are the keystone of insect pest management programs. Such levels are indispensable because they indicate the course of action to be taken in any given pest situation. D. L. Chant, in 1966, termed the study of economic decision levels "bioeconomics" and argued that sensible pesticide use is possible only with an understanding of the insect population level that causes economic damage. Indeed, without such knowledge we risk making absurd economic blunders, like spending more to suppress an insect than the value of the commodity the pest could destroy. Conversely, understanding and properly using economic decision levels in dealing with pests can increase producer profits and conserve environmental quality.

Economic decision levels usually are expressed as number of insects per area, plant or animal unit, or sampling procedure. Less commonly, such levels are given as degree of plant damage or combinations of numbers and damage. The levels are unique in that they have both biological and economic attributes, and they are most often used for management decisions for private concerns, for example, on the farm.

CONCEPTS OF ECONOMIC LEVELS

In 1959, V. M. Stern and colleagues formally proposed the concepts and terminology of bioeconomics that we use today. Specifically, they developed the ideas of **economic damage**, **economic-injury level**, and **economic threshold**, collectively called the **economic-injury level concept**. Although this concept was originally proposed in 1959, some of the ideas expressed had been discussed years earlier (1934) by W. D. Pierce in a particularly farsighted article. Here, Pierce raised questions that became one incentive for developing economic-injury levels (EIL) by asking, "Is all insect attack to be computed as assessable damage? If not, at what point does it become assessable? Is control work warranted when damage is below that point?"

Although fundamental to the concept, such questions may not have been the ultimate impetus for EIL development. The concept actually emerged as an encouragement for more rational use of insecticides. In discussing the EIL concept, Stern and colleagues emphasized the concerns of many persons regarding excessive and other inappropriate uses of insecticides. They highlighted problems of insecticide resistance, residues, and effects on nontarget organisms. These ideas were a critical part of the concept of **integrated control**, a new approach at the time, recommended as a replacement for the overly simplistic strategy of "identify and spray."

Economic Damage and the Damage Boundary

Economic damage was originally defined as "the amount of injury which will justify the cost of artificial control measures." To understand this term, we distinguish between injury and damage. **Injury** is the effect of pest activities on host physiology that is usually deleterious. **Damage** is a measurable loss of host utility, most often including yield quantity, quality, or aesthetics. Therefore, injury is centered on the pest and its activities, and damage is centered on the crop and its response to injury.

As the concept is applied in pest management, economic damage begins to occur when money required for suppressing insect injury is equal to the potential monetary loss from a pest population. The term **gain threshold** has been used to express this beginning point of economic damage. The gain threshold can be expressed as follows:

$$\text{Gain threshold} = \frac{\text{Management costs (\$/acre)}}{\text{Market value (\$/bushel)}} = \text{bushels/acre}$$

Figure 7.1 Diagram relating insect leaf injury to damage of the harvestable produce.

Here, gain threshold is expressed as a unit of measure of the marketable product per a specified land area. For example, if management costs for application of an insecticide are $10 per acre and harvested corn is marketed for $2 per bushel, the gain threshold would be five bushels per acre.

$$\text{Gain threshold} \; = \; \frac{\$10/\text{acre}}{\$2/\text{bushel}} = 5 \text{ bushels/acre}$$

In other words, at least five bushels per acre would need to be saved with an insecticide application for the activity to be profitable.

The gain threshold, therefore, is a very important measure because it is a worksheet standard: our margin for determining benefits of management and establishing decision indices.

Although not recognized by Stern and his colleagues, another important damage level to consider is the **damage boundary**, defined as the lowest level of injury where damage can be measured. The level is reached before economic damage occurs and is a necessary complement to the idea of economic damage. Together, these principles allow Pierce's original questions to be answered. Specifically, no injury level below the damage boundary merits control; and injury estimated to result in economic damage does warrant action.

Economic-Injury Level

The **economic-injury level** (EIL) is defined as the lowest number of insects that will cause economic damage, or the minimum number of insects that would reduce yield equal to the gain threshold. Although expressed as numbers of insects per unit area, as its name implies the EIL is really a level of injury. Because injury is usually difficult to measure in a field situation, however, num-

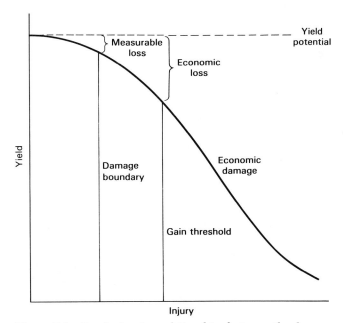

Figure 7.2 Graph showing relationships between the damage boundary and the gain threshold.

bers of insects are used as an index of that injury. For example, it is usually easier to count insect numbers than it is to estimate the area of foliage removed by a pest population or the amount of juices (photosynthates) sucked from plants.

In some instances, particularly when several pest species causing similar injury are present, **insect equivalents** may be considered instead of insect numbers. An insect equivalent is the amount of injury that could be produced by one pest through its complete life cycle. By understanding feeding activity of the various species and comparing these according to a standard, EILs can be developed and decisions made for managing the whole complex. (See Pedigo et al. [1986] for more details on insect equivalents.)

To understand the development of an EIL using conventional insect numbers, consider the previous example of a 5-bushel/acre gain threshold for pest management in corn. If 1 insect/plant causes 1 bushel/acre loss, then the EIL for the pest is 5 insects/plant. In this example, 5 insects/plant potentially could consume enough plant tissue to reduce yields by 5 bushels/acre. Therefore, such an insect population is considered economic, and management activities are justified. Insect populations below this level and whose potential growth will not allow them to reach this level are considered subeconomic; no management is advised.

If management action (insect suppression) can be taken quickly and loss can be averted completely, then the EIL can be expressed as follows:

$$V \times I \times P \times D = C$$

where

V = market value per unit of produce (for example, \$/acre)
I = injury units per insect per production unit (for example, percent defoliation/insect/acre, expressed as a proportion)
P = density or intensity of insect population (for example, insects/acre)
D = damage per unit injury (for example, bushels lost/acre/percent defoliation)
C = cost of management per area (for example, \$/acre)

then

$$P = \frac{C}{V \times I \times D}$$

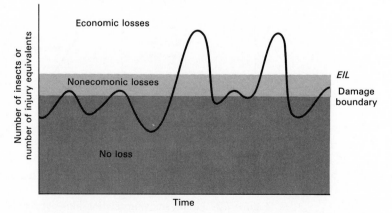

Figure 7.3 Graph showing relationships of a hypothetical insect population with the damage boundary and the economic-injury level.

and

$$EIL = P$$

In instances where some loss from the insect is unavoidable, for example, if injury can be reduced only 80 percent, then the relationship becomes:

$$P = \frac{C}{V \times I \times D \times K}$$

where

K = proportionate reduction in injury (for example, 0.8 for 80%)

With some insect pests, particularly piercing-sucking insects, the separation of the I and D variables presents a problem. This is because the I variable would represent, for plants, photosynthate removed per insect and the D variable would represent yield loss per unit of photosynthate removed. Because these variables would be difficult to measure, a coefficient b representing loss per insect is substituted. These b coefficients are obtained from statistical regression analyses of data by using experimental populations and measuring yield losses. The b coefficient can be obtained from the following expression:

$$Y = a + bx$$

where

Y = yield/area
a = a constant (the y intercept)
b = yield loss/insect
x = number of insects/area

and therefore,

$$EIL = \frac{C}{V \times b} \text{ or } EIL = \frac{C}{V \times b \times K}$$

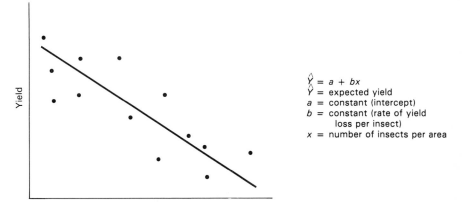

Figure 7.4 A linear-regression used in obtaining b, the loss per insect.

When the relationship between y and x variables cannot be approximated by a straight line (it is curvilinear), a more complex form of the EIL equation must be used.

Economic Threshold

The **economic threshold** (ET) is probably the best-known term and most widely used index in making pest management decisions. The ET indicates the number of insects (density or intensity) when management action should be taken. For this reason, it is sometimes called the **action threshold**. Although expressed in insect numbers, the ET is really a **time parameter**, with pest numbers being used as an index for **when** to implement management. Just as with EILs, ETs also can be expressed in insect equivalents.

If a pest population is growing as the season progresses, growth rates are predicted, and the ET is set below the EIL. By setting the ET at a lower value, we are predicting that once the population reaches the ET, chances are good that it will grow to exceed the EIL. Therefore, it is appropriate for us to take action on an earlier date (before we accrue losses in reaching the EIL).

The relationships between the EIL and the ET are shown in Figure 7.5, where it can be seen that action was taken when a population level exceeded the ET, and the population was forced down before it could reach the EIL. No action was taken at levels below the ET.

From this discussion, it should be clear that the ET is a complex value. It is based on the EIL, a value of economics and potential for injury. However, it also relies on an understanding of population dynamics because potential growth rates need to be predicted. In instances where population dynamics are poorly understood (we cannot predict growth), the ET may be set equal to the EIL or arbitrarily below it. An ET lower than the EIL has the great advantage of giving us time to respond to a problem developing in a field or other habitat. For the previous example with an EIL of 5 insects/plant, we might set the

Figure 7.5 Graph showing the relationship between the economic threshold (ET) and the economic-injury level (EIL) in taking action against an insect population.

ET at 4 insects/plant to obtain added reaction time and avoid some of the early losses.

DYNAMICS OF ECONOMIC-INJURY LEVELS

Economic levels are very dynamic. They change with changes in costs, values, and production environments. An insect pest feeding on a crop at a given time can be expected to have a different EIL when feeding on the crop at another time the same season or in another season. Variations in the levels may be insignificant at times or vary severalfold at other times. The major forces behind change in economic decision levels, as shown in the previous equations, are (1) crop value, (2) management costs, (3) degree of injury per insect, and (4) crop susceptibility to injury.

Although the relationships among these factors are relatively straightforward, complexity is evident when attempting to predict variability in the factors themselves. As shown in Figure 7.6, the primary factors are affected by complex

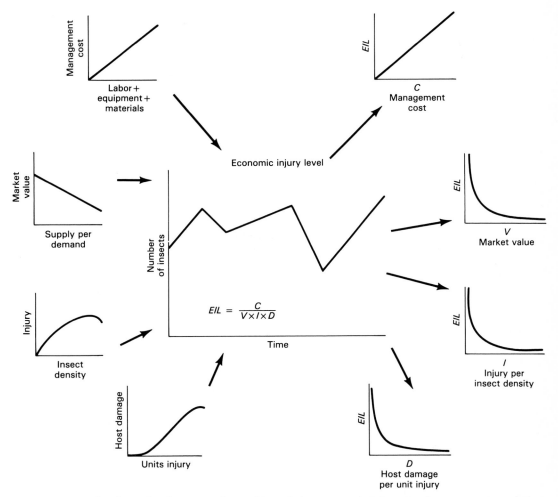

$$EIL = \frac{C}{V \times I \times D}$$

Figure 7.6 Stylized graphs showing relationships of the economic-injury level components and their variables. Primary components (graphs to left) are affected by secondary variables like insect density and labor costs. These variables cause fluctuations in the economic-injury level (EIL) with time (center graph). Resulting responses of the EIL to changes in primary components are shown in graphs at right.

secondary variables like the host-damage/insect-injury relationship. Not shown in the figure are tertiary variables like weather, soil factors, biotic factors, and the human social environment that change the function of the secondary variables. Consequently, the primary factors are not simple constraints; rather, they are complex processes that operate through time. In dealing with these components, entomologists have tended simply to account for the economic aspects and conduct research on the biological components.

Market Value

Of the primary factors, crop value is one of the most variable, and it alone accounts for much of the change in EILs. Crop market values are notorious for fluctuations and unpredictability. For instance, grain prices may fluctuate 70 to 80 percent in a single year, with a profound influence on EILs of insect pests. The relationship between EIL and market value is inverse; as market value increases, EIL decreases and vice versa.

As a general rule, estimates for EIL calculation are based on current or past records of crop value. But these values should be forecasted for the anticipated date of sale to reflect the expected increases or decreases. If, for example, the value of a crop is forecasted to increase from one year to the next, then the EIL for the forecast year should be reduced appropriately.

The quality of a commodity may be of overriding importance in determining market values in many instances. In situations where several specific grades (and prices) exist, the value of the desired grade should be used in the EIL calculations. Similarly, when quality is associated with appearance (cosmetics), the level of control should reflect the desired appearance. A target price, based on the expected quality of production, should be used whenever feasible.

When no clear system of marketing exists for a product, some estimate about the utility of the output to the producer must be made. Forage and pasture crops, for example, often have no established markets but are necessary inputs in animal production. Because most forage and pasture crops are produced for on-farm use, their value depends on the relative contribution that they make to the growth and production of the animal. One way to estimate this value is to determine the substitution price for other, more marketable feeds. To be valid, however, these must be equal in nutritional quality to the on-farm feeds.

As with each of the other primary factors of the EIL, variability in market values cannot be controlled by the producer. In many instances, however, an effective marketing plan can be used to assist in the estimation of this factor.

Management Costs

The cost of suppressing a pest population must be estimated before profitability of an action can be assessed. As management cost increases, net benefit of control decreases. Consequently, EILs must be raised to accommodate the higher gain thresholds.

Most years, management costs tend to be more stable than crop market values. These costs include labor (scouting and application of insecticides), materials (insecticides), and equipment (hydraulic sprayer and sweep nets). Management costs usually change gradually, depending on inflation rates, and therefore are usually predictable.

Because the expenses associated with equipment and labor may equal or exceed the cost of the control material, care should be taken to formulate the best estimate for these application expenses. If, for example, airplane application at

$8.50/acre must be substituted for the usual ground application at $5.50/acre, the additional cost should be reflected with a higher EIL. The variability in cost associated with a particular management measure probably is not significant within a season but could be between seasons. Therefore, most of the uncertainty with this variable in the EIL is caused by differences in costs between management options rather than between seasons. For this reason, management costs for calculating EILs are probably the simplest of the primary factors to estimate.

Degree of Injury per Insect

Injury is a dual-sided phenomenon, governed both by insect and by host populations. The insect aspect concerns a particular act or behavior of individuals

BOX 7.1 BLACK CUTWORM

SPECIES: *Agrotis ipsilon* (Hufnagel) (Lepidoptera: Noctuidae)

DISTRIBUTION: The black cutworm is widely distributed throughout the world, including the United States and southern Canada. Northern regions of the United States must be recolonized each year by migrant moth populations in southern regions where overwintering is possible. Outbreaks within infested regions are often associated with high moisture conditions. Hence, river bottoms or low areas in fields are often very susceptible.

IMPORTANCE: The species is chiefly a pest of seedling plants, causing damage to many species. Most notable plants attacked are corn, cotton, tobacco, and various vegetables. In most instances, damage is seen as a "cutting" of young seedlings, often causing death. This type of damage may be very destructive because several plants can be cut by a single larva. In fact, relatively small populations of cutworms are capable of destroying entire stands of cotton or corn. When seedlings are large enough to avoid cutting, foliar feeding also reduces plant vigor and yield. One practical limitation to controlling black cutworm injury is that larvae live under the soil surface and feed at night. This makes assessment difficult, especially because much damage can be sustained in a short period.

APPEARANCE: Adult moths have a wingspan of about 38 to 50 mm. Forewings are long, narrow, and usually darker than hindwings. In addition the forewings are marked with three characteristic black dashes, or "daggers." Larvae vary from light gray to black. Full-grown larvae are 38 to 50 mm long. Eggs are ribbed and about 0.45 mm high.

LIFE CYCLE: Pupae overwinter in the southern United States and Mexico. In spring, adults emerge, and females oviposit on plant foliage or on the soil surface in the proximity of hosts. In northern latitudes adults are carried north by southerly winds, and females oviposit in the soil near host plants. Larvae hatch in three to six days and move into soil where they remain during the day. At night, larvae move to the surface and feed on young plants. They develop through six stadia (occasionally seven). Depending on temperature, larvae begin to pupate in 25 to 35 days. Pupation occurs in the soil and lasts about 12 to 15 days. Several generations per year are possible, but the spring generation, which places eggs near the young plants (usually the first), is the most damaging.

that causes impairment of a host's ability to survive, grow, and reproduce. The host, as the recipient of the behavior, plays a major role in determining the kind and degree of injury.

Most insect injury is caused by insect **feeding** on host tissues or fluids, although other major causes include **injecting toxins** and **vectoring pathogenic microorganisms**. Chewing and sucking are the most common feeding behaviors, producing injuries to plants such as leaf skeletonizing, leaf mining, stem boring, and fruit scarring.

Some of the most detailed studies of insect injury have been conducted with plants. Here, pests can be placed into at least six different categories: stand reducers, leaf-mass consumers, assimilate sappers, turgor reducers, fruit feeders, and architecture modifiers. These categories include pests that kill plants outright or impair physiological processes.

Stand reducers.   Insects that reduce stand (for example, cutworms, Lepidoptera: Noctuidae) produce an immediate loss in plant biomass and decreased photosynthesis of the crop. Depending on the severity of plant loss, however, compensation in the stand occurs when uninjured plants adjacent to dead plants grow and yield at increased rates. For instance, under some conditions soybean stands reduced to 4 or 5 plants per foot of row may yield as well as stands with 10 plants per foot of row. The total effect of stand reduction is governed by number, timing, and dispersion of plants lost.

Leaf-mass consumers.   Leaf consumption by insects is believed to directly affect absolute photosynthesis per unit of the remaining leaf tissue. Effect of the injury on plant physiology can be accounted for by measuring leaf mass consumed per

Figure 7.7 Black cutworm, *Agrotis ipsilon*, injury to field corn. Such insects can reduce plant stand. (Courtesy Iowa State University Extension Service)

unit land area, timing of leaf consumption, and vertical distribution or location of the defoliation.

Assimilate sappers. These comprise mostly piercing-sucking and rasping insects, which remove plant carbohydrates and nutrients after carbon is taken up and before the plant can convert it to tissue. As mentioned earlier, measuring assimilate removal per pest is a problem, as is the effect of injection of toxic substances during the feeding process, for example, the tarnished plant bug, *Lygus lineolaris*, on peaches.

Turgor reducers. These pests, represented by soil insects and stem feeders, influence plant water and nutrient balances at root and stem sites. Insects like corn rootworms, *Diabrotica* species, prune corn roots, reducing rooting depth and density, and others, like the threecornered alfalfa hopper, *Spissistilus festinus*, girdle soybean stems, thus destroying conductive tissues. Severe instances of reduced water uptake result in decreased plant turgor, followed by reduced expansion of leaves, stems, and fruits, as well as reduced photosynthesis.

Fruit feeders. When insects injure fruit, they cause direct damage to the harvestable produce. Such injury can affect quality (appearance and/or makeup), yield, or both, depending on utility of the produce. Although injury to fruit may seem simple and straightforward to measure, it is not, because yield losses usually are not proportional to percentage loss of reproductive sites. In particular, plant compensation needs to be understood to quantify relationships of losses to this type of injury.

Figure 7.8 Tarnished plant bug, *Lygus lineolaris*, injury (catfacing) to peaches. These pests are assimilate sappers and inject toxins when they feed. (Courtesy Iowa State University Extension Service)

Figure 7.9 Threecornered alfalfa hopper, *Spissistilus festinus*, on a soybean stem. These insects girdle stems, thus destroying conductive tissues and causing plants to lodge. (Courtesy USDA)

Architecture modifiers. Injury from these types of insects change plant morphology to reduce yield. For example, injury caused by the threecornered alfalfa hopper, in addition to destroying conductive tissues, also causes plants to lodge. Lodged plants may die outright or continue to live and grow in a "goose-neck" fashion. This change can reduce physiological yield of the plant, as well as harvestable yield. Other examples of architecture modifiers include the seedcorn maggot, *Delia platura*, which consumes the plumule of soybean seedlings, causing a "Y" plant, and the tunneling of the stalk borer, *Papaipema nebris*, in young corn plants, destroying the growing tip and causing tillering and low yielding or barren plants. In these instances, it is not only the quantity of tissue destroyed but also the quality. Such injury results in potentially drastic changes in subsequent pattern and, usually, rate of plant growth. Because of complexity added by morphological changes in plant structure, this type of injury is difficult to understand and quantify on an injury-per-insect basis.

Injury measurements in EIL calculations. With the majority of insect pests, direct quantification of insect feeding has not been attempted. Most studies have emphasized yield losses, comparing these with observed insect numbers. With data from such studies along with regression analysis, as mentioned earlier, estimates of loss per insect can be made. Before the production system can be modeled, however, quantification of tissue destroyed per insect must be known. Only by quantifying tissue destroyed and understanding physiological responses of hosts to this destruction can we begin to explain yield losses caused by a diversity of pests.

To date, the most detailed studies to quantify tissue consumption by insects

Figure 7.10 Apple maggot, *Rhagoletis pomonella*, injury to an apple. This pest is an important example of a fruit feeder. (Courtesy Iowa State University Extension Service)

has been done for plant defoliators. The simplest approach has been to **visually estimate percent defoliation** of the plant. Then, by estimating the leaf area present and number of insects (of a single species), consumption per insect can be calculated. The major problem with this approach is the error involved in estimating defoliation. Such errors can be great, with a tendency toward overestimation. More accurate studies have been conducted by presenting premeasured leaves to insects and remeasuring leaf area after defoliation. Recent developments of efficient photoplanimeters have made leaf area measurement practical for defoliation studies in the laboratory and the field.

For the purpose of calculation, injury per insect has usually been assumed to have a straight-line relationship with insect numbers. In other words, injury is expected to double with twice as many insects, triple with treble the insects, and so on. However, crowding that occurs at high densities has been shown to reduce injury per insect with some insects because of interference between individuals and (or) relative shortage of food. Although this phenomenon is a concern, for many species the density/injury relationship is straight line at intensities up to and including those in the vicinity of the EIL. Therefore, unless there are other indications to the contrary, insects should be considered in an additive manner in estimating injury for EILs. Such an approach yields conservative estimates in decision making.

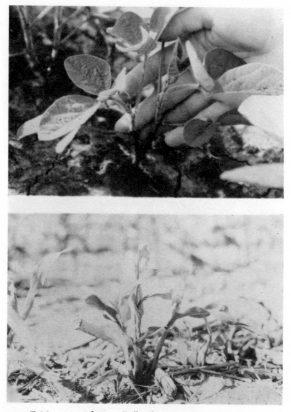

Figure 7.11 A soybean "Y" plant caused by the seedcorn maggot, *Delia platura*, which consumed the growing tip as the plant emerged (top), and tillering of a corn plant caused by stalk borer, *Papaipema nebris*, feeding on the growing tip. Both instances are examples of plant-architecture modification. (Photos by L. Higley (top) and P. Davis)

Crop Susceptibility to Injury

The relationship between injury and crop yield or utility is the most fundamental factor of the EIL. This relationship provides the biological foundation on which economic and practical constraints can be superimposed.

Virtually all theoretical and practical attention to the injury/crop response has been limited to plants. Indeed, our frequent inability to describe these responses with veterinary and medical pests is one reason EILs are usually unavailable or inappropriate for such pests. Therefore, discussion will be limited to injury and plant responses.

In his book, P. G. Fenemore (1982) outlined four major factors involved in the injury/plant-response relationship. They include the following: (1) time of injury with respect to plant growth, (2) type of injury, (3) intensity of injury, and (4) environmental influences on the plant's ability to withstand injury.

Time of injury. The time in a plant's growth cycle when injury occurs has an obvious influence on response to that injury. Usually, seedlings are the most

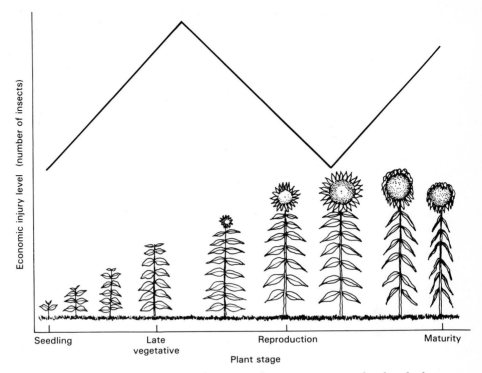

Figure 7.12 Graph showing typical changes in the economic-injury level with changes in plant stage. Many plants are particularly vulnerable during reproductive stages.

susceptible, whereas older plants are better able to tolerate or compensate for injury. Similarly, while yield-producing organs are forming, plants again become very susceptible, but when they mature, injury usually has much less influence (unless fruit is injured directly). The timing of pest injury is most often accommodated in EILs by recognizing a separate yield loss-to-injury relationship (and therefore, separate EIL) for each different stage of crop development.

Plant part injured. The part of the plant injured also influences the plant's response to injury. Usually, a distinction is made between injury to yield-forming organs (**direct injury**), and injury to nonyield-forming organs (**indirect injury**). In general, EILs are calculated for only one injury type, because with only a few exceptions, insects do not produce both direct and indirect injury simultaneously.

Injury types. The common types of injury were discussed earlier. It is important to note that the nature of the injury is fundamental to how a plant responds to different intensities of a particular form.

Intensity of injury. The relationship between the intensity or amount of injury and plant yield is the most important factor in the injury/crop-response relationship. In seeking to explain this relationship, it is of value to consider a generalized response curve of insect injury and plant yield (Figure 7.13a). Not all plants undergo an injury response that includes every portion of the curve, but all potential responses can be described by some part of it.

Specific areas of this curve can be distinguished according to their response rate to injury. In Figure 7.13a, each area of the generalized response curve is described by a different x. These areas and their descriptions are as follows:

x_1: **Tolerance**—no damage per unit injury, yield with injury equal yield without injury, $f(x_1)$ = a constant (zero slope)

x_{2a}: **Overcompensation** (stimulation)—negative damage (yield increase) per unit injury, $f(x_{2a})$ = curvilinear relationship, positive (decreasing) slope

x_2: **Compensation**—increasing damage per unit injury, $f(x_2)$ = curvilinear relationship, negative (decreasing) slope

Figure 7.13 General (a) and specific (b–h) forms of the damage curve of plants. (See text for details.)

x_3: **Linearity**—maximum (constant) damage per unit injury, $f(x_3)$ = linear relationship, negative (constant) slope

x_4: **Desensitization**—decreasing damage per unit injury, $f(x_4)$ = curvilinear relationship, negative (increasing) slope

x_5: **Inherent Impunity**—no damage per unit injury, yield with injury less than yield with no injury, $f(x_5)$ = constant (zero slope)

These responses apply equally to individual plants and to plant stands; however, the responses displayed by a plant and its plant stand at any one time may differ. Usually, individual plants display less of the early portion of the damage curve, but plant stands have a greater ability for tolerance, compensation, and overcompensation. Because EILs usually are developed for plant stands and not single plants, the stand response to injury is of primary importance in calculation.

The last two responses, desensitization and inherent impunity, are not common. When they do occur, injury levels to cause them are much higher than those found with EILs and are unimportant in calculations. However, they could be important in quantifying total crop losses. An example of desensitization (Figure 7.13b) occurs with aphids (Homoptera: Aphididae) when they vector plant pathogens. Inherent impunity (Figure 7.13c) is exemplified by the citrus rust mite, *Phyllocoptruta oleivora*, on citrus. This pest may produce only a slight yield reduction (fresh weight), followed by little further reduction regardless of mite density. This response is shown by the solid line in the figure. But if yield is redefined to include appearance, the damage curve is so radically altered that inherent impunity may not appear at all (dotted line in the figure). Thus, the precise definition of yield, better termed "utility," is important in determining the shape of the damage curve.

Tolerance and overcompensation probably occur more frequently than desensitization or inherent impunity, but these responses may be very small and masked by environmental effects. These occur mostly at low injury levels relative to economic injury and are undetected. Probably all plants display some degree of tolerance to indirect injury. Overcompensation seems not as widespread but, when present, produces a response curve as shown in Figure 7.13d or 7.13e.

More commonly, responses such as in Figures 7.13f, 7.13g, or 7.13h, which include combinations of tolerance, compensation, and linearity, are obtained for calculating EILs. Most plants respond to some level of indirect injury with compensation. Compensation implies that the plant or plant stand is able to prevent the injury from having its maximum influence on yield. In contrast, linearity represents a direct relationship between injury and yield loss. Overcompensation and compensation are primarily limited to indirect injury. An entirely linear response, on the other hand, is expected with direct injury but may also be found with indirect injury.

Environmental effects. Factors in the environment often play an important role in determining how plants respond to injury. Within a given season, environmental factors may influence how long a plant remains susceptible to a specific type of injury. Similarly, between seasons, a plant's response to the same level of injury may be drastically changed. For example, as much as a twofold difference in soybean yield-loss per green cloverworm, *Plathypena scabra*, larva has been

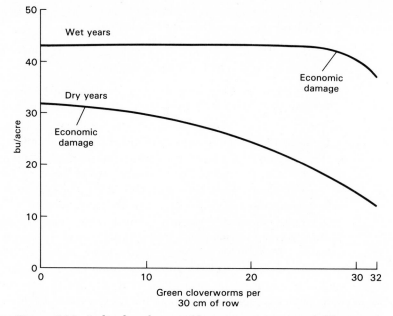

Figure 7.14 Stylized soybean-yield response curves to defoliation of the green cloverworm, *Plathypena scabra,* in years of normal rainfall (wet) and those with below normal rainfall (dry). Note the differences in number of insects required to cause economic damage between the two types of years.

found between growing seasons with different weather conditions. Such extreme variability emphasizes the need to calculate EILs over a range of environmental conditions. A better approach would be to develop EILs for specific sets of weather and environmental factors.

Experimental Techniques to Determine Plant Damage Response

Techniques to determine the type and magnitude of plant response to insect injury vary greatly. The most common approaches can be grouped into four categories: (1) observation of natural insect populations, (2) modification of natural populations, (3) establishment of artificial populations, and (4) injury simulation.

Observation. Observation of natural insect populations is the simplest, but perhaps least precise, approach. The method usually involves making intensity estimates on plants from several fields and correlating these with resulting yield losses in those fields. This approach has been widely used in all types of crops and most often used with perennial or severe pests (pests that are present and cause losses most years).

Modification of natural populations. This approach is also used with perennial and severe pests, except in this instance experiments are designed to produce populations of desired sizes in designated plots. Usually, selective insecticides or selective insecticidal rates are used to create different population densities on plants for analysis. This procedure is more precise than simple observation described before and usually allows insect-crop loss relationships to be determined more effectively. However, the effect of the insecticide on plant physiology must be known and accounted for with this approach.

Figure 7.15 Applying insecticides with a small-plot sprayer. Insecticides may be used to create different-sized insect populations in establishing damage-response curves. (Photo by L. Higley)

If insect numbers are low, the problem in experimentation is to increase numbers, rather than reduce them to desired levels. In this instance, increases to obtain various insect levels in plots may be attempted by using baits, attractant crops, or attractant isolines of the same crop. (Isolines are cultivars that differ in only a single gene and could be expected to yield similarly to the cultivar in question.)

Figure 7.16 Stunting of a glabrous isoline of "Clark" soybeans in response to potato leafhopper, *Empoasca fabae*, injury (center), as compared with the pubescent "Clark" variety (left and right). Such isolines can sometimes aid in establishing damage response curves. (Photo by L. Higley)

Creating artificial populations. This approach can be used when precise control of numbers is sought or numbers in the natural population are low. Most often, this approach involves rearing or collecting the insect, then artificially infesting small plots. The usual procedure is to place feeding larvae or other stage in plots and cover them with cages. In some instances, as with the soybean looper, *Pseudoplusia includens*, the adult stage may be released in cages to produce damaging larval populations. Desired damage levels from the artificial populations then can be achieved by properly timed insecticide applications.

Injury simulation. Simulating injury by an insect is another method used to assess losses by insects. Also known as **surrogate injury**, this approach seeks to mimic insect feeding behavior. It has been applied most frequently with defoliators, where workers have picked leaves or punched holes in them in a real time and rate sequence. The primary advantage of this method is that degree of injury can be precisely controlled and the biology of the crop response followed intensively. However, much preliminary information regarding fidelity of the surrogate damage to actual insect feeding must be gathered before the technique is employed.

When we consider the diverse circumstances involved with insect injury, it is clear that no single method of estimating crop response is appropriate in all instances. The choice of an approach depends on the insect species and crop investigated and the resources available for the investigation.

BOX 7.2 SOYBEAN LOOPER

SPECIES: *Pseudoplusia includens* (Walker) (Lepidoptera: Noctuidae)

DISTRIBUTION: The soybean looper migrates into the southeastern United States each year from Mexico and Central America.

IMPORTANCE: The species is a common and destructive soybean pest in the Gulf coast area of the United States. Larvae defoliate soybean and can cause serious yield losses when outbreaks occur. Usually, populations do not reach high levels until August and September. Damage can be especially serious in areas adjacent to cotton fields. Larvae also attack numerous vegetable crops.

APPEARANCE: Adults are brown with a wingspan of about 38 mm. The forewings display a distinctive silver marking. The hindwings are brown. Larave are green with white stripes running lengthwise along the body, which is about 32 mm long when full-grown. They have black front legs and two pairs of abdominal prolegs (excluding the anal pair). Eggs are greenish white. Pupae are enclosed in silken cocoons.

LIFE HISTORY: Migrating females oviposit up to 600 eggs singly on leaves. After about three days, eggs hatch, and the larvae feed for approximately 18 days (seven molts). Pupation occurs on the underside of leaves, and in about eight days, adults emerge. There can be up to four generations per year.

Figure 7.17 Field cages in alfalfa plots. Insects are introduced into the cages at various levels to establish damage response curves. (Photo by S. Hutchins)

Figure 7.18 Punching holes in leaves with a cork borer to simulate insect defoliation.

CALCULATION OF ECONOMIC DECISION LEVELS

In developing economic indices for pest management decisions, the principal level to estimate is the EIL, because the EIL includes the basic damage potential of a given insect population. It can be used as the ET, or an ET can be determined from knowledge of the EIL and population dynamics. In either situation, the EIL must be known first.

The calculation of EIL for an insect is a continuing process because new values are required with changes in the input variables. Consequently, when market value, management costs, and plant susceptibility change, recalculation is necessary. Additionally, several EILs are usually required for any one season because of crop development and consequent changes in susceptibility.

Basically, the steps required to calculate the EIL are as follows:

1. Estimate the loss per insect
2. Determine the gain threshold as described previously
3. Calculate the EIL as:

$$\text{EIL} = \frac{\text{gain threshold}}{\text{loss per insect}}$$

Of these steps, the first is by far the most difficult. Crude estimates of losses are usually obtained from field observation and experimentation with various-sized insect populations on a crop at specific times. Subsequently, yields are measured and losses caused by the insects are determined.

An example of the steps required in EIL calculation can be shown with the potato leafhopper, *Empoasca fabae*, on soybeans. This insect is very common in the crop and may cause losses when another preferred crop, alfalfa, is clipped. Lacking suitable habitat and food after the alfalfa harvest, leafhoppers migrate into soybeans, where they suck plant juices and can cause a leaf necrosis (hopperburn), especially in young plants.

To estimate the damage per insect, small soybean plots were caged and infested with various leafhopper numbers at three plant growth stages. Yields were measured at season's end, and statistical procedures (regression analyses) determined the loss per insect at each plant growth stage. These values, based on loss caused by one insect per plant, were as follows: early vegetative stage = 1.55 bushels/acre, beginning flowering stage = 0.17 bushels/acre, and beginning pod fill = 0.08 bushels/acre. Loss per insect decreased throughout the season because of plant growth and greater leaf area, which decreased insect intensity.

To calculate the gain threshold, we estimate the cost of applying one pound (actual ingredient) of malathion insecticide by aircraft at $9.50/acre. Further, we predict that we can sell the harvested grain for $4.15/bushel and calculate the gain threshold as:

$$\text{Gain threshold} = \frac{\$9.50/\text{acre}}{\$4.15/\text{bushel}} = 2.29 \text{ bushels/acre}$$

Next, EILs can be calculated for each stage as follows:

$$\text{Seedling stage: EIL} = \frac{2.29 \text{ bushels/acre}}{1.55 \text{ bushels/acre/insect}}$$

$$= 1.48 \text{ (about 2) leafhoppers/plant}$$

Figure 7.19 Statistical regression analysis of soybean yield on number of potato leafhoppers, *Empoasca fabae*, present per plant at three plant stages.

BOX 7.3 POTATO LEAFHOPPER

SPECIES: *Empoasca fabae* (Harris) (Homoptera: Cicadellidae)

DISTRIBUTION: The potato leafhopper occurs throughout the eastern half of the United States and southern Canada in summer. To inhabit this range, however, individuals must migrate annually from an overwintering region in the southern United States. They accomplish this by moving vertically into winds conducive for northern migration. This adaptive movement terminates when precipitation causes "dropout" of individuals.

IMPORTANCE: This species is one of the most important alfalfa pests in the eastern and north central United States. In addition, this pest is capable of feeding on numerous other host species, including soybean and potato. Feeding with piercing-sucking mouthparts results in phloem blockage within the plant and produces an inverted V-shaped yellowing of the leaf (chlorosis). This symptom is referred to as "hopperburn." Injury to plants results in reduced photosynthesis, which translates into shorter, less productive plants. In the case of alfalfa, leaf proteins also are severly reduced. Before hopperburn was associated with potato leafhopper feeding, the

chlorosis was believed to be the symptom of a leaf disease.

APPEARANCE: Adults are green, wedge-shaped and about 3 mm long, with a row of six round white spots along the pronotum. Nymphs, which develop by gradual metamorphosis, are smaller than adults and are wingless. Eggs are about 1 mm long.

LIFE CYCLE: Upon arrival and mating in the northern United States and Canada, females oviposit 60 to 100 eggs over a 30-day period within stems and larger veins of host leaves. Hatching occurs in six to nine days, and nymphs emerge to develop through five stadia. Soon after the adults appear, mating occurs and oviposition soon follows. Approximately 21 days are required for development from egg to adult in warmer summer temperatures. There are several overlapping generations per year, with the exact number depending on time of arrival and temperatures.

$$\text{Beginning-flowering stage: EIL} = \frac{2.29 \text{ bushels/acre}}{0.17 \text{ bushels/acre/insect}}$$

$$= 13.47 \text{ (about 14) leafhoppers/plant}$$

$$\text{Beginning pod-fill stage: EIL} = \frac{2.29 \text{ bushels/acre}}{0.08 \text{ bushels/acre/insect}}$$

$$= 28.63 \text{ (about 29) leafhoppers per plant}$$

Finally, we can use these EILs as our "action levels" (set the ET equal to the EIL), or we may choose to set ETs at levels conservatively below the EILs, say, at 75 percent of the EILs.

USING ECONOMIC LEVELS

As mentioned at the beginning of this chapter, economic decision levels are an important, if not indispensable, part of most insect pest management programs. Of the levels discussed, the ET is the practical operational level recommended to producers for making management decisions. The ET has been described by J. Mumford and G. Norton as a "useable, if not ideal, decision rule"; it has been adopted in many situations.

To date, the most widespread use of the ET has been where **curative** (**therapeutic**) management tactics (mainly insecticides) are used. Here, insect populations are sampled on a regular basis, and when needed, suppressive action is taken. Usually, less value is placed on this decision rule when the pest always causes economic damage, and a preventive management strategy is used.

Implementation Categories

The place and ultimate use of ETs in pest management programs becomes clearer when their state of development is categorized. Accordingly, most of the decision rules currently used can be placed in one of four categories: (1) no thresholds, (2) nominal thresholds, (3) simple thresholds, and (4) comprehensive thresholds.

No thresholds. Decisions in this category were common in applied entomology prior to the late 1950s, when "identify and spray" was the conventional practice. Although using thresholds is considered an advance for most insect problems today, there are certain situations where they are not appropriate:

1. Pest sampling cannot be done economically
2. Practical response to cure a problem cannot be implemented in a timely way
3. Once detected, the problem cannot be cured
4. ET is immeasurably low (some quality losses, disease transmission, rapid growth potential)
5. Populations are intense, where general level of fluctuation is always above the EIL

Management implementation in these situations usually must depend on prevention, rather than cure. When insecticides are used for prevention, scheduled treatments according to calendar date or cultural operation are the rule. More advanced programs may predict treatment dates using degree-day models and computer-based delivery systems.

Nominal thresholds. This category represents decision rules that are declared on the basis of a manager's experience. Historically, these were the first thresholds used and are still the most frequent type implemented. It is common for nominal thresholds to be developed by experienced entomologists of a state's Cooperative Extension Service and to be published in insecticide recommendation bulletins. Such levels, which may be used over broad regions, tend to be static, unchanging with changes in the variables mentioned previously. Although sometimes criticized for not being based on rigorous research, nominal thresholds are an advance over using none at all, because such decision rules tend to be conservative (that is, they err on the side of taking action when it is not needed), and their use has resulted in reduced insecticide applications.

Simple thresholds. This is the type of threshold described at the beginning of the chapter. These levels are calculated on the basis of average responses

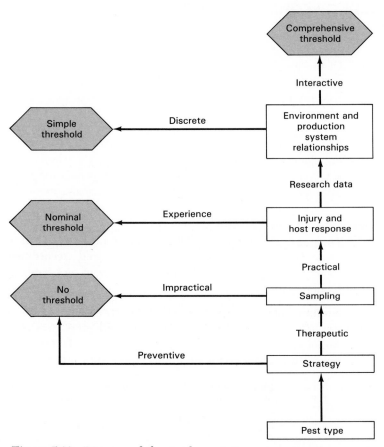

Figure 7.20 Diagram of the implementation categories of economic thresholds.

of hosts to injury caused by an insect. The four major inputs of market value, management costs, tissue destroyed (or damage done) per insect, and yield (or quality) reduction per tissue destroyed are used to make calculations. Such levels may be implemented by using a simple formula to make calculations, consulting tables that consider the input variables, or using software packages with programmable pocket calculators or personal computers. Although these thresholds may be our best current practice, they usually fail to consider possible interactions of several pests and changes in the cropping environment that influence decisions.

Comprehensive thresholds. Decision rules in this category await future development. Only recently have entomological studies been completed that address topics like effects of insect and weed interactions on plant stress and include weather factors in calculating EILs. At the core of establishing truly comprehensive thresholds for crops is an intimate understanding of the host plant and its reaction to combined stressors, both biotic and physical. Such an understanding should include quantification of processes as dry-matter partitioning throughout the growing season and not only a simple measurement of yield, as has been done in the past. It is only in this way that the effects of a **pest community** can be understood and thresholds developed in the context of the entire production system. Opportunities for implementing such thresholds will probably lie with advances in computer-based information delivery systems and the acquisition of on-farm computers.

Limitations of the EIL Concept

Although the economic levels discussed have been used on a practical basis for more than 25 years, they are not without drawbacks. Their limitations relate to the types of pests or injury, management tactics selected, research requirements, and desirability of multiple inputs in making decisions.

Decision levels for management of some types of pests cannot be determined with EILs. Many vectors, medical pests, veterinary pests, and pathogens often do not have a quantitative relationship (or a weak one at best) between damage and injury; consequently, they are not amenable to calculated EILs. Moreover, because the "market value" of human health and life is so high (priceless), it is virtually impossible to put an economic limit on the control of most medical pests.

Economic considerations also limit our ability to use EILs for aesthetic pests. It is very difficult to place a monetary value on the reduction in "aesthetic value" associated with a given type of pest injury. Usually any assigned values are subjective, which greatly limits their usefulness in calculating EILs.

A similar problem exists with forest pests. Pests of a few woody species like fruit trees and Christmas trees may be amenable to EILs; however, forest pests are not easily described by them. Almost all the components of EILs are difficult to estimate for forest insects; accurate market values are a problem because projections often must be made many years in advance; management costs may vary greatly and frequently must include more environmental and social costs than in other pest management programs; and the injury/crop-response relationship may be difficult to determine because the growth of the crop spans many years.

Some pests have a quantitative relationship to yield, but they are still difficult to manage with EILs. Pathogenic microorganisms on plants, although not entomological, are a good example of this. Here, the yield reduction produced by many pathogens is usually quantitatively related to pathogen number. Unfortunately, sampling and quantifying the number or amount of these pathogens is usually impractical. Thus, the question whether pests can be economically sampled is important in determining practicality of EILs. Furthermore, management tactics of these pathogens are more often preventive, not therapeutic; therefore, determining whether or not a pathogen population is at the EIL after infection may not be of significant value if the only management options available must be applied before infection.

In fact, this last example highlights an important limitation of EILs. As mentioned earlier, the EIL concept originally was developed with the objective of reducing insecticide use. Consequently, both the EIL and the ET can be used most appropriately when a single, curative management action can be made. They also are useful with some of the new tactics for killing insects like microbial insecticides and insect growth regulators; however, these new tactics are used similarly to conventional insecticides. Except from an assessment standpoint, EILs are of limited use with preventive measures, such as host-plant resistance and most forms of ecological management. These tactics are implemented as a matter of course, according to historical problems with pests.

Finally, the significant amount of research required and relative unsuitability when several diverse pests (those causing different kinds of injury) occur may present serious obstacles in attempting to employ the EIL concept. But if injuries from different pests produce the same host response and can be placed on a common basis, or if effects of different injuries are additive, the concept may find application for pest complexes.

CONCLUSIONS

The EIL concept has been one of the most important innovations in the development of insect pest management. Although not without drawbacks, EIL decision rules are the most widely accepted and practical tools for their purpose. However, these decision rules are yet to be established for many species and need refinement in others.

Refinement of the EIL concept is possible in the form of comprehensive thresholds, which account for multiple pests as well as other environmental stressors. By necessity, these comprehensive thresholds need to be founded on the total production system of a given agricultural enterprise. In large part, future development depends on our ability to understand how diverse pests cause stresses in plants and animals and on quantification of organismal reactions to these stresses.

Selected References

ANDOW, D. A., and K. KIRITANI. 1983. The economic injury level and the control threshold. *Japan Pesticide Information*. 43:3–9.

BARDNER, R., and K. E. FLETCHER. 1974. Insect infestations and their effects on the growth and yield of field crops: a review. *Bulletin on Entomological Research*. 64:141–160.

BOOTE, K. J. 1981. Concepts for modeling crop response to pest damage. *ASA Paper* 81-4007. American Society of Agricultural Engineering, St. Joseph, Mo.

CAPINERA, J. L. , G. W. WHEATLEY, D. C. THOMPSON, and J. JENKINS. 1983. Computer assisted crop loss assessment: a micro-computer model for estimation of sugarbeet insect effects to facilitate decision-making in pest management. *Protection Ecology*. 5:319–326.

CHIANG, H. C. 1979. A general model of the economic threshold level of pest populations. *United Nations F.A.O. Plant Protection Bulletin*. 27:71–73.

FENEMORE, P. G. 1982. *Plant Pests and Their Control*. Wellington, N. Z. : Butterworths, pp. 125–144.

MUMFORD, J. D., and G. A. NORTON. 1984. Economics of decision making in pest management. *Annual Review of Entomology* 29:157–174.

PEDIGO, L. P., S. H. HUTCHINS, and L. G. HIGLEY. 1986. Economic injury levels in theory and practice. *Annual Review of Entomology*. 31:341–368

PIERCE, W. D. 1934. At what point does insect attack become damage? *Entomological News* 45:1–4.

POSTON, F. L., L. P. PEDIGO, and S. M. WELCH. 1983. Economic injury levels: reality and practicality. *Bulletin of the Entomological Society of America* 29:49–53.

STERN, V. M., R. F. SMITH, R. VAN DEN BOSCH, and K. S. HAGEN. 1959. The integrated control concept. *Hilgardia* 22:81–101.

8 PEST MANAGEMENT THEORY AND PRACTICE

FURNISHED WITH THE KNOWLEDGE of basic insect biology, ecology, and decision making, we appropriately turn our attention to the practicalities of managing insect populations. An understanding of this topic requires a look at the basic nature of pest control.

Pest control is the application of technology, in the context of biological knowledge, to achieve a satisfactory reduction of pest numbers or effects. As such, it has been described as a two-strand science. The technological aspect includes such things as insecticides and equipment used to apply them. Biological knowledge allows us to know where, when, and how to apply the technology.

A simple example of pest control can be made with a common household pest, the German cockroach, *Blattella germanica*. Biological knowledge necessary for pest control includes the following: (1) the species is tropical or subtropical in origin and survives only indoors in temperate locations, (2) it prefers warm, dark, humid environments, (3) it feeds widely on all sorts of plant and animal products, particularly spilled food and cooking debris in the kitchen, and (4) eggs of the species require three or four weeks for incubation. The technology selected for pest control could appropriately be the application of premixed chlorpyrifos sold in a container, complete with a hand-pump sprayer. Satisfactory pest control in this instance usually means complete elimination of the pest from the household. According to biological knowledge, sprays would be applied to dark corners, cracks and crevices, cabinet shelves, undersides of sinks, and other hiding places in the kitchen and bathroom(s). Because the spray kills only nymphs and adults, it would need to be applied two or three times at two-week intervals to kill new nymphs as the eggs hatch. In addition, food spills and debris should be thoroughly removed to limit the pest's food supply and help prevent further introduction and establishment.

Admittedly, this is a very elementary example, but it illustrates the dual nature of pest control. If either biological knowledge or technology is deficient, the control program probably will not succeed.

HISTORICAL HIGHLIGHTS OF PEST CONTROL

Pre-insecticide Era

Agricultural pest problems and actions to alleviate them are nearly as old as the beginnings of crop cultivation. The earliest record of pest control seems to be the use of sulfur (an insecticide still used) by ancient Sumerians about 2500 B.C. In prebiblical times both the Egyptians and Chinese used insecticides formulated from herbs and oils, particularly for protection of seeds and stored grain.

A very important advancement, occurring around 300 B.C., was the recognition of phenology as a science. This led to the principle of timely planting of crops to avoid pest losses. During this time, the Chinese discovered the use of natural enemies to control insects: for example, they placed ants on citrus to reduce pest infestations.

Few developments in pest control were recorded during the Middle Ages (end of the Roman Empire until about A.D. 1000). However, by 1101 the Chinese had discovered the use of soap to control pests, and by the late 1600s, tobacco infusions and insecticides from other herbs, as well as arsenic, were commonly used.

With the expansion of agriculture in the 1700s, Réaumur published on the importance of temperature summation in determining insect phenology. This idea was to be used, almost two centuries later, to predict insect events and determine timing of pest control activities. Also in the 1700s, the concept of plant

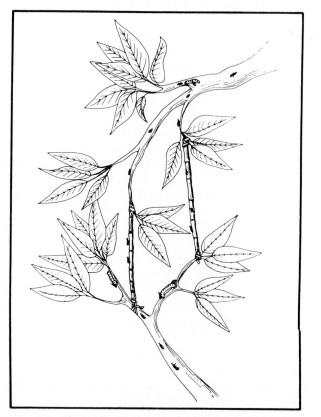

Figure 8.1 The earliest known example of using natural enemies is believed to be in China, where ants were placed in citrus to suppress caterpillars and beetles. Bamboo bridges were tied between branches to encourage ant movement from tree to tree. (Reprinted with permission of Plenum Press from *Introduction to Integrated Pest Management* by M. L. Flint and R. van den Bosch. Copyright ©1981)

resistance to insects was advanced for the Hessian fly, *Mayetiola destructor*, in the United States, as well as the rediscovery or introduction of botanical insecticides (derived from plants).

During the mid-1800s, imperial expansion, accompanied by developing technology, resulted in countries sending citizens and materials far beyond their own borders. Along with this movement, insect species were introduced into new regions. Examples of notable introductions include San Jose scale, *Quadraspidiotus perniciosus*, introduced into the United States from China, and the Colorado potato beetle, *Leptinotarsa decemlineata*, introduced into Europe from the United States. In many countries, significant losses from introduced pests gave impetus to developing inspection and quarantine procedures that would be instrumental in slowing the spread of insect pests around the world.

The late 1800s and early 1900s saw the rapid development of insecticide application equipment. A French company, Vermorel, claims to have invented the first commercial sprayer in 1880, and the first power sprayer was developed in Germany in 1904. These commercial sprayers were used in Europe primarily for pesticide applications on fruit and vine crops. In the United States, the use of insecticides increased rapidly on cotton during this period to combat mounting problems with the boll weevil, *Anthonomus grandis*. In this instance dusts were the primary insecticide formulation applied.

By 1920, the use of airplanes for all sorts of activities was commonplace. Therefore, it is not surprising that they were fitted with a spray boom and pressurized tank and used to apply pesticides. This development occurred in the United States in 1921 and was used in emergency boll weevil pest control programs in 1922.

BOX 8.1 GERMAN COCKROACH

SPECIES: *Blattella germanica* (Linnaeus) (Orthoptera: Blattellidae)

DISTRIBUTION: Although native to Africa, the German cockroach is found throughout the temperate areas of the world, usually in humans' dwellings.

IMPORTANCE: The German cockroach is one of the most common, prolific household cockroaches. Both adults and nymphs are a major annoyance to humans because individuals eat almost anything, excrete on food, and leave a characteristic bad odor. The species is active only at night and prefers warm areas of buildings. The pest is also suspected of being able to vector diseases.

APPEARANCE: Adults, about 12 mm long when full-grown, are yellowish brown with two parallel stripes on the pronotum. The antennae are long and the wings cover the entire abdomen. Nymphs are similar to adults, except they are smaller and lack wings. Oothecae (egg capsules) are light brown and contain 25 to 30 eggs.

LIFE HISTORY: Females carry oothecae for about 21 days, until just before hatching occurs. Nymphs develop for about three months before finally becoming adults. Each female can produce up to five oothecae. There are typically two to three generations per year.

Figure 8.2 Dusting cotton with a horse-drawn power duster in the early 1900s. (Courtesy USDA)

Figure 8.3 Dusting for boll weevils, *Anthonomus grandis*, with an airplane in the early 1920s. (Courtesy USDA)

During the 1920s and 1930s, most insecticides were not very efficient. They were expensive, difficult and hazardous to apply, somewhat phytotoxic (they could kill or injure plants), and by today's standards, they caused low pest mortalities. Therefore, insecticides could not be relied on, by themselves, to achieve pest control. Consequently, applied entomologists developed approaches based on crop sanitation, timing of planting and harvesting, and other environmental manipulations, coupled with the low-key use of insecticides to achieve pest control—it was the dawn of the insecticide era.

Insecticide Era

The insecticide era spans a period roughly from 1939 to 1962. It began with the explosive development of seemingly miraculous insecticides. It ended in a good deal of disillusionment.

The discovery of the insecticidal properties of DDT (*dichlorodiphenyl-trichloroethane*) marked the beginning of this era. Prior to its discovery, the major insecticides in use were formulated from petroleum and coal tar distillates, plants, or inorganic compounds, particularly arsenic. The trend in applied chemistry in the late 1930s was toward synthesis of new compounds, and pest control workers were seeking to synthesize moth-proofing agents. Dr. Paul Müller, a Swiss chemist with Geigy Corporation, seeking to develop a new, purely synthetic insecticide, examined DDT. The obscure compound had been synthesized in 1873 by a German graduate student, Othmar Zeidler, who did not realize its insecticidal properties. After testing, Müller found the compound extraordinarily effective against many insects.

During World War II, DDT was revealed to the Western Allies and was used primarily for mosquito, flea, and lice suppression. Its application saved thousands of lives that would have been lost to malaria and typhus. Indeed, DDT had such an impact on human health that in 1948 Müller received the Nobel prize in medicine for his discovery. After the war, DDT had phenomenal successes against agricultural pests, and a host of other effective insecticides were developed from the chlorination of hydrocarbons.

While the Allies were working with DDT and related compounds, the Germans accelerated their efforts to find suitable replacements for nicotine and other scarce insecticides. Dr. Gerhard Schrader with Farbenfabriken Bayer AG, having worked actively with organophosphates as possible nerve gases, discovered the insecticidal properties of these compounds.

At the close of World War II, insecticide development proceeded at a very rapid pace. Many firms, particularly petroleum companies, became involved in the development of agricultural chemicals. The trend in pest control was toward the testing of new and even more effective insecticides. As this trend continued, agricultural entomologists turned their attention away from multifaceted programs involving sanitation, planting dates, and the like, in favor of a more narrow, purely insecticidal approach to solve pest problems.

The insecticidal approach usually entailed identifying the pest and finding the most "effective" chemical to apply, with little concern for other factors important in the agroecosystem. Spraying was often performed according to a calendar schedule and without knowledge of pest phenology, density, or damage potential. The low cost and high performance of these chemicals fostered the abandonment of earlier combined approaches for suppression. The insecticidal approach relied only on ability to kill insects, and killing, rather than crop protection, became a major preoccupation in pest control.

Figure 8.4 Dusting a child with DDT for typhus control during World War II. (Courtesy World Health Organization)

Such a practice worked well in the short run, but in time, insecticide resistance and other problems began to occur with surprising regularity. In addition, residues of DDT and related compounds were discovered in milk and other foods, as well as in food chains of predator birds and other wildlife.

The realization of unintentional pesticide movement in and effects on ecological systems was brought to the fore in 1962 with the publication of *Silent Spring* by Rachel Carson. In her book, Carson made strong accusations against pesticides, causing a public outcry in the United States for environmentally safe approaches to pest control. This reaction resulted in appointment of the Presidential Science Advisory Committee on Environmental Quality, and a report recommending improved pest control practices was issued in 1965.

Emergence of Pest Management

The concept of pest management grew out of the discontent with the purely insecticidal approach to pest control of the 1950s. Although the term was coined in the early 1960s, pest management had its roots in earlier theories of biological control. Early successes and lasting solutions with devastating pests encouraged the development of biological control, which emphasized the significant impact of effective natural enemies on pest populations.

With many pests, however, natural enemy controls were undependable for satisfactory containment. Subsequently, the concept of **integrated control** was developed, emphasizing the selective use of insecticides so that natural enemies

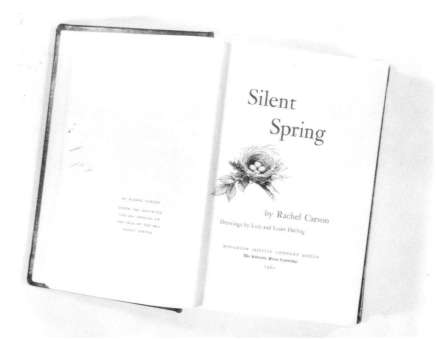

Figure 8.5 The publication of *Silent Spring* in 1962 had a significant influence on the direction of pest control in the United States.

were conserved in the agroecosystem. This "integration" of control techniques was expanded in later years to include other techniques (for example, resistant hosts and sanitation), but natural enemy conservation was always an axiom of integrated control.

Against this background, Australian entomologists L. R. Clark and P.W. Geier clearly outlined the principles of pest management in 1961. They suggested the term "protective population management" or "**pest management**" for their ideas. Pest management differed from earlier approaches in its holistic viewpoint, its synthesis of ideas, and the inclusion of basic population theory in its design.

Since its inception, pest management gained momentum and has become the major pest control philosophy for agricultural pests in developed countries. In the United States, it is supported at all levels of agriculture, including federal (Department of Agriculture, Environmental Protection Agency), and state (university research and extension) agencies, private crop consulting firms, and grower organizations. Pest management was ordered to be used wherever practicable by presidential decree in 1977 and is mandated by California law to be used when feasible.

THE CONCEPT OF PEST MANAGEMENT

Definition and Characteristics of Pest Management

Probably as many definitions of pest management have been suggested as there are authors on the topic. Most descriptions mention three elements: multiple tactics (for example, natural enemies, resistant varieties, and insecticides) used in a compatible manner; pest populations maintained below levels that cause economic damage; and conservation of environmental quality.

Additional characteristics that help us conceptualize pest management were outlined by Geier in 1966. He contended that pest management programs should be (1) selective for the pest, (2) comprehensive for the production system, (3) compatible with ecological principles, and (4) tolerant of potentially harmful species but within economically acceptable limits.

For our purposes, we will consider these foregoing elements and define pest management as *a comprehensive approach to pest control that uses combined means to reduce the status of pests to tolerable levels while maintaining a quality environment.* This definition carries with it several implications that set pest management apart from other pest control approaches. First, it seeks to deal with pests in the context of the whole production system, rather than as a separate collection of individual problems. It proposes that several techniques be employed to alleviate problems, rather than relying on a single tactic (for example, insecticides alone).

The objectives of pest management are also clear from our definition. The main objective of the approach is to reduce pest status. Although reducing status may be achieved by killing pests, killing certainly is not the objective—preventing economic loss is. Indeed, pest status also may be reduced by avoiding or repelling pests or by reducing their reproductive rates.

Another implication of our definition is that pest populations or their effects should be reduced to tolerable levels. In this instance, tolerance means that humans should accept the presence of pest species, although at levels that are not economically important. This aspect admits that complete elimination of pests may not be feasible or even desirable. Particularly, this acceptance of pest presence sets pest management apart from many other approaches to pest control.

Finally, the capstone objective of pest management is the maintenance of a quality environment. This objective clearly refers to conservation, and it includes quality of both cropping and nonagricultural environments (air, water, soil, wildlife, and plant life). With past approaches, environmental quality was largely ignored, excepting the attention paid to crop vigor and soil elements supporting the crop. The principles of pest management emphasize that cropping systems behave similarly to unmanaged or natural systems; that is, they have a diversity of interacting elements like natural insect and weed enemies that influence ecology of the crop environment. By maintaining the quality of this environment, lasting solutions can be achieved. Most approaches that conserve the quality of crop environments also aid in conserving surrounding, nonagricultural environments.

Pest Management Strategies and Tactics

The application of pest management concepts begins with the development of a strategy. A **pest management strategy** is the overall plan to eliminate or alleviate a pest problem. The particular strategy developed depends on the particular life system of the pest and the crop involved.

In addressing problems using pest management, we aim to reduce pest status. Because pest status is determined by both the insect and the crop, our management program may emphasize modification of either or both of these. Therefore, the types of strategies to be followed, based on economics and pest characteristics, are as follows:

1. Do nothing
2. Reduce pest population numbers
3. Reduce crop susceptibility to pest injury
4. Combine reduced population numbers with reduced crop susceptibility

After an appropriate strategy has been developed, the methods of implementing the strategy must be chosen. These methods are usually called **tactics**. Optimally, several tactics are used to implement a management strategy.

Do-nothing Strategy

The "do-nothing" strategy may, at first, seem absurd. Obviously, we "know" when a pest is doing us harm—but do we? It is possible that insect injury only *seems* as if it is causing a loss, when in reality the crop species tolerates the injury without economic damage. The phenomenon of mistaking trivial insect injury for economically significant injury occurs all too often and is usually the result of not assessing population density in relation to the economic threshold. When pest densities are below the economic threshold, "do nothing" definitely is the strategy to follow, otherwise more money is spent on management than is gained in utility; in other words, a net loss occurs from management. The most frequent need for the "do-nothing" strategy usually arises when insects cause indirect injury. This strategy also may be the ultimate one following a successful pest management program. In this instance, only surveillance of the resulting pest population is required.

Obviously, tactics are not implemented in the "do nothing" strategy. However, this does not mean that little activity is involved or that pest suppression is not occurring. Indeed, considerable sampling is required to assure that taking no action is appropriate, and significant pest suppression may have occurred as a result of natural environmental factors.

Reduce-numbers Strategy

Reducing insect numbers to alleviate or prevent problems is probably the most widely used strategy in pest management. This strategy is usually employed in a

Figure 8.6 Weevil (Coleoptera: Curculionidae) injury (notches) to leaves of redbud. Indirect injury often results in no measurable loss in the valued commodity. (Courtesy D. Foster and the Iowa State University Extension Service)

therapeutic manner when densities actually reach the economic threshold or in a preventive manner based on a history of problems.

In seeking to reduce numbers, one of two different objectives may be desirable. Where the pest's long-term average density, the general equilibrium position (GEP), is low compared with the economic threshold (problems are not particularly severe), the best strategy would be only to dampen population peaks. This action would not change the GEP appreciably, yet would prevent economic damage from occurring during outbreaks.

On the other hand, severe pest problems call for more drastic population reductions. With these pests, the GEP lies very close to or is above the economic threshold. What is required for these populations is a general lowering of the GEP so that highest population peaks never reach the economic threshold. The most appropriate method of accomplishing this is to either (1) reduce the environmental **carrying capacity** (the maximum number a given environment will support for a sustained period), or (2) reduce the inherited reproductive and/or survival potentials of populations.

In the first instance we are dealing exclusively with the environment, trying to reduce the favorableness of the habitat. For instance, if numbers of cattle allowed to graze in an area of pasture are reduced, we reduce the carrying capacity of the habitat for populations of cattle tick, *Boophilus annulatus*. The same is true when crops are rotated from a pest's host species to a nonhost species (for example, a legume rotated with a grass) or a field is tilled to eliminate overwintering habitat. In all of these situations, we attempt to reduce the GEP of the pest population over time by reducing the inherent favorableness of the pest's environment.

With the second approach, we do not deal with the environment, but with the characteristics of the pest itself. Depending on genetic makeup, each pest population has a specific ability to reproduce (potential birth rate) under a given set of environmental conditions, as well as a given ability to survive (potential survival rate) under those same conditions. Methods have been proposed and used that reduce those innate abilities (potentials), thus causing the GEP to drop (for example, the sterile-insect technique).

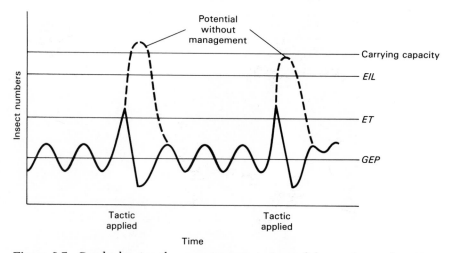

Figure 8.7 Graph showing the management strategy of dampening peaks. EIL, economic-injury level; ET, economic threshold; and GEP, general equilibrium position.

Figure 8.8 Graph showing the management strategy of lowering the carrying capacity of a pest. EIL, economic-injury level; ET, economic threshold; and GEP, general equilibrium position.

Figure 8.9 Graph showing the management strategy of lowering the general equilibrium position (GEP) but not the carrying capacity. EIL, economic-injury level; and ET, economic threshold.

The tactics utilized in the reduction-of-numbers strategy are many and varied. Most of these increase mortality in pest populations by creating or intensifying hazards to insects in their environment. Hazards are increased with tactics such as natural enemies, insecticides, many resistant cultivars, ecological modifications, and insect growth regulators. Other tactics attempt to reduce numbers by reducing reproductive rates in pest populations. Some of these include release of sterilized insects and application of chemicals to disrupt mating activity.

Reduce-crop-susceptibility Strategy

Reducing crop susceptibility to insect injury is often one of the most effective and environmentally desirable strategies available. For this strategy, the insect population is not modified at all. Rather, we rely on changes made in the host plant or animal that render it less susceptible to an otherwise damaging pest population.

The tactics involved in the reduce-crop-susceptibility strategy usually involve elements of host-plant and livestock resistance or ecological management (crop environment manipulations). In the first instance, a certain form of resistance, referred to as **tolerance**, occurs in some cultivars or breeds, which have a greater degree of impunity to insect injury than present in others. When such cultivars or breeds are utilized, insect population numbers are not reduced; however, losses are minimized because the tolerant crop is less affected by a given level of injury. Some of the tactics in ecological management for reducing crop susceptibility are improving plant vitality through fertilization and changing planting dates to upset the synchrony between a pest and a susceptible plant stage.

Combined Strategies

The remaining, yet obvious, strategy is one that combines objectives of all previously mentioned strategies to produce a pest management program with several tactics. In many ways this is the most desirable strategy, when feasible. The diversification resulting from such programs tends to produce a greater degree of consistency in pest control than can be achieved with a single strategy, particularly a single tactic. Past experience has shown that pest control programs using a single tactic are subject to failure when, either gradually or abruptly, that tactic fails. On the other hand, multifaceted programs do not rely on a single tactic, and if one tactic fails, others are present to help moderate losses. As we will see, the use of multiple strategies and tactics is a basic principle in developing insect pest management programs.

Kinds of Pests and Likely Strategies

The appropriate strategy and subsequent complexity of a pest management program is primarily determined by the status of an insect pest in the production system. Four pest types may be designated according to status, and identifying a pest species as one of these is useful in developing an appropriate strategy. The pest types include noneconomic populations (subeconomic pests of some authors), and occasional, perennial, and severe pests.

Subeconomic pests. The subeconomic pest, at first, may seem to be a contradiction in terms. However, these insects are pests in a true sense, although they cause insignificant losses. The GEP with this pest type is far below the economic-injury level, and the highest pest-population fluctuations do not reach that level. As mentioned earlier, attempting to reduce injury from such pests would cost more than the losses they inflict.

Many insects fall within the subeconomic category. An example is the alfalfa caterpillar, *Colias eurytheme*, in central Iowa. This insect is a defoliator of alfalfa and causes direct losses of forage. But densities are low and commodity values modest, so that the most appropriate action is to do nothing about the pest. As discussed previously, the pest management activity involved with this kind of pest is to monitor activity levels for changes that may alter the pest's status. Changes in the pest biology, crop biology, or economics can cause status to change, thereby requiring other strategies.

Another consideration with subeconomic pests is their contribution to overall injury. Should a complex of several subeconomic insect pests occur, management may be necessary when the total injury threatens economic damage. Such a complex sometimes occurs in corn when corn leaf aphids, *Rhopalosiphum maidis*,

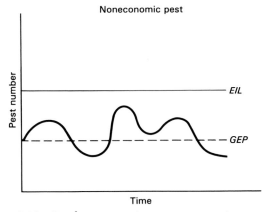

Figure 8.10 Graph representing a noneconomic or subeconomic pest. EIL, economic-injury level; and GEP, general equilibrium position.

BOX 8.2 ALFALFA CATERPILLAR

SPECIES: *Colias eurytheme* Boisduval (Lepidoptera: Pieridae)

DISTRIBUTION: The alfalfa caterpillar is native to North America and is found throughout the United States and southern Canada. However, the pest is most abundant and significant in the southwestern United States.

IMPORTANCE: The species is an important alfalfa pest, especially in the irrigated areas of the southwestern United States. Larvae defoliate host plants and can also attack stems. The pest also feeds on other legumes.

APPEARANCE: Adults typically have yellow to orange wings with a black margin on the edges. Albinism is common among females, which have white wings instead of orange. Wingspans are approximately 50 mm, and the antennae are slightly clubbed. Larvae are about 38 mm long when full-grown, and are green with a pink-and-white stripe running

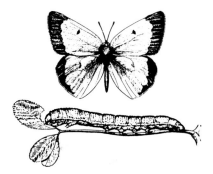

lengthwise along the side of the body. Eggs are spindle-shaped, with numerous ridges.

LIFE HISTORY: In the northern parts of their distribution, pupae overwinter in plant debris; however, all stages overwinter in the southwestern United States. In early spring, adults emerge and females oviposit from 200 to 500 eggs singly underneath leaves. In about three days, eggs hatch, and larvae feed for 12 to 15 days. Pupation occurs, and adults emerge in five to seven days. In the northern United States and southern Canada, there are typically two generations per year. In the southwestern United States, there may be as many as seven generations per year.

feeding on tassels and silks, cover them with honeydew, and western corn root-worm adults, *Diabrotica virgifera*, clip silks, thereby interfering with pollination. In many instances, the activity of only one species is not sufficient for action, but injury from the two, occurring simultaneously, may warrant action.

Occasional pests. The occasional pest is a very common type of insect pest. It has a GEP substantially below the economic-injury level, but the highest fluctuations exceed this level occasionally and usually sporadically. This pest may be present on a crop most years but, more often than not, it does not cause economic damage.

The management approach in this instance usually is to deal with the pest in a therapeutic way. A wait-and-see attitude is assumed, with reliance on early detection, prediction of impending outbreaks, and employment of tactics only when the economic threshold is reached. The objective of this strategy is to dampen outbreak peaks, with no effort toward reducing the GEP. Therefore, pest management programs for pests in this category tend to be less complex than those of more serious pests and may rely on only two or three tactics.

Pesticides often fit well in the strategy for occasional pests. When used judiciously, these can serve for long-term management, because application is infrequent, and consequently, problems with undesirable side effects are minimized.

An example of this type of pest is the green cloverworm, *Plathypena scabra*, a defoliator of soybeans in the midwestern United States. This insect is present

BOX 8.3 CORN LEAF APHID

SPECIES: *Rhopalosiphum maidis* (Fitch) (Homoptera: Aphididae)

DISTRIBUTION: The corn leaf aphid is found throughout the world and is distributed in North America east of the Rocky Mountains. Individuals can be found on corn, sorghum, sugarcane, barley, and several species of wild grasses.

IMPORTANCE: The species is an occasional corn pest during the late growing season. Individuals suck juices from leaves of hosts, causing a mottled yellow appearance. More important, the aphids feed on tassels and silk, coating them with honeydew, which interferes with pollination. Severely infested tassels eventually drop from the plants. The pest also vectors several plant pathogens, including maize dwarf mosaic virus of corn, mosiac disease of sugarcane, and yellow dwarf virus of barley.

APPEARANCE: Individuals are soft-bodied, about 3 mm long, and greenish-blue, with black antennae and appendages. Both winged and wingless forms exist during the growing season.

LIFE HISTORY: Corn leaf aphids migrate to the north central United States each year from southern regions. Individuals in early spring are all wingless females. At certain times in the growing season, winged males and females are produced, which disperse and reproduce. Reproduction is parthenogenetic. Normally, up to nine generations per year occur in the north central United States, and as many as 50 generations per year have been recorded in Texas.

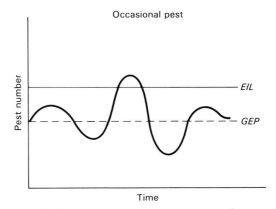

Figure 8.11 Graph representing an occasional pest. EIL, economic-injury level; and GEP, general equilibrium position.

each year in nearly all soybean fields, but unusually heavy moth flights from the South in late spring set the stage for midseason problems with larvae. The insect is rarely a problem in late season, because it is attacked and killed by a disease-causing fungus, an important natural enemy. Efficient grower-based sampling plans, predictive indices, and definitive economic thresholds have been developed for the pest. Tactics of suppression are used only when necessary and can include insect-selective microbial insecticides (*Bacillus thuringiensis*) or low rates of conventional insecticides. In the green cloverworm pest management program, natural enemy conservation is achieved, environmental problems averted, and grower profits sustained.

Perennial and severe pests. Perennial and severe pests cause the most serious and difficult problems in crop production. Only a few insects belong in these categories, and these are often referred to as **key pests**. Problems created by these pests are usually caused by relatively high market value of the crop and/or very dense insect populations. In this category are insects and other arthropods

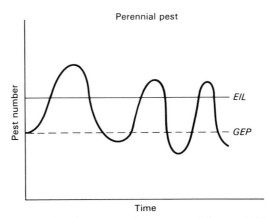

Figure 8.12 Graph representing a perennial pest. EIL, economic-injury level; and GEP, general equilibrium position.

BOX 8.4 GREEN CLOVERWORM

SPECIES: *Plathypena scabra* (Fabricius) (Lepidoptera: Noctuidae)

DISTRIBUTION: The green cloverworm is distributed in North America from the Great Plains eastward and from southeastern Canada to the Gulf of Mexico.

IMPORTANCE: The green cloverworm is considered a serious, yet sporadic, defoliator of soybeans, alfalfa, and other legumes. One of the most widespread of soybean insect pests, green cloverworm populations occasionally cause economic damage in the midwestern and southeastern United States in the form of significant yield losses. Larvae defoliate the host plant by chewing the tissues between the veins of the leaf, leaving a ragged appearance.

APPEARANCE: Adult moths appear triangular when at rest and typically have dark brown wings with lighter markings. But there are numerous color variants, ranging from light brown to black. The moths also have distinctive labial palps pointed forward from the head. Adult males can be differentiated from females in that the eyes of male moths are much larger. The larvae are pale green with two thin white stripes placed horizontally along each side of the body. Three pairs of prolegs (four pair including the anal pair) distinguish these larvae, which are approximately 25 mm long when full-grown. The eggs, which are 0.5 mm in diameter and have longitudinal ridges, are translucent green but turn light brown about 48 hours before hatching. Pupae are brown and lightly wrapped in a silken cocoon.

LIFE HISTORY: Adults overwinter south of 41° North latitude. Near the Gulf coast, feeding activity and reproduction occur year-round. Northern areas are colonized each spring by migrating moths transported by southerly winds. Females oviposit eggs singly on the undersides of leaves. The eggs hatch in about three days, and the larvae undergo six to seven molts in about 14 days. After a 7- to 10-day pupal stage, moths emerge and the cycle is repeated. There are two generations per year in the northern United States and three to four generations per year in the southern United States.

that attack the harvested produce directly and those that almost always occur in great numbers. Some of the worst pests may cause only blemishes on produce, making it unacceptable to consumers and resulting in serious economic losses. An example of losses from blemishes occurs with the citrus rust mite, *Phyllocoptruta oleivora*, which causes a serious russeting of oranges.

Pests in these categories are characterized by a very high average density in relation to the economic-injury level. With the **perennial pest**, the GEP is below, but so close to, the economic-injury level that economic damage occurs most years. Only infrequently do population peaks of this pest type not reach the economic-injury level. **Severe pests** have a GEP that is actually above the economic-injury level, making them a constant problem.

Unlike the strategy for occasional pests, the primary insect pest management strategy for perennial and severe pests is to reduce the GEP of the population.

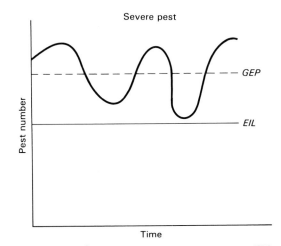

Figure 8.13 Graph representing a severe pest. EIL, economic-injury level; and GEP, general equilibrium position. Note that with this pest and GEP is above the EIL.

In the past, pesticide applications have been used heavily and frequently to accomplish this objective. Such usage has resulted in undesirable side effects and environmental problems. Because of these problems, use of pesticides alone is not a viable strategy for these pests, except as a stop-gap. Instead, more complex programs are required to produce lasting solutions. These programs usually comprise combined strategies involving several tactics to lower the environmental carrying capacity and/or the GEP of the pest population.

Such a program was developed for the spotted alfalfa aphid, *Therioaphis maculata*, in the western United States (California Central Valley). This aphid problem resulted from accidental introductions from the Middle East in 1954, and the species subsequently spread through California and to other states. Early attempts at suppression with insecticides showed success, but resistance (loss of effectiveness) to these chemicals developed rapidly. Particularly in areas with resistant populations, repeated sprays reduced natural enemies, and aphid damage became acute.

Subsequently, an insect pest management program was developed, offering effective protection from losses. The program involved (1) the introduction and establishment of exotic wasp parasites that attack the aphids, (2) strip harvesting (leaving uncut strips on a given harvest date) to conserve and enhance natural enemies, (3) development of useful economic thresholds and implementation of sampling procedures for aphids and natural enemies, and (4) use of insecticides that were less destructive of natural enemies. Later developments of aphid-resistant alfalfa varieties have almost eliminated the need for insecticide sprays in this program. The lasting value of such an integrated approach lies in its reliance on several compatible tactics rather than on a single one.

DEVELOPMENT OF A PEST MANAGEMENT PROGRAM

The practice of insect pest management depends on a series of well-designed programs that apply to important pests in the production system. Each program, sometimes called a "recommendation algorithm," can be visualized as a bridge

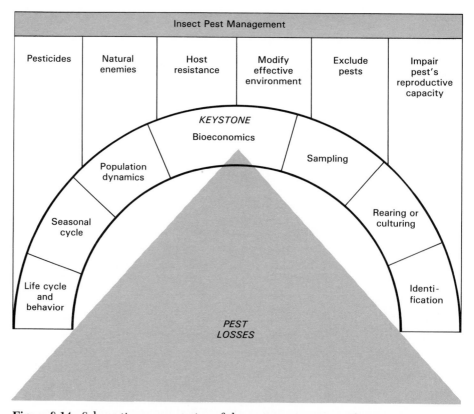

Figure 8.14 Schematic representation of the major components of an insect pest management program.

that can be crossed to avoid significant losses to pests. Such a bridge, shown in Figure 8.14, is composed of a foundation arch (information and techniques), several vertical pillars (tactics), and a road surface (avoidance of losses).

Information and Techniques

The foundation arch of the bridge is made up of several elements and represents the basic information that should be known and techniques needed to manage a pest. The elements to the left of center stand for biological information about a pest; those to the right, for procedures needed to obtain this information.

Biological information includes data on how individuals feed, grow, develop, reproduce, and disperse, as well as habitat requirements. Pest management seeks particularly vulnerable points in the life cycle to exploit. The seasonal cycle is studied along with pest developmental rates to provide predictions of pest occurrence and relationship to biological events of the crop.

Population dynamics is another critical element in the biological information needed for insect pest management. It seeks to explain the forces (natural enemies, weather, food, and habitat) causing population change by quantifying these forces and presenting the numbers as mathematical models. Such models, consisting of component equations, have often been organized to represent an ecological system and have been programmed for computer simulation. Such simulations are believed useful for understanding population dynamics in various environments. Population dynamics information builds on the understanding of behavior, life cycle, and seasonal cycle, and allows prediction of infestations and outbreaks. It also indicates environmental factors that might be exploited to reduce pest status.

The techniques in the foundation arch include species identification, rearing or culturing, and sampling. Identification involves assigning the correct name to a species and knowing its taxonomic classification. Because every species behaves differently, it is critical to make accurate identifications; they are a key to the published literature about a pest. They aid in research and development of pest management programs and serve as a reference for delivering such programs to practitioners. Not only is the ability to rear, culture, or colonize a pest useful for obtaining biological information about a species, but also laboratory and field experimentation with management tactics can proceed more rapidly than by having to depend completely on natural pest occurrence.

Sampling techniques and programs play a particularly important role in insect pest management. As mentioned earlier, such techniques allow the estimation of numbers for population dynamics studies and later serve to assess the population for management decisions. Without precise, economical sampling programs, it is doubtful that an effective pest management program can be designed.

The keystone of the pest management foundation arch is **bioeconomics**. As you will recall, this is the relationship of pest numbers to losses. Bioeconomics ties together biological information of the pest and host crop, sampling programs, and economics. Determination of bioeconomics depends greatly on a knowledge of pest-plant or pest-livestock relationships and how the crop responds to pest damage. Coupling known crop responses with pest densities and accounting for market values and pest management costs allows economic-injury levels and economic thresholds to be developed.

Tactics

The foundation arch of our bridge diagram supports several pillars that represent tactics used in insect pest management. As you might agree, the bridge would not be very stable with only one pillar or tactic. Therefore, the diagram shows that several tactics are used in combination to avoid losses to pests. The tactics shown do not include all those available; however, they are some of the most useful ones.

Putting It All Together: Systems

A basic problem in the design of pest management programs is how to assemble all the elements previously discussed into a comprehensive, understandable set of procedures. In recent years, systems technology has become an increasingly popular method of organizing information about a pest and delivering it to users.

The development and use of systems technology has been greatly enhanced by spectacular growth in the computer industry. Ranging from circuit boards to microchips to software, the idea of "system" has played a central role in computer development, and subsequently has made the word a part of nearly everyone's vocabulary.

Although widely interpreted, a system can be defined as an identifiable set of interrelated elements around which a boundary can be drawn. Your own body can be thought of as a system with inputs and outputs, and it is easily distinguished from its environment. Interrelated elements in this system are organs like the

Figure 8.15 Diagram of a pest management system. (Redrawn from Ruesink, 1982, in Mecalf and Luckmann, eds., *Introduction to Insect Pest Management,* Wiley & Sons)

heart that form subsystems, for example, the circulatory system, that contribute to the functioning of the overall system.

Likewise, a pest management system has interrelated elements like the processes of obtaining information about potential pest problems, decision making, and strategies. W. G. Ruesink presented the pest management system as a diagram with the interrelated elements and associated inputs and outputs shown in Figure 8.15. Here, the boundary of the pest management system is shown with dashed lines, and information for its operation is received from weather agencies, business sources, and sampling of the agroecosystem. Outputs of the system are actions of the grower directed toward the agroecosystem.

At the heart of the pest management system is the so-called recommendation algorithm. An **algorithm** is a precisely defined sequence of rules stating how to produce specific outputs from specific inputs through a given number of steps. Therefore, a recommendation algorithm is the set of steps for managing a pest. Such an algorithm would exist for each insect pest or group of pests important to the agroecosystem. Each recommendation algorithm produces a specific pest management scheme for the inputs received.

An example of a recommendation algorithm is shown in Figure 8.16 for the green cloverworm in soybeans. Although all details are not presented, a number of pest management activities are assembled in an organized manner, requiring inputs on insect population (adults and larvae), weather (degree-day accumulations), plant growth (stage designations), and economics (included in the larval sequential sampling plans). After several steps in the sequence, decisions are made and recommendations (outputs) advanced for a particular set of circumstances.

Such recommendation algorithms can conveniently be programmed for solution by computers. In the United States, there is a growing trend in this direction, with the delivery of pest management recommendations to growers via on-farm personal computers and computer networks sponsored by the State Cooperative Extension Service.

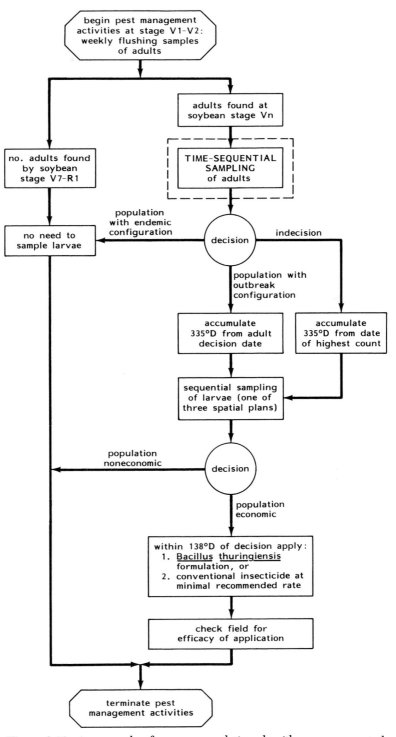

Figure 8.16 An example of a recommendation algorithm as represented by that for the green cloverworm, *Plathypena scabra*, on soybeans.

The advantage of using the systems approach is obvious. It provides a mechanism for integrating complex information from many sources to produce optimal pest management programs—those that are most economically and ecologically sound. In this way systems technology helps to fulfill the objectives set down by the principles of pest management.

Selected References

ALLEN, G.E., and J.E. BATH. 1980. The conceptual and institutional aspects of integrated pest management. *Bioscience* 30:658–664.

CHANT, D.A. 1964. Strategy and tactics of insect control. *Canadian Entomologist* 96: 182–201.

GEIER, P.W. 1966. Management of insect pests. *Annual Review of Entomology* ll:471–490.

JONES, D.P. 1973. Agricultural entomology. In R.F. SMITH, T.E. MITTLER, and C. N. SMITH, eds., *History of Entomology*. Annual Reviews, Palo Alto, Calif: pp. 307–332.

MATTHEWS, G.A. 1984. *Pest Management*. Essex, England: Longman, pp. 165–182.

RABB, R.L., G.R. DEFOLIART, and G.G. KENNEDY. 1984. An ecological approach to managing insect populations. In C.B. HUFFAKER, and R. L. RABB, eds., *Ecological Entomology* New York: Wiley, pp. 697–728.

RUESINK, W.G. 1982. Analysis and modeling in pest management. In R.L. METCALF, and W.H. LUCKMANN, eds., *Introduction to Insect Pest Management*, 2nd ed. New York: Wiley, pp. 353–373.

STERN, V.M., R.F. SMITH, R. VAN DEN BOSCH, and K.S. HAGEN. 1959. The integrated control concept. *Hilgardia* 29:81–101.

9 NATURAL ENEMIES OF INSECTS

UNNOTICED BY THE CASUAL OBSERVER, thousands of insect and mite species attack crops, forests, and livestock but they occur in very low numbers. Most of these species cannot be considered pests at all or are subeconomic pests because their activity does not result in an economic loss. There are many reasons for their low numbers, including unfavorable weather, lack of material requisites (for example, food and nesting sites), and, very often, **natural enemies**.

As discussed previously, natural enemies are living organisms found in nature that kill insects outright, weaken them and thereby contribute to their premature death, or reduce their reproductive potential. A natural enemy usually reduces the subject insect population, the **host** or **prey**, by feeding on individuals, thereby promoting its own population at the expense of the population fed upon. Not only do these natural enemies help prevent some insects from attaining pest status, but they also play a role in reducing the damage potential of significant pests. Established pests, as damaging as they may be, would cause even more damage were it not for the presence of natural enemies.

Virtually all insect populations are affected to a greater or lesser extent by natural enemies. This includes insects from tropical to subarctic regions and those in unmanaged habitats, as well as in agroecosystems. For many species, natural enemies are the primary regulating force in the dynamics of their populations. For this reason alone, it is very important to know what natural enemies are affecting an insect-pest population and to obtain an estimate of their impact. Such information may be the basis for explaining pest population density and predicting outbreaks.

Perhaps even more significant to pest management programs is an understanding of natural enemies so that they can be manipulated to our advantage. Here we would use a strategy that in Perkins's words seeks to use nature as a partial ally, not as a total adversary. Indeed, a basic tenet of pest management is to consider the natural enemy component first in developing a strategy.

The pest management tactic involving purposeful natural enemy manipulation to obtain a reduction in a pest's status is called **biological control** or **biocontrol**. Biological control differs from **natural control** in that the latter may involve agents other than natural enemies, for example, weather or food, and no purposeful manipulation is involved. Some authorities broaden biological control to include all methods that involve living organisms as part of the tactic, like host plant resistance, sterile insect releases, and genetic manipulation. For the purpose of discussion, we consider biological control in the conventional sense, involving only the manipulation of natural enemies.

BRIEF HISTORY OF BIOLOGICAL CONTROL

Biological control is one of the oldest, most effective means of achieving insect control. As already mentioned, the earliest record dates back to fourth-century China (Kwantung Province), where ants were used to suppress pests in citrus. Many other early attempts were made throughout the world using insects to control other insects; birds, lizards, and toads to control insects; and vertebrates to control other vertebrates (mainly rats), all with varying degrees of success.

It was not until 1888 that biological control became firmly established in the United States as a significant method. That year marked the phenomenal success of an introduced predatory insect, the vedalia beetle, *Rodolia cardinalis*, in eliminating the threat to citrus of the cottonycushion scale, *Icerya purchasi*. This scale insect was introduced and became established in the Los Angeles area sometime before 1876. By 1885, it was threatening to wipe out the citrus industry in southern California. C.V. Riley, then the state entomologist for Missouri, examined specimens and determined that the pest had been introduced from Australia. After becoming chief entomologist for the U.S. Department of Agriculture, Riley arranged for an expedition to Australia to collect natural enemies of the cottonycushion scale for possible introduction. Subsequently, another American entomologist, Albert Koebele, traveled to Australia and found two enemies of the scale, a parasitic fly and the vedalia beetle. These were shipped to San Francisco, reared, and sent to Los Angeles. The natural enemies were released on scale-infested trees covered by canvas tents. The vedalia beetle population increased rapidly and was allowed to spread to adjacent trees. With the help of growers and state agencies, the vedalia beetle dispersed throughout the scale-infested areas. Complete economic suppression of the cottonycushion scale had

Figure 9.1 A vedalia beetle, *Rodolia cardinalis*, eating a cottonycushion scale, *Icerya purchasi*. (After Copple and Mertins, 1977, *Biological Insect Suppression*, Springer-Verlag Publishers, with permission)

BOX 9.1 COTTONYCUSHION SCALE

SPECIES: *Icerya purchasi* Maskell (Homoptera: Margarodidae)

DISTRIBUTION: The cottonycushion scale was introduced into the United States (California) about 1868. The species was later introduced into Florida but has never been a serious pest there.

IMPORTANCE: The cottonycushion scale is a citrus pest, but it also attacks shade trees, ornamentals, and other fruit trees. Soon after its introduction, the species became a serious pest, and by 1890, it had killed thousands of trees. With no effective control measures, the scale threatened the entire California citrus industry. The scale was found in New Zealand (where it was a serious pest) and in Australia (where it was not a pest). In Australia, a lady beetle, *Rodolia cardinalis* (Mulsant), commonly called the vedalia beetle, was observed to feed voraciously on the cottonycushion scale, effectively suppressing the populations. Vedalia beetles were subsequently captured and introduced into California in 1888. After 18 months, the vedalia beetle had successfully suppressed the cottonycushion scale population, and currently the species is considered only an occasional pest. The vedalia beetle has been successful against the cottonycushion scale in almost all areas of the world where it has been introduced, making it one of the primary examples of classical biological control.

APPEARANCE: Females are yellow to reddish brown and covered with white to yellow wax-like filaments. They are usually found carrying yellow egg masses. Males are brown and have two wings. Eggs are oval, pale red, and contained in cottonlike masses, with up to 1,000 eggs per mass. Nymphs are minute, about 0.7 mm long, and red with black legs.

LIFE HISTORY: Females oviposit egg masses on trees, and eggs hatch in 3 to 60 days, depending on temperature. Nymphs feed and darken as they mature. Waxy filments are excreted, covering the back and sides of the developing nymphs.

---

occurred by 1889, and that suppression continues even today. Additionally, the fly parasite also became established and is today an important natural enemy of the scale in an area around San Francisco and some parts of southern California. The total cost of this management program was less than $2,000!

Many other successes followed, including the use of insects to control such weeds as the prickly pear cactus in Australia and Klamath weed in the western United States. In a survey of biological control literature to 1980 (Bulletin of the Entomological Society of America vol. 26, pp. 111–114), 602 attempts of classical biological control were found on a worldwide basis. The rate of complete success was 16 percent, and partial success (reduction but not elimination of the pest problem) was 58 percent. In this survey, the most complete successes were obtained with pests in the orders Homoptera (30 percent), Hemiptera (15 percent), Lepidoptera (6 percent), and Coleoptera (4 percent). The degree of success for the amount of money invested in biological control has been particularly impressive and has stimulated much interest in the approach for use in pest management programs.

Figure 9.2 Biological control of prickly pear, *Opuntia* species, by *Cactoblastis cactorum* (Lepidoptera: Pyralidae) in Australia. Top photo is original problem without control. Middle photo is three years after introduction of the natural enemy. Bottom photo is five years after introduction. (After Dodd, 1940, *The Biological Campaign Against Prickly Pear*, Commonwealth Prickly Pear Board)

Figure 9.3 Biological control of Klamath weed by *Chrysolina quadrigemina* (Coleoptera: Chrysomelidae) in California. Flowering weeds before control by the beetle in 1948 (top), area after the release of the beetle showing heavy grass cover in 1950 (middle), and full weed suppression in 1966 (bottom). (After Huffaker and Kennett, 1969, *Canadian Entomologist* 10:425-447)

THE THEORY BEHIND CLASSICAL BIOLOGICAL CONTROL

The theory of biological control is actually no different from that in ecology and population dynamics. As discussed earlier, many environmental factors regulate population density, thereby keeping insects within certain bounds of fluctuation. These include density-independent and perfectly and imperfectly density-dependent factors.

In the context of biological control, we think initially of the pest population as the one being regulated, and natural enemies as imperfectly density-dependent factors responsible for that regulation. The object of biological control is either to introduce natural enemies or to manipulate existing ones so as to cause the pest population to fluctuate (be regulated) at a density below the economic-injury level.

A goal of many of these biological control programs is to establish a self-sustaining system. For example, a natural enemy is introduced in an area with hopes that it will become established, cause the pest to fluctuate below the economic-injury level, and continue to hold the density down indefinitely without further manipulation. Of course, this strategy does not mean total elimination of the pest in an area because achieving that would also mean elimination of the food supply for the natural enemy and, consequently, the natural enemy.

The mechanism for the self-sustaining system, in theory, is based on food supply for and reproductive capability of the natural enemy. Here, an increasing pest population provides plenty of food for the natural enemy, and the natural enemy population expands. As expansion occurs, an increasing proportion of the pest population is destroyed, reducing food for the natural enemy. This food shortage worsens until natural enemy reproduction rates decrease, causing a downward trend in enemy population numbers. As natural enemy numbers fall, less pressure is placed on the pest population, and pest numbers turn upward, followed by an increase in natural enemy numbers.

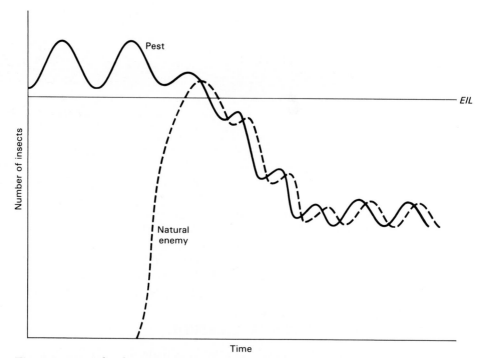

Figure 9.4 Graphical representation of the theoretical relationship between a pest and the introduction of an effective natural enemy.

In reality, the regulating mechanism is usually not so perfectly dependable or smoothly operating; thus regulation by natural enemies is imperfect at best. Therefore, this theoretical description is a rough but reasonable approximation of successful biological control programs, like that of the vedalia beetle and the cottonycushion scale.

In the foregoing discussion, only one aspect of natural enemy response to pest density was mentioned, namely, changes in natural enemy numbers. This aspect has been called the numerical response. Natural enemies may also respond to increasing pest numbers by destroying more pests per natural enemy. This aspect is referred to as the functional response. Numerical responses are usually of more interest, because often they represent the responses most accountable for overall suppression of pest populations.

AGENTS OF BIOLOGICAL CONTROL

Most, if not all, species of insects are preyed upon or serve as hosts for other life forms. These diverse natural enemies of insects include vertebrates, nematodes, microorganisms, invertebrates other than insects, and, perhaps most important, insects themselves. Indeed, other insects may be an insect population's worst enemy. As a group, natural enemies may function as **parasites**, **predators**, or **pathogens**.

Parasites and Parasitoids

A parasite is an animal that lives on or within a larger animal, its host. The parasite feeds on its host, usually weakening it and sometimes killing it. A parasite requires only one or part of one host to reach maturity. Frequently, there are many parasites in a single host. Parasites with the greatest impact on insect populations are insects and nematodes. Mites also parasitize insects but have a lesser impact on host populations.

Parasitoids.   Insects that parasitize other insects and arthropods are most appropriately called **parasitoids**. A parasitoid is parasitic in its immature stages but is free-living as an adult. In all instances, parasitoids kill their hosts, but in some circumstances, the host may live much of its full life before dying. Parasitoids

Figure 9.5   A parasitoid, *Lysiphlebus testaceipes*, ovipositing in an aphid (Homoptera: Aphididae). (Courtesy USDA)

may attack any host stage, but the adult stage is the least frequently parasitized. These natural enemies have been used more frequently in biological control than any other kind of agent.

Six orders of insects including 86 families have been listed as parasitoids. These orders include the Coleoptera, Diptera, Hymenoptera, Lepidoptera, Neuroptera, and Strepsiptera. However, the most important of these by far are the Hymenoptera and Diptera, in that order. The most significant groups of parasitoids in the Hymenoptera include small parasitic wasps in the Ichneumonidae, Braconidae, and Chalcidoidea (several families). In the Diptera, the Tachinidae stand out as the most significant parasitoids.

Parasitoids either penetrate the body wall and lay eggs inside the host or attach eggs to the outside of the host's body. The latter phenomenon requires newly emerged larvae to burrow inside or somehow break through the host's exoskeleton to feed.

Parasitoids are often effective biological control agents because of the following: (1) survival is usually good, (2) only one (or fewer) host is required for complete development of a parasitoid, (3) populations can be sustained at low host levels, and (4) most parasitoids have a narrow host range, often resulting in a good numerical response to host density. Most frequent disadvantages of parasitoids in biological control are the following: (1) host searching capacity may be strongly reduced by weather or other factors, (2) only the female searches, and (3) often the best searchers lay few eggs.

Synchronization is also a difficult problem with some parasitoids. To be effective, the parasitoid life cycle must coincide closely with that of its host before establishment and suppression can occur. Synchronization may be upset by cer-

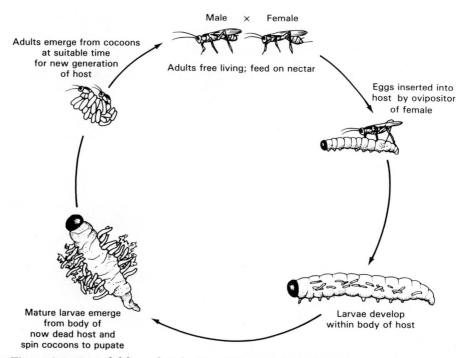

Figure 9.6 Typical life cycle of a parasitic wasp (Hymenoptera). (Redrawn from Finemore, 1984, *Plant Pests and Their Control*, Butterworths Publishing Company)

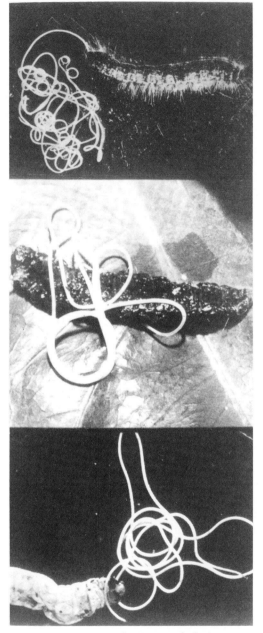

Figure 9.7 Parasitic nematodes (Mermithidae) emerging from Lepidoptera: tent caterpillar (top), mahogany shoot borer (middle), and fall armyworm. [G. C. Papavizas, ed., *Biological Control in Crop Production*, Beltsville Agricultural Research Center (Totowa N.J.: Allanheld, Osmun, & Co.), 1982]

tain environmental conditions, causing the parasitoid to fail to reduce host numbers significantly.

Insect parasitic nematodes. Nematodes are thin, unsegmented roundworms, some of which parasitize insects. Many other species are parasitic on humans, other animals, and plants. In biological control, particular attention has been given to nematodes as parasites of bark beetles (Coleoptera: Scolytidae), grasshoppers (Orthoptera: Acrididae), and black flies (Diptera: Simuliidae). The most important families of the Nematoda affecting insects include the Mermithidae, Neotylenchidae, and Steinernematidae.

Nonbeneficial parasites.   Although parasites and parasitoids of insects are usually thought of as beneficial, some are not, because, like pests, natural enemies have enemies themselves. Parasitoids have parasites (most often, tiny wasps) that may kill them while the parasitoids are feeding on their hosts. This phenomenon has been termed **hyperparasitism**.   Here, the original parasite of the pest is the **primary parasite**, and the hyperparasite is the **secondary parasite**. Primary parasites also are not beneficial to biological control when they parasitize insect predators or plant-feeding insects introduced to suppress weeds. In some situations, the potential of biological control programs can be seriously altered when these detrimental species are present, and they must be considered before major biological control programs are implemented.

Predators

Predators are free-living organisms that feed on other animals, their **prey**, sometimes devouring them completely and usually rapidly. Predators may attack prey both as immatures and adults, and more than one prey individual is required for a predator to reach maturity. Major predators of insects include birds, fish, amphibians (toads, frogs, salamanders), reptiles (lizards, snakes, turtles), mammals (bats, rodents, shrews, and others), and arthropods (bats, insects, mites, spiders). Even some plants are predators of insects, for example, the Venus flytrap which has clustered leaves with blades that close and trap insects for food.

The most important predators in biological control programs have been insects and mites. The habit of predation is very common among insects, and some of our most successful instances of biological control have been with insect predators and insect pests.

It is difficult to rank insect orders with regard to significance in predation because nearly every order has important species. However, with regard to diversity and significance of biological control, the Coleoptera, Neuroptera, Hymenoptera, Diptera, and Hemiptera are outstanding. Some particularly impor-

Figure 9.8 Mosquito fish feeding on a mosquito larva. (After Copple and Mertins, 1977, *Biological Insect Suppression*, Springer-Verlag Publishers, with permission)

tant species groups include ladybird beetles (Coleoptera: Coccinellidae), ground beetles (Coleoptera: Carabidae), lacewings (Neuroptera: Chrysopidae), ants (Hymenoptera: Formicidae), flower flies (Diptera: Syrphidae), damsel bugs (Hemiptera: Nabidae), and assassin bugs (Hemiptera: Reduviidae).

Some predaceous insects like the vedalia beetle have a narrow prey range, feeding almost exclusively on a single species. These predators are **monophagous**. Other predators such as syrphid larvae have a narrow host range, feeding on only a few prey species. These predators are **oligophagous**. As a group, however, predators tend to feed on a wide range of prey and are considered **polyphagous**.

Being polyphagous has both advantages and disadvantages for biological control. It is an advantage that polyphagous predators can survive by shifting to alternate prey when densities of the pest species are low. It is disadvantageous, however, when the polyphagous habit results in lack of responsiveness to changes in pest density; in other words, where the predator does not suppress a growing pest population. But it is difficult to predict the density response of a natural enemy to a given pest, from food specificity alone.

Other characteristics make predators effective biological control agents. These include the following: (1) prey is killed rapidly, (2) often all individuals in the population (males, females, immatures, and adults) search for prey, and (3) synchronization in predator/prey life cycles is not a frequent problem.

Much debate has centered around whether predators or parasitoids are the more efficient biological control agents. Certainly, parasitoids have been used more frequently than predators in attempts at biological control. Ratios of parasitoids to predators used in programs have been estimated at from 2:1 to 4:1. But the most successful type of agent for any one pest problem is almost impos-

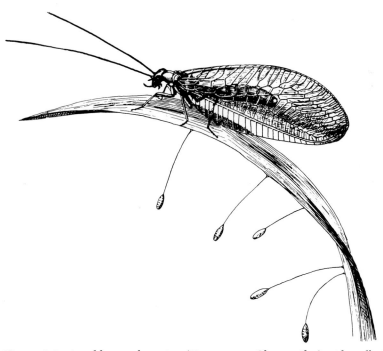

Figure 9.9 A goldeneye lacewing (Neuroptera: Chrysopidae) with stalked eggs. (Courtesy USDA)

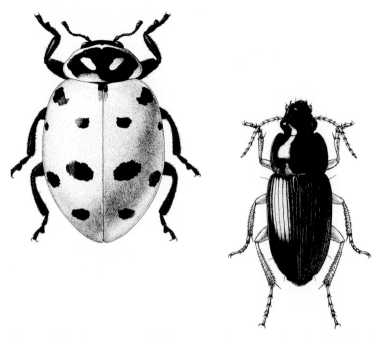

Figure 9.10 Important polyphagous predators: a convergent ladybird beetle, *Hippodamia convergens*, (left) and a ground beetle, *Harpalus pennsylvanicus*. (Courtesy USDA)

Figure 9.11 Wasp parasitoid (Hymenoptera: Braconidae) cocoons on a tobacco hornworm, *Manduca sexta*. The hornworm will die before pupation. (Courtesy USDA)

sible to predict. Success actually depends on the particular agroecosystem, the pest(s) involved, and the environment in which these interact. Consequently, it has usually been best to design management programs that rely on the actions of both parasitoids and predators.

Pathogenic Microorganisms

Just like people and other animals, insects contract diseases, and natural populations are often strongly influenced by disease epidemics. Insect diseases and their symptoms have been recognized as far back as 2700 B.C. in China with the honeybee, *Apis mellifera*, and the silkworm, *Bombyx mori*. Early scientific studies on insect diseases include investigations of white muscardine disease caused by the

BOX 9.2 LADYBIRD BEETLES

SPECIES: numerous coccinellid species (Coleoptera: Coccinellidae)

DISTRIBUTION: Ladybird beetles are widely distributed throughout the world, with many species represented in North America. Throughout their range, they are found on a wide variety of trees, shrubs, weeds, grasses, and cultivated crops.

IMPORTANCE: Ladybird beetles represent one of the most beneficial and recognizable group of insects known to humankind. For most species, both adults and larvae are voracious predators on a wide range of prey, including aphids, scales, mites, and thrips. One of the most common, *Hippodamia convergens* Guerin-Meneville, can consume its weight in aphids daily as a larva and as many as 50 aphids daily as an adult. Lady beetles have been imported to control outbreaks of aphids and scale insects. The most successful of these introductions was in 1888, when *Rodolia cardinalis* (Mulsant), was imported from Australia to control the cottonycushion scale, *Icerya purchasi* Maskell, a pest of California citrus groves. For one hundred years this beetle, commonly called the vedalia beetle, has kept cottonycushion scale populations below economic levels. Two species of Coccinellidae, however, are plant-feeding and considered pests: the Mexican

bean beetle, *Epilachna varivestis* Mulsant, and the squash beetle, *Epilachna borealis* (Fabricius).

APPEARANCE: Adults range from 3 to 8 mm long and are oval. Both males and females are brightly colored, with the most common species red or brown with black spots. A few species have black elytra with red spots. Eggs of most species are orange and deposited in clusters of 10 to 50. Larvae are elongate and usually patched with bright colors.

LIFE CYCLE: Adults overwinter, often aggregating in large masses on plants and then under rocks, bark, leaf debris, and the southern side of structures. In spring, the beetles become active and mate. Females oviposit as many as 1,000 eggs in a one- to three-month period. Incubation time in most instances is two to six days. Most ladybird beetles undergo three larval molts in about 20 days. Pupation lasts from 3 to 10 days, and adults can live from a few months to over a year. Larvae and adults can be seen throughout the summer on the same plant, feeding on aphids and other small insects and their eggs.

fungus *Beauveria bassiana* and pebrine disease caused by the microsporidian *Nosema bombycis* on silkworms. The idea of using a microorganism (a fungus) to control a pest (the sugarbeet curculio, *Cleonus punctiventus*) dates back to the eighteenth century. Today, the science of insect pathology is an established discipline, contributing significantly to the biological control of insects.

The major microorganisms causing diseases in insects include bacteria, viruses, protozoans, fungi, and rickettsiae. They may cause diseases that kill insects outright, reduce their reproductive capabilities, or slow their growth and development. These organisms cause disease epidemics in insect populations naturally, and understanding their action is an important part of predicting the population dynamics in many pests.

Most of these microorganisms have been used or studied for use as so-called **microbial insecticides**. Microbial insecticides are biological preparations that are often sprayed or delivered in ways similar to that of conventional chemical insecticides. Such biological preparations have label directions like insecticides and must be registered with the United States Environmental Protection Agency before they can be sold.

To date, the microorganisms most widely used as microbial insecticides are bacteria and viruses. Fungi, although important in natural population regulation, are strongly influenced by the weather and therefore are not always reliable. Both protozoa and rickettsiae have the disadvantage of being slow to kill insects, if they kill at all. Some bacteria and many viruses kill insects relatively quickly (days to a few weeks rather than several weeks or months) and are rather resilient to extremes in weather.

Of the bacteria, the spore-forming species have been the most important in insect suppression. Particularly, the genus *Bacillus* has been used, some of the

Figure 9.12 Mumified body of a green cloverworm larva, *Plathypena scabra*, killed by a pathogenic fungus, *Nomuraea rileyi*. Conidiophores of the fungus form a white mat over the body surface. (Photo by H. Thorvilson)

Figure 9.13   Microbial insecticides are packaged and labeled similarly to chemical insecticides. This pesticide is a virus preparation for use against *Heliothis* species (Lepidoptera: Noctuidae). (Courtesy Sandoz Incorporated)

most notable examples being *B. popilliae* and *B. lentimorbus*. These bacteria cause milky disease in the Japanese beetle, *Popillia japonica*. Another is *B. thuringiensis*, causing diseases in many species of moths (Lepidoptera), mosquitoes (Diptera: Culicidae), and a few beetles (Coleoptera).

Bacillus popilliae and *B. lentimorbus* spores are ingested by Japanese beetle and other scarab larvae. The spores germinate and penetrate their gut and cause their blood to appear milky. Death follows soon afterward. Both these bacteria are **fastidious** (they must have the host to develop) and therefore are difficult and expensive to propagate. Conversely, *B. thuringiensis* is less fastidious, being propagated on artificial media, and therefore is better suited for commercialization. *Bacillus thuringiensis* also differs in being a crystaliferous bacterium. After ingestion, the proteinaceous crystal, associated with a spore in a sporangium, quickly dissolves in the midgut, causing gut paralysis. The host larva then may be killed outright by the toxic crystal, or the crystal may enhance the penetration of the gut by the spores, which consequently cause a lethal septicemia.

Although bacteria have been demonstrated as effective against many pest species, some insect pathologists believe that viruses are the most promising group of pathogenic microorganisms for biological control. To date, more than 450 viruses have been described from about 500 arthropod species. The majority of these cause diseases in the Lepidoptera, Hymenoptera, and Diptera.

The best-known viruses of insects are the nuclear polyhedrosis viruses (NPV).

Figure 9.14 Above, *Bacillus popilliae* sporangia under a phase-contrast microscope, each with a large spore and a smaller parasporal body. Below, a healthy (left) and *Bacillus popilliae* infected Japanese beetle (Popillia japonica) larva. The metathoracic legs were clipped to produce a droplet of blood. Note that the blood color of the diseased larva on the right is white. (Top, courtesy USDA; bottom, After Copple and Mertins, 1977, *Biological Insect Suppression*, Springer-Verlag Publishers, with permission)

BOX 9.3 JAPANESE BEETLE

SPECIES: *Popillia japonica* Newman (Coleoptera: Scarabaeidae)

DISTRIBUTION: The Japanese beetle was introduced into the United States (New Jersey) from Japan about 1916, probably in perennial plant nursery stock. Since its introduction, the species has spread to almost all states east of the Mississippi River and also portions of Ontario and Quebec in Canada. Two infestations have occurred in California, but the populations were destroyed in costly eradication programs.

IMPORTANCE: The species is only a minor pest in Japan, its native habitat; but in North America it is a serious pest. Larvae live in the soil and feed on the roots of many species of plants. They are especially damaging to pastures, lawns, golf courses, and parks. Grasses are stunted and yellowed under low infestations, but under high infestations (100 to 500 larvae per square yard) grasses turn brown and eventually die. Adults attack leaves, flowers, and fruits of over 275 species of plants, such as fruit trees, shade trees, soybean, corn, flowers, and vegetables.

APPEARANCE: Adults are 8 to 10 mm long and are metallic green with six white tufts of hair along the bronze elytra. Larvae, about 19 mm long when full-grown, are C-shaped white grubs with a brown head and a swollen tail. Eggs, about 1 mm, are white and deposited in 5 to 15 cm of soil.

LIFE HISTORY: Larvae overwinter in soil. In early spring, they move near the surface of the soil and feed on roots. Pupation occurs from late May to early June. In late June, adults begin emerging and are active until late September. Beetles mate during this time, and the females oviposit about 50 eggs in the soil. In about 14 days, the eggs hatch, and the larvae feed until the onset of cold weather. At this time larvae tunnel to about 20 cm in the soil to overwinter. There is usually one generation per year, but during excessively cold and wet conditions, two years may be needed for completion of development.

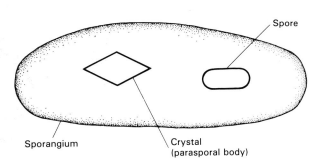

Figure 9.15 Schematic of *Bacillus thuringiensis* sporangium with proteinaceous crystal and associated spore.

Figure 9.16 Cabbage looper (*Trichoplusia ni*) infected with a nuclear polyhedrosis virus (NPV). Healthy larva (top), advanced infection (middle), and dead larva. (Courtesy USDA)

Figure 9.17 Electron micrograph of nuclear polyhedrosis viruses showing packets of virus rods in one polyhedron (bottom) and single rods in the other. (Courtesy USDA)

However, granulosis, cytoplasmic polyhedrosis, and entomopox viruses, among others, also show promise as microbial insecticides. NPV viruses develop in the host cell nucleus where one or several virus rods occur singly or in groups encased in an envelope. The envelopes are occluded (encased) in many-sided crystals called polyhedra. After ingesting the polyhedra, larvae show no outward symptoms for four days to three weeks. At this time, the larval skin darkens, and larvae climb to the highest point on their host plant, where they die. Dead, blackened, larvae may be found hanging from the tops of plants. Subsequently, the integuments of these dead larvae rupture, and millions of polyhedra are released into the environment. Such diseases have been called collectively caterpillar wilt.

The future of microbial insecticides appears to be assured. Increased problems of resistance and environmental contamination with conventional insecticides creates an impelling need for safer alternatives. Although no panacea, these materials may be used similarly to insecticides, and this would make them very adaptable to many established pest management programs.

Also, advances in biotechnology should allow production costs of microbial insecticides to decrease and efficiency to increase. Then these agents would become even more appealing. Among the expectations of biotechnology, we might include modification of fastidious microbial pathogens that allow them to reproduce on artificial media, propagation of host-insect tissue from single cells that can be used as media for fastidious pathogens, and using genetic engineering to transform nonpathogenic microorganisms into insect pathogens. An example of the last objective is recent research to modify a common nonpathogenic soil bacterium, *Pseudomonas fluorescens*, through splicing a gene from *B. thuringiensis*, thereby engineering a pathogenic bacterium that would attack soil insect pests.

THE PRACTICE OF BIOLOGICAL CONTROL

Through the years, many diverse methods have been developed for practicing biological control, but most programs have a common thread. They all seek to reduce pest status by increasing pest mortality, and very often, this is done by increasing existing natural-enemy-to-prey ratios. Attempts at increasing ratios include placing natural enemies in the environment and modifying the environment in such a way as to ecologically promote increases in natural enemy numbers. Another method used to raise ratios is modifying established crop production practices so that natural enemies otherwise destroyed are conserved.

To reflect these practices, many specialists group biological control practices into three categories: **introduction, augmentation**, and **conservation**. Although there is some overlap between these categories, we use them for the sake of discussion. It also should be noted that more than one practice may be used in a single biological control program.

Introduction

Introduction, also known as importation, is often considered the "classical" practice in biological control because most early programs used this approach. The basis for this practice is to identify the natural enemies that regulate a pest in its original location and introduce these into the pest's new location; thus natural enemies are reassociated with their prey and hosts. The hope is that the natural enemies, once introduced, will become established and permanently reduce the pest's general equilibrium position (average population level) below the economic-injury level.

Figure 9.18 Immature stink bugs (Hemiptera: Pentatomidae) sucking body contents from larval prey. (After Copple and Martins, 1977, *Biological Insect Suppression*, Springer-Verlag Publishers, with permission)

One of our most successful examples of this approach has already been mentioned: the introduction of the vedalia beetle to control the cottonycushion scale. Other substantial successes include citrus blackfly, *Aleurocanthus woglumi*, controlled by *Eretmocerus serius*, a wasp parasitoid; woolly apple aphid, *Eriosoma lanigerum*, controlled by *Aphelinus mali*, a wasp parasitoid; purple scale, *Lepidosaphes beckii*, controlled by *Aphytis lepidosaphes*, a wasp parasitoid; and olive scale, *Parlatoria oleae*, controlled by *Aphytis maculicornis* and *Coccophagoides utilis*, both wasp parasitoids.

Spectacular successes from the practice of introduction have occurred in several types of agroecosystems but particularly in those of orchards and forests. Biological control of the alfalfa weevil in alfalfa is an example of the approach in an important forage crop. Here, 11 parasitoids and one predator were introduced in the northeastern United States, most of them in 1959. By 1970, six parasitoids (all Hymenoptera) were widely established and caused significant mortality in egg, larval, and adult stages. By the late 1970s, these parasitoids were responsible for weevil mortality rates of about 70 percent, with about 73 percent of the alfalfa acreage in the Northeast no longer requiring insecticide treatments. Similar introductions by USDA/APHIS, particularly using *Microctonus aethiopoides*, have been made in the Midwest and other parts of the United States, with hopes of achieving similar results.

The pattern for introducing natural enemies in the preceding examples begins with determining the geographical origin of the pest species. If the pest has itself been introduced from another country (that is, it is **exotic**), the introduction of natural enemies from its original location may hold promise. For native pests, this approach usually is not practical. If the species is exotic, foreign exploration is conducted to collect natural enemies. International exploration is aided by such agencies as the Commonwealth Institute of Biological Control, headquartered in Trinidad, West Indies, and the International Organization for Biological Control

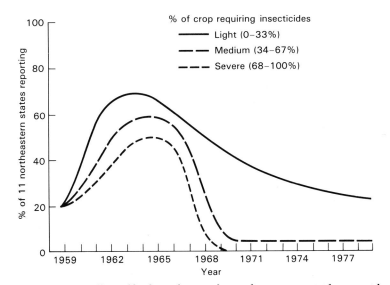

Figure 9.19 Effect of biological control in reducing insecticide use with the alfalfa weevil, *Hypera postica*, in 11 northeastern states. [G.C. Papavizas, ed., *Biological Control in Crop Production*, Beltsville Agricultural Research Center (Totowa, NJ: Allanheld, Osmun, & Co), 1982]

of Noxious Animals and Plants, headquartered in Zurich, Switzerland. Natural enemies are quarantined for study and processing. If approved for release, they are colonized in appropriate agroecosystems, and efforts are made to assist their dispersal. Finally, **evaluations** are conducted to determine natural enemy impact on pest populations.

The value of importing natural enemies has been great. When successful, introduction has been long-lasting and economical. However, certain potential limitations should be mentioned. First, introduction is primarily useful only for exotic pests. In the United States, exotic pests are very important, but they make up only about 39 percent of the 600 most important arthropod-pest species. There are examples that show some success of an exotic natural enemy against a native pest, but it is unlikely that the majority of native pests can be controlled in this way, because introduced enemies must compete with an established complex of native natural enemies, thereby reducing their chances of establishment.

Although the point is debated by some specialists, many others believe that success through introduction also is limited by agroecosystem stability. The majority of successes with introduced natural enemies have been with pests in rather stable systems. These usually include crops of perennials, for example, orchards and forests. Less success has been shown in greatly disturbed systems such as annual row crops, for example, corn, cotton, and wheat. When clean tillage is practiced, the above-ground habitat must be recolonized each year from outside the field. Whereas rapid recolonization has been observed with many native natural enemies, establishment of exotic natural enemies under these circumstances is difficult. However, the status of agroecosystem stability is changing at the present time, with shifts away from conventional tillage toward conservation tillage, for example, no-till and chisel-plow. Such changing practices may have important effects on the potential of classical biological control in annual row crops.

AUGMENTATION

Augmentation is a biological control practice that includes any activity designed to increase numbers or effect of existing natural enemies. These objectives may be achieved by releasing additional numbers of a natural enemy into a system or modifying the environment in such a way as to promote greater numbers or effectiveness.

Releasing additional numbers of a natural enemy is similar to the introduction practices discussed previously, except that augmentive releases are expected to result only in temporary (usually one season or fewer) suppression. Therefore, the ecological result of these is most likely the dampening of pest population peaks, rather than changing the general equilibrium position. Because of their temporary effect, these releases must be made periodically. Periodic releases may be considered as either inundative or inoculative.

Inundative releases. These releases depend on propagation of massive numbers of natural enemies and their widespread distribution. Subsequently, the release of the actual enemies suppresses the pest population, with little or no impact expected from progeny of the released individuals. Massive releases have been attempted in several programs involving natural enemies like *Trichogramma* species, a tiny wasp that parasitizes insect eggs, and general predators like green lacewings, *Chrysoperla carnea*, and ladybird beetles, *Hippodamia convergens*. Several of these polyphagous natural enemies are available commercially (Table 9.1).

Table 9.1. Some Suppliers of Natural Enemies for Use in Insect Pest Management

Supplier	Predators				Parasites		
	Lacewings	Ladybirds	Mantids	Mites	Wasps	Flies	Nematodes
Applied Bionomics P.O. Box 2637 Sidney, B.C. Canada V8L 4C1				X			
Beneficial Biosystems 1603-F 63rd Street Emeryville, CA 94608						X	
Beneficial Insectary 245 Oak Run Road Oak Run, CA 96069						X	
Beneficial Insects Ltd. P.O. Box 154 Banta, CA 95304	X				X	X	
Better Yield Insects 13310 Riverside Drive E. Tecumseh, Ontario Canada N8N 1B2				X			
Bio-Control Company P.O. Box 247 Cedar Ridge, CA 95924	X	X	X		X	X	
Biogenesis, Inc P.O. Box 36 Mathis, TX 78368	X			X	X		
Bio Insect Control 710 S. Columbia Plainview, TX 79072	X	X		X	X	X	
Biotactics, Inc. 7765 Lakeside Drive Riverside, CA 92509				X			
B.R. Supply Company P.O. Box 845 Exeter, CA 93221							X
California Green Lacewings P.O. Box 2495 Merced, CA 95340	X				X	X	
Foothill Ag. Research Inc. 510 W. Chase Drive Corona, CA 91720		X			X	X	
Gurney Seed & Nursery Co. Yankton, SD 57079	X	X	X		X	X	

Table 9.1. (continued)

Supplier	Predators				Parasites		
	Lacewings	Ladybirds	Mantids	Mites	Wasps	Flies	Nematodes
Harmony Farm Supply P.O. Box 451 Graton, CA 95444	X	X		X	X	X	
Henry Field's Seed and Nursery Co. Shenandoah, IA 51602	X	X	X		X	X	
Integrated Pest Management 305 Agostino Road San Gabriel, CA 91776	X	X		X	X	X	
King's Natural Pest Contr. 224 Yost Avenue Spring City, PA 19475	X	X	X	X	X	X	
Kunafin Trichogramma Insectaries Rt. 1, P.O. Box 39 Quemado, TX 78877	X	X		X	X	X	
Lakeland Nurseries Sales Inc. 340 Popular Street Hanover, PA 17331	X	X	X			X	
Mellinger's Nursery 2310 W. South Range Road North Lima, OH 44452	X	X	X			X	
Natural Pest Controls 8864 Little Creek Drive Orangevale, CA 95662	X	X	X	X	X	X	
Nature's Control P.O. Box 35 Medford, OR 97501	X	X		X			
Necessary Trading Co. P.O. Box 603 New Castle, VA 24127	X	X		X	X	X	X
The Nematode Farm, Inc. 3335 Birch Street Palo Alto, CA 94306							X
Organic Pest Control Naturally P.O. Box 55267 Seattle, WA 98115	X	X		X	X	X	
Peaceful Valley Farm 11173 Peaceful Valley Rd. Nevada City, CA 95959	X	X	X	X	X	X	X

Table 9.1. (continued)

Supplier	Predators				Parasites		
	Lacewings	Ladybirds	Mantids	Mites	Wasps	Flies	Nematodes
Richters P.O. Box 26 Goodwood, Ontario Canada L0C 1A0	X	X	X	X	X	X	
Ricon-Vitova Insectaries, Inc. P.O. Box 95 Oak View, CA 93022	X	X		X	X	X	
Spalding Laboratories 760 Printz Road Arroyo Grande, CA 93420					X	X	
Unique Insect Control P.O. Box 15376 Sacramento, CA 95851	X	X	X		X	X	

BOX 9.4 COMMON GREEN LACEWING

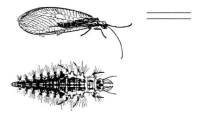

SPECIES: *Chrysoperla carnea* (Stephens) (Neuroptera: Chrysopidae)

DISTRIBUTION: Green lacewings are found as larvae and adults throughout North America on the leaves of many species of weeds, grasses, trees, and shrubs. Adults are especially attracted to lights at night.

IMPORTANCE: Green lacewing larvae are important predators of a wide variety of small insects and their eggs. The larvae are very active and feed on almost any prey their size or smaller. Although many species of adult lacewings are predaceous, common green lacewing adults feed on honeydew. This species, which can be reared relatively easily, is generally tolerant to pesticides and has been used in biological control to manage injurious moth larvae and eggs.

APPEARANCE: Adults have green cylindrical bodies with light green veins on large transparent wings. The antennae are of the long filiform type and the eyes are golden-colored.

Pale white eggs are oviposited at the ends of long stalks that elevate the eggs above plant surfaces. These stalks offer protection to the eggs from predators. The pale larvae, often called aphidlions, have pronounced, hollow sickle-shaped jaws that they use to capture and suck the contents of their prey. When larvae are preparing to pupate, they spin white parchmentlike cocoons from which adults emerge.

LIFE CYCLE: Adults overwinter in dry, dark areas, such as leaf litter, abandoned wasp nests, barns, and stables. The adult life-span is from four to six weeks. Females oviposit from 100 to 200 eggs, which hatch after a three- to six-day incubation. Lacewing larvae undergo three molts, which total approximately 14 days. Prepupal and pupal stages last about eight days.

In many instances, however, inundative releases of insect predators and parasitoids have been unsuccessful. The lack of success is usually ascribed to insufficient coverage over a great enough area or augmentation in environments not supportive of the numbers released. Another reason is the movement of released natural enemies out of the targeted area.

Some of the most successful inundative releases have been with pathogenic microorganisms like viruses and *B. thuringiensis*. Here microbial insecticides have been applied, mostly against caterpillars (Lepidoptera) and mosquitoes (Diptera), that suppress a pest population quickly, much the same as a conventional insecticide. After the initial impact of the "release," little subsequent suppression can be expected. If additional pest outbreaks occur during the season, more applications of the microbes must be made. A number of pesticide and pharmaceutical companies formulate microbial insecticides (Table 9.2).

Table 9.2. Selected Registered Microbial Insecticides

Organism	Trade Name	Manufacturer	Target Pest
		BACTERIA	
Bacillus popilliae + *B. lentimorbus*	Milky Spore	Reuter Labs	Japanese beetle
B. thuringiensis subsp. *kurstaki*	Dipel Bactospeine Thuricide Biobit	Abbott Labs Biochem Corp. Sandoz Inc. Microbial Resources Ltd.	Several moths
B. thuringiensis subsp. *kurstaki* plus beta exotoxin	Javelin	Sandoz Inc.	Armyworm, other moths
B. thuringiensis subsp. *israelensis*	Vectobac Teknar Skeetal Bactimos	Abbott Labs Sandoz Inc. Microbial Resources Ltd. Biochem Corp.	Mosquito, black fly larvae
B. thuringiensis subsp. *galleriae*	Certan	Sandoz Inc.	Wax moth
		FUNGI	
Hirsutella thompsoni	Mycar	Abbott Labs	Mites on citrus
		VIRUSES	
Heliothis virus	Elcar	Sandoz Inc.	Corn earworm, tobacco budworm
Gypsy moth virus	Gypcheck	U.S. Forest Service	Gypsy moth
European pine sawfly virus	Neochek-S	U.S. Forest Service	European pine sawfly
		PROTOZOA	
Nosema locustae	Hopper Stopper Grasshopper Spore	Environmental Technologies Inc. Reuter labs	Grasshoppers

Inoculative releases. These releases differ from inundative releases in that, once they occur, the natural enemy is expected to colonize and spread throughout an area naturally. An inoculation is often made only once in the growing season, and it is the progeny of the released individuals that have the most significant impact on the pest population. Some of the most successful programs utilizing this approach with predators and parasitoids have been reported from field and greenhouse crops in the People's Republic of China and the USSR. In the United States, the largest programs involving inoculative releases have included *Aphytis* species used on about 4,450 hectares against California red scale, *Aonidiella aurantii*; *Chrysoperla* species (lacewings) used on about 2,385 hectares against mealybugs (Homoptera: Pseudococcidae); *Cryptolaemus montrouzieri* (a

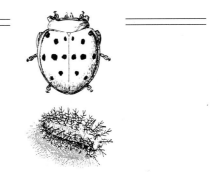

BOX 9.5 MEXICAN BEAN BEETLE

SPECIES: *Epilachna varivestis* Mulsant (Coleoptera: Coccinellidae)

DISTRIBUTION: The Mexican bean beetle is native to the semiarid southwestern United States, with a natural range in and around Mexico. The pest has successfully spread throughout much of the United States since 1920, when it was found in Birmingham, Alabama. Most of the expansion, however, has occurred east of the Mississippi River. An infestation was found in California in 1946.

IMPORTANCE: The Mexican bean beetle is unusual as a pest because it is one of only two plant-eating species of ladybird beetles, which are, as a group, typically beneficial insects. Adults and larvae feed on the underside of leaves and are often economically damaging to bush, pole, lima, and soybean. Leaves injured by Mexican bean beetle are quite distinct, being skeletonized. This appearance results from adults and larvae scraping off the leaf epidermis and soft leaf tissue while leaving the more rigid structured leaf components intact. Eventually, the netting may also die and fall out of the injured area. Larvae are subject to attack by several parasites, including a eulophid wasp, which has been successfully reared and mass-released as a pest management tactic.

APPEARANCE: Adults have 16 black spots arranged in three rows on yellowish bronze elytra. Adults are rounded and convex and about 8 mm long. Eggs are yellow and found in clusters of 40 to 60 on the underside of host leaves. Larvae are pale yellow to orange and have six longitudinal rows of branched spines on their back. Pupae are orange-yellow and remain attached to the underside of leaves by utilizing the last larval skin until adult emergence.

LIFE CYCLE: Adults overwinter among plant residue on the ground. They begin feeding on available foliage in spring, and females oviposit on the undersides of leaves. Eggs hatch in about seven days, and new larvae begin to feed on leaf epidermis of the plant host. There are four stadia, each lasting about six days (depending on temperature). Pupation occurs and lasts about seven days, with total generation time averaging 33 days. There may be one to four generations per year, depending on latitude and temperature.

Figure 9.20 A parasitic wasp, *Pediobius foveolatus*, laying eggs among the body spines of a Mexican bean beetle, *Epilachna varivestis*, larva. Female wasps lay between 10 and 20 eggs, and wasp larvae feed within the host larva. (Courtesy USDA)

ladybird) used on about 5,670 hectares against the black scale, *Saissetia oleae*; and the egg parasite, *Trichogramma* species, used on as many as 100,000 hectares of crops against several lepidopterous pests. Another small program in the eastern United States for soybeans uses an exotic parasitoid, *Pediobius foveolatus* imported from India, against a pest native to Central America and Mexico, the Mexican bean beetle, *Epilachna varivestis*.

Other significant uses of inoculative releases include those against greenhouse (glasshouse) pests in Great Britain, Holland, and other European countries. In particular, planned releases of predatory mites, *Phytoseiulus persimilis*, against plant-feeding mites, and parasitoids, *Encarsia formosa* and *Aphidius matricariae*, against whiteflies (Homoptera: Aleyrodidae) and aphids (Homoptera: Aphididae) have gained credibility in recent years.

Inoculative releases have been particularly useful against Japanese beetle in turf grass. As mentioned earlier, two bacteria are responsible for causing milky disease in this insect. To suppress beetle populations, commercial formulations of spore powder are deposited on the turf surface and washed in artificially or by rainfall. The treatment then causes an epidemic in the local grub (larval) population.

Environmental manipulations. Another method of augmenting numbers of natural enemies besides releasing them involves environmental manipulation.

Here, the crop itself or surrounding crop areas are utilized as "field insectaries" to increase numbers or make existing numbers more efficient. Most practical have been attempts to make requisites (for example, alternate nutrients, nesting habitats, overwintering sites) more available, thereby enhancing natural enemy reproduction. Examples of this approach include providing nesting sites for predaceous wasps, *Polistes* species, and maintaining wild blackberries adjacent to vineyards as food for alternate hosts of a parasitoid, *Anagrus epos*. In the latter instance, the wasp parasitoid increases populations in the wild blackberries and moves into vineyards, thereby suppressing grape leafhoppers, *Erythroneura elegantula*, a pest of imported grapes.

Another important example of environmental manipulation is the planting of cover crops (for example, broadbean, *Vicia faba*) in apple orchards to reduce pest populations. In California, maintaining a cover crop, periodically mowing it, and leaving the mulch was found to lower infestations of aphids, leafhoppers (Homoptera: Cicadellidae), and codling moth, *Cydia pomonella*. Lowered infes-

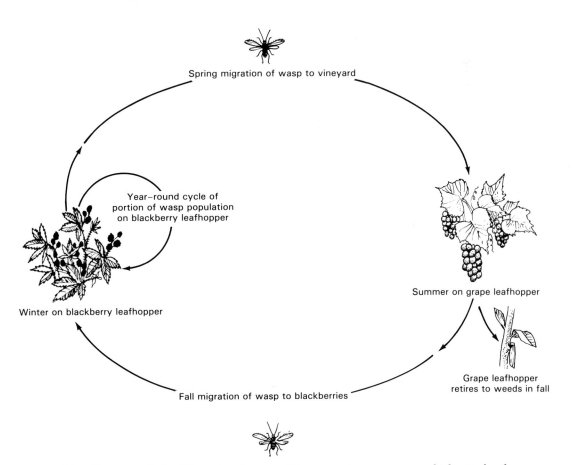

Figure 9.21 Diagram of the life cycle of a parasitic wasp, *Anagrus epos*, which attacks the grape leafhopper, *Erythroneura elegantula*. The parasitoid overwinters in California blackberries, parasitizing eggs of the blackberry leafhopper, *Dikrella cruentata*, when the grape leafhopper is inactive. In spring, some wasps move into vineyards to parasitize the grape leafhopper. (Reprinted with permission of Plenum Press from *Introduction to Integrated Pest Management* by M . L. Flint and R. van den Bosch. Copyright ©1981)

tations were accompanied by increased numbers of surface-dwelling arthropod predators. The cover crop and related activities are believed to interfere with the host-finding ability of the pests, as well as to provide alternate food and habitat for natural enemies.

Although somewhat futuristic, certain chemicals also may play a role in natural enemy augmentation. Presently, substances called **kairomones** are being tested that attract natural enemies and/or stimulate them to become more efficient. For instance, kairomones derived from the scales of corn earworm moths have been shown to stimulate searching behavior of adult *Trichogramma* parasitoids. When these kairomones were applied to field plots, wasp searching increased, and this behavior was followed by increased parasitization rates. It is possible that additional research with kairomones may provide practical substances that enhance the effect of natural enemies already present in agroecosystems.

Conservation of Natural Enemies

Probably the most widely practiced form of biological control is conservation of natural enemies. The objective of conservation is to protect and maintain these existing populations, particularly the insect predators and parasitoids, in an agroecosystem.

+ Disked (no cover crop)

☐ Cover–cropped

Figure 9.22 Differences in leafhopper populations in California apple orchards with and without cover crops. (After Altieri and Schmidt, *California Agriculture*, January-February 1986)

Basically, this approach requires knowledge about all aspects of the natural enemy community, including species present, population numbers, phenology, and impact on pest populations—in other words, the life systems of natural enemies in a given area. With this knowledge, crop production activities, including existing pest management practices, are modified to avoid natural enemy destruction.

Methods to conserve natural enemies may include less frequent mowing of field edges and clipping alfalfa in strips at different times to maintain habitat and alternate food sources for their populations. Perhaps more important as a conservation approach is use of insecticides to avoid significant natural enemy mortality. This method usually includes using chemicals less toxic to natural enemies, reducing numbers of applications, and reducing dosage levels.

CONCLUSIONS

The importance of considering natural enemies of insect pests when developing pest management programs cannot be overestimated. Virtually all insect populations are affected to a certain degree by other organisms in their environment, and some of our most efficient approaches to pest management include using these organisms to our advantage. This can mean simply modifying a crop culture technique to conserve the enemies already present or supplementing their numbers with releases. If we are dealing with an introduced pest, searching for a native natural enemy and introducing it also holds promise.

Although much has been accomplished through biological control, its potential has not been fully developed for most crops. Much remains to be learned about natural enemy ecology and methods of manipulation before this tactic can be fully integrated with other management tactics. With more detailed knowledge, we can look forward to pest management programs that have greater effectiveness and lower costs, and yet are less disruptive of nonagricultural environments.

Selected References

CLAUSEN, C.P. 1940. *Entomophagous Insects*. New York: McGraw-Hill.

COPPEL, H.C. , and J.W. MERTINS. 1977. *Biological Insect Suppression*. New York: Springer-Verlag.

DAY, W.H. 1982. Biological control of the alfalfa weevil in the northeastern United States. In G. C. PAPAVIZAS, ed., *Biological Control in Crop Production*. London: Allanheld, Osmun Publishers, pp. 361–390.

HOY, M., and D.C. HERZOG, eds. 1985. *Biological Control in Agricultural IPM Systems*. Orlando, Fla.: Academic Press.

HUFFAKER, C.B., and P.S. MESSENGER, eds. 1976. *Theory and Practice of Biological Control*. Orlando, Fla.: Academic Press.

MADDOX, J.V. 1982. Use of diseases in pest management. In R. L. METCALF and W. H. LUCKMANN, eds., *Introduction to Insect Pest Management*, 2nd ed. New York: Wiley, pp. 175–216.

NORDLUND, D.A. 1984. Biological control with entomophagous insects. *Journal Georgia Entomological Society* 19 (2nd supp.):14–27.

RIDGWAY, R.L., and S.B. VINSON, eds. 1977. *Biological Control by Augmentation of Natural Enemies*. New York: Plenum Press.

STEHR, F.W. Parasitoids and predators in pest management. In R. L. METCALF and W. H. LUCKMANN, eds. *Introduction to Insect Pest Management*. 2nd ed. New York: Wiley, pp. 135–173.

VAN DEN BOSCH, R., P.S. MESSENGER, and A. P. GUTIERREZ. 1982. *An Introduction to Biological Control*. New York: Plenum Press.

10 ECOLOGICAL MANAGEMENT OF THE CROP ENVIRONMENT

IN ADDITION TO NATURAL ENEMIES, many other aspects of the crop environment greatly influence insect numbers and have potential in pest management. To understand these other factors, we must recount the ecological basis of insect problems.

When crops are produced in an area, usually there is a significant alteration in the species occupying the space, as well as various modifications to soil, water, and topography. Natural flora and fauna may be displaced from the area in favor of mostly exotic plants and animals that are the focus of agricultural production. These crop species are established by clearing the area of undesired species and replacing them with those that produce human food and fiber. The crop species are maintained by high-energy inputs that involve such practices as tillage, fertilization, cultivation, and irrigation.

These practices of crop growth and livestock production are designed to create a favorable environment for the desired species. This environment is usually ecologically simple and also provides a favorable habitat for some insects and other organisms (for example, unwanted plants and microorganisms). These unwanted species may have existed in the area previously or may have been attracted from other areas. With the supply of new requisites or additional amounts of existing requisites, reproductive rates and survival rates of unwanted populations increase, and pest activities result in reduced crop output.

Reduced output does not always occur immediately when a new crop is introduced in an area. At first, producers are preoccupied with the new crop and the cultural activities required to raise it. At this time, there is little concern for potential pests, and for a period, insects may be of little importance. This lag in problems often belies the future importance of injurious populations and may give a false sense of security. As species become adapted to the new environment, pest populations grow, losses increase, and attention must finally be given to alleviating these difficulties.

One of the oldest approaches to dealing with these emerging problems is to modify the established production schemes and techniques to make the crop environment less favorable for the pest, or at least to minimize losses of the commodity in spite of injury.

Ecological management. Historically, entomologists have called modifying the crop environment through changes in production techniques cultural control. Cultural control can be defined as the purposeful manipulation of the environment to reduce rates of pest increase and damage. Cultural control manipulations traditionally involve environmental factors that already exist, as opposed to adding new factors (for example, insecticides or natural enemies). In keeping with the idea of insect pest management and to broaden the notion of cultural control somewhat, we use the term **ecological management** here for the approach.

Ecological management is based on a thorough understanding of pest ecology as it relates to the crop being produced. In general, we must understand a pest's ecological requisites, know how available these requisites are in the agroecosystem, and understand insect behavior in attaining requisites. Major pest requisites include food and appropriate space for feeding, mating, and egg laying, as well as shelter from weather extremes and enemies. When pests are major problems, these requisites may be found totally within the crop area or as supplemented from sources in nearby areas. For example, grasshoppers, *Melanoplus* species and others, find an abundant supply of nutritious food in annual field crops like corn and soybeans but must return to grass sod at the edge of fields for appropriate egg-laying sites. Likewise the European corn borer, *Ostrinia nubilalis*, finds satisfactory food and egg-laying sites in corn fields, but adults seek out waterways and other grassy areas outside the crop for the proper mating environment.

The idea behind ecological management is to find "weak links" in the insect seasonal cycle and exploit them. These weak links may be behavior patterns to complete development, like crawling over the ground surface to reach adequate feeding sites as do the larvae of the stalk borer, *Papaipema nebris*, or locating proper shelter for overwintering, as do adult boll weevils, *Anthonomus grandis*.

Food is often the environmental factor involved in the weak link. Insects vary

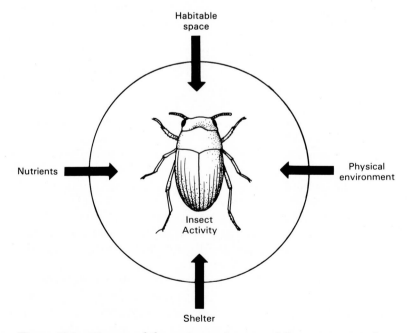

Figure 10.1 Diagram of the major environmental factors to manipulate for reducing pest proliferation and damage.

BOX 10.1 ARMYWORM

SPECIES: *Pseudaletia unipuncta* (Haworth)
(Lepidoptera: Noctuidae)

DISTRIBUTION: The armyworm occurs throughout most of the United States east of the Rocky Mountains. However, this pest has also been found in Arizona, California, and New Mexico.

IMPORTANCE: Larvae of armyworm moths readily feed on corn, oats, barley, and rye. In most years, damage to these crops is insignificant, but in outbreak years large populations can completely devour the original host and "march" to new hosts (hence the term "armyworm"). Outbreaks of this pest tend to occur following a cold, wet spring. Larvae of the first generation are commonly associated with outbreak occurrences. The degree of crop injury is frequently heavier in field borders. Young leaves are completely devoured, but older, less palatable leaves may be only skeletonized, leaving the vein structure.

APPEARANCE: Adult moths are brownish gray, with a distinct white spot in the center of the forewing. The hindwing is a lighter brown and has darker veins; wingspan is about 40 mm. Eggs are laid in masses or rows that are commonly folded in leaves or leaf sheaths. Larvae of the armyworm are green to brown and have a distinctive pattern of longitudinal stripes. A dark stripe can be seen along each side, and a broad stripe is found along the back. This dorsal stripe has a fine, lighter-colored broken line running down the center. The head capsule is pale brown with a green and dark brown mottling.

LIFE CYCLE: Partially grown larvae overwinter in the soil or refuse at the surface and complete their growth the following spring. Larval development continues in early spring and pupation soon follows. Adults emerge, mate, and oviposit within 14 to 21 days. Females lay large masses of eggs, often in the same general location, at night. Hatching occurs within about eight days after deposition, and the neonate larvae begin feeding on leaf tissue. Full-grown larvae are about 40 mm long, then pupate, and a second generation of moths emerges. There may be up to three generations in most regions of the United States, but it is uncommon for successive outbreaks to occur in the same general location. There is continuous breeding in extreme southern locations, where generation overlap is common. This pest is subject to population regulation by numerous larval and egg parasitoids, which often aid in keeping armyworm populations below economic levels.

greatly in their food requirements, with some being quite selective (for example, larvae of the northern corn rootworm, *Diabrotica barberi*, on corn) and others displaying seasonal shifts from one type of food to another (for example, chinch bugs, *Blissus leucopterus*, moving from maturing wheat to corn). Such habits provide potential opportunities for exploiting the insect's seasonal cycle.

In addition to its being vital to the insect, food may be one of the most practical factors in the production scheme that can be manipulated. Modifying the food source may produce profound effects on the pest species, because often it not only provides nutrients but also habitable space for reproduction and other activities.

The physical environment, also critical in insect ecology, may be manipulated through ecological management. Plant culture activities like soil tillage and debris removal can expose insects to intolerable weather extremes and subsequent death. Manipulating temperatures and humidities above or below those tolerated by insects in stored grain are procedures that modify the physical environment to obtain ecological management.

Depending on the objective, most procedures of ecological management can be grouped into one of four categories: (1) reducing the average favorability of the ecosystem, (2) disrupting the continuity of requisite food sources, (3) diverting pest populations from the crop, and (4) reducing the impact of insect injury. One or more approaches may be used in ecological management to keep pest populations "off balance" and prevent them from achieving their full destructive potential.

REDUCING AVERAGE FAVORABILITY OF THE ECOSYSTEM

Ecosystems include interacting biotic and abiotic elements both in the crop and in associated nonagricultural habitats. The level of requisites available to pest species in an agroecosystem determines, to a degree, the average density of the pest and severity of pest problems. Procedures in this category mainly aim at reducing pest density by reducing the average availability of food, shelter, and habitable space. Here, only the level of the requisite is reduced because complete elimination is unwarranted or impractical.

Sanitation

Sanitation is one of the most elementary procedures in reducing favorability of the agroecosystem for pest species. Because many species breed and overwinter in all sorts of debris, removal of debris from habitats can reduce rates of reproduction and survival.

Crop residue destruction and utilization. Destroying or removing crop residues from crop sites is one of the most basic ways to eliminate pest overwintering sites and reduce the spread of infestations. Destruction may be achieved by plowing directly or shredding and chopping with special field implements before plowing. Burning residues in place or raking and scooping them into piles for burning or chemical treatment is another method used to eliminate infestations from residues. Perhaps an even better approach with some residues (for example, dropped fruit) is to feed them to livestock, thereby obtaining an economic benefit. In some operations, livestock can be pastured at appropriate times to consume residues and trample hibernating pests.

Probably the greatest limitations to residue cleanup are cost of destruction and usefulness of residues in soil conservation. If much labor is involved in destruction, the approach quickly becomes impractical for commercial enterprises. For instance, the collection and disposal of dropped fruit is an effective method for reducing infestations of codling moth, *Cydia pomonella*, apple maggot, *Rhagoletis pomonella*, and plum curculio, *Conotrachelus nenuphar*, in small orchards. But it is usually impractical for large commercial orchards. Additionally, the practice of leaving plant residues on the soil surface to reduce wind and water erosion rules out this measure where soil conservation is a top priority.

Figure 10.2 An apple damaged by the apple maggot, *Rhagoletis pomonella*. Collection and disposal of dropped fruit containing such pests can reduce future infestations. (Courtesy Iowa State University Extension Service)

In agronomic crops, some of the most widespread uses of crop residue destruction for insect management have occurred with cotton, corn, and wheat. In particular, cotton production in the southern and southwestern United States emphasizes destruction of infested cotton plants after harvest and before the first killing frost. Texan areawide sanitation programs in cotton have been especially successful. Here, defoliants and mechanical strippers have allowed removal of infested plant parts from the field, and remaining debris is plowed under. Such activities have been particularly important in reducing infestations of the pink bollworm, *Pectinophora gossypiella*, and boll weevil.

With field corn, overwintering European corn borer and southwestern corn borer, *Diatraea grandiosella*, populations have been reduced in previous years by residue shredding and deep plowing techniques. However, production systems emphasizing greater soil conservation in the corn belt region have reduced the current implementation of this approach.

In wheat, the plowing under and disposal of stubble may still be a practical procedure for reducing overwintering populations of important pests like the Hessian fly, *Mayetiola destructor*.

Sanitation also remains a very important pest management procedure in truck farming. Here, efficient producers remove vegetables dropped during packing operations and/or left in fields during harvest. These vegetables are removed and destroyed or fed to livestock to prevent buildup of several Lepidoptera, squash bugs (*Anasa tristis*), some aphids (Homoptera: Aphididae), and other pests. Removing vines and infested tubers also is an important practice in the management of the potato tuberworm, *Phthorimaea operculella*, and the sweet-potato weevil, *Cylas formicarius elegantulus*, by potato growers.

BOX 10.2 BOLL WEEVIL

SPECIES: *Anthonomus grandis grandis* Boheman (Coleoptera: Curculionidae)

DISTRIBUTION: The boll weevil is not native to the United States; rather, it has moved north from its natural range in Central America. It was first reported in the United States in 1894 in Brownsville, Texas. From there, it became established in southern Texas and began to spread throughout the cotton belt at a rate of 40 to 160 miles per year. This pest has continued to spread and adapt to new areas, and authorities expect the pest to eventually occupy the entire cotton belt.

IMPORTANCE: The boll weevil is a major pest on 52 percent of the cotton grown in the United States. Lint and cottonseed losses between 1909 and 1949 averaged $203 million annually. In spite of modern control strategies, cotton growers continue to lose over $200 million per annum because of this pest. Feeding by adults on early flower buds (referred to as "squares") causes the bracts to flare, with eventual square drop. Oviposition into squares also results in shedding. After bloom and during boll (fruit) development, females continue to feed and oviposit within the bolls. At this point in time, however, the boll does not usually shed but larval feeding destroys the lint within the boll. In 1904 the United States Bureau of Plant Industry was charged with demonstrating to farmers how to control this pest. The program was successful and eventually led to the establishment of the Agricultural Extension Service.

APPEARANCE: Adults are about 6 mm long, brown, and hard-shelled. The distinctive snout is half as long as the body. Larvae are legless, about 12 mm long when full-grown, and grublike, with a brown head capsule. Eggs are white, elliptical, and about 0.8 mm long. Pupae are white. A recent taxonomic study has demonstrated that there are three closely related forms of weevils attacking cotton: the boll weevil, the thurberia weevil, *Anthonomus grandis thurberiae* Pierce, and the Mexican boll weevil, *Anthonomus hunteri* Burke and Cate.

LIFE CYCLE: Adults overwinter in ground litter in or around cotton fields, fence rows, woods, or other protected places. Adult diapause period may extend from March to mid-July, with most emergence occurring in June. Adults may disperse several miles in search of host plants. After locating fruiting cotton, adults begin to feed on squares and, less frequently, on leafbuds and growing terminals. Females mate and deposit eggs singly into cavities previously made with their chewing mouthparts. They seal the oviposition site with a frass "nipple," which makes a damaged square or boll easily identifiable. Larvae feed within the square or boll for seven to twelve days and then pupate. Adults emerge within about six days, and disperse to new hosts.

As already mentioned, the frequent removal of dropped fruit can reduce infestations of several insect pests and is recommended for orchards when practical. Also, trees in abandoned orchards or those severely infested by scale insects (Homoptera: Coccoidea) should be removed and destroyed.

Removal of plant debris may also be important in silviculture. Here, it is recommended that slash be destroyed following logging operations, particularly in pine stands of the southern and western United States. Such sanitation procedures help prevent reproduction and buildup of bark beetle populations and reduce

Figure 10.3 Shredding corn stalks with a tractor-drawn shredder. This operation helps reduce European corn borer, *Ostrinia nubilalis*, and southwestern corn borer, *Diatraea grandiosella*, infestations. (Courtesy D. Erbach and USDA)

Figure 10.4 Spreading liquified manure achieves fertilization and allows manure to dry out. The drying helps prevent fly colonization. (Courtesy D. Erbach and USDA)

damage to trees growing near the logged area. The destruction (burning) of elm logs and branches infested with the smaller European elm bark beetle, *Scolytus multistriatus*, and the native elm bark beetle, *Hylurgopinus opaculus*, is another sanitation practice recommended to reduce the spread of Dutch elm disease. Pruning and destruction of insect-infested twigs and branches also have been used as management tactics for European pine shoot moth, *Rhyacionia buoliana*, a pest of pine plantations in the Great Lakes region of the United States.

Elimination of animal wastes.　It goes without saying that good housekeeping is an important aspect of fly and cockroach control in households. Likewise, disposal of animal wastes in and around livestock confinement areas reduces food and habitat for developing fly larvae and is a must if adequate management

BOX 10.3 NATIVE ELM BARK BEETLE

SPECIES: *Hylurgopinus opaculus* (LeConte) (Coleoptera: Scolytidae)

DISTRIBUTION: The native elm bark beetle originated in the United States, and occurs in most of the eastern states from Maine to Virginia, in Mississippi, Kansas to Minnesota, and in eastern Canada. Its range corresponds to that of the American elm tree within the United States. It is mostly monophagous on elm trees but has been reported also to feed on basswood trees.

IMPORTANCE: This pest, in addition to the smaller European elm bark beetle, *Scolytus multistriatus* (Marsham), has played an important role in the near extinction of American elm in the midwestern United States. The beetles cause minimal injury by feeding on foliage, but they transmit the fungal organism *Ceratocystis ulmi*, which is responsible for Dutch elm disease. The disease is present throughout most of the United States and has been reported in Canada. Because of the mobility of elm bark beetles, the disease was able to spread unchecked, eventually killing all but a few American elms in the Midwest.

APPEARANCE: Adults are dull brown and less than 3 mm long. The head is somewhat blunt, and body shape is typical of scolytid beetles. Eggs are small, globular, and pearly white. Larvae are about 3 to 4 mm long and creamy white, with a brown head capsule. They are legless, C-shaped grubs. Damaged trees can be identified by feeding galleries under the bark. Parent galleries consist of two branches diverging from the entry points. Adult galleries extend across the grain, whereas larval galleries usually run with the grain of the wood.

LIFE CYCLE: Larvae and adults overwinter. Adults appear in spring after emerging from bark that was fed upon the previous year. Only dying or injured elm trees are attacked by the beetles, which bore into bark and form galleries. Females deposit eggs in niches along the sides of the galleries. Young larvae begin feeding in the cambial region of trees and form larval galleries independent of adults. The width of the galleries increases with larger larvae and greater consumption. Pupation occurs at the end of larger galleries, followed by adult emergence. The life cycle requires 45 to 60 days for completion, and two to three generations can occur per year.

BOX 10.4 ANGOUMOIS GRAIN MOTH

SPECIES: *Sitotroga cerealella* (Olivier) (Lepidoptera: Gelechiidae)

DISTRIBUTION: This pest is worldwide in distribution. It was first reported in the United States in 1728 and has since spread throughout North America. Larvae attack grains in storage and in the field.

IMPORTANCE: The Angoumois grain moth is the most common of the moths that infest whole grains. Indeed, this species is second in importance only to rice and granary weevils as a stored-grain pest. The pest is of greatest importance in southern North America and in portions of the eastern and central United States. This pest has caused millions of dollars in damage to grain stored in central locations like grain elevators. It can, however, also be a severe household pest by infesting grain products during processing or after purchase.

APPEARANCE: Adults are light grayish brown or straw-colored moths with a satiny luster and a wingspan of about 16 mm. Hind wings are fringed, with long, dark hairs, and they have a point at the tip like a finger, which, distinguishes this pest from clothes moths. Newly laid eggs are white but turn red as they near hatching. Larvae are white, with a brown head capsule. They feed within individual kernels of wheat, corn, or other grains and are about 5 mm long when full-grown.

LIFE CYCLE: Larvae overwinter in the northern United States (two generations per year), but in the Southern United States and in heated buildings, breeding is continuous. Eggs are deposited (in spring if overwintering) on the stored grains or on developing grain heads while hosts are still in the field. Females oviposit an average of 40 eggs but may produce as many as 400 per individual under ideal conditions. Upon hatching, larvae chew into grain and consume the inside portions, usually rendering it inviable and useless. Prior to pupation, larvae prepare exit holes through the seed coats for the moths to exit. Total development may be completed in five weeks under warm conditions.

---

of house flies, *Musca domestica*, and stable flies, *Stomoxys calcitrans*, is to be achieved. Methods here have included scattering pulverized or liquified manure with spreaders to achieve drying and prevent fly colonization. At the same time, the manure fertilizes. With other organic wastes, composting facilities and solid-waste treatment plants turn otherwise useless material into products of value as fertilizers and fuel, while eliminating potential insect habitats.

Efficient storage and processing. Cleaning of facilities and elimination of spillage is of utmost importance in storage of grain and other commodities. Such sanitation, along with maintaining cool, dry atmospheres, is absolutely necessary in preventing infestations of stored-grain pests. Most food processing plants have sanitation programs designed to thoroughly clean bulk storage bins and insure an insect-free environment each time before they are filled. Grain elevators reduce the incidence of infestations by removal of waste grain and chaff from grain augers, elevator equipment, and support facilities. Often heavy-duty vacuum cleaners

Figure 10.5 Sanitation applied to grain-storage facilities includes removing residual grain (top left), cleaning grain bins with a vacuum cleaner (top right), removing chaff from grain-loading equipment (bottom left), and removing weeds around grain bins. (Courtesy University of Nebraska Extension Service)

are used for this purpose, as well as for sanitation in food plants, bakeries, and warehouses. Regularly scheduled inspections and cleaning "field days" are the best approaches to this type of sanitation.

Destruction or Modification of Alternate Hosts and Habitats

Many insect pests have requirements that cannot be met by the crop itself and need to move out or switch to other foods during certain parts of the year. If these alternate hosts and habitats can be destroyed or limited in extent, insect numbers can be reduced.

With many pests of annual crops, individuals leave the unprotected, barren crop area for field borders during the fall. Here, they may find an environment with dense vegetation and other cover in which to spend the winter. Insects sheltered in these areas and in a state of dormancy can successfully survive the adversities of winter and are a source of infestations the next spring. Burning or plowing this habitat may be a practical means of reducing such sources.

One of the most frequently mentioned pests for this kind of management is the sorghum midge, *Contarinia sorghicola*. The most important sources of infestations in sorghums, Sudan grass, and broomcorn in the United States seem

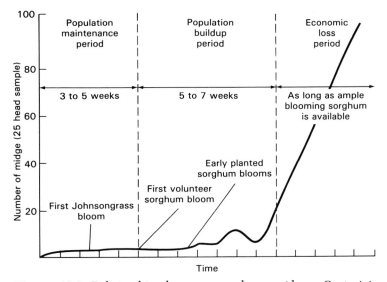

Figure 10.6 Relationship between sorghum midge, *Contarinia sorghicola*, densities and the blooming of Johnsongrass and sorghum. The midge lays eggs in floral spikelets during flowering, and larval feeding prevents seed development. Sorghum is the preferred host, but spring and early summer generations of the midge are maintained on Johnsongrass. (Courtesy Texas Agricultural Experiment Station)

to be from Johnson grass in or along borders of fields. Significant reductions of pest numbers have been reported where weed burning and other forms of pest habitat destruction were used.

Many bugs, like squash bugs and stink bugs, and other garden pests also overwinter in trash and plant cover at the edge of plantings. Elimination of these hibernating habitats can significantly reduce infestations in cucurbits, garden beans, and other vegetables.

Another source of insect infestations is volunteer plants of the crop that remain in turnrows and in parts of a field after harvest. These plants can harbor large numbers of pests at times when pest presence would otherwise be impossible. The destruction of these volunteer plants is particularly important when crop rotation is practiced to eliminate pests. For example, volunteer corn in soybean fields should be removed to prevent corn rootworm adults, *Diabrotica* species, from egg laying and producing larvae that would colonize corn the following season. Other pests where destruction of volunteer plants is recommended for suppression include grasshoppers, Hessian fly, sweetpotato weevil, potato tuberworm, potato aphid (*Macrosiphum euphorbiae*), and the wheat curl mite (*Eriophyes tulipae*).

Although destruction of alternate hosts and habitats may be practical for some insect pests, it also may eliminate important habitat for beneficial insects and wildlife. Therefore, before widespread destruction is implemented, surveys of beneficial organisms should be conducted, after which the benefits of this fauna should be determined and weighed against the benefits of reducing pest numbers. In some instances, a decision against habitat destruction may be the most appropriate one.

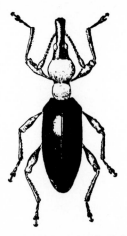

Figure 10.7 The sweet potato weevil, *Cylas formicarius elegantulus,* one of the pests for which destruction of volunteer crop plants is recommended. (Courtesy USDA)

In instances where complete elimination of cover in alternate habitats is ill-advised, it may be possible to replace the vegetation with plantings less favorable to pests. Such an approach succeeded several years ago with the beet leafhopper, *Circulifer tenellus,* in southern Idaho. The beet leafhopper vectors a virus that causes curly top disease in several crops. The source of beet leafhopper infestations was found to be from Russian thistle growing in large range and desert areas of the region. In 1959, over 116,000 acres of this alternate habitat were seeded with perennial range grasses, primarily crested wheatgrass, to replace the thistle. This program was an outstanding success, providing a substantial reduction in beet leafhopper and, subsequently, a reduction of curly top incidence.

Tillage

Tillage operations are a major cultural activity in many plant production schemes, serving primarily for seed bed preparation and weed control. Also, it is often the method of choice for the elimination of plant refuse and the destruction of alternate pest habitats in insect management. In addition to these benefits, tillage changes the physical environment of insects inhabiting the soil.

Changes in the soil environment can be very influential on insects, because a large majority (perhaps over 90 percent) of terrestrial species spend some part of their lives in soil or on the soil surface. Tilling the soil for plant culture modifies soil texture, moisture, temperature, and other characteristics in ways that may be either beneficial or detrimental to insects. Therefore, modification of tillage operations to suppress insects must be based on a knowledge of soil community ecology, as well as the acceptable limits of good agronomic practice.

Often timing and tillage depth are the major modifications made to manage insects. Tillage may be conducted in the fall or early winter and/or in the spring before planting. With land that is fallowed, it may be conducted several times during the summer. The frequency of tillage varies from a single pass with a one-way disk, often used in dryland spring wheat production, to repeated cultivations and harrowing, used in seedbed preparation for vegetable crops. The depth of soil affected by the tillage procedure may be from 15 to 30 centimeters, attained with

Figure 10.8 Map showing migration routes of the beet leafhopper, *Circulifer tenellus*, from breeding locations into sugarbeet cropping areas. (Courtesy USDA)

moldboard plows, to a shallow stirring, created with harrows. In California and many areas of the southern United States, subsoiling, or the disturbance of hard soil beneath normal plow depth, is another tillage method that has been used. Newer tillage systems employed in soil conservation include such procedures as fall chisel plow, till plant, and no till.

If tillage is deliberately used for reducing insect infestations, it should be timed with the insect's life cycle. The objective is to move insects when they are helpless (usually meaning immobile) from a favorable location to an unfavorable one. Therefore, timing is usually determined by the period of the life cycle when either pupation or dormancy occurs, and depth is recommended by the location of this helpless stage.

As an example of timing and depth, ecological management of the European corn borer can be considered. This insect undergoes diapause as a full-grown larva in corn residue (stalks and earshanks) on or above the soil surface. Chopping

Figure 10.9 Wheat stem sawfly, *Cephus cinctus*, and injured wheat. Spring plowing has been recommended as an ecological management tactic to reduce infestations of this pest. (Courtesy USDA)

and shredding, followed by deep plowing with a moldboard plow in the spring, is recommended. With this approach, some mortality occurs from exposure to low winter temperatures above ground; mortality is added by the chopping and shredding process; and many of the remaining survivors are entombed by the plowing process. Plowing alone also may be used but is not as effective.

In addition to reducing European corn borer populations, spring plowing is recommended to reduce infestations of wheat stem sawfly, *Cephus cinctus*, in spring wheat. Moreover, it is used to aid in destruction of grasshopper eggs and to eliminate early weeds that are food for newly emerged nymphs. In this instance, the grasshopper eggs are moved from soil to the surface, where they are subject to desiccation and predation by other insects and birds. In vineyards, spring plowing is recommended to cover overwintering cocoons of the grape berry moth, *Endopiza viteana*, under grape trellises and thereby to reduce infestations. Yet another example of spring plowing for management has been reported for infestations of grass grub, *Costelytra zealandica*, in New Zealand. Here, plowing in September and October (spring in the southern hemisphere) disrupts pupal cells in the soil that cannot be reformed, causing insect mortality.

Fall plowing is often most appropriate for insects that are attracted to an undisturbed surface for egg laying and/or overwinter in soil. For insects like the corn earworm, *Heliothis zea*, and several species of cutworms (Lepidoptera: Noctuidae) that undergo diapause in soil during winter, fall plowing moves them to the soil surface, where they are exposed to winter temperatures and vertebrate predators like birds and mice.

Some tillage operations are not always of benefit to pest management and actually may encourage pest problems. For instance, tillage of fallowed land during August and September in Canada fosters successful oviposition by the pale western cutworm, *Agrotis orthogonia*, a pest of small grains. Avoiding tillage

Figure 10.10 Soil ridges thrown under a grape trellis in the spring to help prevent emergence of the grape berry moth, *Endopiza viteana*. (Courtesy USDA)

during these months promotes crusting of the soil surface and prevents females from penetrating soil to lay eggs. In another instance, parasitoids of the cereal leaf beetle, *Oulema melanopus*, a pest of small grains, can be severely affected by tillage operations, which have little effect on the pest.

Irrigation and Water Management

Irrigation is a primary plant culture activity in many regions, and little emphasis has been placed on its use in preventing insect problems. An exception to this statement is the recent technology called **chemigation** that uses the irrigation system to dispense insecticides and other pesticides over the area. However, this is not an ecological management approach.

Water management is practiced as a form of ecological management in the muckland Everglades agricultural area of Florida. Here, fallow land is flooded prior to the production of winter vegetables. Of approximately 300,000 hectares, about 20 percent is flooded annually to suppress soil pests and to conserve and stabilize organic soils in the region. A primary target of pest management in this operation is immature wireworms (Coleoptera: Elateridae), which attack germinating plants and underground plant parts. To be effective against this pest, production areas are flooded for a period of six weeks or more.

In another instance, sprinkler irrigation has been effective in suppressing certain foliage-feeding insects. The potato moth, *Phthorimaea operculella*, is effec-

---

BOX 10.5 CEREAL LEAF BEETLE

SPECIES: *Oulema melanopus* (Linnaeus) (Coleoptera: Chrysomelidae)

DISTRIBUTION: The cereal leaf beetle is distributed throughout Europe, the Middle East, and North Africa. The species was first detected in the United States (Michigan) in 1962 and is currently found in the eastern United States and southern Canada.

IMPORTANCE: Both adults and larvae attack young leaves of grasses, including wheat, oats, barley, rye, and corn. Adults feed on grain shoots, whereas larvae consume tissues between leaf veins, giving the injured plants a silvery appearance. The pest is capable of vectoring the corn pathogen maize chlorotic mottle virus. Quarantines have been initiated to aid in preventing movement of the pest.

APPEARANCE: Adults are 4 mm long and have a metallic bluish black head and elytra with red legs and prothorax. Larvae are pale yellow with a black head. Often, larvae are covered with fecal excrement, giving them a dark appearance. Eggs are initially yellow but darken to black before hatching. Pupae are yellow but change to adult coloration before emergence.

LIFE HISTORY: Adults overwinter in plant debris on the ground, and in early spring, they begin feeding on wild grasses. They eventually move to cultivated fields. Females oviposit on leaves, and hatching occurs in about five days. Larvae feed for about 10 days and then pupate under 50 mm of soil. After 20 to 25 days, adults emerge, feed for a short time, then aestivate until fall. The beetles become active again in fall and locate a suitable overwintering site. A typical life cycle takes 46 days, and there is only one generation per year.

Figure 10.11 Overhead irrigation may be used to dispense insecticides (chemigation) and may be useful in suppressing some insect populations; even without chemicals. (Courtesy L. Hodgden and Dow Chemical Company)

tively suppressed by frequent overhead irrigation of potatoes in New Zealand. Here, it is believed that the moist conditions may deter egg laying and cause mortality of newly eclosed larvae before they tunnel into the plants.

Of course, water management can also be used to grow more vigorous plants and thereby reduce losses. This aspect, however, is not a means of reducing pest numbers or decreasing the average favorability of the pest environment.

Another very significant use of water management in insect suppression has been with mosquitoes. In the United States, the control of water levels in water impoundments and other sources contributes significantly to reductions in nuisance mosquitoes, as well as those transmitting human pathogens. In addition, thousands of hectares of tidal marsh area have been diked and ditched to reduce mosquito breeding habitats, and thereby have reduced mosquito incidence. Modifications to some marsh areas are being questioned, however, because of the trade-off with eliminating habitat for certain fish and wildlife.

DISRUPTING CONTINUITY OF PEST REQUISITES

Insect pest problems originate with the establishment of agroecosystems and the creation of environments favorable to certain species. These problems continue because of the uninterrupted supply of requisites provided at appropriate times and places relative to the seasonal cycle of these species.

Within the limits of good cultural practice, the supply of requisites can sometimes be interrupted to reduce infestations of pests. In this instance, we are usually dealing with the crop species itself, manipulating its presence in time and space for the elimination of insect requisites. Such would reduce insect reproductive rates or cause pests to search elsewhere for their needs.

Figure 10.12 Mosquitoes reproduce well in marshy, vegetation-choked swamps (top). Swamps can be cleared and water impounded to reduce mosquito populations in an area (bottom). (Courtesy USDA)

Reduce Continuity in Space

With this approach, time is held constant. Our perspective is the cropping layout over a given area during a season for annual crops or over the life of a single planting for perennial ones. The focus is on crop planning and where to place crop plants, relative to each other and in relation to other crops and unmanaged habitats.

Crop spacing. The location of plants within a crop has always been of great interest to agriculturists concerned with maximum production. Too few plants or too many plants result in less than optimal yields and have been the subject of detailed study. Although much is known about crop spacing in relation to weed pests, little is known about this aspect in relation to insect infestations.

Spacing can affect relative growth rate of plants and the development of environments favorable to insect population growth. Plants placed close together create closed canopies that aid in insect movement, a situation that may be favorable for pests as well as natural enemies. Therefore, the detriments and benefits of spacing may vary from pest to pest.

In the instance of a spruce weevil, *Thylacites incanus*, open sunlit areas seem to be a requirement for survival. If trees are placed well apart, eggs in the open areas receive enough sunlight and sufficient temperatures to hatch. A management approach to reduce weevil infestations involves establishing plantings of young trees as close as possible to achieve early closing of the canopy. In this way, the continuity of open areas is eliminated.

Soybean production is another instance where a closed canopy is important in reducing some insect problems. Here, it has been found that corn earworms prefer open canopies and therefore colonize late planted and wide-row soybeans most readily. Early planting in narrow rows causes canopy closure earlier and helps to reduce infestations. Moreover, the high humidities of closed soybean canopies have enhanced epidemics of the pathogenic fungus *Nomuraea rileyi*, an important natural enemy of several defoliating caterpillars.

In yet another instance, shading has been important with the citrus rust mite, *Phyllocoptruta oleivora*, in Florida citrus. Here, groves planted in hammocks under oak and palm trees, which are shaded in the morning, have much lower mite populations than those grown in unshaded areas. Presently, however, priorities involving grove equipment maneuverability have encouraged hedging and clearing, producing less shaded environments and more mite problems.

Crop location. In selecting the location for plantings, it is prudent to consider other crops and environments near the proposed site. Many insects can move quickly from one field to the next and between botanically related crops to attain their requisites. Considerations should also be given to overwintering pests that may move out of woodlots and unmanaged habitats adjacent to crop planting.

In planning crop location, field size and sources of insects that will colonize the proposed crop should also be considered . The nature of insect colonization of a field varies, with many uncontrollable factors like wind and windbreaks, as well as source location. Less mobile and weak flying species, overwintering or initially feeding outside the field proper, cause injury distributed in a definite edge pattern (the edge effect). Active insects that are strong fliers invade and quickly produce injury throughout the field. Pests producing edge effects tend to cause disproportionately greater damage in small fields versus large ones because there is a greater edge-to-area ratio in the smaller fields. An example of this is the stalk borer, *Papaipema nebris*, which overwinters in field borders but rarely damages conventionally tilled corn more than 10 rows from the field's edge. Here, small fields or cultivated strips near waterways or terraces suffer greater damage than do larger fields. For highly mobile pests, field size usually makes little difference.

As a rule of thumb, it is wise to locate dissimilar crops adjacent to one another to moderate pest movement. For instance, locating soybean adjacent to corn is

Figure 10.13 Corn in edge rows (left) is injured by stalk borers, *Papaipema nebris*, that move in from grassy terraces (right). Person in foreground is measuring stalk borer movement with a linear pitfall trap.

a good choice. Because of botanical dissimilarity (a legume next to a grass), few insect species would find requisites in both, and neither crop usually supports pest infestations of the other. On the other hand, planting soybeans next to alfalfa (a legume next to a legume) is not a good idea, as it invites injury (particularly early season in edge rows) from potato leafhoppers, *Empoasca fabae*, which may move into soybean when alfalfa is clipped for hay.

Likewise, seed potatoes are not grown adjacent to main potato plantings to limit aphid infestations. Such a tactic is used to avoid potato viruses transmitted by aphids, which is important in seed certification programs.

Other situations where certain adjacent plantings are not recommended include small grains next to corn because of potential chinch bug problems in corn; cotton and soybeans next to wheat because of southern masked chafer, *Cyclocephala immaculata*, problems in cotton and soybeans; and late planted potatoes next to early planted potatoes because of potato tuberworm, *Phthorimaea operculella*, problems in the late plantings.

Upset Chronological Continuity

The general idea of this approach is to create a gap, in time, in the insect food source. Here, our perspective is a single location (space is held constant). Activities to achieve this goal are implemented at different times in the growing season, from one growing season to the next, or, for perennials, from one planting to the next.

Crop rotation.   Probably the most important method of providing discontinuity in a pest's requisites is to rotate from one crop to another in alternate plantings. The practice of crop rotation actually developed from a desire to improve soil structure and fertility. Benefits in pest management have been mostly coincidental to this main purpose.

This method has been used widely with a number of different crops and pest species. Rotation schemes work best, however, when three factors are present: (1) the pest has a narrow host range, (2) eggs are laid before the new crop is planted, and (3) the feeding stage is not very mobile. The latter factor is characteristic mostly of pests that feed underground on plant roots. Insects having these characteristics are usually purged from an area when a crop lacking food requisites is planted in place of a crop that has them. Such insects cannot move out of the area to obtain requisites and will die by the time the rotation has completed a full cycle. Rotation cycles that involve two or three crops are common.

One of the best examples of eliminating insect pests by crop rotation has been with corn rootworms, *Diabrotica virgifera virgifera* (western) and *Diabrotica barberi* (northern), on corn in the midwestern United States. Here, the development of corn production technology in the 1950s and 1960s supported the practice of corn monocropping, the continuous production of the same crop in one area. Monocropping, in turn, promoted the spread of corn rootworms in the corn belt.

Rootworm larvae emerge in the spring from eggs laid in corn fields in mid- to late summer of the previous season. They are monophagous, feeding on roots of developing corn that was planted into their habitat. A two-year rotation cycle of corn and soybeans has been used effectively to eliminate the problem, because larvae cannot survive in the soybean environment. Recently, however, certain strains of the northern corn rootworm have developed in southeastern South Dakota, northwestern Iowa, and southern Minnesota that can remain dormant in the egg stage for two years. With the development of these strains, a two-year rotation has become ineffectual, and like nematodes that encyst (*Heterodera* species) and many plant-pathogenic microorganisms, rotations with longer cycles and involving more crops will be required to manage these new rootworm strains.

Rotations also have been successful for arthropods like white grubs (Coleoptera: Scarabaeidae) and other insects feeding on roots in grass sod. For these insects, legumes (such as alfalfa) are planted and maintained in the rotation cycle for at least two years.

Whitefringed beetles, *Graphognathus* species, also can be managed with rotations. These weevils do not fly, and females become fecund after feeding on peanuts, soybeans, and velvetbean. When corn and small grains are planted in a rotation cycle, the weevils feed on them but do not obtain proper nutrition, and females lay few eggs. These grass crops are not greatly injured by the beetles, and populations of the pest can be reduced with the rotation and effective weed control.

Some mites also are effectively managed by rotations. The winter grain mite, *Penthaleus major*, on wheat, oats, and barley can be suppressed by rotating to nongrass crops every third year.

In almost all instances, rotations involving botanically similar crops should be avoided. For example, corn following grass sod may suffer serious damage from wireworms. But this is not to say that rotations of dissimilar crops are always safe. For example, wireworm problems can occur in potatoes when this crop is grown after red or sweet clover.

Crop fallowing. For many years fallowing practices have been used in dry regions to allow increases of soil moisture and fertility for growing crops. For

BOX 10.6 CORN ROOTWORMS

SPECIES: *Diabrotica* species (Coleoptera: Chrysomelidae)

DISTRIBUTION: Three species of *Diabrotica*, the northern corn rootworm (*D. barberi* [Smith and Lawrence]), the western corn rootworm, (*D. virgifera virgifera* LeConte), and the southern corn rootworm (*D. undecimpunctata howardi* Barber) inhabit different regions of the United States. The northern corn rootworm once solely inhabited the corn belt states. The western corn rootworm, once restricted to Colorado, Kansas, and Nebraska, has spread northward and eastward to inhabit most of the corn belt states. The southern corn rootworm is found mainly in the South and lower Midwest, but because of migration it can be found throughout the eastern United States and into Canada. There are many regions where one of more of the species inhabit the same location.

IMPORTANCE: The three species of beetles, referred to as rootworms because of larval feeding habits, are major corn pests. Adults damage corn during silking by clipping corn silks and reducing the chances for complete fertilization of the plants. This results in yield loss when corn ears are not completely filled with kernels. Larvae, or rootworms, cause the greatest degree of damage to corn plants. They damage corn by feeding on, and then tunneling in, the roots of growing plants. This injury reduces the amount of food available for plant growth and ear development, which consequently lowers yield (often up to 30 percent or more). In addition, the damaged root system provides less support for the plant during wind and rain storms. This injury often causes lodging of plants and difficulty in harvesting.

APPEARANCE: Adults of the northern corn rootworm are about 6 mm long and yellow to pale green (no striping or spots). Adults of the western corn rootworm are 6 to 8 mm long and yellow, with a black stripe on the outside of each elytron. Adults of the southern corn rootworm are yellow to pale green, with three rows of black spots on the elytra. Larvae of all species are about 6 to 7 mm long, slender, legless, and white, with a brown head capsule and terminal segment.

LIFE CYCLE: Northern and western corn rootworm species have a similar life cycle. Adults oviposit eggs in the soil near corn plants during the fall. The eggs overwinter and hatch the following spring, and larvae begin feeding on the young roots of corn. They develop through three stadia and pupate in the soil. Upon emergence, adults begin feeding on corn silks and oviposit in the soil. Only one generation per year occurs for these two species. The life cycle of the southern corn rootworm is similar except that more than one generation is typical, and it overwinters as adults or eggs in the southern part of its range.

instance, in western Kansas winter wheat is often planted in fields that had been idled and kept weed free the previous season. Without vegetation growing on it, the fallowed land is able to store a significant share of a season's precipitation. This stored moisture along with precipitation of the next season is able to support a biennial wheat crop. Fallowing may also be practiced as part of set-aside programs

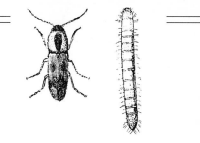

BOX 10.7 WIREWORMS

SPECIES: numerous species in several genera (Coleoptera: Elateridae)

DISTRIBUTION: The term "wireworm" describes the larvae of adult click beetles (Family Elateridae). Because the various genera of plant-feeding elaterids are widespread, wireworms are also cosmopolitan and are found in almost any kind of soil.

IMPORTANCE: Wireworms are polyphagous and feed reaidly on numerous hosts while in larval form. They injure crops by devouring seeds in the soil, cutting off small underground stems and roots, and boring into larger stems, roots, and tubers. Normally, the crop is planted in soil already containing wireworms, and the pests simply begin feeding on the new food source. This is especiallly true in cases where prairie or sod is newly cultivated. No crop is known to be entirely immune to wireworm attack, and the estimated loss to farmers annually runs into the millions of dollars.

APPEARANCE: Adult click beetles are hardshelled, 7 to 25 mm long (depending on species), brown to black to gray, with a tapered body at each end. The name "click beetle" refers to the clicking sound emitted by adults as they flip their body to right themselves. Larvae (wireworms) are slender, cylindrical, 7 to 35 mm long, hardened, and yellow to brown. Adults and larvae vary in details of appearance, depending on species.

LIFE CYCLE: Wireworms have varied life histories depending on species, but some aspects of development are similar. Adult females lay eggs singly in the soil, 2 to 15 cm deep during the spring or early summer. Hatching occurs in 14 to 30 days, and young larvae move within the soil to locate a food source. Heavy mortality occurs during the larval stage if environmental conditions are too harsh or food is too scarce. Larvae feed in the soil for one to five years, with overlap of generations. Pupation occurs in the soil, and adults leave the soil and complete the life cycle by mating and ovipositing. Survival of larvae is increased by females seeking grassy areas (indicative of the extensive root mass) and laying eggs there.

in the United States for the reduction of commodity surpluses and for weed control.

Although fallowing is not always practical, it can be effective against certain insect pests. Ecologically, fallowing creates a hiatus in the supply of requisites for pests and may be a way of purging an area of difficult pests. Among other pests, wireworms are said to have been reduced by summer fallow practices in parts of Canada. With these insects, two or three fallows may be required to reduce severe infestations. Where more than one planting of the same crop occurs in a single growing season, short fallow periods between plantings may be adequate to reduce infestations. In California and Florida, it is sometimes customary for early and late plantings of celery to overlap in time. Here, aphids infesting early plantings have been a problem in the later plantings when they migrate and transmit celery mosaic virus. By employing a celery-free period to eliminate the overlapping presence of a host, aphid infestations and the mosaic disease have been effectively suppressed.

A somewhat similar procedure to fallowing also occurs with livestock. Here, cattle and other animals are moved from one pasture to another, leaving the first pasture idle for a period of time. Such has been called **pasture spelling**, and it has the same effect on less mobile livestock parasites as fallowing has on some plant pests. In the United States, the spelling or rotation of cattle from pastures was particularly important against the cattle tick, *Boophilus annulatus*, a vector of pathogens that cause bovine piroplasmosis. Elimination of cattle in fields for a few months caused ticks to starve and allowed the cleanup of infested pastures. Other parasites, such as the lone star tick, *Amblyomma americanum*, in the southwestern United States, have a wider host range and are not as easily managed by pasture spelling because they may survive on birds and other wildlife in the area.

Disrupting crop and insect synchrony. Part of the reason that insects are pests on given crops is because the insect seasonal cycle is synchronized with the seasonal cycle of these crops. If crop phenology (timing of biological events like emergence, flowering, fruiting, and seed maturity) can be changed to be asynchronous with such insect events as egg laying and larval development, insect numbers and/or injury can be reduced.

Changes in crop phenology may sometimes be accomplished by planting alternate varieties of the same crop, changing planting dates, or both. With some crops, like clover, clipping may modify reproductive phenology.

Modifying planting dates has been one of the most widely used methods to cause asynchrony between crops and pest insects. One of the classic examples of this approach is with the Hessian fly and the planting of winter wheat. Winter wheat is planted and emerges in the fall as do Hessian fly adults. These adults live only three or four days, and females lay eggs on the leaves of early-sown wheat. Eggs hatch in the fall, and larvae develop into pupae ("flaxseed" stage), which overwinter. Ecological management can be practiced by understanding the adult emergence period and delaying wheat planting until all adults have died. In so doing, the emerging wheat escapes infestation by the fall generation. Estimates of fly-free dates for planting are available for most northern states and are especially recommended when fly-resistant cultivars are unavailable.

In another example, modifying planting dates of kale (a type of cabbage) can reduce larval infestations of cabbage looper, *Trichoplusia ni*, on the crop. The pest is most abundant late spring to early fall in the southern United States, and numbers on kale can be held to manageable levels by early planting of the spring crop and late planting of the fall crop. With this plan, the crop matures too early in the spring and too late in the fall to be affected by peak populations of the pest.

Another horticultural crop, tomatoes, have also been shown to benefit from altering planting dates. In North Carolina, by planting tomatoes early, ripening occurs before corn matures, and injury by the corn earworm is minimal. Late-planted tomatoes can be heavily infested by moths that move from the succeedingly less attractive corn into the fruiting tomatoes.

In a somewhat different situation, the formation of clover seed can be delayed by spring grazing or clipping. This has been done to avert damage by the clover seed midge, *Dasineura leguminicola*, in the northwestern United States, where clover is grown for seed. In this instance, the delayed phenology causes seeds to

Figure 10.14 Fly-free dates for planting wheat in the central United States to escape egg laying by the Hessian fly, *Mayetiola destructor*. (Courtesy USDA)

set after adults of the first fly generation have died. By the time second-generation flies are ready to oviposit, the seeds are too far advanced to serve as a host.

Timing of livestock production activities like dehorning and castration also may be modified to prevent problems from insects. Here, transmission of disease-producing organisms by flies is reduced by conducting these and other such procedures at cooler times of the year, when adults of the species are not active.

DIVERTING PEST POPULATIONS AWAY FROM THE CROP

In many instances, it may not be practical to modify the crop itself, or modifications of existing environments may not prove very effective for pest management. Ineffectiveness may be prevalent when insects have a high dispersal rate.

As another approach to habitat management, it is sometimes possible to take advantage of insect dispersal abilities and/or their preferences for one host over another. With this procedure, we would attempt to divert the insect from a crop to be protected by presenting it with a more favorable substitute. Among the most common ways of diverting insects are **trap cropping** and **strip harvesting**.

Trap Cropping

Trap cropping usually involves planting small areas of a crop or other species near the protected crop. The favorability of this alternate environment (the trap) compels the pest to move into the trap and stay away from the protected crop. Depending on the seasonal cycle, the insects are left to develop in the trap or are killed in place with an insecticide. Even if an insecticide is used, costs and environmental impact are reduced, because entire fields are not treated. In trap cropping, the trap may be a different species than the crop or the same species planted at a different time.

An instance showing feasibility of this technique has been developed with soybeans in the Mississippi delta region of the southern United States. Here, populations of bean leaf beetle, *Cerotoma trifurcata*, can be suppressed by planting 5 to 10 percent of a field with an early maturing soybean variety (group V) about 10 days to three weeks earlier than the main planting. The main planting of the field is of a late maturing variety (groups VI or VII). Early in the season, vegetative growth of the early planting attracts overwintered bean leaf beetles, which seek food and lay eggs of the first generation. Treatments of insecticides are made in the trap 7 to 10 days after emergence of first-generation adults to prevent infestations in the main planting. Later in the season, at soybean pod fill, southern green stink bugs, *Nezara viridula*, are also attracted to and concentrate in the trap. Subsequently, these may be treated with insecticide, if necessary. The net result of using this approach can be a reduction in both treatment costs and insecticide residues in the environment. An added benefit can be conservation of the insect natural enemies in the agroecosystem.

Planting strips of alfalfa in cotton fields in the western United States is another

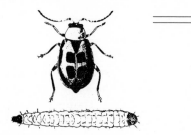

BOX 10.8 BEAN LEAF BEETLE

SPECIES: *Cerotoma trifurcata* (Forster) (Coleoptera: Chrysomelidae)

DISTRIBUTION: The bean leaf beetle is widely distributed throughout the United States. Populations tend to be more abundant in the southeastern states but also occur regularly in the midwestern and eastern regions.

IMPORTANCE: Bean leaf beetles feed on several host species but are commonly found in soybeans, pea, cowpea, and several other bean species. Injury is predominately leaf consumption by adult beetles, but larvae also injure plants by feeding on roots and nodules. In most instances, adults do not cause economic damage by themselves, but their defoliation (rather symmetrical round holes) contributes to the feeding injury of other defoliating species. Adults also have been observed feeding on the mature bean pods. The consequence of this pod injury is poorly understood but can be seen as round necrotic areas on the pod directly over the seeds.

APPEARANCE: Adults are about 7 mm long and yellow, tan, or red. They usually have black spots on the elytra and have a black wing margin. This species is polymorphic, with exact color, number, and arrangement of spots varying among individuals. Eggs are orange, lemon-shaped, commonly laid in clusters of 15 to 30, and found in or on soil near host plants. Larvae are white with a dark head and are distinctively segmented. Pupae are found in the soil within an earthen chamber.

LIFE CYCLE: Adults overwinter in the soil or under crop refuse in and around the field. Beetles become active in spring and begin feeding and laying eggs in or on the soil surface near stems of host plants. Hatching occurs within about 11 days, and larvae feed on plant nodules or root tissue through three stadia. Pupation occurs underground, and adults emerge and feed on available foliage. The complete life cycle requires 35 to 55 days, and one to three generations per year may occur, depending on temperature and latitude.

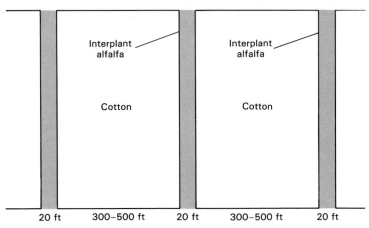

Figure 10.15 Schematic drawing of restricted alfalfa plantings among cotton to trap populations of western lygus bugs, *Lygus hesperus*. (Reprinted with permission, from V. Stern, 1981, in D. Pimentel, ed., *Handbook of Pest Management*, volume I, CRC Press)

example of trap cropping. This approach has given effective management for western lygus bugs, *Lygus hesperus*, in both experimental and commercial fields. With this technique, 5- to 10-meter strips of alfalfa are planted in every 100 to 120 meters of cotton. The alfalfa effectively draws the bugs out of the cotton, and they remain in the alfalfa to feed. In this way, blossom abortion caused by bug feeding is significantly reduced in the cotton, and the alfalfa receives little injury. The technique has been found most effective with low to moderate populations of plant bugs.

Strip Harvesting

Strip harvesting is similar to trap cropping except that a trap is created in a main crop by harvesting different areas at different times. With this approach, insects in the crop are not forced to search for requisites in adjacent crops.

Again, the main example of this approach is the western lygus bug, with cotton and alfalfa. In California, cotton and alfalfa may be produced adjacent to each other, and alfalfa fields have been the source of western lygus bugs. Movement of the bugs into cotton is particularly heavy after alfalfa fields have been clipped for hay. To reduce this movement, alfalfa can be harvested in alternate strips. When a strip is cut, strips on either side of it are half-grown. Consequently, instead of leaving the field and moving to cotton, as they do in conventional operations, the bugs remain in alfalfa fields, moving over from harvested to unharvested plants. This approach is most applicable for forage crops, which are harvested several times each growing season.

REDUCING THE IMPACT OF INSECT INJURY

The purpose of this technique is to manage losses. We realize that insects are present in the crop and injury is being sustained. Instead of focusing on the insects themselves, we consider the crop and modify cultural techniques to minimize

Figure 10.16 Graph showing increases in densities of western lygus bugs, *Lygus hesperus*, in cotton following complete clipping of adjacent alfalfa. (Reprinted with permission, from V. Stern, 1981, in D. Pimentel, ed., *Handbook of Pest Management*, volume I, CRC Press)

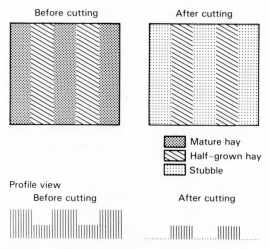

Figure 10.17 Schematic drawing of alfalfa strip harvesting to reduce movement of western lygus bugs, *Lygus hesperus*, into adjacent cotton. (Reprinted with permission of Plenum Press from *Introduction to Integrated Pest Management* by M. L. Flint and R. van den Bosch. Copyright ©1981)

losses from the injury. Such an approach often is used in conjunction with other pest management tactics.

Modify Host Tolerance

The modification of plants and animals to be more tolerant of insect injury is often achieved genetically. New forms of plants or animals are bred that yield well in spite of insect attack.

Host tolerance also can be modified in a cultivar or breed by nongenetic means. Most crops sustain a wide range in yield loss to a given degree of injury, depending on the vigor of crop growth. For instance, soybeans grown during periods of drought in the midwestern United States can suffer up to twice the yield loss from a given amount of defoliation as those in years with normal precipitation. Therefore, good production practices like proper irrigation, fertilization, and weed control can have a significant influence on the vigor of the crop and consequently, the amount of damage sustained from insect injury. By producing a vigorous crop, it is sometimes possible to overcome a degree of the damage that otherwise would be sustained from a pest population.

Although vigorous growth of plants and animals may reduce losses in many situations, there are documented instances where such growth attracts greater numbers. For example, leafhoppers (Homoptera: Cicadellidae) have been found most abundant on rice fertilized with high rates of nitrogen, as well as mites (Acari) and aphids on an array of well-fertilized crops. This does not mean that we should limit crop fertility, only that we should be cognizant of possible pest increases and be prepared to deal with them if they occur.

Modify Harvest Schedules

With many crops, harvest time can vary within certain acceptable limits. When this is the case, harvest dates can sometimes be modified to avoid some types of insect losses. As a rule, crops injured by insects should be harvested at the earliest possible date.

In forage crops like alfalfa, damage from the potato leafhopper, *Empoasca fabae*, and the alfalfa weevil, *Hypera postica*, can be minimized by early clipping. If significant numbers of the potato leafhopper are present, clipping is done at the very-early-bloom or late-bud stage. For the alfalfa weevil, decisions on early clipping versus insecticide spraying are based on numbers of larvae and plant height (Table 6. 2). If the crop is near harvest stage, clipping early may produce better economic returns than spraying weevils and delaying harvest.

Early harvest also is recommended for insect pests of other crops. One is the wheat stem sawfly, which tunnels in wheat stems, causing plant lodging and harvest loss. In South Dakota, some growers clip infested wheat before it lodges and windrow it. After drying, the windrowed plants are threshed with a special attachment on the combine, and harvest losses are reduced. The loss of yield from corn ears that drop early because of European corn borer tunneling also is reduced by early harvesting. However, early harvested corn usually has unacceptably high moisture content and must be dried before it is stored.

In forestry, logging operations are scheduled during specific times of the year to minimize damage from wood-boring insects. Timber cut from late summer to early winter is least likely to be damaged by wood borers. Losses from bark beetles, *Ips* species, in pine are least likely if logging operations are completed during the fall or winter months, thus avoiding the period of greatest beetle activity.

Figure 10.18 A grain-drying system for corn. Early harvesting to reduce losses from dropped grain usually requires drying facilities. (Courtesy D. Erbach and USDA)

Figure 10.19 Chances of timber loss from some forest pests are reduced if logging operations are completed during fall and winter months. (Courtesy USDA)

CONCLUSIONS

Ecological management is one of the oldest, least expensive, and most ecologically compatible tactics for solving insect problems. To use this approach, weak links in the insect seasonal cycle are identified and exploited. Usually, food sources or physical factors of the crop environment are made unfavorable for insects through manipulating conventional production practices.

Figure 10.20 Moldboard plowing (top) and corn plants emerging in a no-till environment. Conventional methods of sanitation like deep plowing are not possible in many conservation tillage systems. (Courtesy D. Erbach and USDA)

In most instances, the approach is a preventive tactic that anticipates problems before they occur and attempts to avoid or minimize their impact. It serves as a baseline procedure that is compatible and can be integrated with many other pest management tactics. As a rule, it is most effective with insects that have a narrow host range, a low rate of dispersal, and/or complex requirements in the seasonal cycle.

The major limitation of ecological management is that the most effective procedures to manage pests may not be practical in terms of crop production objectives. Although insect pests are an important part of crop production, other priorities like greater efficiency, higher yields, and soil conservation often dictate the confines in which ecological management can operate.

Additionally, new technologies are leading to changes in crop production and making older approaches to ecological management less viable. Particularly, the development of selective herbicides has permitted the refinement and acceptance of no tillage or reduced tillage practices. Such practices save fuel, time, and labor, and encourage soil conservation. They also eliminate such options as sanitation and tillage for insect pest management.

Will such changes in crop culture cause significant changes in the status of insect pests? Probably so. For some pests, status will increase, but it may decrease with other pests. In either case, pest managers will need to accept these technologies and understand the ecology of new crop environments. Where pest infestations increase, a search for other ways to make crops less favorable for insects will be needed.

Selected References

FENEMORE, P.G. 1982. *Plant Pests and Their Control.* Wellington: Butterworths of New Zealand. pp. 167–172.

FLINT, M.L., and R. VAN DEN BOSCH. 1981. *Introduction to Integrated Pest Management.* New York: Plenum Press, pp. 46–48.

GENUNG, W.G. 1974. Flooding in everglades soil pest management. *Proceedings Tall Timbers Conference on Ecological Animal Control by Habitat Management,* Feb. 28–March 1, Gainesville, Fla., No. 6, pp. 165–172.

HANSON, C.H., S.G. TURNIPSEED, N.T. POWELL, J.M. GOOD, and D.L. KLINGMAN. 1979. Cultural and preventive practices. In W. B. ENNIS, JR., ed., *Introduction to Crop Protection.* Madison Wisc.: American Society of Agronomy and Crop Science Society, pp. 91–101.

MUMA, M.H. 1970. Preliminary studies on environmental manipulation to control injurious insects and mites in Florida citrus groves. *Proceedings Tall Timbers Conference on Ecological Animal Control by Habitat Management,* Feb. 26–28, Tallahassee, Fla., No. 2, pp. 23–40.

NATIONAL ACADEMY OF SCIENCES. 1969. *Insect Pest Management and Control. Principles of Plant and Animal Pest Control,* vol. 3. U.S. National Academy of Science Publication 1695,. pp. 208–242.

PEDIGO, L.P., R.A. HIGGINS, R.B. HAMMOND, and E.J. BECHINSKI. 1981. Soybean pest management. In D. PIMENTEL, ed., *CRC Handbook of Pest Management in Agriculture.* vol. 3, Boca Raton, Fla.: CRC Press, pp. 417–537.

RABB, R.L., G.K. DEFOLIART, and G.G. KENNEDY. 1984. An ecological approach to managing insect populations. In C.B. HUFFAKER and R.L. RABB, eds., *Ecological Entomology.* New York: Wiley, pp. 697–728.

RALSTON, L.H., and C.E. MCCOY. 1966. *Introduction to Applied Entomology*. New York: Ronald Press, pp. 102–107.

SAILER, R.I. 1981. Extent of biological and cultural control of insect pests of crops. In D. Pimentel, ed., *CRC Handbook of Pest Management in Agriculture*, vol. 2. Boca Raton, Fla.: CRC Press, pp. 57–67.

STERN, V. 1981. Environmental control of insects using trap crops, sanitation, prevention, and harvesting. In D. PIMENTEL, ed. *CRC Handbook of Pest Management in Agriculture*, vol. 1. Boca Raton, Fla.: CRC Press, pp. 199–207.

TEETES, G.L. 1981. The environmental control of insects using planting times and plant spacing. In D. PIMENTEL, ed., *CRC Handbook of Pest Management in Agriculture*, vol. 1. Boca Raton, Fla.: CRC Press, pp. 209–221.

11 CONVENTIONAL INSECTICIDES

INSECTICIDES AND OTHER PESTICIDES are some of the most important chemicals used for the well-being of human populations. They are indispensable in maintaining high levels of health, nutrition, and quality surroundings. They rank with medicines in their impact on our existence.

In agricultural production, pesticides are a regular component of most systems, and their development has given rise to entirely new ways of growing crops. Indeed, we could not consider using such energy and soil conserving approaches as no-till cropping without them. Moreover, the quantity and quality of our food and fiber production could not be maintained without substantial pesticide inputs.

Pesticide use has increased phenomenally since the development of modern synthetic poisons in the 1940s. In the United States alone, over 600 million kilograms of pesticides are produced annually, with retail sales of more than $5 billion. Of these sales, it is estimated that agriculture accounts for 72 percent, and industry and government about 21 percent. Home and garden use accounts for only about 7 percent. A study of household, lawn, and garden use has indicated that 91 percent of American households use some kind of pesticide. This study reveals the pervasiveness of these chemicals in our society.

Insecticides are particular kinds of pesticides for killing insects and other invertebrates. They account for about 36 percent of the pesticide quantity used in agriculture, but they represent over 50 percent of that used for home and garden purposes. Furthermore, over 400 insecticidally active compounds with thousands of uses are registered with the United States Environmental Protection Agency (EPA).

Increasing insecticide use since World War II is the result of several advantages these chemicals have over other techniques for insect control. Of utmost importance is the effectiveness of insecticides. They can treat an insect problem while it is in progress, reducing numbers to insignificant levels. Insecticide action is rapid; it usually takes effect within hours and alleviates a problem within a few days. In particular, rapid action allows this tactic to cure a problem by dampening population peaks before major losses are sustained. Insecticides also are economical, at least in the short run, compared with many other pest management tactics. In terms of only the crop, cost/benefit ratios regularly are on the order of four to five dollars return for every dollar invested in insecticide applications.

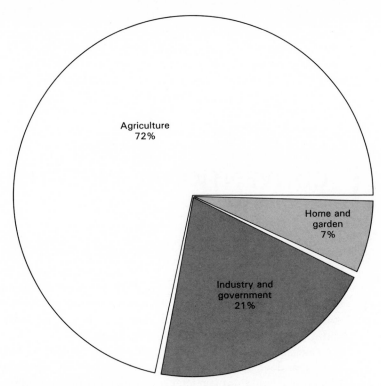

Figure 11.1 Proportionate use of pesticides by groups in the United States. (Based on EPA data)

Finally, insecticides offer ease of application. Persons with minimal experience or training can apply them effectively.

Unfortunately, insecticides have important disadvantages. Some problems occur within the cropping system itself from frequent use. Insecticide resistance, pest resurgence, and pest replacement can occur following repetitive applications of these compounds. Therefore, we may see decreasing effectiveness with their use. Moreover, insecticides may have a negative impact on nontarget species like honey bees, fish, and wildlife both within and outside agroecosystems. Destruction of these organisms is a cost often not included when calculating cost/benefit ratios. Furthermore, insecticides present risks to the user. Many insecticidal compounds are highly toxic to humans and can injure or kill when applied improperly or when accidents occur.

If we consider the alternatives, the benefits of insecticides probably outweigh the risks. But the goal of insecticide use to obtain maximal benefits with minimal risks has not yet been achieved in most situations. Therefore, it is important that a good understanding of these chemical tools and the consequences of their use be acquired if this approach is to be improved.

INSECTICIDE NAMES AND FORMULAS

The word **insecticide** literally means killer of insects, derived from the Latin suffix, *cida*, or "killer." The term "pesticide" is broader, denoting a killer of pests in general. In addition to insecticides, other common pesticides include **acaricides** (mite and tick killers), **herbicides** (weed killers), **fungicides** (fungus killers), and **nematicides** (nematode killers). Other designations of pesticides have even more specific names. For example, some insecticides may be called aphicides when used for aphids and termiticides when used for termites. Legally, these are all classed as **economic poisons**, substances used for controlling, preventing, destroying, repelling, or mitigating any pest.

In most instances, insecticides are chemicals. Therefore, although fly swatters are used to kill insects, they are not called insecticides. In this chapter, we deal mostly with conventional or traditional insecticides, which are chemical compounds with wide use that kill insects quickly. Most insecticides are nerve poisons. Nonconventional insecticides include **microbial insecticides** and **insect growth regulators**, whose action is usually slower, and death occurs by other means than those of conventional insecticides.

Insecticide Nomenclature

Insecticide nomenclature is the formal process by which insecticides are named. Insecticides are designated by three names: the **approved common name**, the **trade name**, and the **chemical name**. In the United States, common names for insecticides are officially selected by the Entomological Society of America and approved by the American National Standards Institute and the International Organization for Standardization. The trade name (also the proprietary name and brand name) is given by the manufacturer or formulator of the insecticide. The chemical name provides a description of the insecticide's structure and is formed by following the "Definitive Rules for the Nomenclature of Organic Chemistry," which are developed by the International Union of Pure and Applied Chemistry.

The following is an example of the names for a widely used insecticide:

> common name: carbaryl
> trade name: Sevin®
> chemical name: 1-naphthyl N-methylcarbamate

Note that the trade name has a **registered trademark** superscript, indicating that a patent exists or is pending and that the rights of the patent holder are protected by law. More than one manufacturer may hold patent rights to a compound; therefore it is not uncommon to see more than one trade name for a single insecticide. In publications, the symbol is used the first time the insecticide is mentioned and not after that.

An understanding of the equivalence of these names is important, because insecticide recommendations may be presented either as common names or as trade names. The U. S. Department of Agriculture (USDA) recommends insecticides by common names to avoid bias toward manufacturers. The state Cooperative Extension Service may publish either trade names or common names, usually depending on the name most frequently used for a compound. Equivalences of insecticide names for many compounds are given in Appendix 3, and more complete lists of pesticides can be found in publications like the *Farm Chemicals Handbook*.

Chemical Formulas

Chemical formulas represent the composition of chemical compounds. They describe the components of a single molecule and are important in our characterization of insecticides.

Both molecular formulas and structural formulas are used to describe insecticidal compounds. Molecular formulas use standard symbols of the elements and numbers to show the kind and quantity of atoms in a molecule. For example, H_2O symbolizes water and indicates that two atoms of hydrogen are bonded by an atom of oxygen. On the other hand, the structural formula is more like a map, showing where atoms are located relative to each other.

Molecular   Structural
Formula   Formula
H_2O H — O — H

One of the most common structural components in insecticide chemistry is the benzene ring, also called the phenyl ring when other chemical groups are attached to it. This six-carbon ring with six hydrogens is presented as a hexagon with double bonds. For brevity, the benzene or phenyl ring is usually reduced to a simple hexagon with bonding symbols.

Benzene

Because molecules are actually three-dimensional structures, the structural formula does not give a complete representation; it is only two-dimensional. Therefore, *stereo*, or three-dimensional, formulas are sometimes given so that depth and true spacial form of a compound can be perceived.

Dieldrin   Stereo Dieldrin

SURVEY OF COMMON INSECTICIDES

Insecticides have been grouped or classified in several ways, according to application or chemical composition. If application is the main focus, they may be grouped according to the site of insect encounter. In some of the established literature, insecticides are listed under the headings stomach poisons, contact poisons, and fumigants.

Stomach poisons enter the insect body through the gut and are fatal only after they are eaten. It is mainly the oldest insecticides that fall within this grouping, and few true stomach poisons are used today. One such stomach poison still used is boric acid (H_3BO_4), for use against cockroaches and other crawling insects in the household.

However, there are also modern insecticides whose main encounter with the insect is through the gut. These are so-called **systemic insecticides**. Systemics are taken up and translocated within plants and animals. Insects feeding on the protected host contact the insecticide through the gut, and susceptible individuals are killed. Systemics in plants mostly kill piercing-sucking insects; larger chewing insects usually are not affected. Piercing-sucking pests are killed more readily because they normally receive a greater insecticide dose than chewing insects feeding on the same plant. In livestock, systemics are often used against internal parasites like cattle grubs, *Hypoderma* species. Systemic insecticides also have properties of contact insecticides and are not usually thought of as true stomach poisons.

Contact poisons are the major group of modern insecticides. They usually enter the body when the insect walks or crawls over a treated surface. Here, the insecticide is absorbed through the body wall. If the treated surface is a food source like a leaf or blossom, these poisons may also enter the digestive tract and be absorbed through it. Grouping insecticides by manner of entry begins to

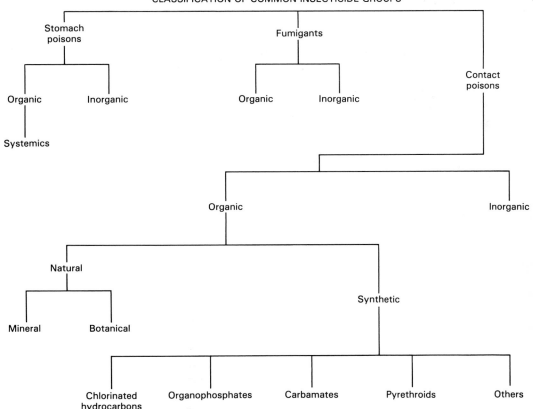

Figure 11.2 Classification of common insecticide groups. Initial divisions are based on site of encounter.

break down when multiple entry sites are encountered. Still, the primary entry site of contact poisons is from the environment and through the body wall.

A fumigant's contact site is the tracheal system. **Fumigants** are insecticides that become gases at temperatures above 5°C. These insecticides are applied to enclosures and to soil. Being volatile, they enter the tracheal system, circulate, and subsequently are absorbed by body tissues.

Insecticides are further grouped or classified according to their nature and source. Therefore, some insecticides are **inorganic**, lacking carbon atoms, and others are **organic**, possessing carbon atoms. Most modern insecticides are organic, and these are even further subdivided as natural or synthetic. Natural insecticides are produced by refining natural substances. These consist mainly of **botanical insecticides**, made from plants, and mineral oils, obtained by refining petroleum. Botanical insecticides have wide usage, but oils are used mainly for suppression of fruit-tree insects and mosquito larvae. Most insecticides used today are synthetics, manufactured by chemically joining elements or simple compounds.

The most precise method of grouping insecticides is according to their chemical makeup. Several classes of compounds can be designated according to **active ingredients** (chemicals responsible for the toxic effect). The four major classes are **chlorinated hydrocarbons**, **organophosphates**, **carbamates**, and **pyrethroids**. These major classes are addressed first; then we discuss several classes of lesser importance.

Chlorinated Hydrocarbons

Chlorinated hydrocarbons are the oldest major insecticide class, having been the first widely used synthetic organic insecticides. All insecticides of this group contain chlorine, hydrogen, and carbon. Occasionally, these insecticides also contain oxygen and sulfur.

DDT and relatives. DDT is the most famous, perhaps infamous, of all insecticides. Its history has already been discussed, and today most uses have been cancelled in the United States (official cancellation January 1, 1973). The primary reason for cancellation of DDT (and several other chlorinated hydrocarbons) is its characteristic stability and fat solubility. Insecticides of the other major

DDT

1,1,1-trichloro-2,2-bis(*p*-chlorophenyl)ethane

TDE (DDD)

1,1-dichloro-2,2-bis(*p*-chlorophenyl)ethane

Methoxychlor

1,1,1-trichloro-2,2-bis(*p*-methoxyphenyl)ethane

classes are relatively unstable; they are broken down in animals by enzymes and in the environment by microorganisms, heat, and/or ultraviolet light.

However, DDT is broken down slowly, sometimes over several years, and this stability provides active residues for animal uptake. Sublethal doses of the chemical are ingested by animals and, not being amenable to metabolism, are stored in body fat. Having been used widely for many years, DDT accumulated in animals that fed on residue-laden plants. Consequently, dairy cows fed DDT-treated hay produced milk with high residues in the milk fat. Ultimately, humans ingested DDT through milk and other animal fats and stored these residues in their own tissues. Indeed, until recently Americans had an average of 10 parts per million of DDT in their body tissues. This concentration has fallen since the cancellation of most DDT uses.

Similar problems occurred with predators in many ecosystems. Here, invertebrates that consumed plants and detritus with DDT residues stored concentrations in body fat. These invertebrates were fed upon by rodents, reptiles, amphibians, fish, and other small insectivores, further concentrating the residues in their tissues. These primary predators, in turn, were eaten by yet other top predators in the food chain, such as hawks and owls, which received yet higher insecticide concentrations. This phenomenon, known as **biomagnification**, was believed partly responsible for the decline of such predators as ospreys, falcons, eagles, seagulls, pelicans, and others. At least in some instances, the mode of decline was caused by thin egg shells and reproductive failure of the populations. Other chlorinated hydrocarbons that have caused biomagnification are DDE (a metabolite of DDT), TDE, dieldrin, aldrin, HCH and its isomers, mirex, heptachlor, and endrin.

The very characteristics of DDT that resulted in its cancellation also made it an effective insecticide. Because of stability, DDT's insecticidal activity lasts considerably longer than many of the insecticides that have replaced it. Consequently, where one application had to be made previously, several may be needed now.

In addition to its importance in medicine (suppression of mosquitoes, lice, and

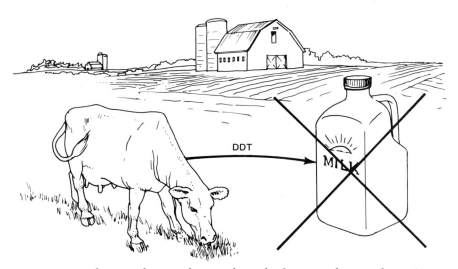

Figure 11.3 Schematic drawing showing the path of DDT to dairy products. (Courtesy EPA)

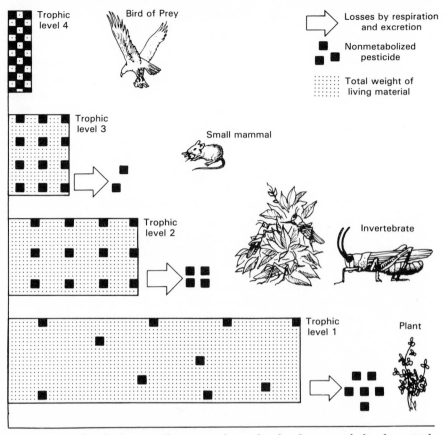

Figure 11.4 Graph showing the accumulating levels of nonmetabolized pesticide residue as trophic levels are ascended. The size of the bars in the graph represents total biomass at a trophic level. Such is an example of biomagnification. (Redrawn from Flint and van den Bosch, 1981, *Introduction to Integrated Pest Management*, Plenum Press)

fleas), DDT was used widely in agriculture and reached a peak of production in the United States in 1961. It was effective against many insects, with the notable exception of grasshoppers (Orthoptera: Acrididae), the boll weevil (*Anthonomus grandis*), the Mexican bean beetle (*Epilachna varivestis*), and aphids (Homoptera: Aphididae). Unfortunately, the overwhelming effectiveness of DDT and its exceptionally low cost contributed to overuse and, subsequently, to its demise. It is still used in some parts of the world but at a fraction of former levels.

Insecticides closely related to DDT are TDE, ethylan, and methoxychlor. Methoxychlor was probably the most important of these other materials, having been used against flies after they had become resistant to DDT and as a spray to eliminate pests from empty stored-grain bins.

HCH and Lindane. HCH, earlier called BHC, was developed as an insecticide by French and British entomologists, beginning about 1940. This insecticide has a wider spectrum (kills more insects) than DDT and is effective against aphids. HCH has a disagreeable musty odor and flavor and its taste can be detected in plants and plant products that come in contact with the insecticide.

HCH (BHC)

1,2,3,4,5,6-hexachlorocyclohexane

HCH comprises five isomers, only one of which is highly active. This isomer, the gamma isomer, was isolated, manufactured, and sold directly as the insecticide lindane. Lindane is odorless and volatile. It was used widely as a household fumigant, dispensed by placing insecticide pellets in electric wall vaporizers.

Presently, all registered products of HCH have been eliminated, as has lindane as a household fumigant. Lindane has been used mainly in sprays for commercial ornamentals and livestock, in dog shampoos, and as a human parasiticide.

Cyclodienes.   The cyclodienes were developed after DDT and HCH, beginning about 1945. The major compounds in this group are aldrin, dieldrin, chlordane, heptachlor, endrin, mirex, endosulfan, and chlordecone.

Chlordane

1,2,4,5,6,7,8,8-octachloro-3a, 4,7,7a-tetrahydro-4,7-methanoindane

Heptachlor

1,4,5,6,7,8,8-heptachloro-3a,4,7,7a-tetrahydro-4,7-methanoindene

Aldrin

1,2,3,4,10,10-hexachloro-1,4,4a,5,8,8a-hexahydro-1,4-*endo-exo*-5,8-dimethanonaphthalene

Endosulfan (Thiodan®)

6,7,8,9,10,10-hexachloro-1,5,5a, 6,9,9a-hexahydro-6,9-methano-2,4,3-benzodioxathiepin 3-oxide

Mirex

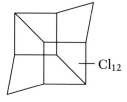

dodecachlorooctahydro-1,3,4-metheno-1H-cyclobuta[cd] pentalene

Chlordecone (Kepone®)

decachlorooctahydro-1,3,4-metheno-2H-cyclobuta[cd] pentalen-2-one

Dieldrin

endo-exo

1,2,3,4,10,10-hexachloro-6,7-epoxy-
1,4,4a,5,6,7,8,8a-octahydro-1,4-*endo-endo*-
5,8-dimethanonaphthalene

Endrin

endo-endo

1,2,3,4,10,10-hexachloro-6,7-epoxy-
1,4,4a,5,6,7,8,8a-octahydro-1,4-*endo-endo*-
5,8-dimethanonaphthalene

These insecticides are persistent chemicals, stable in soil and relatively so in sunlight. Therefore, many were used in great quantities against such soil insects as corn rootworms (*Diabrotica* species), wireworms (Coleoptera: Elateridae), and cutworms (Lepidoptera: Noctuidae). Most of the cyclodienes have higher levels of mammalian toxicity than DDT and are more dangerous to apply. Growing ineffectiveness from insecticide resistance and problems with residue uptake in harvested produce have caused the elimination of most cyclodienes in agriculture. Cyclodienes also have been eliminated for use in termite control.

Polychloroterpenes. Polychloroterpene insecticides are almost exclusively agricultural chemicals. There are only two materials in this group, strobane and toxaphene. Until a few years ago, toxaphene was the single most widely used insecticide in agriculture. It is prepared by the chlorination of camphene, a derivative of pine tree materials. Toxaphene was used widely, particularly in cotton, where it was mixed with other insecticides (especially methyl parathion) to give greater effectiveness. At its peak, toxaphene accounted for over 40 percent of insecticide quantity in cotton.

The polychloroterpene insecticides are easily metabolized by birds and mammals, and storage in body fat is low. Even though not highly toxic to other animals, toxaphene is a potent fish poison. Most uses of toxaphene have been cancelled, as well as all uses of strobane.

Toxaphene

chlorinated camphene containing 67 to 69 percent chlorine

Organophosphates

The organophosphate insecticides were developed in Germany during World War II as a substitute for nicotine, an insecticide used against Colorado potato beetle, *Leptinotarsa decemlineata*. The discovery of the insecticidal properties of this group was associated with other German studies on related chemicals, the so-called "nerve gases" (sarin, soman, and tabun).

Organophosphates (OP) are derived from phosphoric acid and are some of the most toxic insecticides. As opposed to the chlorinated hydrocarbons, they are unstable (in the presence of light) and quickly break down into nontoxic compounds. Breakdown can occur within a few hours or days, as compared with months or years for many of the chlorinated hydrocarbons. For this reason and because they are effective, OPs have replaced the chlorinated hydrocarbons in many programs. In fact, they are perhaps the most widely used group of insecticides today.

OPs are characterized as having different alcohols attached to their phosphorus atoms, and the various phosphorus acids produced are termed esters. These esters have different combinations of oxygen, carbon, sulfur, and nitrogen, and OPs formed from them can be divided into three groups of derivatives: **aliphatic, phenyl,** and **heterocyclic**.

Ortho Phosphoric Acid

Aliphatic derivatives. The aliphatic derivatives are compounds with straight carbon chains, as opposed to those with ring structures. The oldest and one of the most toxic aliphatic derivatives is **TEPP**. TEPP was used for fly control in dairy barns because of effectiveness and short residue (breakdown in 12 to 24 hours).

TEPP

tetraethyl pyrophosphate

The most widely used aliphatic is malathion, one of the safest and most effective of all OPs. Malathion has been used for all types of agricultural purposes and is effective against many insects. It is safe for home use, including most garden and household pests, and does not harm pets when used according to directions. It is commonly prescribed by physicians for head, body, and crab louse (Anoplura: Pediculidae) problems. Also, it is formulated as powders or dips for flea (Siphonaptera) control on pets and mange mites (Acari) on livestock. Ultra low volume sprays, those applied as fine particles in concentrated form, also have been found effective and economical for management of such insects as grasshoppers on rangeland.

Malathion

O,O-dimethyl-*S*-1,2-di(carboethoxy)
ethyl phosphorodithioate

In addition, several aliphatics are plant systemics, including dimethoate, disulfoton, oxydemetonmethyl, and dicrotophos. Most of these insecticides are applied to the soil and are taken up by plants and translocated to above-ground plant parts. They are effectively used for many aphids, leafhoppers (Homoptera: Cicadellidae), and other piercing-sucking insects. Dimethoate also is an effective contact poison against mites during outbreaks on soybeans and other crops.

Dimethoate (Cygon®)

O,O-dimethyl-*S*-[2-(methylamino)-
2-oxoethyl] phosphorodithioate

Oxydemetonmethyl (Meta Systox R®)

S-2-[(ethylsulfinyl)ethyl]*O,O*-dimethyl
phosphorothioate

Dicrotophos (Bidrin®)

O,O-dimethyl*O*-1-methylvinyl-
N,N-dimethyl carbamoyl phosphate

Disulfoton (Di-Syston®)

O,O-diethyl*S*-2-[ethylthio)ethyl]
phosphorodithioate

Other common aliphatics include trichlorfon, methamidophos, and acephate. Trichlorfon has been used in crop insect management and around farms for fly control. It is one of the few modern insecticides that is insect-selective, killing some insects but leaving others (many natural enemies) unharmed. Both methamidophos and acephate are more recent aliphatics used widely in agriculture and especially for management of vegetable pests.

Trichlorfon (Dylox®)

$$(CH_3O)_2\overset{\overset{\textstyle O}{\|}}{P}-\overset{\overset{\textstyle OH}{|}}{CH}CCl_3$$

dimethyl (2,2,2-trichloro-1-hydroxyethyl)
phosphonate

Methamidophos (Monitor®)

$$\begin{array}{c} CH_3O \\ \\ CH_3S \end{array}\!\!\!\overset{\overset{\textstyle O}{\|}}{P}-NH_2$$

O,S-dimethyl phosphoramidothioate

Acephate (Orthene®)

$$\begin{array}{c} CH_3O \\ \\ CH_3S \end{array}\!\!\!\overset{\overset{\textstyle O}{\|}}{P}--NH-\overset{\overset{\textstyle O}{\|}}{C}-CH_3$$

O,S-dimethyl acetylphosphoramidothioate

Three other aliphatics that have been important as soil insecticides include phorate, terbufos, and ethoprop. Phorate is an older OP that is economical and effective against corn rootworms. Terbufos is a newer compound also widely used for protection of monocropped corn from corn rootworms. Both phorate and terbufos also have plant-systemic properties. Ethoprop is used similarly but is not systemic.

Phorate (Thimet®)

$$\begin{array}{c} C_2H_5O \\ \\ C_2H_5O \end{array}\!\!\!\overset{\overset{\textstyle S}{\|}}{P}-S-CH_2-S-C_2H_5$$

O,O-diethyl S-2-(ethylthio)methyl
phosphorodithioate

Terbufos (Counter®)

$$(C_2H_5O)_2-\overset{\overset{\textstyle S}{\|}}{P}-S-CH_2-S-C(CH_3)_3$$

S-[[(1,1-dimethylethyl)=thio]methyl]
O,O-diethyl phosphorodithioate

Ethoprop (Mocap®)

$$C_2H_5O-\overset{\overset{\textstyle O}{\|}}{P}\!\!\!\begin{array}{c} S-C_3H_7 \\ \\ S-C_3H_7 \end{array}$$

O-ethyl S-S-dipropyl phosphorodithioate

Phenyl derivatives. Phenyl OPs differ basically from aliphatic OPs in having a phenyl ring, which has one of the hydrogens displaced by a phosphorus moiety and others (one or more) displaced by CH_3, Cl, CN, NO_2, or S. Stability of phenyl OPs is greater, and residues usually last somewhat longer in the environment than the aliphatics.

One of the older, most widely used phenyls is parathion. The first form of parathion, ethyl parathion, was introduced in 1947 and early use was against aphids. It is very toxic to humans, and its use diminished with the introduction in 1949 of the somewhat less toxic methyl parathion. Methyl parathion has a broad range of toxicity to many insect pests and has several agricultural applications. It is particularly important for difficult-to-kill insects and is sometimes mixed with other insecticides to increase overall effectiveness. Methyl parathion is too toxic for normal use around the home.

Ethyl Parathion

O,O-diethyl O-p-nitrophenyl phosphorothioate

Methyl Parathion

O,O-dimethyl O-p-nitrophenyl phosphorothioate

Much less toxic is stirofos, a household insecticide used for livestock parasites. Its toxicity is similar to that of malathion.

Stirofos (Gardona®)

O,O-dimethyl O-2-chloro
1-(2,4,5-trichlorophenyl)
vinyl phosphate

Some of the phenyl OPs also are used for animal systemics against cattle grubs. These include famphur and fenthion, which can be used on beef cattle and nonlactating dairy cattle. Usually, they are poured on the animal and are absorbed through the skin.

Famphur (Warbex®)

$C_{10}H_{16}NO_5PS_2$

O-[4-[(dimethylamino)=sulfonyl]phenyl] O,O-
dimethyl phosphorothioate

Fenthion (Baytex®)

O,O-dimethyl O-[3-
methyl-4-(methylthio)phenyl
phosphorothioate

Other phenyl OPs used on field crops include profenophos and sulprofos. Yet others, for example, isofenphos and fonofos, have been particularly useful against soil insects in both field and vegetable crops.

Profenofos (Curacron®)

O-(4-bromo-2-chlorophenyl)-*O*-ethyl-
S-propyl phosphorothioate

Sulprofos (Bolstar®)

O-ethyl S-propyl *O*(4-methylthio)
phenyl phosphorodithioate

Fonofos (Dyfonate®)

$C_{10}H_{15}OPS_2$

O-ethyl S-phenyl ethylphos =
phonodithioate

Heterocyclic derivatives. Heterocyclic OPs, like phenyl OPs, have ring structures but differ in having one or more carbon atoms displaced by O, N, or S. Also, structural rings in this group may have three, five, or six atoms. Compounds of this group are the most stable and long-lasting of the OPs. They have complex molecules that break down into many products. This characteristic makes residues difficult to measure in the laboratory, rendering these compounds somewhat more limited for use on food for human consumption.

One of the most common heterocyclics is diazinon, a moderately safe insecticide contained in many household and garden sprays. In the 1960s, it was used as a soil insecticide against corn rootworms but was replaced after resistance to the chemical developed. It is still useful for household, garden, and ornamental problems, and sometimes for flies in stables and around pets' quarters. Another, older heterocyclic OP is azinphosmethyl. It has been used extensively as both an insecticide and an acaricide on cotton.

One of the most useful heterocyclics is chlorpyrifos. It is a relatively safe insecticide to apply and has become a standard in the structural pest control industry. Marketed as Dursban®, it is effective against cockroaches (Orthoptera: Blattoidea), termites (Isoptera), and many other household pests and can be applied on pets. Marketed under the trade name Lorsban®, it has many uses in field and forage crops. For example, Lorsban is a major replacement for the cancelled cyclodienes formerly used against black cutworms, *Agrotis ipsilon*, on corn.

Two other heterocyclic OPs with wide use in agriculture include methidathion and phosmet. These have uses on field, forage, fruit, and nut crops against a variety of insects and mites. Phosmet also is used against two important weevils, the boll weevil on cotton and the plum curculio, *Conotrachelus nenuphar*, on fruit trees.

Diazinon

O,O-diethyl*O*-(2-isopropyl-4-methyl-6-pyrimidinyl) phosphorothioate

Azinphosmethyl (Guthion®)

O,O-dimethyl*S*(4-oxo-1,2,3-benzotriazin-3(4*H*)-ylmethyl) phosphorodithioate

Chlorpyrifos (Dursban®, Lorsban®)

O,O-diethyl*O*-(3,5,6-trichloro-2-pyridyl) phosphorothioate

Methidathion (Supracide®)

O,O-dimethyl phosphorodithioate *S*-ester with 4-[mercaptomethyl]-2-methoxy-Δ²-1,3,4-thiadiazolin-5-one

Phosmet (Imidan®)

N-(mercaptomethyl)-phthalimide *S*-(*O,O*-dimethylphosphorodithioate)

Carbamates

The carbamates are broad-spectrum insecticides that have had wide application in agriculture. They were developed by the Geigy Corporation in 1951 but, because of initial problems, were not practical until 1956. These insecticides are produced from carbamic acid and are similar in environmental persistence to the organophosphates. A distinct limitation of carbamates in pest management is their toxicity to Hymenoptera, including both pollinators and parasitoids.

Two carbamates widely used in agriculture are carbaryl and carbofuran. Carbaryl is the oldest of the effective carbamates, introduced in 1956. It has low toxicity to humans and, therefore, is a common insecticide for use in home lawns and gardens. Its greatest usage, however, is probably in fruit production, where it both kills insects and serves as a fruit-thinning agent. Carbofuran is used widely as a soil insecticide for suppression of nematodes (Nematoda), corn rootworms, and other soil pests and is a plant systemic. It is highly toxic to humans and should be handled carefully. Most recently, carbofuran has become increasingly ineffective when used continuously in the same soil. This phenomenon is seemingly caused by increases in microorganisms capable of quickly degrading the compound after application. When used for soil insects, carbofuran should be rotated with a soil insecticide of another class, for example, an organophosphate.

Carbaryl (Sevin®) Carbofuran (Furadan®)

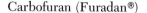

1-naphthyl N-methylcarbamate

2,3-dihydro-2,2-dimethyl-7-benzofuranyl methylcarbamate

Other common carbamates used in agriculture are aldicarb, trimethacarb, and methomyl. The first two are soil insecticides, and methomyl has been particularly effective against caterpillars (Lepidoptera) on vegetables. Aldicarb is an important

Aldicarb (Temik®)

2-methyl-2-(methylthio)propionaldehyde
O-(methylcarbamoyl) oxime

Methomyl (Lannate,®Nudrin®)

Trimethacarb (Broot®)

methyl N-[(methylcarbamoyl)oxy]thioacetimidate

3,4,5-trimethylphenyl methylcarbamate, 75%
2,3,5-trimethylphenyl methylcarbamate, 18%

systemic for controlling nematodes, as well as insects. It is one of the most toxic and, therefore, dangerous to handle of all insecticides. This high toxicity has greatly limited aldicarb's usefulness. Under certain circumstances, aldicarb has been discovered in shallow groundwater, and its use must be monitored very carefully.

A remaining carbamate, propoxur, is used by professionals against cockroaches. It is particularly effective against species in restaurants and homes that have become resistant to organophosphate insecticides.

Propoxur (Baygon®)

o-isopropoxyphenyl methylcarbamate

Pyrethroids

Pyrethroids are the fastest developing group of modern insecticides. They are replacing many older insecticides because of their great effectiveness and safety of application.

Pyrethroids are not new insecticides; the first, allethrin, was developed in 1949. Allethrin was synthesized to duplicate the insecticidal activity of a natural product, cinerin I, a component of the botanical insecticide **pyrethrum.** The term pyrethroid comes from "pyrethrum," and "-oid" means something that resembles something else. In other words, these insecticides resemble pyrethrum.

Allethrin (Pynamin®)

2-methyl-4-oxo-3(2-propenyl)-2-
cyclopenten-1-yl 2,2-dimethyl-3-
(2-methyl-1-propenyl)cyclopropane-
carboxylate

However, pyrethroids have certain advantages over their namesake. They are highly toxic to insects at very low rates. In addition to quick knockdown ability, there is less recovery of poisoned insects, compared with the natural product. Although pyrethrum is too expensive to be used in agriculture, modern pyrethroids compare favorably in price at low application rates with organophosphates and carbamates. Finally, the modern pyrethroids are broken down by ultraviolet wavelengths more slowly than is pyrethrum. Consequently, many pyrethroids are effective on plant foliage for four to seven days.

The pyrethroids often are categorized into generations of development, of which there are four. Allethrin belongs to the first generation, and several compounds including resmethrin belong to the second generation. But the most widely used compounds today are from the third and fourth generations.

Third-generation pyrethroids. These include fenvalerate, introduced in 1972, and permethrin, introduced in 1973. These insecticides are used in many crops, including cotton, corn, and soybeans, where they are effective against several aboveground insect pests. Rates of application in these crops are exceptionally low, for example, 0.11 kilogram or less of actual ingredient per hectare (0.1 pound or less per acre), compared with many organophosphates and carbamates, for example, 1.1 to 2.3 kilograms of actual ingredient per hectare (1 to 2 pounds per acre).

Fenvalerate (Pydrin®)

cyano (3-phenoxyphenyl) methyl 4-chloro-α-(1-methyl-ethyl) benzeneacetate

Permethrin (Ambush®, Pounce®)

m-phenoxybenzyl ($\pm$)-*cis, trans*-3-(2,2 dichlorovinyl)-2,2-dimethylcyclopropanecarboxylate

Fourth-generation pyrethroids. These pyrethroids are even more potent insect poisons than are the third-generation chemicals. Application rates for these may be only one-tenth those of the third-generation pyrethroids for equal effectiveness. Much research is being conducted with fourth-generation com-

Cypermethrin (Ammo®)

cyano(3-phenoxyphenyl) methyl 3-(2,2-dichloroethenyl)-2,2-dimethylcyclopropanecarboxylate

Flucythrinate (Payoff®)

(±)cyano(3-phenoxyphenyl)methyl
(+)-4-(difluoromethoxy)-α-(1
methylethyl)benzeneacetate

Fluvalinate (Mavrik®, Spur®)

N-[2-chloro-4-(trifluoro=
methyl)phenyl]DL-valine, cyano=
(3-phenoxyphenyl)methyl ester

Deltamethrin (Decis®)

cyano(3-phenoxyphenyl) methyl
3-(2,2-dibromoethenyl)-2,2-dimethyl=
cyclopropanecarboxylate

pounds for greater usefulness and even more efficacy. Some of the new, truly exciting insecticides in this category are cypermethrin, flucythrinate, fluvalinate, and deltamethrin.

Botanicals

Botanical insecticides are derived directly from plant or plant products. With the possible exception of sulfur, these have been used longer than any type of insecticide. Many botanicals are of interest to organic gardeners because they are "natural" products. However, as a group they are no safer than synthetics available to the lay public. Botanicals are expensive to extract from plants, and for this reason, they are generally impractical in commercial agriculture. The use of these materials reached a peak in 1966 and has steadily declined since. Presently, there are five botanicals that have any notable use.

Pyrethrum. Pyrethrum was already mentioned with regard to the pyrethroids, and it is by far the most widely used botanical. This insecticide is extracted from flower petals of *Chrysanthemum* species grown in Kenya, Ecuador and other countries. Pyrethrum is made up of four compounds: pyrethrins I and II and cinerins I and II.

Pyrethrin I	Cinerin I
$R_1 = -CH_3$	$R_1 = -CH_3$
$R_2 = -CH_2CH=CHCH=CH_2$	$R_2 = -CH_2CH=CHCH_3$

Pyrethrin II	Cinerin II
$R_1 = -\underset{\underset{O}{\|\|}}{C}-OCH_3$	$R_1 = -\underset{\underset{O}{\|\|}}{C}-OCH_3$
$R_2 = -CH_2CH=CHCH=CH_2$	$R_2 = -CH_2CH=CHCH_3$

This insecticide is commonly contained in household aerosol sprays because it has a wide spectrum and rapid knockdown characteristics. Unless certain additives (synergists) are included, however, many insects are only stunned by the material and soon recover to full activity. Also, pyrethrum is available as a spray concentrate or dust for use on fruit trees, ornamentals, vegetables, and flowers. When it is used, fruits and other edibles can be harvested and eaten shortly after treatment. This insecticide breaks down quickly in the presence of sunlight.

Nicotine. Nicotine is obtained from extractions of tobacco leaves and was used as an insecticide as long ago as 1690. A favorite garden spray, Black Leaf 40®, has as its active ingredient 40 percent nicotine sulfate. Nicotine is an alkaloid, as are, for example, caffeine, quinine, morphine, LSD, cocaine, and strychnine.

Nicotine Sulfate

3-(1-methyl-2-pyrrolidyl) pyridine sulfate

Nicotine sulfate is very toxic to insects, as well as to humans. In fact, it is the most dangerous of the botanicals to apply. This material is sold as sprays because dusts are too dangerous to use. It is used mainly in small gardens against piercing-sucking insects and mites. Because of toxic residues, food cannot be harvested for at least seven days after treatment.

Rotenone. Rotenone is probably the second-most used botanical. It is extracted from the roots of the legumes *Derris* species, grown in Malaysia and the East Indies, and *Lonchocarpus* species, grown in South America. Rotenone has been applied as an insecticide since 1848 and as a fish poison by South American aborigines at least since 1649.

Rotenone

1,2,12,12a, tetrahydro-2-isopropenyl-8,9-dimethoxy-
[1]benzopyrano-[3,4-b]furo[2,3-b][1]benzopyran-
6(6aH)one

Today, the insecticide is sold as a spray or dust for both chewing and piercing-sucking insects, mainly in garden and fruit crops. Another major use is as a fish poison for reclaiming lakes and ponds. Here, it is applied to eliminate all fish, including rough species, before restocking with desirable game fish.

Ryania. Ryania is a botanical extracted from the stem and roots of *Ryania speciosa*, a shrub grown in Trinidad. Like nicotine, the active ingredient in this material is an alkaloid; however, it has low toxicity to humans. Ryania has been used most widely against caterpillars on fruit trees, particularly codling moth, *Cydia pomonella*, on apples. Also, it is effective against such insects as European corn borer, *Ostrinia nubilalis*, on corn, as well as many other insects in home gardens and fruit trees.

Sabadilla. This botanical is extracted from the seed of *Schoenocaulon officinale* and contains alkaloids as the active ingredients. Sabadilla has low toxicity to humans; however, it is known to cause eye irritation and sneezing in some individuals. This insecticide probably is the least used of the botanicals mentioned but is effective against most garden pests except aphids and mites.

Fumigants

Fumigants are highly volatile pesticides, most of which contain one or more of the halogen gases: Cl, Br, or F. Effective fumigants have high penetrating ability and kill all stages of insects in enclosures, including eggs. Good fumigants are not corrosive and do not affect the quality of commodities that are treated. They desorb quickly from treated surfaces and leave no harmful residue.

Fumigants are used to eliminate all pests in grain elevators, greenhouses, homes, and warehouses and also in packaged products like beans, grain, breakfast cereals, and dried fruits. They are also used against soil insects, nematodes, and pathogenic microorganisms. With soil applications, plastic sheets sometimes are laid down after application to reduce rate of evaporation. The greatest hazards of using these insecticides are flammability of the gas and possible accidental poisoning of humans.

There are over 15 fumigants registered for use in the United States. Only a few of the most common ones are mentioned here.

Paradichlorobenzene and naphthalene. These are some of the most common fumigants used in the home. They are produced as solids that emit gas at a slow rate. In households they are used as moth balls or flakes for protection against clothes moths, *Tineola* species. Also, they are used as soil fumigants.

Dichloropropene and dichloropropane. These fumigants are often mixed together as a soil fumigant against both insects and nematodes. Because it can damage plants and sometimes cause off-flavoring of some (potatoes), the mixture is applied well in advance of planting.

Methyl Bromide	CH_3Br
Chloropicrin	Cl_3CNO_2
Sulfuryl Fluoride (Vikane®)	SO_2F_2

Vapam

$$CH_3NHC\overset{\overset{\displaystyle S}{\|}}{-}S-Na$$

Telone $\quad ClCH = CH - CH_2Cl$

D-D® $\quad CHCl = CHCH_2Cl$ plus

$$\underset{ClCH_2CHCH_3}{\overset{\overset{\displaystyle Cl}{|}}{}}$$

Ethylene Oxide

$$\underset{H_2C \overline{\qquad} CH_2}{\overset{O}{\diagup \diagdown}}$$

Naphthalene (crystals)

p-dichlorobenzene (PDB crystals)

Methyl bromide. This is a highly volatile insecticide widely used as a general fumigant. It is fairly stable and nonflammable and is very toxic to insects and some mites. Methyl bromide penetrates well and desorbs quickly. It has been particularly useful in the fumigation of mills, granaries, and warehouses. Also, it is frequently used to fumigate quarantined plants before they are shipped.

Chloropicrin. This compound is the active ingredient of "tear gas" used by police and is also frequently added to odorless fumigants as a safety precaution. When in a mixture, as with methyl bromide, the fumigant's presence repels users, and the chance of accidents is reduced. Chloropicrin also is a fumigant in its own right, being effective against insects, fungi, nematodes, and weed seeds.

Oils

Refined petroleum oils have long been an important source of insecticides used in management of mosquito larvae and fruit-tree pests. They are also used as carriers of other insecticides.

Because petroleum oil is **phytotoxic** (kills or injures plants), it must be highly refined before it can be applied to plants. The lower the viscosity (resistance to flowing) and distillation (boiling point) range, the less phytotoxic the oil. Light oils probably have less toxicity to plants because they volatilize more quickly than heavier oil and do not remain on the plant surface as long. Also, the amount of unsaturated hydrocarbons in an oil is another factor in its phytotoxicity. Unsaturated hydrocarbons are unstable and combine to form compounds toxic to plants. U. R. ratings (percent unsulfonated residue from laboratory tests) are a measure of these unsaturated hydrocarbons. Heavy oils (dormant oils) have U. R. ratings of 50 to 90 percent, and light oils (summer oils) have U. R. ratings of 90 to 96 percent.

Oils are usually emulsified with water for application. **Summer oils** are the most highly refined and can be applied to trees in full foliage without problems of phytotoxicity. The summer oils are usually not as toxic to insects as dormant oils but can be useful against mites and scale insects (Homoptera: Coccoidea) on citrus. **Dormant oils** are less refined and, to prevent phytotoxic effects, are applied when no foliage is present. These are applied to fruit trees and ornamentals during periods of mild weather in late winter before budswell occurs. They are applied mostly to kill partially grown scale insects and eggs of aphids and mites that are overwintering on branches and twigs.

The advantages of oil applications are many. They are inexpensive, usually result in good coverage, are simple to mix, and are safe to warm-blooded animals. Also, few insects have become resistant to these. Some disadvantages of use include phytotoxicity, instability in storage, and ineffectiveness against certain pests.

Other Insecticides

Although the insecticide groups discussed make up the vast majority of those used, others are notable for specific purposes. These are mentioned briefly.

Formamidines. Formamidines are a group of insecticides that gained acceptance in the 1970s, mainly for use against organophosphate- and carbamate-resistant pests. They are quite effective against eggs and young caterpillars of several moth species, as well as against most stages of mites and ticks.

An early formamidine, chlordimeform, is one of the most widely used in this group. It was removed from the market in 1976, when laboratory studies indicated cancerous tumor formation in strains of mice that fed on it over their lifetime. It was brought back on the market in 1978 for use on nonfood crops, primarily cotton, and is strictly regulated. Another formamidine being developed is amitraz.

Chlordimeform (Galecron®, Fundal®)

N'-(4-chloro-*o*-tolyl)-*N*,*N*-dimethylformamidine

Amitraz (Baam®)

N'-(2,4-dimethylphenyl)-*N*-[[2,4-dimethyl-
phenyl) imino]methyl]-*N*-methylmethanimidamide

Thiocyanates. These insecticides can be distinguished by their creosotelike odor, and they are relatively safe for use around the home. They give very quick knockdown of flying insects and are sometimes found as aerosol sprays for application around pets and farm animals. Their use may decline with the development of synthetic pyrethroids.

Lethane 384®

2-(2-butoxyethoxy)ethyl thiocyanate

Thanite®

isobornyl thiocyanoacetate

Dinitrophenols. Dinitrophenols have a broad range of toxicity for many kinds of organisms, and therefore they have been developed as herbicides, fungicides, and insecticides. They are considerably toxic to humans. The oldest chemical of this group is DNOC (dinitrocresol); it has been used as an ovicide (egg killer), herbicide, fungicide, and blossom-thinning agent. Today, DNOC is used mainly for killing all plants in an area. Another dinitrophenol, dinoseb, is used as a dormant spray against insects and mites on fruit.

Dinitrocresol (DNOC)

Dinoseb

4,6-dinitro-*o*-cresol, sodium salt

2-*sec*-butyl-4,6-dinitrophenol

Organosulfurs and organotins. The compounds of these groups are often used as acaricides, effective against many species of plant-feeding mites. Organosulfurs have sulfur as their central atom but are more toxic to mites than sulfur alone. Of significance to pest management is the property of these compounds to suppress mites without causing ecological upsets in insect populations. Tetradifon is one of the older organosulfurs used.

Tetradifon (Tedion®)

p-chlorophenyl 2,4,5-trichlorophenyl sulfone

The organotins are relatively new acaricides, many of which are selective in suppressing mites. However, some serve a dual purpose as fungicides as well as acaricides. Two of the most common organotins are cyhexatin and fenbutatin-oxide. Fenbutatin-oxide is particularly effective against mites on citrus, deciduous fruits, and ornamentals, as well as in greenhouses.

Cyhexatin (Plictran®)

tricyclohexylhydroxytin

Fenbutatin-oxide (Vendex®)

hexakis (2-methyl-2-phenylpropyl)distannoxane

Inorganics. Some of the oldest insecticides used in agriculture are inorganic compounds. Most were used in the early part of the century and have been replaced by the much more effective organic compounds.

Some of the inorganics that remain in use include sulfur, sodium fluosilicate, and cryolite. Sulfur is both a contact poison and stomach poison and is applied as a dust against mites and some fungi. Liquid lime-sulfur, made by boiling mixtures of sulfur and freshly slaked or hydrated lime, is used on fruit trees as a fungicide and pesticide against mites, aphids, and scale insects. Sodium fluosilicate is an insecticide used in cockroach and grasshopper baits, and cryolite is somewhat effective against a number of insects in truck crops. Both sodium fluosilicate and cryolite are stomach poisons, and as such, they usually do not have a strong effect on natural enemy populations.

CHEMICALS USED WITH INSECTICIDES

When applying insecticides, often there are other chemicals added to give desired results. Some of these increase the toxicity of the insecticide directly and are called **synergists**. Others, in general called auxiliaries or **adjuvants**, are added to improve adhesion, mixing, surface tension, or smell, or serve to carry the insecticide.

Synergists

Synergists are chemicals that are not toxic to insects by themselves but, when added to another substance, make an insecticide much more toxic. They are often added to insecticides in ratios of 8:1 to 10:1 (synergist : insecticide).

Synergists were initially developed and used to increase the effectiveness of pyrethrum. Indeed, most aerosols with pyrethrum and some of the pyrethroids for household use today are enhanced with a synergist. In testing, synergists have also been found to increase the toxicity of some chlorinated hydrocarbons, organophosphates, carbamates, and other types of insecticides. In most

Piperonyl Butoxide   MGK 264®

α-[2-(2-butoxyethoxy)ethoxy]-4,5-methylenedioxy-   N-(2-ethylhexyl)-5-norbornene-2,3-dicarboximide
2-propyltoluene

instances, synergists function by inhibiting the **mixed-function oxidases**, enzymes that metabolize foreign substances.

Some of the most common synergists are piperonyl butoxide, sulfoxide, and MGK 264®. Many of these are used to prepare sprays for home and garden and stored-grain and dairy cattle facilities. Because of their cost, synergists are not frequently used for spraying broad cropping areas.

Solvents

Many of the organic compounds used for insecticides are insoluble in water. Before they can be turned into spray concentrates or aerosols, they must be dissolved. The solvent chosen depends on the planned use of the material. Such factors as solvency, phytotoxicity, animal toxicity, combustibility, odor, and cost are major considerations. Some examples of solvents used to dissolve insecticidal compounds include carbon tetrachloride, kerosene, and xylene.

Diluents

Diluents are substances combined with concentrated insecticides as carriers and are necessary to obtain proper coverage of treated surfaces. Diluents can be either liquid or solid.

Liquid diluents of insecticides are usually water or refined oils. When water is used, it is necessary to add wetting and dispersing agents for proper suspension of the insecticide. Emulsifying agents are required where oil solutions are used with water.

Solid diluents are used to formulate insecticide dusts or granules. Here, coarse to finely ground particles of the diluent serve as a carrier of the insecticide. Common solid diluents include organic flours (for example, soybean flour) and minerals (bentonite clay, talc, volcanic ash).

Surfactants

The term **surfactant** is a euphemism for several agents that aid or enhance the surface-modifying properties of a pesticide formulation. It is derived from the words "*surf*ace *active agent.*" Surfactants improve emulsifying, wetting, and spreading properties of the mixture.

Usually, liquid insecticides, oils, and insecticides in water-insoluble solvents are formulated and applied as water emulsions. Emulsions are suspensions of microscopic droplets of one liquid in another. These can be formed by vigorous agitation but not very effectively. For effective suspensions, detergentlike materials are added to the insecticide formulation to enhance the mixing. In most instances, when the insecticide and emulsifier are added to water, the oil carrier disperses immediately and uniformly, giving a milky appearance. A thoroughly emulsified liquid has efficient wetting and spreading properties, as opposed to a crude suspension.

Stickers

The amount of insecticide residue adhering to a treated surface is partly governed by the wetting and spreading properties of the mixture. When emulsions and wettable powders are applied, however, stickers are sometimes added to retain the active ingredient on a surface longer than otherwise possible. Such materials as casein, gelatin, and vegetable oils have been used as insecticide stickers or adhesives. Also, water- or rain-resistant formulations containing **latex** have been used to extend residues, for example, in Sevin XLR®.

Deodorants

Deodorants are materials added to insecticides to mask unpleasant odors. Many insecticides like the thiocyanates, pyrethrum, and several organophosphates have strong odors that can be offensive. This is particularly unacceptable when formulated for home use. Therefore, various substances like cedar oil, pine oil, or flower scents are added to insecticide concentrates to disguise odor.

INSECTICIDE FORMULATIONS

The beginning stages of insecticide production involve the manufacture of **technical-grade material**. This material is a relatively pure form of the insecticidal compound that comprises the **active ingredient** (AI) in the final insecticide mixture. Before sale, however, the technical-grade material is mixed with auxiliaries to make the insecticide convenient to handle and easy to apply. Some of the auxiliaries may also be active against insects; however, many of them are inert. **Inert ingredients** have no direct effect on pests. The mixture of active and inert ingredients for killing insects is called an "insecticide formulation." Some formulations are ready for use out of the container, but others must be diluted with water or oil.

There are many kinds of formulations available on the market, including liquids and solids, and a few are prepared for release of the active ingredient over a period of time. Only the most widely used formulations are discussed here.

Liquid Formulations

Liquid formulations usually are sold in small cans and bottles, medium-sized pails, or large drums. If mixing is required, these formulations are the most convenient to use.

Emulsifiable concentrates (EC or E). It has been estimated that more than 75 percent of all pesticides are applied as sprays. The great majority of these sprays are applied as water-based emulsions using emulsifiable concentrates. With this kind of formulation, a detergentlike material is added to break up the insecticide into microscopic droplets, producing a milky liquid. ECs often contain 240 to 1,920 grams of active ingredient (AI) per liter (two to eight pounds per gallon).

Solutions (S). Solutions are also liquid concentrates that may be used directly or require diluting. Those used directly from the container are low concentrates, usually containing fewer than 240 grams AI per liter (two pounds per gallon). Most low concentrates are solutions in highly refined oils, and many have a convenient atomizing pump. Solutions are used mainly as household sprays, mothproofers, livestock sprays, and space sprays in barns. High concentrates usually contain 1,920 grams or more AI per liter (eight pounds per gallon). If dilution is required, oil is usually the diluent. A special kind of high-concentrate solution is the **ultra low volume** *concentrate* (ULV). ULV formulations are applied without dilution with special aerial or ground equipment to produce an extremely fine spray. Volumes delivered by ULV are 0.6 liters to 4.7 liters per hectare (two quarts or fewer per acre), as compared with more than 20 times that amount for

Figure 11.5 Types of liquid pesticide formulations.

conventional high-volume sprays. ULV application is particularly effective with certain insecticides, because it reaches the insects in a concentrated dose. It is also economical, because large areas can be treated without refilling the sprayer.

Flowables (F or L). In some instances, insecticidal compounds can be made only as a solid or semisolid. To give them some of the mixing advantages of ECs and solutions, they are wet-milled with a clay diluent and water. This leaves the technical material finely ground and wet, with a puddinglike consistency. This formulation then can be mixed with water for spraying. Flowables must be constantly agitated to prevent the insecticide from coming out of suspension and settling to the bottom of the spray tank.

Aerosols (A). Aerosols are the most frequently used formulation of insecticides around households. Insecticides used in aerosols are dissolved in volatile petroleum solvents. The solution then is pressurized in a can by a propellant gas like carbon dioxide or fluorocarbons. When sprayed, the solution is atomized and quickly evaporates, leaving microsized droplets suspended in air. Aerosols are sold in push-button or total-release containers. Although easy to use, aerosols have a low percentage of active ingredient and, therefore, are expensive. Most uses are in households, greenhouses, and around barns.

Liquified gas (LG or F).   Several fumigants when placed under pressure turn into a liquid. These are stored in metal bottles under pressure and are released into structures like grain bins, into the soil by injection, or under tarps. Other insecticidal compounds remain liquid at normal atmospheric pressure but turn into a gas after they are applied. They are not stored under pressure and vaporize after they are placed in soil or in enclosures.

Dry Formulations

Dry formulations are usually sold in paper cans or bags, which may be lined with plastic. Some are used directly from the container, but others require a diluent.

Dusts (D).   Dusts are some of the oldest and simplest formulations of insecticides. They are prepared by milling the insecticidal compound into a fine powder. This powder is diluted with a dry diluent like organic flour or finely

Figure 11.6 Types of dry pesticide formulations.

ground mineral. The percentage active ingredient in dusts usually ranges from 1 to 10 percent. Dusts are often easy to use in small areas because they can be shaken directly on a surface from the container or puffed into cracks and crevices with an applicator. However, dusts are the least effective, least economical insecticide formulations for outdoor use. This is because of wind-caused drift and poor rate of deposit on foliage and other surfaces. Sometimes only 10 to 40 percent of the insecticide reaches a crop when dust is applied by airplane, as opposed to 60 to 99 percent with an EC spray. Moreover, dusts are perhaps the most toxic formulation to honey bees and parasitic Hymenoptera. These characteristics make dusts rather poor pest management formulations for outdoor use.

Granules (G). Granular formulations are prepared by applying liquid insecticide to coarse particles of a porous material. These particles may be formed from corncobs, walnut shells, clay, or other materials. The insecticide is absorbed into the granule, coats the outside, or does both. The amount of active ingredient in granular formulations ranges from 2 to 40 percent. Because of the size of the granular particle, this formulation is much safer (not inhaled) to apply than dusts or even ECs. Granulars are most often used against soil insects and may be placed directly with or beside the seed at planting time, banded over the top, or broadcast over an area. When dropped over plants, granules accumulate in leaf whorls. This type of application is useful against such insects as European corn borer larvae, which feed at the whorl before boring into the plant. As might be expected, proper coverage may be a problem with this formulation.

Wettable powders (WP or W). Wettable powders look like dusts while in the container but are formulated to be mixed with water and sprayed on surfaces. A surfactant added to the dust allows wetting during the mixing processes. A particle suspension results when water is mixed. WPs are much more concentrated than dusts, containing 15 to 95 percent active ingredient. Like flowables, frequent agitation is required to keep the insecticide in suspension. WPs usually cause less phytotoxicity than ECs, but they are more abrasive to spray pumps and nozzles. WPs should never be used without dilution.

Soluble powders (SP). Unlike wettable powders, soluble powders dissolve in water, forming a true solution. Some are packaged in bags that look like plastic and dissolve when thrown into water. Some agitation is needed to get SPs into solution, but after they are dissolved, additional agitation is not needed. Usually, SPs are formulated with 50 percent or more active ingredient and always require dilution.

Poisonous baits (B). This type of formulation combines an insect-edible or other attractive substance with the insecticide to improve effectiveness of treatment. Dried and pulverized fruit and other materials are often used to draw insects to a spot where they ingest or simply crawl across the insecticide. Baits can be used in buildings or outside for agricultural pests. Usually, active ingredient concentrations are very low in baits, on the order of 5 percent or less.

Slow-release formulations (SR). Along with the development of environmentally unstable insecticides have come problems of short-term effectiveness and the increased expense of several applications. Therefore, ways to extend the life of organophosphates and other chemicals have been the goal of considerable research. A degree of success in extending insecticide activity has been achieved through the development of slow-release insecticides. The first significant development of a slow-release formulation occurred in 1963 with the introduction of the Shell No-Pest Strip®. With this formulation, the volatile organophosphate, dichlorvos, was embedded into strips of polychlorovinyl resin. The polychlorovinyl resin slows the rate of volatilization of the insecticide, allowing it to kill most flying and some crawling insects in the vicinity. Other developments of the original "fly strip" idea now can be seen in ear tags and ear tape for fly control on livestock and flea collars for dogs and cats.

Another form of slow-release formulation has been in microencapsulation of insecticides. This formulation is accomplished by incorporating the insecticide in a permeable covering. This process forms microscopic spheres or microcapsules that permit the insecticide to escape at a reduced but effective rate. The life on an insecticide formulated in this manner is often two to four times that of an emulsifiable concentrate. A commonly used example of this formulation is Penncap M®, a microencapsulated form of methyl parathion.

Yet another slow-release formulation is a paint-on type. Here, the insecticide is dissolved in a volatile petroleum solvent, along with certain plastics and lacquers. This formulation may be "painted" on directly or blended with latex or oil-base paints for application. When the solvents evaporate, the insecticide remains in the thin transparent film or the actual paint. Over a period of time the insecticide escapes or "blooms," causing the surface to be active against crawling insects.

INSECTICIDE TOXICITY

Toxicity of a substance refers to its inherent poisonous potency under a given set of laboratory conditions. All insecticides are poisons, and the degree of toxicity varies greatly among them. To understand toxicity and how to use insecticides efficiently and safely, we must first understand their mode of action.

Insecticide Modes of Action

Insecticide **mode of action** involves all the anatomical, physical, and biochemical responses to a chemical, as well as its fate in the organism. Of greatest focus in this discussion are the most fundamental physiological steps that result in the death of insects.

All insecticides block metabolic processes in insects, but this is done in different ways by different compounds. According to their mode of action, the major groups of the most frequently used insecticides are (1) nerve poisons, (2) muscle poisons, and (3) physical toxicants.

Nerve poisons.   By far most conventional insecticides act as nerve poisons. These affect the insect nervous system mostly as narcotics, axonic poisons, or synaptic poisons.

Narcotics' mode of action is mostly physical. Many fumigants (particularly those containing Cl, Br, and F) are narcotics, inducing unconsciousness in insects. These narcotics are fat-soluble and lodge in fatty tissues, including nerve sheaths and lipoproteins of the brain. An important characteristic of these insecticides is that their action is reversible; if a human being shows early symptoms of poisoning, removing him or her from the source of the fumigant allows recovery. This also means that poisoned insects can recover if the fumigated space is ventilated too quickly.

Axonic poisons act primarily by interrupting normal axonic transmission of the nervous system. Remember from Chapter 2 that the axon of the nerve

Figure 11.7 Schematic representation of the interruption of axonic transmission of a nerve cell by cyclodiene and pyrethroid insecticides.

cell is an elongated extension of the cell body that transmits nerve impulses to other cells. These impulses are electrical and arise from the flow of positive sodium and potassium ions through the cell membrane, creating a wavelike action potential (the impulse). Subsequently, the action potential is followed by a resting potential. All the chlorinated hydrocarbons and pyrethroids are believed to disrupt normal transmission along the axon. Although many details are unknown, some of these like the cyclodienes and pyrethroids are believed to induce changes in axonic membrane permeability, causing repetitive discharges. Such discharges eventually result in convulsions, paralysis, and death.

Synaptic poisons act by interrupting normal synaptic transmission of the nervous system. You may recall that a synapse is the junction between a neuron and another cell, including a muscle, gland, or sensory receptor cell or another neuron. In the central nervous system, the chemical acetylcholine forms and transmits an impulse across the synapse (gap) to the next cell. Subsequently, acetylcholine is broken down to form acetic acid and choline in the presence of an enzyme, acetylcholinesterase. Therefore, acetylcholinesterase clears the system before another nerve transmission occurs. Both organophosphates and carbamates inhibit or tie up acetylcholinesterase. Inhibition causes a buildup of acetylcholine and a malfunction of the transmitting system. Rapid nerve firing occurs, producing symptoms of restlessness, hyperexcitability, tremors and convulsions, paralysis, and death. One major difference in the action of these two insecticidal groups is that inhibition of acetylcholinesterase by carbamates is reversible but not so with organophosphates.

Other synaptic poisons include nicotine, nicotine sulfate, and the formamidines. Both nicotine and nicotine sulfate poison by mimicking acetylcholine at the synapse; the receptors cannot distinguish between acetylcholine and nicotine. This phenomenon results in symptoms similar to the inhibition of acetylcholinesterase. Formamidines cause insect death by affecting enzymes involved with other chemical transmitters (including norepinephrine and octopamine) at certain nerve synapses.

Figure 11.8 Schematic representation of the interruption of normal synaptic transmission between nerve cells by organophosphate and carbamate insecticides. These insecticides inhibit acetylcholinesterase, which causes a buildup of acetylcholine and subsequent rapid nerve firing.

Muscle poisons. Muscle poisons have a direct influence on muscle tissue. A common example of their action is the botanical, ryania. In this insecticide the alkaloid, ryanodine, disrupts the excitable membrane of muscle. This disruption results in up to a tenfold increase in oxygen consumption, followed by flaccid paralysis and death. A similar mode of action is seen in another botanical, sabadilla.

Physical toxicants. Physical poisons block a metabolic process by physical, rather than chemical, means. The most frequent occurrence of physical poisoning is with oils. When refined oils are applied to the surface of water, mosquito larvae and pupae suffocate because their spiracles become clogged with the insecticide. Clogging of spiracles occurs also in scale insects when dormant oils are applied to trees. Other physical toxicants against insects include inert, abrasive dusts that adsorb waxes from the insect cuticle. Such action results in increased water loss, desiccation, and ultimately death of the insect from dehydration. For example, boric acid, in addition to being a stomach poison, acts as an abrasive dust in killing insects.

Toxicity to Humans

Insecticides poison humans and insects similarly. A given product's **hazard** is the danger that injury will occur to people coming into contact with it. The degree of hazard depends on the toxicity of the active ingredient and the chance of exposure to toxic amounts of the product. According to the risks of insecticide use, two types of human poisoning can be distinguished, acute and chronic.

Acute poisoning. **Acute poisoning** causes illness or death from a single dose or exposure. This type of poisoning is of particular concern to persons involved directly in the manufacture and application of pesticides, because they are at greatest risk. Acute poisonings also occur among nonprofessionals, however, usually as a result of accidents, ignorance, suicide, or crime. Of all accidental poisonings in the United States, only about 1 percent are believed caused by pesticides and fertilizers. Regretfully, a substantial percentage of these poisonings occur in children under age five.

Antidotes are available for treatment of victims of organophosphate and carbamate insecticide poisoning. After first aid measures have been taken (explained on insecticide label), physicians may administer **atropine** and 2-PAM. Atropine is often used for acute poisoning, and 2-PAM is given later (with organophosphate poisoning) to speed long-term recovery. Poison control centers are also designated in each state to advise treatment for accidental poisoning. An emergency toll-free number for the National Drug and Poison Information Center is 800-845-7633 (800-922-0193 in South Carolina).

Chronic poisoning. **Chronic poisoning** occurs from long-time exposure to low levels of a toxicant. This type of poisoning is often revealed only after several weeks of exposure and is of special concern to the general public. The main worry is that food will contain residues capable of causing sickness or death after repeated consumption over time.

Experiments with laboratory animals fed sublethal doses of pesticides over time have shown that several types of maladies are possible. Some of the most insidious effects are cancer (**carcinogenic** effects), genetic damage to future generations (**mutagenic** effects), and birth defects in offspring of exposed pregnant females (**teratogenic** effects). From animal studies, pesticide tolerances are set (by the

EPA) that are sufficiently low so daily consumption of a particular food will not cause significant risk of such diseases.

Tolerances for residues allowed by law are established from an acceptable daily intake (ADI). The ADI is the level of a residue to which daily exposure over a lifetime will not cause appreciable risk. This level is set at 100 times less than the level where no observable effects of the poison can be detected (NOEL).

Previous, continuing studies of pesticide residues on food in the United States have shown that actual pesticide levels in a typical diet are far below the ADI and established tolerances. Therefore, it is unlikely that today's public is exposed to much risk from chronic poisoning associated with food consumption.

Estimation of toxicity to humans.   The estimation of an active ingredient's inherent toxicity to humans is required before insecticides (and other pesticides) can be registered and sold. This estimate is made by the manufacturer using prescribed laboratory tests with white rats or rabbits. Toxicity in this instance is defined as the dose that will kill 50 percent of the test animals to which it is administered. This lethal dose is expressed as milligrams of toxicant per kilogram (mg/kg) of body weight and is referred to as the LD_{50}.

Toxicity is established by feeding (oral), skin application (dermal), and inhala-

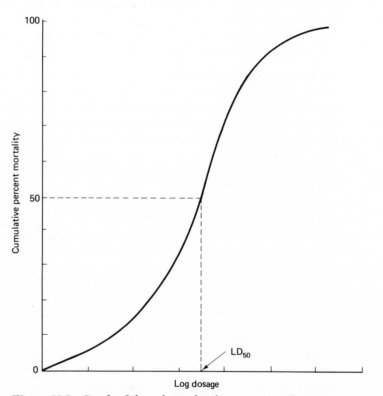

Figure 11.9 Graph of the relationship between cumulative percent mortality and log dosage of a pesticide. The LD_{50} represents the dosage required to kill 50 percent of a population of laboratory animals, usually white rats or rabbits. (Reprinted with permission of Macmillan Publishing Company from *Insects in Perspective* by Michael D. Atkins. Copyright ©1978 by Michael D. Atkins)

tion (respiratory) tests. LD_{50}s are determined from oral tests with white rats and dermal tests with rats or rabbits. These tests are used to rank insecticides for toxicity, and subsequently insecticide labels carry information on this toxicologic ranking. Another value, the LC_{50}, is the lethal concentration given as milligrams or cubic centimeters per animal. It is used most frequently in environmental studies where it is also expressed as parts per million (ppm) or parts per billion (ppb) in the medium (usually water) that kills 50 percent of the test organisms.

Figures 11.10 and 11.11 show toxicities of some common insecticides. Note that the lower the LD_{50}, the greater the toxicity. Factors determining toxicity to humans include dose, length of exposure, and route of absorption. But size of the dose is probably of greatest single importance. Estimates of the dosage required to kill a human being can be extrapolated from LD_{50} values of laboratory rats and are based on a person's weight. To convert LD_{50} values from milligrams per kilogram to ounces per 100 pounds, divide the mg/kg value by 625.

Pesticide law in the United States requires that labels be placed on all products. An important part of this label are **signal words** that give the user some idea of the toxicity of the material. A knowledge of these words is very useful and is required for pesticide applicator certification. There are four categories of insecticides (pesticides in general) based on their potential hazard, each with its own signal word. All categories must also bear the words "Keep Out of Reach of Children."

Category I. Highly toxic—signal words **Danger—Poison**, accompanied by a skull and crossbones symbol. Pesticides in this group have oral LD_{50}s of 50 mg/kg or fewer—a few drops to one teaspoon. Dermal LD_{50}s in this group are fewer than 200 mg/kg.

Category II. Moderately toxic—signal word **Warning**. Pesticides in this group have oral LD_{50}s within the range of 50 to 500 mg/kg—one teaspoon to two tablespoons. Dermal LD_{50}s in this group are between 200 and 2,000 mg/kg.

Category III. Slightly toxic—signal word **Caution**. Pesticides in this group have oral LD_{50}s within the range of 500 to 5,000 mg/kg—one ounce to one pint. Dermal LD_{50}s in this group are between 2,000 and 20,000 mg/kg.

Category IV. Low toxicity—signal word **Caution** applied here also. Pesticides in this group have oral LD_{50}s over 5,000 and dermal LD_{50}s over 20,000.

INSECTICIDE LAWS AND REGULATIONS

In the United States, federal laws have been enacted to protect pesticide users, consumers of treated products, domestic animals, pets, and the general environment. The first federal law to do this was the Federal Insecticide Act of 1910. This law, governing insecticides and fungicides, was designed mainly to protect farmers and other users from substandard or fraudulent products.

The next pesticide law was established in 1938, as an amendment to the Pure Food Act of 1906. The amendment forbade the presence of harmful residues of pesticides, primarily the old stomach poisons like lead arsenate and Paris green, on food. It also required the addition of coloring to pesticides to prevent consumers from mistaking them for food products like flour.

Federal Insecticide, Fungicide, and Rodenticide Act

The Federal Insecticide Act of 1910 was superseded by the Federal Insecticide, Fungicide, and Rodenticide Act of 1947. This law, called FIFRA, extended coverage of the law to include rodenticides and herbicides and required that all

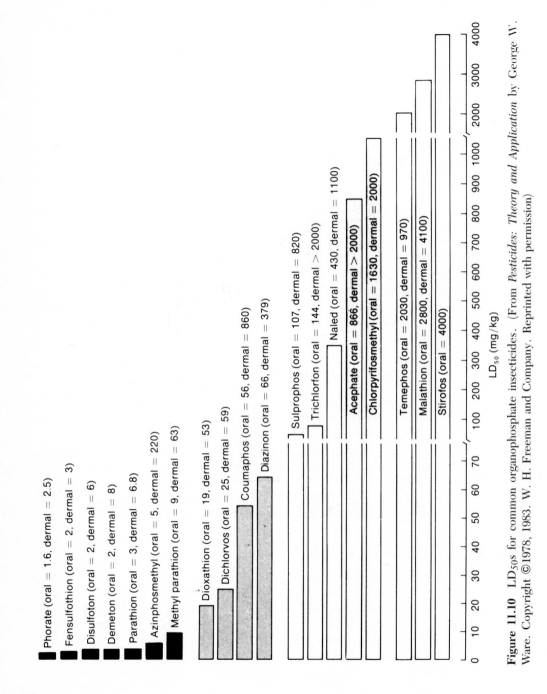

Figure 11.10 LD$_{50}$s for common organophosphate insecticides. (From *Pesticides: Theory and Application* by George W. Ware. Copyright ©1978, 1983. W. H. Freeman and Company. Reprinted with permission)

The figure shows LD$_{50}$ (mg/kg) values for the following organophosphate insecticides:

Phorate (oral = 1.6, dermal = 2.5)
Fensulfothion (oral = 2, dermal = 3)
Disulfoton (oral = 2, dermal = 6)
Demeton (oral = 2, dermal = 8)
Parathion (oral = 3, dermal = 6.8)
Azinphosmethyl (oral = 5, dermal = 220)
Methyl parathion (oral = 9, dermal = 63)
Dioxathion (oral = 19, dermal = 53)
Dichlorvos (oral = 25, dermal = 59)
Coumaphos (oral = 56, dermal = 860)
Diazinon (oral = 66, dermal = 379)
Sulprophos (oral = 107, dermal = 820)
Trichlorfon (oral = 144, dermal > 2000)
Naled (oral = 430, dermal = 1100)
Acephate (oral = 866, dermal > 2000)
Chlorpyrifosmethyl (oral = 1630, dermal = 2000)
Temephos (oral = 2030, dermal = 970)
Malathion (oral = 2800, dermal = 4100)
Stirofos (oral = 4000)

LD$_{50}$ (mg/kg)

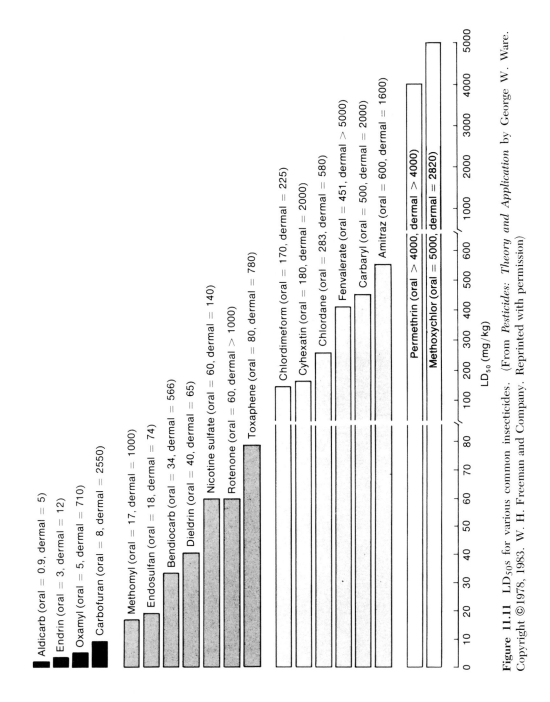

Figure 11.11 LD$_{50}$s for various common insecticides. (From *Pesticides: Theory and Application* by George W. Ware. Copyright ©1978, 1983. W. H. Freeman and Company. Reprinted with permission)

The chart displays LD$_{50}$ values (mg/kg) for various insecticides:

- Aldicarb (oral = 0.9, dermal = 5)
- Endrin (oral = 3, dermal = 12)
- Oxamyl (oral = 5, dermal = 710)
- Carbofuran (oral = 8, dermal = 2550)
- Methomyl (oral = 17, dermal = 1000)
- Endosulfan (oral = 18, dermal = 74)
- Bendiocarb (oral = 34, dermal = 566)
- Dieldrin (oral = 40, dermal = 65)
- Nicotine sulfate (oral = 60, dermal = 140)
- Rotenone (oral = 60, dermal > 1000)
- Toxaphene (oral = 80, dermal = 780)
- Chlordimeform (oral = 170, dermal = 225)
- Cyhexatin (oral = 180, dermal = 2000)
- Chlordane (oral = 283, dermal = 580)
- Fenvalerate (oral = 451, dermal > 5000)
- Carbaryl (oral = 500, dermal = 2000)
- Amitraz (oral = 600, dermal = 1600)
- Permethrin (oral > 4000, dermal > 4000)
- Methoxychlor (oral = 5000, dermal = 2820)

LD$_{50}$ (mg/kg)

pesticides for sale in interstate commerce be registered with the USDA . It did not cover products manufactured and sold within state borders. Primarily, the act of 1947 required adequate labeling and safety of the product, if label directions were followed.

In 1954, the Miller Amendment to the Food, Drug, and Cosmetic Act (1906, 1938) further strengthened the laws concerning pesticide residues on food. This amendment clearly set tolerances on all pesticides, which previously had been rather vague. In 1958, another amendment to the Food, Drug, and Cosmetic Act was passed that included the Delaney Clause. This clause disallowed any cancer-causing chemical (carcinogen) on food for human consumption.

A later amendment to FIFRA in 1959 extended coverage to include other chemicals in the category of economic poisons, including nematicides, plant regulators, defoliants, and desiccants. (Other poisons and repellents of vertebrates and invertebrates not previously included subsequently have been added in coverage.) FIFRA was amended again in 1964 to improve safety through specific label requirements.

As can be seen, both FIFRA and the Federal Drug and Cosmetic Act operated together to achieve consumer protection of treated products, as well as that for the pesticide applicator. Until December 1970, these laws were administered jointly by the USDA and the Food and Drug Administration (FDA). With growing concerns of environmental quality, the Environmental Protection Agency (EPA) was established, and all responsibility for pesticide registration and establishing tolerances on food was transferred to this agency in 1970. The EPA remains in charge today, and the FDA is responsible for enforcing pesticide tolerances on food established by the EPA.

Federal Environmental Pesticide Control Act

A significant change in FIFRA was enacted in 1972 with the Federal Environmental Pesticide Control Act (FEPCA), also known as FIFRA Amended, 1972. This is fundamentally our present law, comprising the following major provisions:

l. Use of any pesticide in a manner inconsistent with the label is prohibited.
2. All pesticides must be classified for general use or restricted use.

An Act

To amend the Federal Insecticide, Fungicide, and Rodenticide Act, and for other purposes.

Be it enacted by the Senate and House of Representatives of the United States of America in Congress assembled, That this Act may be cited as the "Federal Environmental Pesticide Control Act of 1972".

AMENDMENTS TO FEDERAL INSECTICIDE, FUNGICIDE, AND RODENTICIDE ACT

Sec. 2. The Federal Insecticide, Fungicide, and Rodenticide Act (7 U.S.C. 135 et seq.) is amended to read as follows:

"SECTION 1. SHORT TITLE AND TABLE OF CONTENTS.

"(a) Short Title.—This Act may be cited as the 'Federal Insecticide, Fungicide, and Rodenticide Act'.

"(b) Table of Contents.—

Figure 11.12 The present pesticide laws in the United States are stipulated in the Federal Environmental Pesticide Control Act of 1972.

3. Restricted pesticides may be applied only by an applicator certified by the state in which that person works.
4. The states will certify pesticide applicators.
5. Farmers and other private applicators can be fined and/or imprisoned upon criminal conviction for a knowing violation of the law. Additional penalties apply to registrants, applicators, dealers, and other distributors.
6. All pesticide products must be registered by the EPA, including those shipped in interstate and intrastate commerce.
7. For registration, a manufacturer must prove that, when used as directed, a product will be effective; not injure humans, crops, livestock, or wildlife or damage the environment; and not produce illegal residues on food or feed.
8. Pesticide manufacturing plants must be registered and inspected by the EPA.

Several other amendments have been added to the law from 1975 through 1981, mostly to clarify the intent of the law. Two important amendments allow the efficacy provision to be waived and permit the use of pesticides at lower than labeled dosages. Also, certain other environmentally safe deviations from the label are allowed under the new provisions.

The EPA is responsible for interpreting and implementing FEPCA, through duly processed regulations that carry the force of law. Regulations are developed in consultation with those persons most affected by them.

Pesticide Label Regulations

The pesticide label is the printed information on or attached to the pesticide container; federal regulations stipulate exactly the information to be included on it. The label is a license to sell a product; to abide by the law, its directions must be followed.

Examples of general-use and restricted-use labels are shown in Figures 11.13 and 11.14, respectively. Label requirements and a key to label information is presented as follows:

1. Trade name
2. Manufacturer's name and address
3. Net contents (total quantity)
4. EPA registration number
5. EPA formulation manufacturer establishment number
6. a. Ingredients statement
   b. Pounds per gallon (for liquids)
7. Precautionary statements
   a. Child hazard warning, **Keep Out of Reach of Children**
   b. Signal word—**DANGER, WARNING, or CAUTION**
   c. Skull and crossbones and word **POISON** in red
   d. Statement of practical treatment in case of accident
   e. Referral statement
8. Side- or back-panel precautionary statements
   a. Hazards to humans and domestic animals
   b. Environmental hazards
   c. Physical or chemical hazards
9. Box showing words **RESTRICTED-USE PESTICIDE**
   a. Statement of pesticide classification
   b. Statement on misuse
10. Statement on how soon field can be reentered
    a. Applicator category (limits of use to certain categories of commercial applicators)
    b. Instructions for storage and disposal
    c. Directions for use

Figure 11.13 Elements of a general-use pesticide label as required by law. See text for explanation of numbers.

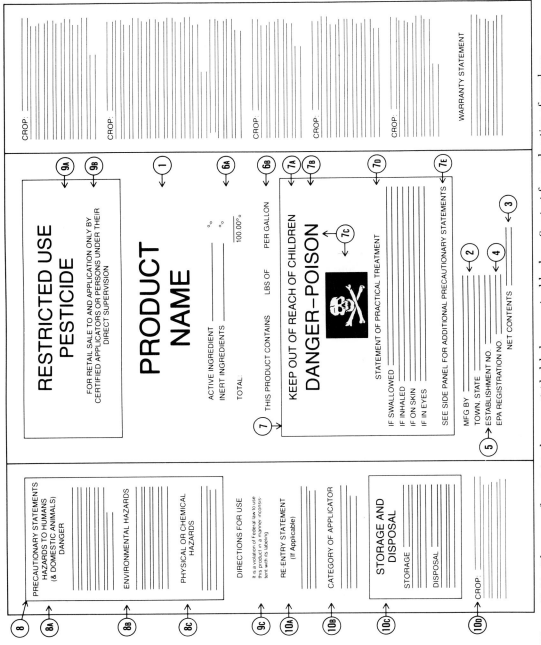

RESTRICTED USE PESTICIDE

FOR RETAIL SALE TO AND APPLICATION ONLY BY CERTIFIED APPLICATORS OR PERSONS UNDER THEIR DIRECT SUPERVISION

PRODUCT NAME

ACTIVE INGREDIENT _____ %
INERT INGREDIENTS _____ %

TOTAL. _____ 100.00%

THIS PRODUCT CONTAINS _____ LBS OF _____ PER GALLON

KEEP OUT OF REACH OF CHILDREN

DANGER–POISON

STATEMENT OF PRACTICAL TREATMENT

IF SWALLOWED _____
IF INHALED _____
IF ON SKIN _____
IF IN EYES _____

SEE SIDE PANEL FOR ADDITIONAL PRECAUTIONARY STATEMENTS

MFG BY _____
TOWN, STATE _____
ESTABLISHMENT NO. _____
EPA REGISTRATION NO _____
NET CONTENTS _____

PRECAUTIONARY STATEMENTS
HAZARDS TO HUMANS
(& DOMESTIC ANIMALS)
DANGER

ENVIRONMENTAL HAZARDS

PHYSICAL OR CHEMICAL HAZARDS

DIRECTIONS FOR USE

It is a violation of Federal law to use this product in a manner inconsistent with its labeling

RE-ENTRY STATEMENT
(If Applicable)

CATEGORY OF APPLICATOR

STORAGE AND DISPOSAL

STORAGE _____

DISPOSAL _____

CROP: _____

CROP: _____

CROP: _____

CROP: _____

CROP: _____

CROP: _____

WARRANTY STATEMENT

Figure 11.14 Elements of a restricted-use pesticide label as required by law. See text for explanation of numbers.

401

Applicator Certification

According to the provisions of FEPCA, all individuals applying restricted-use pesticides must be certified. Restricted-use pesticides are so designated by law because of their inherent toxicity or potential hazard to the environment. The basic divisions of applicators approved to handle these pesticides are private and commercial. The general distinction between these divisions is whether or not the individual is applying the pesticide for his or her personal use (private) or is in some way compensated for applying it (commercial). Training programs on pesticide application in preparation of certification are administered by the Cooperative Extension Service of most states. Testing for certification is often conducted by a state's department of agriculture.

The EPA has established ten categories of commercial applicator certification: (1) agricultural pest control (plant and animal), (2) forest pest control, (3) ornamental and turf pest control, (4) seed treatment, (5) aquatic pest control, (6) right-of-way pest control, (7) industrial, institutional, structural, and health-related pest control, (8) public health pest control, (9) regulatory pest control, and (10) demonstration and research pest control. Some states have additional categories.

Examination for commercial-applicator certification emphasizes competency in the following areas: (1) label and labeling, (2) safety, (3) environment, (4) pests, (5) pesticides, (6) equipment, (7) application techniques, and (8) laws and regulations. Manuals and other study materials for exam preparation are available from the state Cooperative Extension Service.

Regulating Pesticides

The EPA regulates approximately 45,000 pesticide products, which involve roughly 1,400 active ingredients. The regulated community includes all users and commercial suppliers. The latter includes about 30 major pesticide producers, 100 companies that market active ingredients, 3,300 product formulators, and 29,000 distributors.

Regulation of new pesticides.   Under FEPCA, the EPA is charged with registering new pesticides. The EPA requires manufacturers to show that the new material, when used according to label directions, will not present unreasonable risks to human health or the environment. The law requires the EPA to take into account economic, social, and environmental costs and benefits in making decisions. Most registration decisions made each year are for new formulations containing active ingredients that are already registered with EPA or new uses of existing products.

The EPA bases registration decisions for new pesticides on its evaluation of test data provided by the applicants. Required studies include testing to show whether a pesticide has the potential to cause adverse effects in humans, fish, wildlife, and endangered species. Potential human risks include acute reactions like toxic poisoning and skin and eye irritation. Also, long-term risks like cancer, birth defects, or reproductive system disorders are assessed. Data on how a pesticide behaves in the environment (**environmental fate**) also are required so that threats to ground or surface water can be detected. In addition, if a food-crop application is involved, the applicant also must petition EPA for a tolerance and submit appropriate data so that a safe and realistic level can be established.

New active ingredients typically attain full registration within six to nine years and require from $2.4 million to $4 million of capital investment. Total development costs are between $50 million and $70 million.

The registration process begins with an experimental use permit (EUP) to allow premarket field testing of the pesticide. Subsequently, manufacturers typically submit an application for registration. This application must include appropriate health and safety data from experimental use, after which other data may be requested by the EPA. Registration is granted after study shows the new pesticide meets federal regulations.

The EPA also evaluates several other types of special registration submissions. Here, under section 18, the EPA may temporarily authorize state or federal agencies to combat emergencies with pesticides not permitted by usual registrations. Under another provision of the law, section 24(c), "special local need" registrations that take effect immediately can be granted directly by a state, but the EPA can override these registrations within 90 days.

Regulating existing pesticides. In addition to regulating new pesticides, the EPA monitors and evaluates human and environmental risks from already registered pesticides. To insure that these registered materials meet current scientific and regulatory standards, the law requires their "reregistration."

The reregistration program conducted by the EPA has been underway for several years. For reregistration, an originally registered pesticide must meet the same criteria that apply to new pesticides. Here, the reregistration application data are reviewed, and conditions of reregistration, if any, are determined. If existing data largely satisfy registration requirements, no further study is needed, and the reregistration will be granted. If data are inadequate for analysis, additional studies are conducted by the manufacturer and the requested data submitted. Most older pesticides do not have enough existing data for reregistration.

If a registered pesticide shows evidence of posing unacceptable personal or environmental risks, the EPA can conduct a Special Review. A Special Review is a formal review process for obtaining risk/benefit information on a pesticide. Special reviews can be initiated at any time against any pesticide the EPA believes meets or exceeds specified risk criteria. Once a pesticide is placed on the Special Review list, the manufacturer must produce data that show its product does not cause adverse effects or risks. Subsequently, a Special Review team, made up of individuals from the federal and state governments, university personnel, and others, reviews the data and considers input from various user groups. The outcome of the Special Review may be to (1) retain current registration, sometimes with label changes, (2) retain current registration but with restricted-use classification and/or geographical restrictions, or (3) issue a notice of intent to cancel registration.

Cancellation notification is given to the manufacturer when a product is allegedly causing unacceptable risks. In these proceedings, the manufacturer has 30 days to request a hearing. This hearing is a formal, courtlike process before an administrative law judge. During the hearing, the manufacturer attempts to show that the product does not cause the alleged ill effects. During these proceedings, which may last two or more years, it is legal to manufacture, distribute, sell, and use the pesticide in question. As a result of cancellation hearings, any of the registration outcomes of the Special Review may be possible, or registration is cancelled.

If the EPA administrator finds that continued registration of a pesticide presents an imminent hazard to human health, however, he or she may call for suspension of the pesticide. Here, a manufacturer can request a suspension hearing within five days after notification. The hearing is begun immediately after request to determine if the pesticide is indeed an imminent hazard. If imminent hazard is proven, cancellation proceedings are begun, and a suspension order stops manufacture, distribution, and sale of the pesticide during the proceedings. Suspension hearings take about six months, during which time the pesticide may be marketed.

If continued marketing during the suspension hearing constitutes too great a risk, the EPA administrator may call for an emergency suspension. This is the most drastic step the EPA can take against a pesticide and is used when both an imminent hazard and an emergency exist. Here, an emergency suspension order is issued to the manufacturer, without prior notification, calling for cessation of manufacture, distribution, and sale. An expedited suspension hearing may be requested but manufacture, distribution, and sale are not allowed during the hearing. The emergency suspension order is accompanied by a notice to begin cancellation proceedings.

Suspension orders are interim steps taken to quickly remove products from the market when they are found hazardous. Full cancellation proceedings that weigh the risks and benefits of product use are required to permanently eliminate a pesticide.

USING INSECTICIDES

Insecticides are some of the most potent, dependable substances that can be employed to manage insect pests. Indeed, many of the world's most serious insect problems can be contained only by these important compounds. Because of their broad-spectrum nature, however, they can be hazardous to humans and can cause undesirable side effects, both in agricultural and nonagricultural ecosystems.

Most often, the cause of undesirable side effects is the way insecticides are used. Side effects cannot be avoided totally, but they can be minimized by proper use, which includes applying insecticides as effectively as possible and having utmost regard for human and environmental safety.

Effective Use

Using insecticides effectively begins with a proper strategy. With agricultural pests, insecticide use is a curative tactic that should be applied only after pest status has been assessed. Assessment involves accurate identifications of the pest species present and estimates of its population level. Moreover, information on other potential pests, as well as natural enemies, in the agroecosystem should be a part of the overall assessment. If assessment shows that one or more pests will likely exceed the economic-injury level, and no other tactic is practical, an appropriate insecticide should be chosen.

Choosing an insecticide. The choice of the most appropriate insecticide for a given pest situation depends on several factors including effectiveness, cost, formulations available, and equipment required. When choosing an insecticide, a place to start is recommendations published by the state Cooperative Extension Service. Such publications are frequently organized according to crop and are oriented toward local needs and conditions. They are usually updated frequently so that users can take advantage of the latest knowledge and developments from research.

Often, several alternatives are presented in extension recommendations. The most appropriate from among these may depend on the particular equipment required for application. If ULV sprays are suggested and only low-pressure, high-volume sprayers are available, the ULV compound is obviously eliminated in the choice. In other instances, alternatives can be eliminated on the basis of time remaining until harvest. Here, some insecticides will be eliminated because unacceptable residues would be present on the crop at the expected harvest time. Also, if several pests are present that require suppression, some insecticides will not be effective against all. This will require cross-referencing of the recommendations to see if possible choices are effective against the entire complex of pests. Subsequent elimination of impractical alternatives also should be made after considering such factors as probable effects on natural enemies and wildlife.

The elimination process is continued until only one or a few acceptable choices are left. Final product acceptance then can be determined according to cost, availability, and convenience.

Choosing a dosage.   Most insecticide recommendations list a dosage range for application, as does the insecticide label. As a general rule in insect pest management, least is best. To avoid insecticide resistance and other population counter-responses in agricultural applications, it is often best to reduce populations to insignificant levels, rather than trying to eliminate them completely. This can be achieved most easily by keeping dosage levels as low as possible. Also, remember that EPA regulations allow pesticides to be used at rates lower than those stated on the label. Reducing dosage also cuts costs.

However, exceptions to the low-dosage rule do apply. For instance, where exotic pests have been newly introduced and occupy a limited area, higher doses for complete elimination may be necessary. Additionally, certain mites and a few insects have been shown to be reproductively stimulated by sublethal doses of pesticides. When this phenomenon, **hormoligosis**, is expected, highest legal doses may be necessary. In yet other instances, highest dosages may be prescribed when attempting to avoid certain types of insecticide resistance.

Timing of applications.   No factor is more important in efficacy and environmental safety than proper timing of insecticide applications. The timing of applications is determined from characteristics and status of the target pest(s) and by conditions of the environment.

Insect life cycles and seasonal cycles greatly determine the general time when applications are made. Susceptible stages are usually targets of insecticide programs. For instance, sprays timed to kill small grasshopper nymphs usually are more effective than those applied against large nymphs and adults. In another instance, insecticide applications against second-generation European corn borer on corn must be timed to expose the peak occurrence of small larvae before they bore into the stalk.

Another important aspect of timing is to expose the population to an insecticide before significant injury has occurred. It may be consoling to kill insects that have injured a crop; however, unless significant damage can be avoided, it is economically unwise to do so. Usually, early surveillance and following recommended economic thresholds provides this proper timing information, which indicates estimates within a time period, perhaps three or four days, when application should take place.

The weather and concern with other production activities will probably determine when insecticides can be applied in this time period. To reduce drift and to concentrate most material in the target area, applications should be made under calm conditions. Windy days should be avoided and, when possible, applications should be made early in the morning or late in the evening before wind speed picks up. By avoiding periods from about 10 A. M. to 4 P. M. , especially during

Figure 11.15 A typical spraying system and common types of equipment for applying liquid formulations of pesticides.

Low-pressure sprayer

Air blast sprayer

Fixed-wing aircraft

Helicopter

Figure 11.15 (continued)

hot weather, both insecticide drift and evaporation can be reduced. Of course, if rain is occurring or imminent, additional limitations are placed on the application time. In addition, attention to production activities like cultivating, calving, and irrigating may take priority and further limit timing.

Coverage and confinement of applications. Thorough coverage of the target area is another very important factor for insecticide efficacy. Good coverage requires that the insecticide mixture be adequate to reach all pest feeding sites. In some instances, as with some foliage feeders, insects concentrate in the sunny portions of the upper canopy. For these pests, airplane sprays with low volumes (for example, 15 to 20 liters per hectare) may give adequate coverage. However, when insect pests are found throughout the plant canopy or concentrate in lower strata, sprays of greater volume (for example, 100 or more liters per hectare) applied with ground equipment may be required.

Sometimes, special sprayers are used to obtain effective coverage. An example is the air-blast sprayer. With this sprayer, a blast of air propels the formulation, allowing it to penetrate dense tree canopies and reach leaf and branch surfaces. Such sprayers have been used effectively in orchards and for shade trees.

Restricting the application to the target area is an important prerequisite to good coverage. Good coverage cannot be achieved when significant drift occurs or when residues are quickly washed off surfaces and carried away in runoff. Such actions reduce exposure levels for target insects and can cause destruction of fish and wildlife populations in surrounding areas.

Tractor–mounted
granular applicator

Broadcast, tractor–mounted
applicator

Broadcast, tractor–pulled
applicator

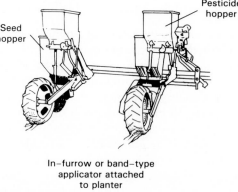

Pesticide
hopper

Seed
hopper

In–furrow or band–type
applicator attached
to planter

Figure 11.16 Common types of equipment for applying pesticide granules.

Choosing certain formulations may help keep insecticides within the target area. For instance, granular formulations contain heavy particles that fall to the surface and remain in an area. Conversely, dusts have small particle size and do not adhere well, making applications difficult to contain in the target area. However, even though larger particles may keep an insecticide in place, coverage on a surface decreases with particle size, thereby moderating the use of granulars and large-droplet sprays. Adding sticking agents or selecting formulations with stickers included in them also may help keep an insecticide on target. For example, formulations like Sevin XLR that contain binding and sticking agents can extend active residues several days by keeping the active ingredient, carbaryl, on plant leaves.

Figure 11.17 Parts of a typical nozzle and common nozzle types.

Choice of application equipment also can affect containment of insecticides. In comparisons of airplane and ground sprays in cotton, high-clearance sprayers with three nozzles per row deposited an average of 82 percent of the spray on target, versus only 54.5 percent for aerial sprays. But certain factors like wet fields, canopy closure, and problems of soil compaction may make ground applications impractical.

Using Insecticides Safely

Personal safety with insecticides begins with a thorough reading of the label. The label gives explicit precautions and steps to be taken for mixing and applying an insecticide. It also gives information on disposal of containers and explains what to do if an accident occurs.

Any time insecticides or other pesticides are handled, protective clothing should be worn. This clothing includes a long-sleeved shirt and long-legged trousers or coveralls. Trousers should be worn outside of boots. Unlined neoprene gloves should be worn in most instances, with shirt-sleeves worn over the gloves. Other clothing for safe use includes a wide-brimmed hat, neoprene boots, and goggles or a face shield.

Clothing worn while handling pesticides should be laundered frequently and by itself. When it is soiled by an insecticide concentrate, the clothing should be safely destroyed.

Lungs and other parts of the respiratory system are much more absorbent of pesticides than skin. An approved respiratory device must be worn if the pesticide

Figure 11.18 The use of insecticides for personal safety begins with a thorough reading of the label. (Courtesy EPA)

Cartridge respirator

Canister respirator

Self–contained breathing apparatus

Protective clothing

Goggles

Figure 11.19 Types of clothing and protective equipment used in applying pesticides. (Courtesy EPA)

label so directs. Several types of respirators can be worn, and specific information as to type is also given in label directions.

Taking time to use pesticides safely is one of the most important investments that any user can make. Religiously following label directions and having respect for personal and environmental hazards is the only acceptable policy in using pesticides. It is also the law.

CONCLUSIONS

It is difficult to imagine a technology that would produce the amount of food and fiber and maintain the level of public health that we have today without pesticides. But their use presents a dilemma. Speaking on pesticides, M. J. Dover comments, "As their hazards become more apparent, so does the need to use them. Although designed to kill, they are often life-savers. Although increasingly costly, they bring economic benefits. And while they have opened up many possibilities for improving agriculture and public health, they have closed others, making us extremely dependent on them for our continued survival."

The challenge ahead for pest management is to reduce our dependency on pesticides.

Selected References

DOVER, M.J. 1985. *A Better Mousetrap—Improving Pest Management for Agriculture*. Washington, D.C.: World Resources Institute, Study No. 4.

FRONK, W.D. 1985. Chemical control. In R.E. PFADT, ed., *Fundamentals of Applied Entomology*. New York: Macmillan, pp. 203–229.

———1985. Insecticide application equipment. In R. E. PFADT, ed., *Fundamentals of Applied Entomology*. New York: Macmillan, pp. 230–246.

LARSON, L.L., E.E. KENAGA, and R.W. MORGAN. 1985. *Commercial and Experimental Organic Insecticides*. College Park, Md.: Entomological Society of America.

PLAPP, F.W., Jr. 1981. The nature, modes of action, and toxicity of insecticides. In D. PIMENTEL, ed., *CRC Handbook of Pest Management in Agriculture*. Boca Raton, Fl.: CRC Press, vol. 3, pp. 3–16.

United States Environmental Protection Agency. 1985. *Suspended, Cancelled and Restricted Pesticides*. Third Revision. Washington D.C.: United States Environmental Protection Agency.

———1986. *Pesticides Fact Book*. Washington, D.C.: United States Environmental Protection Agency, Pub. A–107/86–003.

WARE, G.W. 1983. *Pesticides—Theory and Application*. San Francisco: Freeman.

WATSON, T.F., L. MOORE, and G.W. WARE. 1975. *Practical Insect Pest Management*. San Francisco: Freeman, pp. 57–105.

12 PLANT RESISTANCE TO INSECTS

ONE OF THE MOST PROMISING ways to reduce dependence on pesticides in agriculture is to plant insect-resistant crops. Planting resistant cultivars when available is one of our most effective, economical, and environmentally safe tactics.

The concept of using host resistance to our advantage comes from the knowledge that most plants and animals are resistant to most potential insect attackers. Certain physiological, morphological, and/or behavioral characteristics inherited by organisms are a core of defense against species that would otherwise attack them.

These defenses are the result of natural selection. In nature, crosses that produce highly susceptible plants and animals are not repeated because resultant progeny do not survive to reproduce. However, although resistant to most attacks, even these well-adapted survivors are susceptible to a few forms with the ability to overcome their defenses. Therefore, attackers are able to sustain their populations at the expense, though not usually the complete elimination, of the host. These successful attackers on wild hosts are equivalent to what we would call pests on domesticated hosts.

Just as wild plants and animals have degrees of susceptibility to enemies, so do domestic species. From the gene pool of a crop species, certain crosses produce **phenotypes** (visible expressions of the crosses) that vary from complete susceptibility to high levels of resistance. Susceptibility is the underlying cause of a pest problem, and the degree of it in a species forms the basis of developing useful, pest-resistant types.

By definition, pest resistance is any inherited characteristic of a host that lessens the effects of attack. From an evolutionary standpoint, such characteristics are preadaptive traits that have allowed the organism to overcome the pressures of insects and disease organisms and thereby increased the organism's chances of survival and reproduction. The technology of plant and animal breeding is aimed, in part, at discovering these preadaptive traits and using them to develop pest-resistant cultivars and breeds.

Breeding for pest resistance is only one of many objectives of total breeding programs. Indeed, usually yield and quality are the two most important aims, with pest resistance an important complementary objective. Breeding for pest

413

resistance may emphasize any type of pest, including pathogenic microorganisms, nematodes, vertebrates, weeds, and arthropods.

Breeding programs emphasizing resistance to insects and other arthropods are some of the most important in agriculture. Presently, breeding for insect resistance in plants is far more advanced than that for domesticated animals. Seemingly, this is because of the lower costs, shorter time, and greater opportunity for hybridization involved in plant breeding. Because of its significance in plant pest management and lesser importance in animal production, only plant resistance is discussed here.

BRIEF HISTORY

Some of the earliest observations of plant resistance to insects were recorded in the late eighteenth and early nineteenth centuries. As early as 1792, "Underhill" variety wheat resistant to the introduced Hessian fly, *Mayetiola destructor*, was documented in the United States by J. N. Havens. This is generally considered the earliest documented report of an insect-resistant plant variety. Somewhat later, in 1831, "Winter Majetin" apples were reported resistant to the woolly apple aphid, *Eriosoma lanigerum*.

The first dramatic example of the value of plant resistance against insects occurred in the late 1800s. An insect species, the grape phylloxera, *Daktulosphaira vitifoliae*, was inadvertently introduced into French vineyards and

---

BOX 12.1 WOOLLY APPLE APHID

SPECIES: *Eriosoma lanigerum* (Hausmann) (Homoptera: Aphididae)

DISTRIBUTION: The woolly apple aphid is distributed throughout the world and is found on apple, elm, mountain ash, pear, and hawthorn.

IMPORTANCE: Woolly apple aphids feed on sap from large root knots, underground portions of trunks, and wounds on trunks and branches. Primary injury, however, is caused by root feeding, which causes stunting of growth. Under severe infestations, trees may die. An efficient wasp parasitoid, *Aphelinus mali* (Haldeman), has successfully suppressed populations of the woolly apple aphid in practically all the areas in which the wasp has been released, including the Pacific Northwest of the United States. Here, chemical controls are now generally unnecessary.

APPEARANCE: Adults and nymphs are red to purple and covered with bluish white, cotton-like wax filaments. Winged and wingless forms appear during the year.

LIFE HISTORY: Eggs overwinter on elm bark; however, a number of wingless individuals also overwinter on apple trees. In spring, eggs hatch and wingless females begin parthenogenetically reproducing at a rapid rate on elm trees. Winged individuals develop in early summer and disperse to other plant hosts like apple. Reproduction continues throughout the summer, and in fall winged individuals appear again, mate, and oviposit overwintering eggs.

spread across Europe. The effect was catastrophic, and by 1880, the pest threatened to wipe out the French wine industry. Earlier, it had been found that American grapes were resistant to the phylloxera, and this knowledge led to the grafting of susceptible European grapevine scions to resistant American rootstocks. This produced vines resistant to grape phylloxera and yet allowed quality European wine grapes to be grown. Satisfactory insect control occurred by 1890, thereby saving the French wine industry. An American entomologist, C. V. Riley, was awarded a gold medal by the French government for making this recommendation, and resistant grapevines continue to be the major means of combating grape phylloxera.

Although the grape phylloxera example is often cited to illustrate the value of plant resistance, there is more to this story. Ironically, plants introduced from North America carried the pathogen *Plasmopara viticola* (the causal agent of downy mildew), to which European grapes were highly susceptible. The American root stock saved French vineyards from grape phylloxera but exposed them to an even more dangerous risk. The devastating epidemic of downy mildew that followed threatened not only French wine production but also wine production throughout Europe. Ultimately the development of Bordeaux mixture (lime plus copper sulfate), an early fungicide active against the pathogen, saved European vineyards. Thus, the grape phylloxera/downy mildew story offers an important lesson—the need to consider the whole pest complex and the implications of any management strategy.

Modern work with plant resistance to insects was pioneered by R.H. Painter and colleagues at Kansas State University, beginning in the 1920s. Painter's

Figure 12.1 Grape phylloxera, *Daktulosphaira vitifoliae*, female (upper right) and galls on terminal roots of grape caused by the pest. Gall at lower right is enlarged to show size relative to the insect (Courtesy USDA)

thorough, methodical studies paved the way for later achievements, and his book, *Insect Resistance in Crop Plants*, published in 1951, was the first on this topic. Today, plant breeding for insect resistance is conducted by universities, seed industries, governments, and numerous national and international institutes. Moreover, resistant varieties have become a crucial element in the success of many insect pest management programs.

INSECT AND HOST-PLANT RELATIONSHIPS

To understand the mechanisms of plant resistance to insects, it is important to understand some of the basic relationships between these organisms. Here, such relationships, commonly referred to as insect/plant interactions, are usually couched with heavy emphasis on insects and plant response to attack.

The Insect Aspect

The insect aspect of the insect/plant interaction is often described as a series of steps, in time and space, that lead to suitability of a plant for the insect. The major steps usually recognized by authorities include (1) finding the general habitat, (2) finding the host plant, (3) acceptance of the plant as a proper host, and (4) sufficiency of the plant for survival and successful reproduction of the insect population.

Finding the general habitat. Insects locate the general area of the host by means usually unrelated to the plant. Physical stimuli such as light, wind, gravity, and perhaps temperature and humidity may help orient dispersing insects to the overall location of the host. This step is most important when a species does not reside in an area year-round, as with many migrating forms.

Finding the host plant. Once in the general area the insect next must find a proper host. Most insects rely on vision and/or smell to locate a host plant. Remote factors in locating the plant include color, size, and shape. Much of the information on color in host finding is limited mainly to aphids (Homoptera: Aphididae) and whiteflies (Homoptera: Aleyrodidae), which are attracted to yellow-green surfaces. Color usually cannot be used in plant resistance because changing it affects fundamental physiological processes. In a few instances, however, red cultivars of cotton, cabbage, and oats have been shown less attractive to insects, and yet they have retained good agronomic characteristics. In addition to color, some insects, like fruit flies, *Ragoletis* species, are known to associate shape and size of trees in locating hosts.

Once insects are in contact with the plant, short-range stimuli arrest further movement. These stimuli are both physical, exciting tactile receptors and chemical, exciting chemoreceptors on tarsi, antennae, and mouthparts.

Acceptance of the plant as a proper host. Subsequent to host finding, insects may take test bites, as with some caterpillars, to confirm host recognition. Continuous feeding seemingly is governed by the stimulation from various chemicals. In a monophagous insect, the silkworm (*Bombyx mori*), a series of substances are perceived in mulberry leaves that seem to mediate biting, swallowing, and continued feeding. Feeding to satiation then follows in the presence of appropriate chemicals.

Figure 12.2 Yellow-green (upper) and gray-green garden hoses connected together. Aphid color preference is suspected by their clustering on the yellow-green hose only. (Courtesy D. Foster and Iowa State University Extension)

Major physical factors involved in acceptance of a host may include such factors as leaf and stem toughness, leaf surface waxes, and pubescence (density and types of hairs). These factors may be important in relation to feeding and/or oviposition.

Sufficiency of the plant for requisites. Sufficiency of the plant as a host is finally determined during feeding. If nutrients are adequate and no toxicity occurs, the insect completes development within a normal time period and becomes an adult. Also, sufficiency is indicated in normal adult longevity and fecundity (the production of male and female gametes).

The Plant Aspect

As the supplier of physical and chemical stimuli, the plant itself becomes an important participant in the insect and host-plant relationship. Both morphological and physiological characteristics of a plant elicit given insect responses.

Morphological characteristics. Plant morphological features may produce physical stimuli or bar insect activity. Variations in foliage size, shape, color, and presence or absence of glandular secretions may determine degree of acceptance or utilization by insects. Pubescence and tissue toughness sometimes limit insect mobility and feeding.

Physiological characteristics. Physiological characteristics influencing insects usually involve chemicals that are the products of plant metabolism. Such chemicals are the result of primary and secondary metabolic processes.

Primary metabolic processes in plants produce substances to catalyze reactions, build tissues, and supply energy. The plant requires inorganic ions and produces enzymes, hormones, carbohydrates, lipids, proteins, and phosphorus compounds for energy transfer. Together, these **primary metabolites** promote growth and reproduction of the plant. For insects, some of these primary metabolites are feeding stimulants, nutrients, and toxicants. Other primary metabolites are inert as far as an insect is concerned.

Secondary metabolic processes in plants seem to be coincidental to primary metabolism. The chemicals produced, **secondary metabolites**, vary widely among plants and are believed nonessential in primary metabolism. Some of these secondary metabolites are thought to have arisen as mechanisms for chemical defense against plant eating. They may be stored in any convenient place in the plant structure and, often, are exuded from outer layers of plant tissues. Here, they may be sensed by insects and function as **token stimuli**. A token stimulus elicits a response initially and afterward has no effect.

This relationship between plant chemical stimuli and insect response is a form of chemical communication between these organisms. Such chemicals are called **semiochemicals**. Among semiochemicals are **pheromones**, which promote communication between members of the same species, and **allelochemics,** which promote communication between members of different species. Metabolites in plants that stimulate responses are kinds of allelochemics.

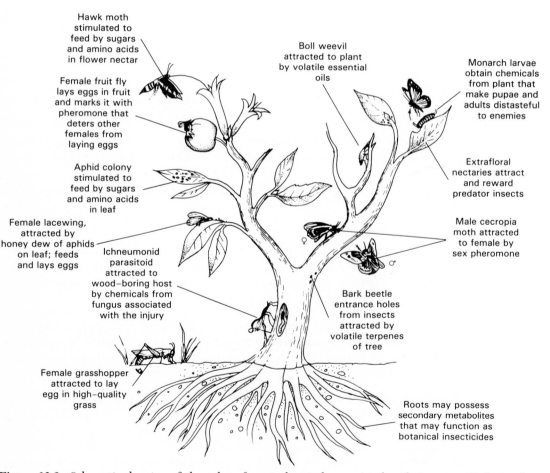

Figure 12.3 Schematic drawing of the roles of semiochemicals associated with insects. (Redrawn from Hagen, Dadd, and Reese, 1984, in Huffaker and Rabb, *Ecological Entomology*, Wiley and Sons)

Allelochemics can be subdivided further into allomones and kairomones. **Allomones** are mostly defensive chemicals, producing negative responses in insects and reducing chances of contact and utilization. They include repellents, oviposition and feeding deterrents, and toxicants. Conversely, **kairomones** are advantageous to an insect, promoting host finding, oviposition, and feeding. They include attractants, arrestants, excitants, and stimulants.

Host-plant Selection

Host-plant selection by insects usually involves both primary and secondary metabolites. Some theorists emphasize secondary metabolites in this process; however, many contend that both play an important role. According to M. Kogan, host-plant odor or taste for insects comes from nutrients and odd compounds that are intertwined as complex sensorial inputs. These inputs are interpreted by the insect's central nervous system to determine whether a given plant is a host. The genetically programmed "correct" signal supports a lasting host association.

In breeding programs for plant resistance, it is important to understand the nature of insect/plant relationships and their expression in host selection. Such understanding gives insights as to the underlying causes of the susceptibility/immunity gradient and enhances efficient development of resistant cultivars.

MECHANISMS OF RESISTANCE

Resistant cultivars function in many different ways to reduce the effects of insect attack. Various steps are required in host selection by insects. For normal insect growth and development to occur, certain requisites, available in proper amounts and at specific times, are necessary. Resistant cultivars, by one means or another, do not supply these requisites and thereby interrupt the normal host-selection process. In some instances, the mechanism of resistance involves new allomones or increased levels of existing ones; in others, it may be based on reduced levels of kairomones. Also, physical factors may be involved.

Most authorities consider true plant resistance as being primarily under genetic control. In other words, the mechanisms of resistance are derived from preadapted inherited characters. Therefore, the expression of these characters always occurs, although they can be mediated by environmental conditions.

The most widely accepted classification of genetic resistance modes in plants is that of R. H. Painter. These modes or mechanisms include **nonpreference, antibiosis,** and **tolerance.**

Nonpreference

Nonpreference refers to plant characteristics that lead insects away from a particular host; it includes activities of both plant and insect. Another, perhaps more appropriate term suggested for nonpreference is **antixenosis.** *Xenosis* is Greek for "guest"; therefore, antixenosis means against or expelling guests.

With nonpreference, normal insect behavior is impaired in such a way as to lessen chances of the insect's using a plant for oviposition, food, or shelter. Nonpreference can be expressed in a cultivar through either allelochemic or morphological characteristics.

Allelochemic nonpreference. This form of nonpreference is common among plants, sometimes causing them to be totally rejected by insects. Allelochemic nonpreference occurs with insects like the spotted cucumber beetle, *Diabrotica*

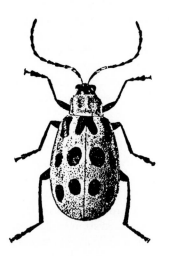

Figure 12.4 Spotted cucumber beetle, also referred to as the southern corn rootworm, *Diabrotica undecimpunctata howardi,* adult (top) and larva. Beetles are attracted to cucurbits and incited to feed by cucurbitacins. (Courtesy Illinois Natural History Survey)

undecimpunctata howardi, and other *Diabrotica* species on cucurbits. In this insect/plant relationship, cucurbitacins (a class of tetracyclic terpenes) produced by cucurbits act as attractants and feeding incitants for the beetles. Cultivars that lack or display low levels of specific cucurbitacins attract fewer beetles and receive much less damage than those with the feeding requisite.

Morphological nonpreference. This form of nonpreference results from plant structural characteristics, which disrupt normal behavior by physical means. In an example of this, the corn earworm, *Heliothis zea,* an important pest of many field and garden crops, prefers to oviposit on pubescent surfaces. Experiments involving cotton genotypes that lack hairs have shown that these genotypes suffer much less damage by many insect species because of lower than normal rates of oviposition.

Feeding activity also is diminished in many instances by morphological factors, including pubescence, tissue characteristics, and gummy exudates. As examples, husk tightness in corn resists damage from corn earworms, stem density of pith and node tissues in wheat resists damage by the wheat stem sawfly, *Cephus cinctus,* and hard, woody stems with closely packed vascular bundles in cucurbits resist feeding of the squash vine borer, *Melittia cucurbitae.*

Use of nonpreference. In a practical sense, the use of some nonpreference characteristics may be limited by given cultural environments. Many cultivars may show nonpreference if alternate hosts are in the vicinity, but in the absence

Figure 12.5 Squash vine borer, *Melittia cucurbitae*, adult (top), pupa (bottom left), and larva in stem. Hard, woody stems of cucurbits resist feeding by this insect. (Courtesy USDA)

of alternate hosts, the nonpreference may break down. Here, the insect species does not particularly prefer the cultivar, but if nothing else is available, it will accept it. Because of the widespread practice of monocropping and the breakdown of resistance, allelochemic nonpreference is not a primary goal in plant breeding programs.

On the other hand, forms of morphological nonpreference that impair feeding behavior are very important and may be a first line of defense against many pests. This is partly because morphological nonpreference provides long-lasting effectiveness, compared with most chemically based resistance; that is, insect populations have a difficult time overcoming this form of resistance.

Antibiosis

By far, antibiosis is the most widely sought after objective of plant breeders. This mechanism usually impairs an insect's metabolic processes and often involves consumption of plant metabolites. As with nonpreference, both insect and plant factors are involved in the antibiosis mechanism.

Allelochemics frequently are associated with antibiosis. Some of the best-documented allelochemics include the cyclic hydroxamic acids in corn (DIM-BOA), gossypol and related compounds in cotton, steroidal glycosides in potato, and saponins in alfalfa.

Quantity and quality of primary metabolites also may be important in conferring antibiosis. Particularly significant in this regard are imbalances of sugars and amino acids that result in nutritional deficiencies for insects feeding on the plant. For example, pea cultivars with low amino acid levels and increased sugar content show resistance to the pea aphid, *Acyrthosiphon pisum*, and rice cultivars deficient in asparagine (an amino acid) cause reduced fecundity in the brown planthopper, *Nilaparvata lugens*.

Symptoms of insects affected by antibiosis include the following:

1. Death of young immatures
2. Reduced growth rate
3. Increased mortality in pupal stage
4. Small adults with reduced fecundity
5. Shortened adult life-span
6. Morphological malformations
7. Restlessness and other abnormal behavior

Figure 12.6 Alfalfa susceptible to the alfalfa weevil, *Hypera postica*, (left) compared to resistant plants. (Courtesy USDA)

Tolerance

Unlike nonpreference and antibiosis, only a plant response is involved in tolerance. Here, the plant has the ability to give satisfactory yields in spite of injury levels that would debilitate nonresistant plants. This is the least dramatic resistance mechanism, and some plant scientists do not consider it a form of resistance.

Many factors are involved in plant tolerance, yet the overall mechanisms are poorly understood. Known components of this form of resistance include general vigor, compensatory growth in individual plants and/or the plant population, wound healing, mechanical support in tissues and organs, and changes in photosynthate partitioning.

An important advantage of tolerance is that it places no selective pressure on insect populations, as do nonpreference and antibiosis mechanisms. Without selection pressure, variants do not develop that can overcome the resistance. Its disadvantage is that insect populations may be allowed to sustain epidemics in an area, with these causing problems in other crops. Also, producers are wary of recommendations that allow large populations of seemingly injurious species to build up. Another, perhaps more important disadvantage is that tolerance is more strongly affected by environmental extremes than are other forms of resistance.

An example of tolerance is found in several corn genotypes that have the ability to repair and replace roots fed upon by the western corn rootworm, *Diabrotica*

Figure 12.7 Injury to corn roots from rootworm, *Diabrotica* species, feeding and remaining root volume can be measured by pounds of pull on a beam and scale. Corn hybrids are often evaluated for rootworm tolerance in this way. (Courtesy J. Tollefson and Iowa State University)

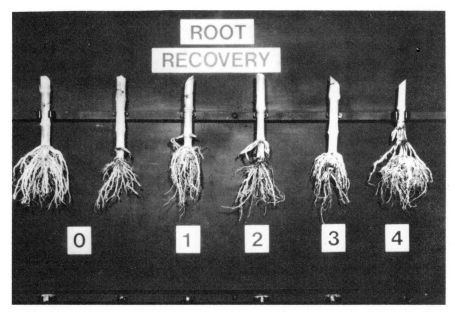

Figure 12.8 Degree of tolerance to rootworm, *Diabrotica* species, feeding is often evaluated by digging and washing roots, then rating them accordingly to numerical indices of root recovery. Root systems to the left show little or no recovery, but those to the right show considerable growth of secondary fibrous roots in response to the injury. (Courtesy J. Tollefson and Iowa State University)

virgifera. Such tolerance allows the plants adequate water and nutrient uptake and anchorage despite heavy feeding. Surprisingly, tolerant genotypes developed greater root volume with rootworm feeding than without.

Apparent Resistance

Apparent resistance, sometimes called ecological resistance or pseudoresistance, is not considered true resistance by many plant breeders. This is because expression of apparent resistance relies more heavily on environmental conditions than on genetics.

The characteristics of this resistance are temporary, and cultivars involved are potentially susceptible. Apparent resistance is important in insect pest management, but its use must be carefully synchronized with prevailing environmental conditions for effectiveness. The three types of apparent resistance recognized by most authorities include **host evasion**, **induced resistance**, and **host escape**.

Host evasion.   With host evasion, the plant passes through a susceptible stage quickly or at a time such that its exposure to potentially injurious insects is reduced. Often, host evasion is accomplished by planting early maturing varieties. A good example is the planting of fast-fruiting, short-season cotton varieties in Texas to provide a long, host-free period for populations of the boll weevil, *Anthonomus grandis*, and pink bollworm, *Pectinophora gossypiella*. The same varieties also give considerable evasion from populations of corn earworm and tobacco budworm, *Heliothis virescens*.

Sometimes evasion with early maturing varieties is confused with true resistance. To test early varieties for true resistance, they can be planted later than usual and inspected for late-season injury.

Induced resistance.   Induced resistance is a form of temporary resistance derived from plant condition or the environment. Here, factors like fertilization or changes in soil moisture levels may make plants more tolerant of insects than under other circumstances. For instance, nitrogen and potassium levels are known to affect aphid populations on plants. High nitrogen levels usually allow increases in survival, but the opposite may occur for high levels of potassium. Providing a proper balance of these nutrients in fertilizers has been suggested as a means of inducing resistance to aphids.

Recently, some attention has been given to the role of **phytoalexins** in inducing plant resistance to insects. Phytoalexins are phenolic compounds produced by plants when they become diseased or are attacked by insects. These compounds enable plants, once fed upon, to resist further damage by the pests. The mechanism involved results from an accumulation of allomones triggered by the injury or some other environmental factor. For example, phytoalexin production in laboratory soybeans has been induced by inoculation with a fungus, *Phytophthora megasperma* var. *sojae*. Subsequently, phytoalexin levels in cotyledons increased and functioned as feeding deterrents against larvae of the Mexican bean beetle, *Epilachna varivestis*.

Host escape.   This category explains the lack of infestation of susceptible plants in a population of otherwise infested plants. The principle of host escape recognizes that the presence of an uninfested plant may not mean that it is resistant and emphasizes that escapes occur in most plant populations, even with heavy insect infestations. The reason for escapes is rarely understood.

GENETIC NATURE OF RESISTANCE

Plant breeding activities aimed at developing resistance rely heavily on a knowledge of genetic background for the resistance. Such knowledge provides a quantitative basis for designs to recombine genes and select for proper characters. It also allows the identification of stable resistance factors that are least likely to be overcome by a pest population.

Epidemiological Types of Resistance

This classification of resistant types has been advanced by plant pathologists to express effectiveness and stability against a population of pests. Effectiveness

---

BOX 12.2 HESSIAN FLY

SPECIES: *Mayetiola destructor* (Say) (Diptera: Cecidomyiidae)

DISTRIBUTION: The Hessian fly was introduced into the United States from Europe, supposedly in straw bedding material used by Hessian soldiers during the Revolutionary War. The species is currently distributed throughout the wheat-producing areas of the world.

IMPORTANCE: The Hessian fly is one of the most destructive wheat pests. Larvae (maggots) feed between the leaf sheath and stalk, causing stunted and weakened plants. Injury results from salivary gland secretions from larvae that enter the plants, interfering with normal metabolism. The first generation in the spring stunts wheat and interferes with normal seed production, whereas the second generation, in the fall, weakens the stems, killing many plants before spring arrives. During outbreaks, 20 to 30 larvae may infest a single plant. In 1915, Hessian flies caused over $100 million damage to wheat in the United States. Currently, with the advent of "fly-free sowing dates," resistant wheat varieties, and new cultural practices, the pest causes about $16 million in damage to wheat each year. The species also attacks barley and rye but to a lesser extent.

APPEARANCE: Adults are black, about 2.5 mm long, and resemble mosquitoes. Gravid females have red abdomens. Larvae are red when newly eclosed, but eventually turn white. A green stripe (visible gut contents) runs down the back of the legless larvae. Eggs are 0.5 mm long and red. Puparia are called "flaxseeds," which they resemble. They are dark brown and 3 to 5 mm long.

LIFE HISTORY: Larvae overwinter on plant debris, volunteer wheat, and winter wheat. In early spring, larvae pupate and after 14 to 21 days, adults emerge. The time of adult emergence varies with latitude. Adults do not feed; they mate and live only two to three days. Females oviposit from 30 to 485 eggs in grooves of young wheat leaves. After 3 to 15 days, eggs hatch, and larvae feed at the leaf sheath base for four to six weeks. Larvae then become dormant in puparia during a portion of summer. In fall, adults emerge and females oviposit. The eggs hatch and larvae feed for a short time, then overwinter as puparia (flaxseeds). There are typically two generations per year; however, owing to climatic conditions, three or more generations may result.

---

and stability of resistant varieties are determined by the plant genes that confer resistance and the insect genes that allow the resistance to be overcome.

The gene-for-gene relationship.   In many pest populations, there are individuals with **virulent genes**, which allow a pest species to overcome resistance and once more attack a plant. Here, one or more virulent genes may be present that allow an individual pest to overcome the effects of one or more plant genes responsible for resistance. This principle has been called the **gene-for-gene** relationship.

In the gene-for-gene relationship, plant cultivars are resistant because they have a resistant allele at a gene locus that corresponds to an avirulent (susceptible) allele at an equivalent locus in the insect. Even though the resistant cultivar is effective against most insects in the population, an occasional insect may have a virulent allele instead of the normally avirulent allele. For example, the resistance gene in the host plant may code for a protein toxic to the insect, and a corresponding virulent gene in the insect may code for an enzyme that detoxifies the toxic protein of the plant. This circumstance allows virulent individuals to attack the otherwise resistant plant, and over a period of time, the virulent genotype can replace the avirulent genotype. Eventually, the effectiveness of the resistant cultivar would decrease.

Different populations of an insect species that vary in their virulence to a cultivar are referred to as **biotypes**. Some species of insects like the Hessian fly, *Mayetiola destructor*, are known to have several biotypes. The term biotype is frequently used for certain insect populations that overcome plant resistance.

To date, most insect biotypes have developed among aphid pests. Prominent examples include spotted alfalfa aphid, *Therioaphis maculata*, on alfalfa; greenbug, *Schizaphis graminum*, on wheat; corn leaf aphid, *Rhopalosiphum maidis*, on sorghum and corn; and pea aphid, *Acyrthosiphon pisum*, on peas and alfalfa.

Figure 12.9   Corn leaf aphids, *Rhopalosiphum maidis*, on corn leaf. To date, most insect biotypes have developed among aphid species. (Courtesy Iowa State University Extension)

Other homopterans with biotypes include grape phylloxera on grapes and brown planthopper on rice.

However, one of the best understood and most famous examples of biotypes is with a dipteran, the Hessian fly. To date, nine biotypes of this pest have been discovered.

Vertical and horizontal types of resistance.   Judging from the range of effectiveness of a resistant plant variety, J. E. van der Plank recognized two types of resistance, vertical and horizontal. In entomological terms **vertical resistance** refers to cultivars with resistance limited to one or a few pest genotypes. **Horizontal resistance** describes cultivars that express resistance against a broad range of genotypes.

Some authorities have argued against vertical resistance in breeding programs because of the potential development of biotypes. But it has been successful in many instances, as with the Hessian fly on wheat, and is easier to incorporate into new varieties than horizontal resistance. To manage insects by using vertical resistance, it has been suggested that resistant plant genes need to be identified and incorporated into varieties that are held, then released when biotypes appear.

Improvement in crops with horizontal resistance is a building process based on stepwise accumulation of genes with favorable additive effects. At present, the only known way to accumulate these favorable genes is by selective breeding over several generations, involving genetic recombinations (recurrent selection).

Horizontal resistance has low heritability and is difficult for plant breeders to incorporate. Nevertheless, some success with insects has been possible, for example, cultivars of corn with high resistance to both generations of the Euro-

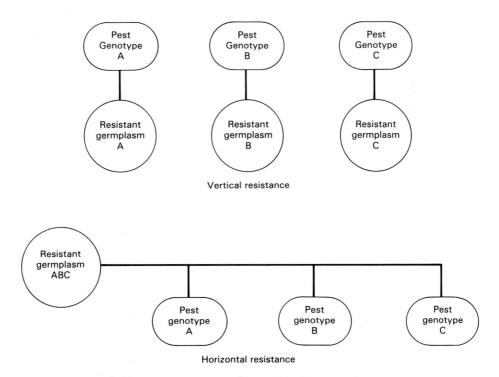

Figure 12.10 Schematic representation of vertical and horizontal resistance.

pean corn borer, *Ostrinia nubilalis*. Furthermore, horizontal resistance may be the most desirable type of resistance to use in pest management because of its stability.

Resistance Classes Based on Mode of Inheritance

Classes of plant resistance also can be distinguished according to the mode by which the resistance is inherited. In this regard, P. R. Day recognized three major resistance categories: oligogenic, polygenic, and cytoplasmic.

Oligogenic resistance. Oligogenic resistance is also called "major-gene resistance" and is conferred by one or only a few genes. This type usually produces vertical resistance against insects and may be inherited through dominant or recessive genes. Single-gene dominant resistance has been incorporated into varieties of such crops as apple, cotton, raspberry, rye, rice, and sweet clover. Single-gene recessive resistance can be found in corn lines resistant to the western corn rootworm and wheat resistant to the greenbug. Wheats resistant to Hessian fly probably should also be considered oligogenic; in this instance, however, resistance is conferred by a series of dominant or partially dominant genes, as well as by several recessive genes. Resistance is often considered oligogenic in this instance because a well-understood gene-for-gene relationship exists between resistance genes in wheat and the corresponding virulence genes in the Hessian fly.

Polygenic resistance. Polygenic resistance is conferred by many genes, each contributing to the resistance effect. For this reason, it is also called "minor-gene resistance." Resistance inherited through the polygenic mode is usually very complex and may be associated with such quantitative traits as plant vigor and yield. Horizontal resistance is usually polygenic. An example of polygenic resistance, as already mentioned, occurs in corn varieties resistant to the European corn borer.

Cytoplasmic resistance. Cytoplasmic resistance is conferred by mutable (capable or liable to mutation) substances in cell cytoplasm. Cytoplasmic inheritance is maternal because most cytoplasm of the zygote comes from the ovum. Although cytoplasmic inheritance is very important in resistance to pathogenic microorganisms, it has not been a factor in resistance to insects.

FACTORS MEDIATING THE EXPRESSION OF RESISTANCE

Although resistance is governed primarily by genetics, physical and biotic elements of the environment often influence its expression. Indeed, abnormal deviations of environmental factors can have profound effects on the performance of many resistant cultivars.

Physical Factors

Weather, soil, plant architecture, and cultural practices are some of the most important influences on the plant's physical environment. These factors can affect plant resistance by influencing such elements as temperature, light intensity, and soil fertility. Changes in these elements cause fundamental changes in plant

Figure 12.11 Colorado potato beetle, *Leptinotarsa decemlineata*, showing striped adult, larvae (center), pupa (lower left), and eggs (far left and far right). Shading has been found associated with reduced levels of steroidal glycosides in potato leaves. These compounds are known to retard development of the beetle. (Courtesy USDA)

physiological processes and can alter levels of allelochemics or cause imbalances in basic nutrients.

Temperature. Abnormally high or low temperatures for a period of time may cause loss of resistance. As an example, exceptionally low temperatures have caused the loss of resistance of some alfalfa genotypes to spotted alfalfa aphid and pea aphid and of some sorghum genotypes to greenbug. Loss of resistance to the Hessian fly has been found in some wheats at temperatures above 18°C.

Light intensity. Shade-induced loss of resistance has been found in several instances. These include wheats resistant to the wheat stem sawfly, *Cephus cinctus*; sugar beets resistant to the green peach aphid, *Myzus persicae*; and potatoes resistant to the Colorado potato beetle, *Leptinotarsa decemlineata*. With potatoes, shading was found associated with reduced levels of steroidal glycosides in leaves. These substances are known to retard feeding and development of the beetle.

Soil fertility. Changes in soil-nutrient levels also may mediate the expression of resistance in some plants, but little is known about the mechanisms involved. In an example of this phenomenon, clones of alfalfa resistant to the spotted alfalfa aphid were found to have reduced resistance if deficient levels of calcium or potassium or excess levels of magnesium or nitrogen were present. In the same example, resistance was increased by deficiency of phosphorus.

Biological Factors

Just as physical factors can influence expressions of resistance, so can biological ones. The most significant biological factors are the selection of biotypes and changes in resistance with plant age.

Biotypes. Increases in the proportion of resistance-breaking biotypes has already been discussed. When resistant cultivars are grown widely, selection pressure is imposed by these hosts on the insect population. When capable, the insect population responds with genotypes having virulence to overcome the resistance. As time passes, the virulent genotypes with superior fitness increase in number, displacing the avirulent types. The result is a situation of growing ineffectiveness of the resistant cultivar.

Biotypes occur in all types of pests. In plant pathology biotypes are called "races," or in instances of plant-parasitic nematodes, "pathotypes."

G. E. Russell distinguishes among four main types of genotypic variants in relation to resistant cultivars. These are included among the following:

1. True resistance-breaking biotypes that can attack previously resistant varieties. Here, there is a definite gene-for-gene relationship between the pest insect and the resistant plant. Most Hessian fly biotypes are examples of true resistance-breaking biotypes.
2. Unusually vigorous variants that have high reproductive potentials on all plant genotypes, with no definite gene-for-gene relationship. An example of this occurs with biotypes of the cabbage aphid, *Brevicoryne brassicae*, on rape.
3. Geographical variants that may or may not be true resistance-breaking biotypes. These have been observed among biotypes of the Hessian fly and, seemingly, are produced by factors other than widespread planting of resistant varieties.
4. Mistaken identities believed to be biotypes but, rather, are different species for which resistance was never developed. Examples here are often difficult-to-identify forms like some nematodes.

It is important to distinguish among these forms of "biotypes" in developing a proper counter-response to the problem.

The time required for breakdown in resistance may be only a few years, as for example, brown planthopper resistance in rice. However, it has not been uncommon for a resistant cultivar to remain effective for eight to ten years. During this period, time is available to search for new resistant genes and incorporate them into new varieties, anticipating the next appearance of another biotype. As an example of excellent stability, some cotton cultivars in Africa with vertical leafhopper resistance have remained effective for over fifty years.

Other than aphids, the development of biotypes has occurred infrequently in most insect groups. The frequent occurrence of aphid biotypes is partly the result of aphid reproduction, which has a parthenogenetic mode during parts of the seasonal cycle. Consequently, only one individual mutant, capable of feeding on a resistant variety, can produce a new biotype. Even including aphids, however, resistance-breaking insect biotypes have occurred much less frequently than have those among fungal and bacterial plant pathogens.

Plant age. Physiological responses in plants vary with age, and these can lead to changes in the expression of cultivar resistance. For instance, resistance in corn to the European corn borer results from the presence of the cyclic hydroxamic acid **DIMBOA** (2,4-dihydroxy-7-methoxy-1,4-benzoxazine-3-one). DIMBOA levels are highest early in the growing season (midwhorl growth stage), thus offering maximum resistance to first-generation corn borers. DIMBOA levels decline as the season progresses (more rapidly in susceptible than in resistant cultivars), and most commercial cultivars have very little resistance to second-generation corn borers. However, research indicates that second-generation resistance is possible, and development of this resistance objective is continuing.

The expression of resistance corresponding to plant age also has been reported in alfalfa resistant to spotted alfalfa aphid and some tobacco resistant to green peach aphid.

Figure 12.12 Corn susceptible to sheath and collar feeding by second-generation European corn borer, *Ostrinia nubilalis*, (top) compared with the resistant hybrid (bottom) under development. (Courtesy, W. Guthrie, and USDA)

STEPS IN DEVELOPING RESISTANT VARIETIES

It is beyond the scope of this discussion to describe the many activities and details of resistant-plant development. But a brief overview of the major steps involved may be of value in obtaining a perspective of the process. According to W. M. Tingey (1986), these major steps include the following:

1. Review pest problems of the crop with an interdisciplinary group composed of entomologists, plant breeders, plant pathologists, and other crop production specialists.
2. Determine the economic need for management of insect pests and define the nature of plant damage. Review the biology and ecology of the target pests and the crop host.
3. Collect adapted and wild germplasm for evaluation of resistance.
4. Determine methodology appropriate in mass screening for resistance and initiate the screening program.
5. Intermate parental germplasm and select resistant progeny. Recycle germplasm for recombination as long as reasonable progress is made each generation. Initiate selection for desired properties.

6. Initiate studies of plant traits conferring resistance, their influence on pest performance, mechanisms of reducing damage, and inheritance.

7. Determine the permanence and stability of resistance by evaluation of advanced genotypes in diverse cultural and environmental conditions.

8. Evaluate the influence of resistance traits on key predators and parasitoids of the cropping system.

9. Determine the vulnerability of advanced resistant germplasm by using infested versus noninfested comparisons.

10. Determine the economic value of resistance in elite germplasm by using infested versus noninfested comparisons.

11. Multiply seed of elite resistant germplasm and release to public and commercial organizations.

12. Monitor postrelease performance of resistant cultivars to determine the appearance of host-specific biotypes and varietal reaction to other commercial production hazards and practices.

13. Publicize the economic and environmental benefits of insect-resistant cultivars to grower groups using demonstration plots, press releases, and other extension information devices.

14. Continue breeding and selection programs for intensified levels and alternate modalities of resistance.

15. Publish research findings to familiarize the scientific community with progress and to establish a formal reference base for justification of continued financial support.

Crop resistance programs are based on the involvement of specialists from several disciplines. With plant resistance to insects, the programs require input from plant geneticists, plant breeders, and entomologists. The actual crossing, genetic analysis, and evaluation of agronomic characteristics are the responsibilities of plant scientists. Entomologists contribute by identifying resistance sources, characterizing mechanisms of resistance, and performing laboratory and field assays on resistance sources.

SUCCESSFUL USES OF INSECT-RESISTANT CULTIVARS

Since the first major successes of grapes resistant to grape phylloxera in the late 1800s, many other outstanding accomplishments in plant resistance to insects have been achieved. Major developments have taken place particularly in grain and forage crops, including wheat, rice, corn, sorghum, sugarcane, beans, barley, and alfalfa (Table 12.1). Progress, with varying degrees of success, has also occurred in vegetables, flowers, and other field crops. According to a survey of G. F. Sprague and R. G. Dahms, over 100 cultivars resistant to over 25 insect species had been released to commercial production by 1972. Many more could be added to this total today.

Many specific examples of success with plant resistance against insects could be mentioned; in the United States, however, five stand out as having provided significant monetary gains to users. These include wheats resistant to Hessian fly and wheat stem sawfly, corn resistant to the European corn borer, alfalfa resistant to the spotted alfalfa aphid, and barley resistant to the greenbug.

Resistance to Hessian Fly

The search for wheat resistant to the Hessian fly was begun in 1914 in Kansas. An early variety, Kawvale, developed from a soft-wheat variety, Indiana Swamp, was released as resistant in 1928. Subsequently, Kawvale was superseded by

Table 12.1. **Summary of Inheritance of Resistance and Presence of Resistant-breaking Biotypes by Insect and Crop.**

Inheritance	Insect	Crop	Biotypes
Oligogenic	greenbug	sorghum	Yes
		wheat	Yes
		barley	Yes
	aphids	raspberry	Yes
	planthoppers	rice	Yes
	leafhoppers	cotton	No
	pea aphid	alfalfa	No
	wheat stem sawfly	wheat	No
	Hessian fly	wheat	Yes
	European corn borer	corn	No
Polygenic	cereal leaf beetle	oats	No
	stem borers	rice	No
	corn leaf aphid	corn	?
	corn earworm	corn	No
	spotted alfalfa aphid	alfalfa	No
	aphids	Brassicae	Yes

SOURCE: Modified from Russell (1978).

Figure 12.13 Hessian fly, *Mayetiola destructor,* on a wheat stem. (Courtesy Kansas Agricultural Experiment Station)

higher-quality and more resistant varieties like Pawnee and, later, Ponca. In Kansas, growing resistant varieties has resulted in the virtual disappearance of the insect, except where susceptible varieties are present. Annual savings from growing resistant varieties has been estimated at $238 million, produced from reductions in fly populations of 95 percent.

Resistance to European Corn Borer

The release of corn varieties resistant to the European corn borer began in the 1940s, and by the 1950s, workers in Iowa and Minnesota had developed more than 40 insect-resistant lines and hybrids. Subsequently, many other varieties have been bred and released that have significantly reduced the threat of this introduced pest.

One of the more popular varieties, CI31A, has shown good resistance to the European corn borer in the United States, as well as Romania, Canada, Yugoslavia, Hungary, and the USSR. Although many resistant cultivars do not show activity against second-generation borers, one line, B52, offers a degree of

Figure 12.14 Leaf feeding from larvae of first-generation European corn borer, *Ostrinia nubilalis*, on susceptible corn (top), compared with a resistant hybrid. (Courtesy W. Guthrie, and USDA)

resistance. B52 has been successfully transferred to sweet corn, where second-generation injury is particularly important. However, heavy infestations are able to seriously impede development of the primary ear of B52 and other resistant lines, signifying the need for higher levels of resistance. Two other cultivars, B86 and BS9C5, also are resistant to both generations of borers, and B86 has been successfully transferred to popcorn, which suffers injury from both generations.

Today, many commercial field corn hybrids possess genes for resistance to the European corn borer. It has been estimated that resistant cultivars are grown on about one-third of the corn area of the United States, resulting in annual savings of approximately $150 million.

Resistance to Spotted Alfalfa Aphid

The spotted alfalfa aphid was introduced into the United States probably about 1954. Shortly after the introduction of this insect, the variety Lahontan, along with three of its five parental clones, were discovered to have resistance to the insect in Kansas. Further work allowed the development of the variety Cody, which possessed even higher levels of resistance and was well adapted to Kansas growing conditions. Other varieties developed with good resistance and adapted

---

BOX 12.3 SPOTTED ALFALFA APHID

SPECIES: *Therioaphis maculata* (Buckton) (Homoptera: Aphididae)

DISTRIBUTION: The spotted alfalfa aphid is native to the Middle East and was introduced into the United States (New Mexico) in 1954. The species spread rapidly, and by 1957, it was distributed over most of the United States.

IMPORTANCE: The species is a significant alfalfa pest, especially in the southwestern United States. In just two years, from 1954 to 1956, the pest caused $81 million in damage to alfalfa. Both adults and nymphs inject toxic salivary secretions and suck juices from leaves and stems, causing disruption of phloem flow. This injury causes leaves to become curled and yellowed. Individuals typically feed on lower leaves, then move up the plant. With large infestations, plants are severely stunted and often die.

APPEARANCE: Individuals are approximately 2 mm long and yellow, with four to six rows of black spots (short spines) on the abdomen. Alate (winged) individuals have clear wings with smoky-gray veins.

LIFE HISTORY: Eggs overwinter in the northern part of the species range, whereas in the South, the aphids are parthenogenetic the entire year. Females produce active young at rates of one to eight per day, depending on temperature. Each female can produce up to 140 offspring during her life-span. Nymphs reach the adult stages in one to four weeks (four instars), depending on climatic conditions. In colder regions, winged males and females appear in fall, mate, then females oviposit. There are 20 to 30 generations per year.

Figure 12.15 Spotted alfalfa aphids, *Therioaphis maculata* (top), compared with the larger pea aphid, *Acyrthosiphon pisum*, on an alfalfa leaflet.

to regional growing conditions include Mopa in California and Zia in New Mexico. Reselection within existing varieties is continuing, and yearly savings from growing resistant cultivars are estimated at $60 million.

Resistance to Wheat Stem Sawfly

One of the most dramatic instances of success in host-plant resistance has occurred with the wheat stem sawfly. This insect was a very destructive pest of wheat grown in the western United States and Canada until the use of fly-resistant wheat, beginning in 1946. At that time the solid-stemmed resistant variety, Rescue, was developed and released in Canada. Subsequently, the variety has been grown extensively in Canada and areas of the northern United States. Rescue has continued to offer protection against wheat stem sawfly since its introduction, and biotypes have not been known to develop. However, relatively low-quality agronomic characteristics have prompted recommendations of Rescue's rotation with susceptible varieties in some instances. Nevertheless, savings of $4 million per year have been estimated because of the use of plant resistance.

Figure 12.16 Greenbug, *Schizaphis graminum*, injury to susceptible (left) and resistant sorghums. (Reprinted by permission of John Wiley and Sons, Inc. from *Breeding Plants Resistant to Insects* by F. G. Maxwell and P. R. Jennings. Copyright ©1980 John Wiley and Sons)

Resistance to the Greenbug

Much plant-breeding activity has been invested in developing greenbug resistance in both wheat and barley. More than 7,000 wheat genotypes have been screened by the USDA in Kansas and Oklahoma, and eight spring wheat cultivars have shown substantial levels of resistance. Triticale, a hybrid cross of wheat and rye, also is significantly resistant.

However, barleys have shown the highest levels of resistance. These include the American varieties Dicktoo and Will, as well as other varieties developed in China, South Korea, and Japan. Although resistance in barley is controlled by a single dominant gene, biotypes have not been a major problem. The savings from growing greenbug-resistant barley in the United States are estimated at one-half million dollars annually.

USE OF PLANT RESISTANCE IN INSECT PEST MANAGEMENT

In insect pest management, plant resistance is used as a preventive, and most resistant cultivars have been developed for severe pests where economically significant populations are probable. When varieties providing nonpreference or antibiosis are grown, the carrying capacity of the agroecosystem is reduced and the general equilibrium position of the pest population drops. The use of tolerant varieties does not reduce environmental carrying capacities; instead, loss is managed by insuring yield output. In the first two instances selective pressure is placed on the pest population and in the latter it is not.

Plant Resistance as the Sole or Primary Tactic

Resistant plants have been used as the primary management tactic where losses are heavy and economics will not allow such curative measures as pesticides. In dealing with most plant diseases, it is the only feasible approach because curative tactics are usually absent and other preventive measures generally lack effectiveness.

Considerable success using resistant varieties as the sole or primary tactic against several insects has shown the approach worthy of continued development. But because of the threat of resistance-breaking biotypes (particularly with vertical resistance), continual monitoring of the pest's status is necessary, as well as dynamic breeding programs to provide timely supplies of new resistant varieties when needed. The use of horizontal resistance seems not to present these dangers; however, continued pest population assessment is still advisable.

Plant Resistance Integrated with Other Tactics

A basic principle of insect pest management is that long-lasting solutions to pest problems depend on several harmonious tactics. Plant resistance can serve as an excellent component tactic, because it is usually compatible with pesticides and sometimes with biological agents.

One of the most important combinations is plant resistance and insecticides. Quite often, levels of plant resistance are not sufficient, by themselves, to avoid economic losses from insect pests. Combining plant resistance with well-timed low dosages of insecticides sometimes can achieve adequate suppression while reducing otherwise high insecticide inputs. Such was the case with using resistant sweet corn (471-U6 x 81-1) and insecticides against the corn earworm in Georgia. Here, the resistant cultivar plus insecticide resulted in about 93 percent protection, compared with 86 percent for an insecticide-treated susceptible cultivar and 78 percent for the untreated resistant cultivar.

Although plant resistance and natural enemies are not always compatible, integrating them has been effective in certain instances. Theoretical studies have indicated that even small reductions in the growth rate of a pest population feeding on a resistant plant would allow significant increases in the action of a natural enemy. This has been established with resistant varieties of sorghum and barley against greenbugs. Here, resistant cultivars enhanced the depressive influence of a parasitoid, *Lysiphlebus testaceipes*, on greenbugs severalfold compared with that on susceptible plants.

The reason for complementary and sometimes synergistic effects of plant resistance and natural enemies is not always understood. In some instances secondary plant metabolites may be involved. Here, some of the chemicals are incorporated into body odors of the pest, and this body odor is detected by natural enemies to locate their hosts. In other instances, natural enemies may be attracted directly by plant volatiles, where subsequently they find their prey or host. Moreover, resistance may prolong the susceptible host stages, giving parasitoids more time to locate and parasitize pests.

It should be mentioned, however, that some characteristics of resistant plants, either morphological or chemical, may be deleterious to natural enemies. This situation could significantly alter the ultimate utility of such cultivars. Much research toward understanding plant-insect-natural enemy relationships, so-called tritrophic interactions, is currently underway (see Boethel and Eikenbary [1986]). This research should enhance our ability to design new cultivars with greater resistance and more environmental compatibility.

CONCLUSIONS

Plant resistance has many advantages as a primary tactic in insect pest management strategies. Among the most important are effectiveness, selectivity against the pest, relatively long stability, compatibility with other tactics, and human and environmental safety. In addition, resistant varieties can be adopted into crop production schemes easily and economically, resulting in both short-term and long-term gains.

Although the gains of developing and using resistant plants far outweigh the disadvantages, some of the most important limitations of the approach should be mentioned. Two of these are time required for development and problems with biotypes.

The time required for the development of new resistant varieties may be as long as 15 to 20 years, as has occurred with some wheats. Even more time may be required with woody plants. This time is exceptionally long compared with the six to nine years required for insecticide registration. However, the development of plant resistance can be reduced substantially by increasing the rate of funding for research activities. Also, new techniques in selection and breeding are being developed that are expected to shorten this time considerably.

The problem of biotypes is another worry, particularly with the use of vertical resistance, but, as mentioned, this side effect has been much more important with pathogenic microorganisms than with insects. In most instances, entomologists and plant breeders have been able to overcome the hazards of this problem with aggressive, dynamic breeding programs.

Other limitations of plant resistance to insects sometimes include poor agronomic characteristics of resistant cultivars and conflicts of resistant characters between different pest species. In the latter instance, a trait resistant to one insect may be attractive to another. For example, leaf pubescence of soybean leaves is the primary mechanism for resistance to the potato leafhopper, *Empoasca fabae*. However, the hairy surface also stimulates egg laying by the green cloverworm, *Plathypena scabra*. In instances of conflicting characters, the relative impact of one pest must be weighed against others and decisions made accordingly. In the example, the destructive potential of the potato leafhopper is much greater than that of the green cloverworm, making resistance to the more important pest preeminent.

Looking toward the near future, we generally concede that plant resistance to insects will continue to rely heavily on identification of sources of resistance and traditional plant breeding. Some authorities, however, like W. M. Tingey (1981) envision molecular engineering of the plant for optimum resistance expression and stability as the way of the future. We will look forward to seeing whether or not this type of approach, biotechnology, will meet all the expectations of it.

Selected References

AINSWORTH, G.C. 1981. *Introduction to the History of Plant Pathology*. New York: Cambridge University Press.

BOETHEL, D.J., and R.D. Eikenbary, eds., 1986. *Interactions of Plant Resistance and Parasitoids and Predators of Insects*. New York: Wiley.

DAY, P.R. 1972. Crop resistance to pests and pathogens. In *Pest Control Strategies for the Future*. Washington, D.C.: National Academy of Sciences, pp. 257–271.

GALLUN, R.L., and G.S. KHUSH. 1980. Genetic factors affecting expression and stability of resistance. In F.G. Maxwell and P. R. Jennings, eds., *Breeding Plants Resistant to Insects*. New York: Wiley, pp. 63–85.

KOGAN, M. 1982. Plant resistance in pest management. In R.L. Metcalf and W.H. Luckmann, eds., *Introduction to Insect Pest Management*. 2nd ed. New York: Wiley, pp. 93–134.

MAXWELL, F.G., and P.R. JENNINGS, eds., 1980 *Breeding Plants Resistant to Insects*. New York: Wiley.

NATIONAL ACADEMY OF SCIENCES. 1969. *Insect Pest Management and Control, Principles of Plant and Animal Pest Control*, vol. 3. U.S. National Academy of Sciences Publication 1695, pp. 64–99.

NORRIS, D.M., and M. KOGAN. 1980. Biochemical and morphological bases of resistance. In F.G. Maxwell and P.R. Jennings, eds., *Breeding Plants Resistant to Insects*. New York: Wiley, pp. 23–61.

PAINTER, R.H. 1951. *Insect Resistance in Crop Plants*. New York: Macmillan.

RUSSELL, G.E. 1978. *Plant Breeding for Pest and Disease Resistance*. London: Butterworths.

SPRAGUE, G.E., and R.G. DAHMS. 1972. Development of crop resistance to insects. *Journal of Environmental Quality* 1:28–34.

TINGEY, W.M. 1981. The environmental control of insects using plant resistance. In D. Pimentel, ed., *CRC Handbook of Pest Management in Agriculture*. Boca Raton, Fla.: CRC Press, vol. 1, pp. 175–197.

———.1986. Techniques for evaluating plant resistance to insects. In J.R. Miller and T.A. Miller, eds., *Insect Plant Interactions*. New York: Springer-Verlag, pp. 251–284.

VAN DER PLANK, J.E. 1963. *Plant Diseases: Epidemics and Control*. New York: Academic Press.

13 CHEMICAL MODIFIERS OF DEVELOPMENT AND BEHAVIOR

SINCE THE EARLY 1960s much activity has been focused on the search for safer and more effective insecticides. The goal has been to discover and develop compounds that are truly selective, that is, to find a means of effectively curing an insect problem without exposing applicators to hazards or causing undue harm to nontarget organisms in the environment.

The search for new insecticidal compounds has caused investigators to delve deeply into insect physiology for innovative ways to upset normal life processes. Most conventional insecticides affect the insect's nervous system, which functions similarly to our own and that of other animals. This characteristic gives these chemicals broad powers to poison and, consequently, presents important safety risks. To obtain selectivity, disruptions of processes that are unique to the Insecta (or at least to the Arthropoda) seem to offer the greatest potential. Presently, the unique life processes receiving the most attention involve growth and development as well as behavior.

Significant advances in chemical technology in the last two decades have allowed the discovery, identification, and synthesis of specific chemicals that regulate or mediate growth and development and species behavior. These chemicals, not previously identified, potentially provide new means of reducing an insect population. The mode of action with some of these chemicals is to cause premature death from abnormal molting or metamorphosis. Yet others may be used to repel insects from a source or attract them for our advantage.

Because these chemicals operate on systems that are different from those of warm-blooded animals, they are believed safe to use in most situations. Although a few such agents have been developed and are already in use on a limited scale, many are still in experimental stages. Therefore, we must consider them potentially useful, realizing that if they succeed at all, widespread application may be some time away.

It is important to understand the nature and possible applications of these chemicals, even though they are prospective. Such understanding will provide a better grasp of pest management objectives and allow realistic evaluations of future developments.

DISRUPTING NORMAL GROWTH AND DEVELOPMENT

Of the potential chemicals discussed, those that alter normal growth and development are considered by many authorities to hold the most promise. These chemicals have been called **insect growth regulators** (IGRs), a name in keeping with the familiar **plant growth regulators** (PGRs) that influence plant growth and phenology. IGRs are also known as biorationals, or third-generation insecticides, to reflect their environmental safety and advanced development (first-generation insecticides are the stomach poisons, and second-generation insecticides are the contact poisons).

IGRs are like conventional insecticides in many ways. They are used to kill or sterilize insects in a given area and can be applied with insecticide equipment to obtain suppression. Because they act on growth and development, however, IGRs are effective only when immature insects are exposed to them. Therefore, timing of applications becomes even more important and limiting than with conventional insecticides. Moreover, because their mode of action requires some time, pest presence and sometimes levels of injury must be tolerated longer than with most conventional insecticides.

The Basis for IGR Development

IGRs affect insects mainly by disturbing normal activity of insect endocrine systems. To understand their mode of action, we must rely on an understanding of developmental physiology.

Functions of the principal growth hormones. Recall that hormones produced by endocrine glands initiate and regulate molting and metamorphosis. The principal hormones involved in these life processes include brain hormone, ecdysone, and juvenile hormone. A review of glands supplying these hormones and their sites of action is presented in Figure 13.1. In the figure, neurosecretory cells of the back of the insect brain (in the location of the pars intercerebralis) secrete brain hormone that links environmental stimuli with other hormone systems. When brain hormone is present in the blood, the prothoracic gland is stimulated to secrete ecdysone, which causes the insect to molt. The body form of the insect after the molt is determined by the concentration of juvenile hormone from the corpora allata in the blood.

Insects with gradual metamorphosis experience a gradual reduction in juvenile hormone concentration, which causes subsequent nymphal stages to appear more and more adultlike. At the time of final molt, juvenile hormone is absent, and the adult emerges. The larvae of insects with complete metamorphosis have groups of cells called **imaginal disks** with the potential of expressing adult characters. Concentrations of juvenile hormone inhibit the growth of these disks, and the insect retains its larval form in subsequent molts. When a larva reaches full growth, juvenile hormone production is drastically reduced, and its level in the blood declines. This causes the insect to take the pupal form with the next molt. The juvenile hormone level declines still further during the pupal stage, allowing active growth of the imaginal disks and resulting in the adult form. In young adults of many species, the corpora allata increase activity once more, producing levels of juvenile hormone that stimulate ovary development and egg yolk production.

In addition to their metamorphic and reproductive functions, juvenile hormones also can have important influences on diapause, behavior, and communication (through pheromone production). Moreover, in social insects they are known to regulate caste differentiation and have special influences on morphogenesis.

Experimental modification of growth hormones. Significant modifications in the balance among the principal hormones causes aberrations and malformations

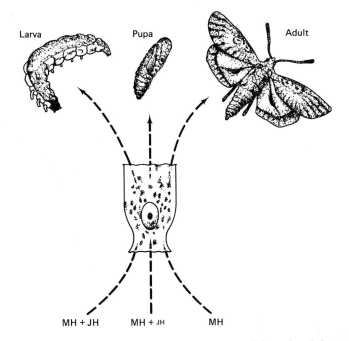

Larva   Pupa   Adult

MH + JH   MH + JH   MH

Figure 13.1 Diagrammatic representation of the role of the principal hormones in complete metamorphosis. MH symbolizes molting hormone and JH symbolizes juvenile hormone. Symbol size represents relative concentrations. (Reprinted with permission of Macmillan Publishing Company from *Insects in Perspective* by Michael D. Atkins. Copyright ©1978 by Michael D. Atkins)

of body structures. Several effects of hormone modification have been demonstrated experimentally. In studies where the brain was removed from immature insects, production of brain hormone ceased and further molting was suspended. When active brains from other individuals were implanted into these immatures, molting resumed. Removal of the corpora allata from young immatures caused cessation of juvenile hormone production, followed by premature pupation to form dwarfed adults. When active corpora allata from young immatures were implanted into full-grown immatures, the titer of juvenile hormone increased, and the insects remained in an immature stage with the next molt. When active corpora allata have been implanted into pupae, monstrosities developed, with mixtures of pupal and adult characteristics. More advanced studies with extracts containing juvenile hormone from male cecropia moths, *Hyalophora cecropia*, showed that the material could penetrate the insect cuticle and that topical applications to immatures could produce lethal developmental derangements. In instances where insects survived to reach adulthood, they were sterile.

Hormone mimics. In other laboratory observations, it was found quite by chance that some plant compounds produce results in insects similar to increasing the concentration of juvenile hormone. While attempting to rear the bug *Pyrrhocoris apterus* in cages with a paper towel lining, researchers obtained extra-large nymphs that were unable to molt into adults. By process of elimination, they found that the paper used, made from balsam fir, caused this reaction in the insects. Later, it was discovered that the "paper factor" resulted from the presence of a chemical, now called **juvabione**, that is structurally similar to juvenile hormone. With continued research, it has been shown that such compounds, called hormone mimics, are present in a variety of plants where they may function in part as defensive mechanisms against insect herbivores. For example, many ferns have high concentrations of these chemicals and are fed upon by only a few species of insects that have overcome this defense.

Synthetic hormones. With the discovery of hormone functions, particularly those controlling metamorphosis and reproduction, isolation and identification of natural hormones were undertaken in serious detail. Glandular and other extracts from many insect species produced a variety of compounds; however, the discovery of significant juvenile hormone activity in the frass (excrement) of the yellow mealworm, *Tenebrio molitor*, provided a breakthrough. With this discovery, the active compound **farnesol** was identified, and this eventually culminated in the development of the first synthetic hormone useful in pest management.

Other potential IGRs. In attempting to upset hormone balances, several possibilities exist. Potentially, concentrations in insect blood could be either increased or decreased to cause disruption. Increasing levels of juvenile hormone analogs or mimics has seemed more easily achieved than attempting to decrease it. However, a problem with agricultural pests is that adding juvenile hormone also keeps the insects in an immature and potentially injurious stage longer than normal. Therefore, if a method of shutting off juvenile hormone production could

Figure 13.2 Normal adult yellow mealworm, *Tenebrio molitor* (left) and abnormal adult treated topically with low doses of a juvenile hormone mimic (right). The abnormal adult retains a number of pupal structures along with the adult structures. The pupal skin of the abnormal individual was removed artificially because normal emergence would not have occurred. (Courtesy USDA)

be found, it would be more useful because it could prevent damage initially, as well as reduce populations in the long run.

With the knowledge that some plants produce compounds that mimic juvenile and molting hormones, researchers tested plant extracts to see if they may also produce antagonists that inhibit juvenile hormone. Subsequently, one such antagonist was discovered in the bedding plant, *Ageratum houstonianum*. Two compounds were identified and named precocene I and precocene II. The name precocene was given to these compounds because they induce precocious metamorphosis and because of their chromene chemical structure. In contact feeding and fumigation studies, these so-called antijuvenile hormones were found to induce premature metamorphosis in insects with gradual metamorphosis but not in those with complete metamorphosis. In many species with complete metamorphosis, however, the adults became sterilized after exposure. The mode of action of these compounds seemingly is lethal activation within the corpora allata, thus destroying the glands.

Several other materials, prototypes shown to interfere with juvenile hormone synthesis, have been investigated by researchers of the pesticide industry in an attempt to develop practical materials. These include FMev (fluoromevalonate), piperonyl butoxide, ETB (4-[2-[(*tert*-butylcarbonyl)oxy]butoxy]benzoate), and EMD (ethyl (*E*)-3-methyl-2-dodecanoate), among others. But none has been shown to be sufficiently active for practical purposes. Recently, it was demonstrated that some allylic alcohols offer promise as antijuvenile hormone agents against Lepidoptera.

As can be seen, development of antijuvenile hormones for actual use in pest management remains tentative. However, because of the diversity of chemicals that show juvenile hormone disruption, there is still hope that discoveries of practical materials will be forthcoming.

Finally, another potential IGR that has been considered is ecdysone. Like juvenile hormone mimics, ecdysones are found in plants where they also may play a part in insect-plant relationships. But ecdysones are steroidal compounds with complex molecules, and many tests have produced only low-level responses to application. Together, these characteristics have tended to discourage development. Most recently, however, azadirachtin, an ingredient in the seeds of neem trees (a subtropical species), has been found to disrupt molting and has been formulated as a liquid and a dust from ethanol extracts. This potential IGR, a new botanical insecticide, presently is being considered for registration on nonfood crops.

Practical IGRs

Currently, only four IGRs are registered for actual use. Three of these are juvenile hormone analogs, including **methoprene**, **kinoprene**, and **hydroprene**. According to G. B. Staal, certain conditions are required before such compounds will be effective. These include one or more of the following:

1. Larvae of the pest species do not cause damage; only the adults are perceived as harmful or as a nuisance (mosquitoes, flies, fleas).
2. Where individual insects or low numbers are not harmful; only large populations that expand through several short population cycles are significant (greenhouse Homoptera, stored product pests, and mushroom pests).
3. Environments where applied have a high degree of protection against breakdown factors, particularly sunlight (most applications).

4. Environments that are enclosed and thus protected against massive reinfestations (stored products, greenhouses).
5. Sedentary insect targets (low mobility) that are not likely to resurge quickly because of reinfestation (scale insects, mealybugs).
6. Special instances where the safety of the compound and the lack of other control methods are the prevalent considerations (pharaoh ant, *Monomorium pharaonis*).

In other words, juvenile hormone analogs are presently most practical where instantaneous control is not needed and residues are protected by enclosures. Furthermore, they are most appropriate with insects having short life cycles and where low numbers can be tolerated.

The fourth IGR registered is diflubenzuron. Although classed with the IGRs, it is not a juvenile hormone analog and is not limited by the conditions mentioned for the juvenile hormone analogs.

Methoprene. Methoprene is an IGR with good activity against many Diptera, Siphonaptera, and Coleoptera and some Lepidoptera and Homoptera. It is sold under the trade name Altosid® for suppression of mosquito larvae and horn flies, *Haematobia irritans*. In the latter instance, Altosid is a food additive for cattle, through which it passes with sufficient integrity to affect developing maggots in cow dung. It is used mainly with cattle in feedlots. Other uses for methoprene include the following: Coleoptera and Lepidoptera in stored tobacco (Kabat®); sciarid flies in mushrooms (Apex®); flea larvae indoors (Precor® and others); Homoptera on houseplants (houseplant mist); pharaoh ants indoors (Pharorid®); and leaf-mining flies (Agromyzidae) in vegetable and flower crops (Minex® IGR). Methoprene is also used to enhance silk production in silkworms (*Bombyx mori*).

Methoprene (Altosid®)

isopropyl (2E-4E)-11-methoxy-3,7,
11-trimethyl-2,4,-dodecadienoate

Kinoprene. Kinoprene is a strong, highly selective juvenile hormone analog effective against Homoptera. Sold under the trade name Enstar®, it affects all life stages, including eggs of whiteflies (Homoptera: Aleyrodidae) and mealybugs (Homoptera: Pseudococcidae). Because it lacks environmental stability, it has been used most successfully with ornamental plants in greenhouses.

Kinoprene (Enstar®)

2-propynyl (*E,E*)-3,7,11-trimethyl-2,4-dodecadienoate

Hydroprene. Hydroprene is a recent IGR and is sold under the trade name Gencor®. This IGR is related to methoprene and used indoors against cockroaches. In this instance, exposure of nymphs to the material causes them to become sterile adults. Outward symptoms of sterile individuals are twisted and deformed wings. Hydroprene may offer a new solution for pest species that have become resistant to conventional insecticides.

Hydroprene (Gencore®)

ethyl (*E,E*)-3,7,11-trimethyl-2,4-dodecadienoate

Diflubenzuron. This compound, sold as Dimilin®, has greater environmental stability than the juvenile hormone analogs mentioned and is registered for use against the gypsy moth, *Lymantria dispar*, in forests and boll weevil, *Anthonomus grandis*, in cotton. Moreover, registration is forthcoming for a variety of pests on several crops. In tests, this IGR has been shown effective against pests of forests, woody ornamentals, fruits, vegetables, mushrooms, cotton, soybeans, and citrus. It is also very useful against nuisance and medical pests like flies, gnats, midges, and mosquitoes.

Diflubenzuron (Dimilin®)

1-(4-chlorophenyl) 3-(2,6-difluorobenzoyl)urea

The mode of action of diflubenzuron is quite different from the juvenile hormone analogs. Diflubenzuron acts on larvae of most insects by inhibiting chitin synthesis and thus affects the integrity of the insect exoskeleton. Exposure causes improper attachment of the new cuticle during molting and produces a cuticle that lacks some of the layers that normally occur. Most larvae die from ruptures of the new malformed cuticle or from starvation.

Compatibility of IGRs with Other Tactics

It is too early to tell if the IGRs will be compatible with other pest management tactics against insects. Certainly, the juvenile hormone analogs satisfy many of the important selectivity requirements and would not be expected to strongly impact arthropod enemies. However, diflubenzuron, having broad activity against insects, may reduce many natural enemy populations, as would conventional insecticides.

In either case, much development remains before these potential compounds can be widely integrated with other tactics. Lack of environmental stability is the most serious weakness for the use of present juvenile hormone analogs in agriculture. Also, the slower activity, even with diflubenzuron, may always preclude their use in certain crops where blemish-free produce is a priority.

The great advantage of IGRs is safety. They all display low toxicity to warm-blooded animals, which makes them safe to apply and, it would seem, environmentally compatible.

Figure 13.3 Pupa of a butterfly (Lepidoptera) in normal condition (left) and abnormal condition. Abnormal insect was treated with diflubenzuron during the last stage of the larva. In the abnormal pupa, the abdomen has a pupal form, and the head and thorax have larval forms. (Courtesy Philips-Duphar, B. V., Amsterdam, Holland)

MODIFYING BEHAVIOR PATTERNS

So far in this discussion, we have dealt with means of disrupting physiological processes within the insect's body, specifically, endocrine-related processes. Other advanced pest management tactics attempt to mediate processes outside the body to change insect behavior patterns.

As discussed in previous chapters, insects are involved in many interactions with individuals of their own species and others in the environment. Such interactions involve complex and predictable behavior patterns. For example, mating involves complex intraspecies (within a species) behavior, and host finding and natural enemy avoidance include intricate interspecies behaviors.

In an elementary sense, insect behavior patterns that result in many interactions can be related to degrees of attraction and repellency, for example, males being attracted to females or individuals being repelled by a noxious plant. Associated behaviors where shorter distances are involved include stimulation and deterrence. These are sometimes more appropriate terms to use when we speak of attraction or repellency.

Attraction, repellency, stimulation, and deterrence are often mediated by chemicals from the exocrine systems of the interacting individuals. As mentioned earlier, these mediating chemicals are called **semiochemicals**. They can be further subdivided into **pheromones**, involved in intraspecific communication, and **allelochemics**, involved in interspecific communication.

In insect pest management, some of the newest and as yet mostly potential tactics involve manipulating insect behavior through the application of semiochemicals in the insect's environment. Experimentally, such applications have been used to disrupt insect perception and compel individuals to behave according to our desires.

Tactics Involving Insect Attraction

The idea of using attractants in pest control is not a new one. As far back as the early 1900s, food lures or baits treated with insecticide were used for suppression of a variety of agricultural pests. Such poisonous baits serve to attract insects, entice them to feed, and kill them outright. This approach is still used widely in households and other situations where exposure to broad applications of an insecticide presents unacceptable risks.

From a tactical standpoint, attractants can be aimed at attracting insects to a source where they are killed, as with poison baits, or simply used to draw them away from a valued commodity to one that is expendable, such as a trap crop. However, the attract-and-kill approach is by far the most common. Another notion is to attract insects with a chemical so compelling as to divert them from performing a vital activity, specifically, mating. Theoretically, a successful application in this way would reduce numbers by lowering reproductive rates.

Although the tactic of attraction is not new, some of the agents involved are. Therefore, in addition to traditional food lures, potent attractants like pheromones, totally unknown in the past, increasingly are being considered. Moreover, new formulations and methods of dispensing attractants have spurred a renewed interest in these compounds.

Use of pheromones in attraction. Although observations of individual insects attracting other insects of the same species were reported hundreds of years ago, it was not until 1959 that A. Butenandt and colleagues first reported the isolation and identification of a pheromone. This pheromone (*trans, cis*-10, 12-

BOX 13.1 SILKWORM

SPECIES: *Bombyx mori* (Linnaeus) (Lepidoptera: Bombycidae)

DISTRIBUTION: This species, the only member of the Bombycidae, is native to Asia. The insect has been artificially reared for centuries in silk production. After years of rearing, it is now mostly a domestic species and probably would not survive in nature. Many different varieties of silkworms have been developed for silk production, and rearing occurs in many countries, including the United States.

IMPORTANCE: The process of obtaining silk produced by silkworms, called sericulture, has been practiced for more than 35 centuries. For over 2,000 years only the Chinese knew the origin of silk (discovered by Lotzu, Empress of Kwang-Ti, about 2697 B.C.). Anyone who tried to reveal the secret of silk production outside of China was quickly executed. It was not until A.D. 555 that two monks, sent as spies to China, discovered the secrets and returned some silkworm eggs to Europe in a hollow staff. Today sericulture is practiced extensively in Japan, China, Spain, France, and Italy and has a commercial value from $200 to $500 million annually (around 100 billion larvae are cultured per year). Silk is formed when full-grown larvae spin cocoons in preparation for pupation. In this process, approximately 914 meters of silk are produced by a single larvae. For commercial purposes, pupae are killed prior to emergence because eclosion will break the valuable silk fibers. Approximately 3,000 cocoons are required to make one pound of commercial silk.

APPEARANCE: Adult moths are creamy white and have several faint brown lines across the forewings. They have a wingspan of about 50 mm and have heavy, hairy bodies. Larvae are naked but possess a short anal horn. Pupae are large and elliptical. Eggs are yellowish white and semispherical.

LIFE CYCLE: Pupae allowed to survive to maintain the colony emerge as moths that do not feed and that live only two to three days. In this period a single moth will lay 300 to 500 eggs. Colonies are maintained at 25°C, and larvae molt at 6, 12, 18, and 26 days after eggs hatch. During larval growth, mulberry leaves are supplied for food, with each larva consuming about 90 grams of foliage. Approximately three days are required to produce the silk cocoon.

---

hexadecadien-1-ol), isolated from silkworms, was a sex attractant involved in successful matings of the species. Since this first report, hundreds of pheromones have been identified in many organisms, ranging from algae to primates. Such advancements have been possible with the many significant developments in analytical chemistry.

Although sex pheromones are the most widely discussed, insects possess several other types. A categorization of insect pheromones is as follows:

1. **Sex pheromones.** These substances often produced by females attract males for mating, but they also may be produced by males to attract females. They seem to be the most highly developed in Lepidoptera and are frequently produced by

eversible glands at the tip of the abdomen. The release of sex pheromones is a complex physiological process, often associated with sexual maturity and environmental stimuli such as photoperiod and light intensity. Female sex pheromones are usually received by sensory sensillae on male antennae, and males search upwind, following the odor corridor of the female.

2. **Alarm pheromones.** These pheromones are common in social insects such as ants and bees. They elicit attack or retreat behavior. For instance, the remains of a sting apparatus of a honey bee left in a victim's body releases an alarm pheromone that attracts other bees and stimulates them to sting also.

3. **Trail-marking pheromones.** These chemicals are produced by foraging ants and termites to indicate sources of requisites to other members of the colony.

4. **Aggregation pheromones.** These pheromones are prominent in some species of beetles and cause insects to aggregate or congregate at food sites, reproductive habitats, hibernation sites, and the like. They are particularly well known in bark beetles, *Ips* species and *Dendroctonus* species, where they are involved in tree attacks.

5. **Epideictic pheromones.** Epideictic pheromones, also called spacing pheromones, elicit dispersal away from potentially crowded food sources, thereby reducing numbers. Hence, they are one of the few pheromones that serve to repel, rather than attract. They are also produced by bark beetles, as well as other Coleoptera, Lepidoptera, Diptera, Homoptera, Orthoptera, and Hymenoptera.

Of these types, the majority of research, to date, has been done with sex pheromones because of their potential usefulness in pest management. Presently, they have been used or perceived for use in three different ways: (1) in sampling and detection, (2) to attract and kill, and (3) to disrupt mating.

Pheromones in sampling and detection. The use of sex pheromones as attractants in traps is one of the oldest practical applications of semiochemicals in pest management. Presently, either sex or aggregation pheromones may be employed to monitor insect activity, with detection, phenology, and relative density information being obtained.

Pheromone traps are used frequently to gain information about pests for making tactical decisions. The first insect caught can serve as the beginning point for the accumulation of degree days, or catches over a period of time are useful in predicting population peaks or egg hatching times. Such predictions are useful in deciding if insecticides are necessary and, if so, when they should be applied.

Some of the most extensive uses of pheromone traps for making pest management decisions have occurred in apple orchards. Here, strategically placed pheromone traps continually monitored have reduced reliance on pesticide sprays and increased grower profits, as compared with calendar-spray schedules. For example, in a New York pest management program, growers sampled five major lepidopterous pests with pheromone traps and applied pesticides according to trap data, natural enemies, and weather information. Compared with other growers, those following the trapping regimen spent $60 less per hectare, without decrease in quality or quantity of fruit, and pesticide inputs (insecticides and acaricides) were reduced by 50 percent. In the Netherlands such programs in apple orchards allowed the total number of seasonal sprays to be reduced by two or three.

Pheromone traps are used regularly to monitor codling moth *(Cydia pomonella)*, redbanded leaf roller *(Argyrotaenia velutinana)*, and other leaf rollers in deciduous fruit; boll weevil, pink bollworm *(Pectinophora gossypiella)*, and corn earworm *(Heliothis zea)* in cotton; black cutworm *(Agrotis ipsilon)* in corn; and California red scale *(Aonidiella aurantii)* in citrus. In addition to these exam-

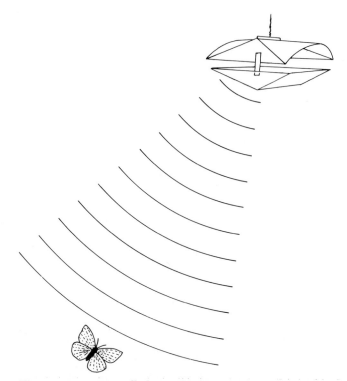

Figure 13.4 A so-called wing trap used to monitor Lepidoptera and other insects. Some models have reusable plastic tops and disposable sticky-coated bottoms. Insects are lured inside by pheromones and stick to the bottom of the trap. Counts are made at regular intervals.

ples of pest monitoring for decision making, pheromone traps are used to detect the presence of a harmful species, for example, the gypsy moth, as it invades new geographical areas. Pheromone traps are particularly valuable for detection because they can detect presence when pest numbers are very low.

In passing, it should be mentioned that chemicals other than pheromones are also valuable as attractants for survey and detection. For instance, mixtures

Figure 13.5 Redbanded leafroller, *Argyrotaenia velutinana*, pupa (top left), adults (top right), and larva. Pheromone traps are used regularly to monitor this species and other Lepidoptera in decidous fruit. (Courtesy USDA)

Figure 13.6 Japanese beetle, *Popillia japonica*, trap baited with phenethyl propionate and eugenol. (Courtesy USDA)

of geraniol and eugenol and those of phenethyl propionate and eugenol have been used by APHIS (USDA) to detect infestations of Japanese beetle, *Popillia japonica*. Additionally, methyl eugenol has been used to detect oriental fruit fly, *Dacus dorsalis*, and mango fruit fly, *Dacus zonatus*, as well as trimedlure to monitor the activity of Mediterranean fruit fly, *Ceratitis capitata*. Several other chemicals are also used to form an early warning network for action against many exotic and potentially harmful species.

Pheromone traps and traps with other attractants provide inexpensive, effective monitoring for many pests. However, the interpretation of trap catch is sometimes difficult, particularly when attempting to estimate relative density. This is because numbers caught are not always directly proportional to actual population size. Some of the factors to consider in utilizing pheromone traps and interpreting their catch include the following: (1) attractiveness of the pheromone, (2) pheromone concentration and release rate, (3) trap design (including color), (4) trap placement, (5) trap durability, and (6) area of trap influence.

Pheromones used in attract-and-kill programs. With this approach, insects are attracted to a source and killed by one means or another. Techniques in attract-and-kill programs range from entanglement in sticky materials to outright killing with insecticides or pathogenic microorganisms. The conventional approach to attract-and-kill has been to use traps lined with sticky material, but recent advances have allowed slow-release formulations of small particles that emit both a pheromone and an insecticide. In theory, when sex pheromones are used, pheromone sources affect half the population (for insects with a 1:1 sex ratio). If a large portion of one sex are attracted and killed, those individuals are eliminated, mating success is reduced, and numbers in the next generation fall.

When pheromone traps are used in attract-and-kill programs, the approach is often referred to as mass trapping. Traps used in mass trapping are often the same as those used in sampling and detection programs. They may or may not have an active killing agent. Operationally, the distinction between mass trapping and sampling usually is in the number of traps used. For mass trapping, trap densities as great as 100 per hectare may be utilized, compared with 5 to 10 per hectare in sampling. Mass trapping, or trap out, has been used experimentally in fruit tree, field, and forest crops, as well as with stored product and household pests. In most instances, trapping alone has not been sufficiently effective to obtain adequate suppression. Tests with the boll weevil showing promise involved pheromone traps placed in strips of cotton to aggregate the beetles. After aggregation was accomplished, the strips were treated with an insecticide. Trapping has also shown promise for use against some stored grain pests, like the khapra beetle, *Trogoderma granarium*, as well as with dermestid beetles in stored products. In the latter instance when combined with spores of the protozoan *Mattesia trogodermae*, mortalities of 100 percent have been observed. To date, one of the largest mass trapping programs has been conducted with a very destructive bark beetle, *Ips typographus*, in Norway and Sweden.

The newest and potentially most promising approach to attract-and-kill is the application of slow-release particles. With this technique, small particles are applied over an area by airplane. Each particle is a dispenser for both pheromone and insecticide, which are released gradually. Insects are attracted and exposed to a contact insecticide when they attempt to mate with the particles.

One such type of dispenser is the Hercon® flake, produced by the Hercon Division of Health-Chem Corporation. Hercon flakes are multilayered plastic

Figure 13.7 Hercon® controlled-release dispenser (3 mm²) containing pheromone and permethrin (a pyrethroid insecticide). Both insecticide and pheromone are in each dispenser or flake, thereby being protected from degradation. The insecticide enhances mating disruption because male insects are attracted to the flakes and killed or sexually impaired after contact. (Redrawn from Hercon brochure)

laminates, about 3 mm^2 (1/8 in.^{2}), that contain pheromone alone or a pheromone-insecticide mix in the inner layer or reservoir. The outer layers of the flake serve as a protective barrier but allow the mixture to diffuse into the air. As the material is dissipated into the air, a replacement quantity automatically moves outward from the reservoir to maintain the desired surface concentration and to give long-lasting effectiveness. The flakes are applied somewhat like a conventional insecticide with the use of a special device, the Hercon/Schweizer aerial applicator, attached to an airplane. This unit automatically combines the flakes with a special sticker (Phero-Tac™) that causes them to adhere to foliage as they land. A product marketed using this system is Disrupt®/Lure N Kill® PBW. It contains gossyplure, a synthetic pheromone, and permethrin, a pyrethroid insecticide, and has been used against pink bollworm in California and Arizona with reasonable success.

Mating disruption by air permeation. The final method proposed for the use of pheromones has been called the confusion, or decoy, method. This approach attempts to permeate the air with sex pheromone. Theoretically, if this is done, insects entering the area could not locate mates emitting natural pheromone because the synthetic pheromone permeates the whole environment. This would seemingly cause a reduction of reproductive rates and achieve crop protection without the use of insecticides. This basic idea was one of the earliest suggestions for the use of pheromones in pest management.

The first preliminary field test demonstrating the potential of this approach was conducted in 1967 with the cabbage looper, *Trichoplusia ni*. In this test, pheromone concentrations were shown to thwart males from being lured to female moths. Since this time, many studies have been conducted to apply the approach but with little success. These have included work with fruit, vegetable, field, forest, and stored product pests. In these studies, pheromone dispensing seems to have been the greatest obstacle to success. Subsequently, tiny dispensers, as mentioned previously, have been developed, and these hold greatest promise for future development. In addition to the Hercon flake, small, hollow, polymeric fibers have been developed as dispensers by Conrel Corporation and other firms. These chopped hollow fibers, combined with a sticker, are also applied by airplane. Some other dispensers that have been evaluated include hollow fiber tapes (parallel arrays of hollow fibers), rubber tubes, and microcapsules.

Both flakes and fibers have been commonly used to permeate atmospheres for mating disruption of several insects. In most instances, these approaches have shown degrees of success only when population levels are low. Therefore, air permeation may find a use only during certain parts of the growing season or in some years. For example, Disrupt PBW (gossyplure without insecticide) is recommended for use against the pink bollworm early in the season when populations are low and cotton plants are small. Small plants are particularly important because more point sources (flakes) are produced per leaf surface than with large plants. For adequate management, early-season pink bollworm suppression with pheromones is usually supplemented by later applications of conventional insecticides.

Although the confusion method has received much attention over the past 20 years, progress toward implementation has been very slow. Indeed, demonstrations of unqualified successes with the approach are few. This would lead us to believe that the use of pheromones for this purpose will have little impact on

Figure 13.8 Hercon/Schweizer aerial applicator for applying Hercon flakes. (Courtesy Health-Chem Corporation)

Figure 13.9 Application of tiny dispensers to plant foliage permeates the air with sex pheromone. The objective of applying the pheromone is to confuse male insects in their attempts to find mates. The approach has been named the confusion, or decoy, method. (Courtesy Health-Chem Corporation)

Figure 13.10 Hercon Disrupt® products available for pink bollworm management on cotton. (Courtesy Health-Chem Corporation)

future pest management programs. Conversely, their usefulness as attractants for pest monitoring and insecticide baits has been proven, and we would expect continued development of these tactics.

Use of traditional baits. Traditional baits or food lures also may be considered for attract-and-kill programs. Baits have been used for decades, and as mentioned in Chapter 11, some insecticides are sold commercially in bait formulations.

Some of the earliest food lures included wheat bran. Here, the bran was moistened with water, and stomach poisons or botanicals were added as the killing agent. Other bait carriers have included hardwood sawdust, fresh horse manure, ground corncobs, and seed hulls. Additives like molasses, chopped oranges or other fruit, amyl acetate, sugar, and honey were often included with these basic ingredients for attraction. With species that are attracted to fat, for example, some ants (Hymenoptera: Formicidae), ground fatty meat or plant products high in vegetable oil have been added to the insecticide carrier. Liquid bait sprays with sugar and molasses added to the insecticide solution have been commonly used with many species of flies.

In more modern programs, fiberboard squares impregnated with methyl eugenol and naled insecticide (an organophosphate) were used to lure and kill male oriental fruit flies, thereby eradicating the insect on the Pacific island of Rota. Bait sprays of yeast protein and malathion (an organophosphate) have been successful in helping eliminate introductions of the Mediterranean fruit fly from

Florida. This bait spray has also been used frequently in continuing campaigns against the pest in California.

Insect Repellents

Repellents, strictly speaking, are chemicals that cause insects to orient their movements away from a source. Allied materials that do not cause movement away but do prevent feeding or oviposition by insects are called **deterrents**. Both repellents and deterrents have been important in specialized areas of pest management. In concept, repellents are conventionally viewed as chemicals applied to surfaces to drive insects away. Deterrents, on the other hand, are usually thought of as natural plant constituents that are most important in host-plant resistance. It should be realized, however, that many synthetic chemicals, including some insecticides, can have deterrent or irritant properties (for example, chlordimeform and pyrethroids) for certain insects and other arthropods. Actual repellents will be emphasized in this discussion.

Repellents are usually volatile chemicals that express their activity in the volatile phase. A strong repellent will be sensed by insects a few centimeters distant, causing them to fly or crawl away. Less active repellents may allow insects to alight or touch the surface before being repelled.

Insect repellents based on such natural substances as wood smoke, oils, pitches, tars, various earths, and camel urine were used by humans hundreds of years ago to ward off insects and other arthropods. In the early 1900s, repellency of several essential oils, including oil of citronella, was discovered, and these were widely used as mosquito repellents. Before World War II, there were only four principal repellents, but significant advances were made after the beginning of the war to protect military personnel, particularly in the tropics. Subsequently, the early repellents were replaced by more effective synthetic compounds.

The advantages of using repellents are twofold. They often have low toxicity, and therefore they can be used safely on humans, plants, and domestic animals. Additionally, effective repellents protect the desired source, and because insects are not killed directly, undesirable side effects like resistance to the chemical are not likely. The limitations of repellents include the need to completely cover all susceptible surfaces with repeated applications and the possibility of increasing infestations on nearby untreated surfaces.

Traditional repellents. Presently this group of repellents is the most widely used. They include chemicals applied mostly for protection of humans and human possessions. Insect repellents to protect the body are aimed mostly at repelling mosquitoes, biting flies, fleas, ticks, and chiggers (larval trombiculid mites). These repellents are produced in several formulations, including aerosols, creams, lotions, grease sticks, powders, suntan oils, and clothes-impregnating laundry emulsions. Most repellents last only a few hours because they evaporate, are absorbed by the skin, are abraded by clothing, or are diluted by perspiration. Some of the most common repellents today are **DEET** (N,N-diethyl-*m*-toluamide, Delphene®) for biting flies and mosquitoes and **benzyl benzoate** for ticks and chiggers.

Repellents for human possessions include chemicals to protect fabrics and wood products. Protection of woolens against clothes moths (Lepidoptera: Tineidae) and carpet beetles (Coleoptera: Dermestidae) has been a particularly important problem that can be handled effectively with repellents. Colorless

DEET (Delphene®)

N,N-diethyl-*m*-toluamide

Benzyl Benzoate

benzyl benzoate

dyestuffs like Mitin FF® (a sulfonate) can be applied to wool during the dyeing operation, and these can effectively protect the fabric over its lifetime. Sodium aluminum fluosilicate has also been applied to woolens as an insect repellent and is not removed by dry cleaning.

Several chemicals are also used as repellents or feeding deterrents against wood-damaging insects. For instance, creosote is a repellent to many insects and is commonly used to protect such products as telephone poles and railroad ties. Synthetic chemicals like pentachlorophenol are also used widely as surface treatments or pressure impregnants of telephone poles and foundation timbers to deter termite feeding.

The use of traditional repellents with crop and domestic animal protection has been limited. One reason is the high cost of repeated treatments and another is the necessity of complete coverage. However, one exception is pine oil from wood pulp residue. Pine oil serves as a repellent and/or deterrent for bark beetles (Coleoptera: Scolytidae) when applied to individual trees.

Plant allomones as repellents. You may recall that allomones are natural substances from plants and animals that produce a response in the receiving species favorable to the emitting species. Most allomones that serve as repellents are usually considered from the aspect of host-plant resistance; in other words, some resistant plants emit repellents naturally, and this may be the primary mode of nonpreference. But the use of pine oil, mentioned earlier, could also be considered application of a plant allomone for protection; however, it is applied exogenously.

The exogenous application of plant allomones for repellency and deterrence has not been widely practiced. Although many of these chemicals have been identified as showing activity, they suffer from the same limitations as traditional repellents. It is possible that future research with these plant compounds will allow them to be used on a limited scale, as with wood products and high-value shade trees.

Epideictic pheromones as repellents. Another group of semiochemicals that may become future repellents are epideictic pheromones. Because these substances reduce population density of a species in an area, it stands to reason that when applied to a surface, there is potential for repellency. To date, epideictic pheromones have been used in this way against only two insect groups, *Dendroctonus* bark beetles and *Rhagoletis* fruit flies.

With bark beetles, various pheromones have been dispensed to limit their landing on trees. Here, 3,2-MCH dispensed near freshly felled Douglas fir reduced attacks as much as 96 percent and lowered subsequent larval population as much as 91 percent. Other tests with granular and liquid formulations of the pheromone also showed effective results against other bark beetles on felled Sitka spruce trees. Similar encouraging results were obtained in Switzerland, where oviposition-deterring pheromones were sprayed on cherry trees to reduce fruit damage. Although promising, much work with these materials remains before they can be used in practical pest management.

Integration of Behavior Modification with Other Tactics

In general, chemicals that attract and repel are well suited to be combined with other pest management tactics. As mentioned, one of the most effective combinations is an attractant integrated with a killing agent. This agent most often is an insecticide, but some success with pathogenic-microorganism combinations indicates that a potential also exists here.

Another potentially important combination are special attractant kairomones to increase populations of natural enemies in crops for enhanced biological control. For example, green lacewings, *Chrysoperla carnea*, can be attracted by applying artificial honeydew (an aphid exudate) to plants. This promotes egg laying in the desired location and potentially increases predation of plant pests. Parasitoids may also be managed in future years by the addition of kairomones to attract them to target areas, where a stronger than usual depressive influence may be gained. Although this approach is largely tentative, significant research is being conducted to achieve practical implementation.

Plant resistance and use of semiochemicals is yet another potentially viable combination for insect pest management. By understanding the role of semiochemicals in conferring plant resistance, we can become more efficient in selecting genotypes with higher levels of resistance than has been obtained in the past. In addition, synthetic allomones similar to the plant's natural products may be employed to artificially enhance a resistant plant's overall effect on a pest population.

CONCLUSIONS

Using chemicals to modify insect development and behavior represents a unique approach to the management of insect populations. These chemicals offer several advantages of which the foremost is environmental and human safety. For this reason, much research is being directed toward their development. These are still largely prospective compounds, however, and only time will tell whether or not they will be widely adopted. To succeed, they will have to measure up to the standards of other established control procedures. In other words, they must be effective in suppressing insect populations, be compatible with other management tactics, and be comparable in cost with conventional insecticides.

Selected References

BERRYMAN, AA. 1986. *Forest Insects—Principles and Practice of Population Management*. New York: Plenum Press, pp. 182-188.

BOWERS, W.S. 1982. Endocrine strategies for insect control. *Entomologia Experimentalis et Applicata* 31:3–14.

JACOBSON, M. 1986. The neem tree: natural resistance par excellence. In M.B. Green and P.A. Hedin, eds., *Natural Resistance of Plants to Pests—Roles of Allelochemicals*, ACS Symposium Series 296. Washington, D.C.: American Chemical Society, pp. 220-232.

KUBO, I., and J. KLOCKE. 1986. Insect ecdysis inhibitors. In M.B. Green and P.A. Hedin, eds., *Natural Resistance of Plants to Pests—Roles of Allelochemicals*, ACS Symposium Series 296. Washington, D.C.: American Chemical Society, pp. 206-219.

KYDYDONIEUS, A.F., and M. BEROZA, eds., 1982. *Insect Suppression with Controlled Release Pheromone Systems*. Boca Raton, Fla.: CRC Press, vol. 2.

METCALF, R.L., and R.A. METCALF. 1982. Attractants, repellents, and genetic control in pest management. In R. L. Metcalf and W. H. Luckmann, eds., *Introduction to Insect Pest Management*, 2nd ed. New York: Wiley, pp. 279-314.

NATIONAL ACADEMY OF SCIENCES. 1969. *Insect Pest Management and Control. Principles of Plant and Animal Pest Control*, vol. 3. U.S. National Academy of Sciences Publication 1695, pp. 290–323.

NORDLUND, D.A., R.L. JONES, and W.J. LEWIS. 1981. *Semiochemicals—Their Role in Pest Control*. New York: Wiley.

QUISTAD, G.B., D.C. CERF, S.J. KRAMER, B.J. BERGOT, and D.A. SCHOOLEY. 1985. Design of novel insect antijuvenile hormones: allylic alcohol derivatives. *Journal Agricultural Food Chemistry* 33:47–50.

STAAL, G.B. 1982. Insect control with growth regulators interfering with the endocrine system. *Entomologia Experimentalis et Applicata* 31:15–23.

14 STERILE-INSECT AND OTHER GENETIC TACTICS

THE USE OF STERILIZED INSECTS and manipulations of pest genetics have developed gradually since about 1916. These tactics are pest-selective, aimed mostly at reducing insect populations by lowering reproductive potentials, and they include some of the most innovative and unusual procedures in insect pest management. Here, pest insects are used against members of their own species to reduce population levels, and for this reason, the approach often is called **autocidal control**.

Autocidal control received major impetus from the work of E. F. Knipling, a USDA entomologist, in the late 1930s. It began as a strategy of replacing normal matings in a population with infertile ones, in effect, inducing sterility. Fundamentally, the sterility principle aims at flooding a population with sterile males, which seek out and mate with normal females. Such matings result in inviable (unfertilized) eggs, and with continued sterile-male releases, the population declines. As decline occurs, the ratio of sterile to normal males increases until virtually no normal males remain. At this point, the population becomes extinct for lack of progeny.

This sterile-insect technique evolved into yet other ideas that emphasized genetic manipulations of the pest population. Whereas the sterile-insect technique involves competition between sterile and normal males, other approaches include competition between gametes (usually sperm) of mutant and normal insects. Collectively, these approaches have been termed **genetic control**. Genetic control focuses on altering genetic makeup in such a way as to (1) produce sterility of progeny, (2) reduce fecundity, or (3) reduce survivorship in otherwise favorable environments.

Although the sterile-insect technique is treated separately from genetic control in this discussion, many authorities group the tactics together as genetic control. The rationale for a combined grouping is that genetic elements—gametes—are involved in both approaches. Distinguishing between them is appropriate, however, because successful use of the sterile-insect technique has been demonstrated, whereas most other genetic methods are still largely theoretical.

THE STERILE-INSECT TECHNIQUE

The sterile-insect technique developed out of studies of the screwworm fly, *Cochliomyia hominivorax*, a parasite of cattle primarily in tropical and subtropical regions of the New World. Knipling and coworkers noted that adult females mated only once (were monogamous) to fertilize their eggs. Using this knowledge, these workers postulated that if males could be sterilized without impairing their mating behavior, releases into the wild population would result in infertile matings and isolated populations could be eliminated.

BOX 14.1 SCREWWORM

SPECIES: *Cochliomyia hominivorax* (Coquerel) (Diptera: Calliphoridae)

DISTRIBUTION: Before eradicaion, the screwworm was found throughout Texas, New Mexico, Arizona, Mexico, and Central and South America. Through migration (female flies travel up to 300 miles) and transport by humans, the species has been found as far north as Canada, but the flies cannot successfully overwinter in northern areas and die with the onset of cold weather. Presently the species is found in Central and South America.

IMPORTANCE: The screwworm, commonly referred to as primary screwworm or obligate screwworm, is a serious pest of livestock and wildlife. The species infests warm-blooded animals when they develop open wounds, on their bodies as with dehorning, castration, or fly and tick bites. Adult female flies oviposit on wounds, and developing larvae (maggots) feed directly on flesh and wound exudate. The odor from infested animals attracts more female flies, which lay additional eggs on the wounds. Moreover, secondary screwworms, such as *Cochliomyia macellaria* (which usually oviposit on carcasses), will oviposit on the primary screwworm-infested animals.

Infested animals abstain from feeding and may die if not treated. Before substantial control of primary screwworm was initiated, livestock losses were between $50 million and $100 million per year in the southwestern United States. In 1937, entomologist E. F. Knipling developed a method to sterilize male flies. Male pupae were sterilized in the laboratory with Cobalt-60, and the newly emerged males were released from aircraft into the environment. Sterilized males, by advantage in numbers, would outcompete normal wild males for females. Because females mate with only one male, they produced infertile eggs. The sterile insect technique (SIT) has continued to be an effective method for controlling primary screwworms.

APPEARANCE: Adults are metallic blue to bluish-green, with an orange to brown head and three dark stripes on the top of the thorax. Larvae (maggots) are white and about 16 mm long when full-grown. Eggs are 1 mm long and white. Puparia are dark brown and about 10 mm long.

LIFE HISTORY: Adults are active year-round in warm climates; in cool climates, however, pupae hibernate in soil for about two months, until warmer weather. Adults feed on manure juices and wound exudates. Females oviposit from 10 to 500 eggs in about 16 hours, and the maggots feed gregariously on live flesh and fluids of the open wound. After four to nine days (three stadia), larvae drop to the ground and pupate in soil. Adults emerge in seven to 54 days, depending upon temperature. Flies mate two to five days after emergence. Each female can oviposit up to 3,000 eggs.

Table 14.1. Trend of a Hypothetical Population with Each Female Producing Five Female Offspring Each Generation (Sex Ratio 1:1)

Generation	Number Females per Unit Area	Total Number in Population
1	1,000,000	2,000,000
2	5,000,000	10,000,000
3	25,000,000	50,000,000
4	125,000,000	250,000,000
5	625,000,000	1,250,000,000

Theoretical Background

The theory behind this approach can be illustrated by a number of simple models (tables), as was done by Knipling, to show the behavior of population numbers alone and when subjected to sterile-insect releases. In these models, a hypothetical population, having a 1:1 (female to male) sex ratio and growing fivefold each generation, is convenient for presentation. This exponentially growing population is shown in Table 14.1. (This simplistic pattern is not prevalent in the environment because of density-dependent factors and randomly operating density-independent factors. It is used here for explanatory purposes only).

If the same population is subjected to releases of 9 million sterile males each generation, as was done in Table 14.2, numbers are essentially eliminated (eradicated) after the fifth generation. The reason for eradication is that every time a sterile male mates with a female, she is prevented from laying fertile eggs. Therefore, in generation one, only 10 percent (100,000) of the 1 million females mated with a normal male, and even though these females increased their numbers fivefold, the population still began a downward trend. Continued inundations of the same number of sterile males caused increasing rates of decline because of increases in the sterile-to-normal-male ratio. In other words, a **reproduction penalty** (genetic load of geneticists) was placed on the population so great that it could not be overcome, and numbers dropped. As they dropped, the method increased in effectiveness.

A primary principle of the sterile-insect technique is shown from this simple example: the initial number of sterile releases must be great enough to overcome reproduction and start a downward trend in the population. The higher the initial sterile-to-normal-male ratio, the quicker is pest eradication. Had only 4 million sterile males been released each generation in the example, the population would still have grown, the method would have become less effective with time, and the

Table 14.2. Trend of the Hypothetical Population Subjected to Sterile-insect Releases

Generation	No. Females without Releases	No. Sterile Males Released	No. Females with Releases	Ratio Sterile to Normal Males	No. Fertile Females
1	1,000,000	9,000,000	1,000,000	9:1	100,000
2	5,000,000	9,000,000	500,000	18:1	26,316
3	25,000,000	9,000,000	131,579	68:1	1,907
4	125,000,000	9,000,000	9,535	944:1	10
5	625,000,000	9,000,000	50	180,000:1	0

Table 14.3. Trend of the Hypothetical Population when Sterile Releases Are Compared with an Insecticide Treatment

Generation	No. Females with No Treatment	No. Females with Sterile Releases 9:1	No. Females with Insecticide— 90% Kill
1	1,000,000	1,000,000	1,000,000
2	5,000,000	500,000	500,000
3	25,000,000	131,579	250,000
4	125,000,000	9,535	125,000
5	625,000,000	50	62,500
6	3,125,000,000	0	31,250

pest would not be eliminated. Theoretically, the initial ratio of sterile-to-normal males must be greater than the pest's net increase potential from generation to generation to be effective. In other words, if the population is increasing by a factor of five each generation, the ratio would have to be greater than 5:1 for effectiveness. In practice, however, the ratio may need to be double the increase factor to compensate for environmental and genetic factors.

The value of the sterile-insect technique can be viewed, at least theoretically, by comparing the hypothetical example with management using another tactic, insecticide applications. Here, it is assumed that an insecticide is applied each generation to a pest population, causing a 90 percent reduction in numbers. The results, shown in Table 14.3, indicate that even though numbers are reduced by the insecticide, a residual population still remains after generation five. If applications are suspended at that time, we would assume the pest population would rebound. On the other hand, the population was completely eliminated with the sterile-insect technique, and the program could have been terminated.

An important principle from this example is that tactics such as insecticide applications tend to become less efficient as numbers are reduced, but the sterile-insect technique becomes more efficient with small numbers. Conversely, when pest populations are large, insecticides are most efficient. Therefore, an effective strategy is to integrate insecticide applications with sterile-male releases in management programs. Here, it is suggested that an insecticide be applied to cause initial number reduction, followed by successive sterile-male releases. As Table 14.4 shows, pest numbers with this integrated approach are reduced very rapidly, compared with the sterile-insect technique alone.

Table 14.4. Trend of the Hypothetical Population When Insecticides Are Used, Followed by Sterile-insect Releases

Generation	No. Sterile Males Released	Without Insecticides		With Initial Insecticide Application—90% Kill	
		No. Females	Ratio Sterile to Normal Males	No. Females	Ratio Sterile to Normal Males
1	9,000,000	1,000,000	9:1	100,000	90:1
2	9,000,000	500,000	18:1	5,495	1,638:1
3	9,000,000	131,000	68:1	17	529,412:1
4	9,000,000	9,535	944:1	0	
5	9,000,000	50	180,000:1		
6	9,000,000	0			

A mathematical model of the sterile-insect technique presented by E. S. Krafsur and coworkers is as follows:

$$W_{n+1} = \left[\left(W_n \times \frac{W_n}{W_n + (R \times C)} \right) + M \right] \times I$$

where,

W = number of wild individuals of a sex (the first W is egg-laying females and the others are males)

n = generation number

R = number of sterile males released

C = competitiveness (if released males are equally competitive to wild males, this value equals 1.00)

M = number of immigrant females inseminated with fertile sperm and unwilling to mate

I = effective rate of population increase

In this equation, if R is constant over successive generations, there would be an accelerated decline of population numbers toward eradication as the R: W ratio progressively improved. Eradication would be prevented if (1) M is not zero, (2) I increases because of density-dependent factors, or (3) C declines because released males consistently fail to find localized wild populations.

Circumstances for application. Several pest situations are envisaged for use of the sterile-insect technique: (1) against well-established pests when they reach low points in their density cycles (this could be within a season or between seasons); (2) against newly introduced pests or established pests spreading into new areas; (3) with other tactics like insecticides and cultural procedures, which precede sterile-insect releases; and (4) against isolated populations like those on islands and other such situations.

Sterilizing Insects in a Natural Population

The idea of sterilizing insects in a natural population was advanced several years ago when it was found that certain chemicals, called **chemosterilants**, had potential in this regard. Subsequently, it was discovered that these chemicals presented unacceptable human health and environmental hazards, and this discouraged further advancement of the approach. Even though not practical presently, future developments could change the acceptability of the method, and therefore a brief explanation of this idea is appropriate.

Sterilizing insects in a natural population should not be confused with the sterile-insect release procedure. Although sterility is involved in both instances, the mechanism of suppression of each is quite different. Whereas mating is required for suppression in sterile-insect release programs, population reproductive rates are reduced directly by the sterilization of individuals in the natural population. Here, both males and females of the population are sterilized, which in regard to reproductive capacity is the same as killing them; they will not add individuals to the next generation. Moreover, the sterilized individuals, still being active, can mate with individuals of the population that were not sterilized, further reducing the reproductive potential. This latter phenomenon behaves similarly to the release method and, with this approach, has been called the **bonus effect**.

To achieve this bonus effect, the sterility technique must not eliminate mating behavior and competitiveness of the sterilized insects. Indeed, unless a high

Table 14.5. Trend of the Hypothetical Population Comparing Application with an Insecticide, Sterile-male Release, and Application of a Chemosterilant

Generation	No. Females with Insecticide— 90% Kill	No. Females with Sterile Releases 9:1	No. Females with Chemosterilant— 90% Sterility
1	1,000,000	1,000,000	1,000,000
2	500,000	500,000	50,000
3	250,000	131,579	2,500
4	125,000	9,535	125
5	62,500	50	6
6	31,250	0	0

level of competitiveness is retained, sterilizing feral (wild) individuals generally would have the same effect on reproduction as killing them outright with an insecticide. Conversely, fully competitive sterile insects theoretically would be capable of reducing reproduction in normal individuals of the population in proportion to the sterile-insect numbers. For instance, if 90 percent sterility in the population is achieved, that percentage would be similar to mortality from an insecticide. However, unlike the insecticide, the sterilized individuals would further reduce reproduction in the 10 percent normal females by 90 percent. A hypothetical example of the rapid population reduction that could be achieved with an effective chemosterilant, along with a comparison with insecticides and sterile-male releases, is shown in Table 14.5.

The method of sterilizing feral insects in their environment has been tested extensively on a worldwide basis. Most research on this approach has been directed toward the house fly, *Musca domestica*, and several mosquitoes (Diptera: Culicidae) in isolated areas. In house fly studies, chemosterilants have been combined with highly selective baits to maximize exposure of the target organisms and reduce risk to other species. With pilot studies, using the chemosterilant TEPA in a granular bait (67 percent cornmeal, 15 percent sugar, 15 percent powdered milk, 2.5 percent powdered eggs, and 0. 5 percent TEPA), house flies were virtually eliminated around a dump on Bahia Honda Key in Florida during a five-week period. Another even more ambitious program with the house fly on Grand Turk Island in the West Indies caused large reductions in the fly population; however, the program was considered only moderately successful because final eradication was not achieved.

The advantages of sterilizing individuals in the natural population are very clear. With the bonus effect, much higher levels of effectiveness can be achieved than with direct mortality agents like insecticides. In addition, many of the obstacles of the sterile-insect release technique are eliminated, particularly the expensive and sometimes formidable task of rearing large numbers of insects for release. Yet the problem remains of finding safe, efficient sterilizing agents that can be applied in the environment. Until this problem is solved, we would not expect much use of the concept.

Methods of Sterilization

The concept of autocidal control is sound in principle, but its application relies greatly on acceptable procedures to sterilize insects. Without these, as seen in

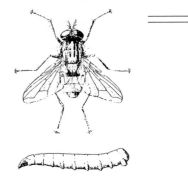

BOX 14.2 HOUSE FLY

SPECIES: *Musca domestica* (Linnaeus) (Diptera: Muscidae)

DISTRIBUTION: The house fly is a cosmopolitan species, found throughout the world. Populations are particularly prevalent at locations in or near human dwellings, where they are well adapted to survive.

IMPORTANCE: House flies are very annoying to both humans and animals. In addition, house flies are known to be carriers of disease organisms. For example, typhoid fever, anthrax, cholera, diarrhea, and dysentery have been associated with mechanical transmission by house flies. They also serve as intermediate hosts for parasitic tapeworms (helminths).

APPEARANCE: Adult flies have sponging mouthparts that they use to pick up liquids or solids that they liquify. Larvae (maggots) are tapered exteriorly and are creamy white. Full-grown larvae are about 12 mm long. Puparia are shorter than larvae and more robust. They are dark brown and about 6 mm long when mature. Eggs are pearly white, elongate, and about 1 mm long. Larvae are found in manure, rotting food, and other fermenting vegetable matter.

LIFE CYCLE: Most house flies overwinter in the northern United States as larvae or pupae, but some adults probably survive in heated buildings. After maturing in spring, female flies oviposit 2 to 21 egg masses in any kind of animal excrement or decaying matter. Each egg mass contains about 130 eggs, which hatch within 24 hours. Maggots feed in decaying material for three to seven days, move to the margins of the food source, and pupate. Pupae in puparia develop in three to seven days, and adults emerge to begin the next cycle. The entire life cycle requires only 10 to 12 days under ideal temperatures but may take considerably longer if cooler temperatures occur. The reproductive potential is tremendous. It has been calculated that the offspring of a single gravid female would easily produce 2×10^{20} flies in a one-year period, filling 2,000 cubic miles. Indeed house flies usually reproduce to the limit of available food resources. This suggests that sanitation is one of the best methods of limiting house fly populations.

---

the concept of sterilizing natural populations, progress toward application can be agonizingly slow or halted completely. During the development of the approach, both ionizing radiation and chemicals have been found to cause insect sterility, but the former has been used more successfully.

Ionizing radiation.   The sterilizing effects of X-rays on insects was observed as early as 1916 with adult cigarette beetles, *Lasioderma serricorne*, and was the first form of radiation investigated. These and other investigations with X-ray–caused mutations in pomace flies, *Drosophila melanogaster*, eventually led to the discovery by R. C. Bushland that pupae of the screwworm, when irradiated near adult emergence, could result in competitive, sterile adults.

After World War II with the development of the atom bomb, it became much more efficient to work with manufactured isotopes, primarily those producing gamma rays. Further studies showed that there was little difference between certain X-rays and gamma rays in treating insects, and subsequently, most programs have utilized gamma radiation.

High-energy radiation exerts its genetic effects by chromosome breakage and point mutations of DNA. After irradiation, gametes are produced, but they carry dominant lethal mutations so that the zygotes formed die early in development.

The main objective of irradiation is to sterilize insects without greatly affecting their ability to live, mate, and carry on otherwise normal life functions. Because species vary in the radiation dose required to achieve this objective, considerable laboratory study must be conducted with various doses on different insect stages to establish the appropriate procedures for sterilization.

In many instances, the pupal stage is the most appropriate for irradiation. Usually being less active than larvae and adults, pupae are simple to handle and can be sterilized with relatively low radiation doses after most adult structures are formed. Doses required for sterility in adults are often about the same as those in older pupae, but adults are usually more difficult to manipulate. Both eggs and larvae are radiosensitive, and the radiation levels required to induce

---

BOX 14.3 FRUIT FLY, POMACE FLY, OR VINEGAR FLY

SPECIES: *Drosophila melanogaster* (Meigen) (Diptera: Drosophilidae)

DISTRIBUTION: The species is found throughout the world, in both urban and rural areas.

IMPORTANCE: *Drosophila melanogaster* is the most extensively studied insect, and one of the best genetically understood organisms in the world. The species is particularly attractive to geneticists because it has very large chromosomes in the nuclei of larval salivary gland cells, allowing researchers to gain information regarding gene mapping, chromosome structure, and gene expression. The flies are ideal laboratory subjects because they have short life cycles (10 to 14 days), are easy to rear, and mutations occur often in relatively small populations. Thomas Hunt Morgan, the first geneticist to receive the Nobel prize, experimented with the species in the early part of the twentieth century, revealing detailed information on the genetics of *Drosophila* embryology.

APPEARANCE: Adults are yellow to brown, and 3 to 4 mm long. Males have three bands on the top of their abdomen, whereas females have five bands. Larvae (maggots) are white and about 5 mm long when full-grown. Eggs are about 0.5 mm long, with two projections at one end which prevent the egg from sinking into the decaying fruit. Puparia are brown.

LIFE HISTORY: Females oviposit in decaying fruit or fermenting vegetable matter. After about two days, eggs hatch, and the larvae feed for about four to five days (three stadia). Larvae then move out of the food material and attach themselves to a dry surface before pupating. Adults emerge in about three days. Generation time is only 10 to 14 days.

sterility often cause premature death or abnormal adults. In some organisms, it has been shown that higher levels of radiation are required to sterilize females than males. Sterilizing doses also vary greatly among insect species but overall are much greater than those causing sterility in mammals.

Chemosterilization. The ability of some chemicals to sterilize insects as well as some of the drawbacks of this method have already been mentioned. When applied before the onset of meiosis, chemosterilants prevent gamete production. Knowing this, we are not surprised to learn that such chemicals prevent all types of cell reproduction and have received much development in cancer therapy (chemotherapy). Indeed, chemotherapy research probably gave impetus to the suggestions of chemical use in insect sterilization.

Chemosterilants can be divided into four basic groups. These include the alkylating agents, phosphorus amides, triazines, and antimetabolites. Presently, the alkylating agents represent the largest class of chemosterilants, and their effects are similar to those of X-rays and gamma rays. These agents cause multiple dominant lethal mutations or severely injured genetic material in the sperm or egg.

Bisulfan (Alkylating Agent)

1,4-butanediol dimethane-sulfonate

Alkylating agents are unstable in the environment and degrade rapidly. Possible contamination of food and water with even small residues, however, makes crop applications presently unfeasible. Safe applications are possible only under laboratory conditions, where these materials may be used to sterilize insects in sterile-release programs. Such an approach has been attempted with the boll weevil, *Anthonomus grandis*, by incorporating busulfan, an alkylating agent, into the diet of mass-reared beetles and releasing these into test areas. Although these attempts have met with only moderate success, a newer chemosterilant, penfluron (not an alkylating agent), has shown some promise.

It is unfortunate that chemosterilants are dangerous to humans and wildlife. Most present compounds, particularly the alkylating agents, are strongly carcinogenic or mutagenic and thereby create undue risk if applied over broad areas. Therefore, we would not expect them to be registered for widespread use. But the goal of achieving safe chemosterilants, although formidable, is a worthy one.

Sterile-insect Release Programs

Much effort has been directed toward developing the sterile-insect release technique for practical use. The USDA has been particularly responsible for many studies and pilot programs that have advanced the concept, and the agency has been the primary force behind large-scale implementations.

A development from these areawide programs has been the so-called concept of **total population management** (TPM). The TPM approach attempts to use all available means to eradicate a pest over a broad area. This objective is not new, but until the development of the sterile-male technique, it had been considered nearly impossible.

Pest eradication has been realized in only a few instances where sterile-insect releases were made, but successful degrees of suppression have been achieved in several other instances. Some of the most important examples of the technique follow.

Screwworm eradication and suppression.   One of the most dramatic successes with the sterile-insect technique was achieved with eradication of the screwworm fly from the Florida peninsula. As a parasite of livestock, this fly lays up to 400 eggs in wounds, and developing larvae feed on tissues and enlarge the open wound. Such feeding attracts still other flies to oviposit, and death of the parasitized animal may occur in as few as ten days. Before the sterile-male-release program, livestock losses were estimated at $20 million annually in the southeastern United States and $50 to $100 million in the southwestern states. During a particularly severe outbreak in 1935, 1.2 million cases of infestation and 180,000 livestock deaths were reported in the United States.

Use of the sterile-male technique against this insect was conceived in 1937. After two years of research, often with negative results, field trials on the island of Sanibel (off the west coast of Florida) gave encouraging results. This test prompted a larger pilot program in 1954 on the larger, more isolated island of Curacao, off the coast of Venezuela. Here, 400 sterilized males per square mile per week were released over a period of about three and a half months (four to five generations). The program resulted in eradication from the island and demonstrated the potential of the technique.

An even larger program, financed jointly by the USDA and the livestock industry in the Southeast, was implemented in 1958. Here, up to 50 million sterile flies of both sexes were produced each week, and more than 2 billion were released over an 18-month period. The area involved 85,000 square miles, including Florida and parts of Georgia and Alabama. In this program, more than

Figure 14.1 Wound of a calf infested with screwworm, *Cochliomyia hominivorax*. (Courtesy USDA)

40 tons of ground meat were required each week to rear the flies, and 20 aircraft were used to release them.

This campaign resulted in complete eradication of screwworm populations from Florida. Subsequent recurrences of the pest from importation of infested animals have been quickly eradicated by treatment of the animals and additional sterile-fly releases. The cost of the Florida program was estimated at $10 million, but it has been credited with saving southeastern livestock producers many times that amount.

With eradication achieved in Florida, attention was turned to screwworm infestations in the southern parts of Texas, New Mexico, Arizona, and California. In these instances, flies were migrating northward into the United States from Mexico and producing spotty but significant infestations. To eliminate this migration, the strategy of creating a **fly-free barrier zone** was advanced. Subsequently, a fly-rearing plant was established in Mission, Texas, capable of producing over 150 million flies per week. Beginning in February 1962, sterile flies were released along the border between Texas and Mexico. In 1964 the releases were extended to include the Mexican border of Arizona and California. These latter releases created a fly-free zone 2,000 miles long by 300 to 500 miles deep and served as a "barrier" to northward movement into the United States. Using this strategy, fly infestation reports in the United States dropped from over 50,000 in 1962 to about 150 by 1970.

Unfortunately, the screwworm populations did not remain low. Infestation frequency began a slow increase in 1971 and reached an alarming 95,642 reports by 1972, despite continued fly releases. Release rates were increased as much as fivefold, but infestations remained at an unacceptable level. Later, it was

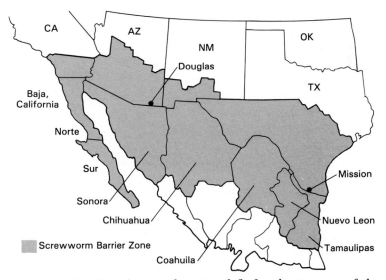

Figure 14.2 Map showing the original fly-free barrier zone of the Southwestern Screwworm Eradication Program in 1962. Mission, Texas, was the original location of the fly-production facility. Both Mission and Douglas, Arizona, were packaging and dispersal centers. Northern Mexico frontier states were Baja California, Sonora, Chihuahua, Coahnila, and Tamaulipas. (Redrawn from Krafsur, et al., 1987, *Parasitology Today,* volume 3: 131-137, with permission)

Figure 14.3 An insect container used in aerial release of sterilized screwworms. (Courtesy USDA)

hypothesized that new strains of laboratory flies were inadvertently produced that were less active and competitive for mates than feral flies. Following modified procedures and repeated efforts, the effectiveness of the program was renewed by 1977.

In 1981 the Mission, Texas, rearing facility was closed and replaced completely by a more efficient plant in Tuxtla Gutierrez, Mexico, opened in 1976. The Mexican plant is capable of producing up to 500 million flies per week.

Along with increased rearing capabilities, strategies were developed to gradually push the fly-free barrier zone south to the Tehuantepec Isthmus, where a shorter barrier (only 140 miles) was necessary. This objective was established by a joint Mexico–United States agreement in 1972, and the program was initiated in 1976. Among the new tactics used in this program have been the application of poison baits (pellets of the attractant swormlure-2 and the insecticide dichlorvos), followed by releases of new-strain competitive flies into the reduced populations.

Case reports since this joint program was initiated indicate that the original barrier zone (southwestern United States and northern Mexico) became effective in 1979, and eradication south to the Isthmus of Tehuantepec was declared in 1984. In 1986, declining reports of cases from continued releases indicate that Mexico should nearly be free of screwworm.

Figure 14.4 Laboratory-confirmed screwworm reports, with data logarithmically transformed [$\log_{10}$ (cases $+$ 1)]. Reports from Texas (top), Arizona and New Mexico (center), and northern Mexico (bottom). The arrow in the northern Mexico graph indicates an episode of sabotage where unirradiated flies were believed deliberately released. The dotted line shows the trend without this episode. No native screwworms have been detected in the northern Mexico zone since August 10, 1985, but two cases were intercepted in San Luis Potosi in November 1985 in animals transported from southeastern Mexico. (Redrawn from Krafsur, et al., 1987, *Parasitology Today*, volume 3: 131-137, with permission)

Figure 14.5 Mass-rearing screwworms, in large vats. The diet is a mixture of ground meat, citrated beef blood, and water, which is piped into the vats. Racks containing the vats move along a monorail. (Courtesy USDA)

Figure 14.6 When full-grown, mass-reared screwworm larvae migrate from vats to a trough. Subsequently, they fall through a funnel and grating into a water channel under the flooring. The water transports them to a separator and dryer. (Courtesy USDA)

Figure 14.7 Accumulations of full-grown screwworm larvae are placed in trays of sawdust where they pupate. Trays are emptied onto a machine that screens out sawdust; pupae and remaining larvae are then placed on a moving belt. Pupae remain on the belt for packaging, and larvae are driven off by lights into a collecting trough at the edge of the belt to be returned to a sawdust tray for pupation. (Courtesy USDA)

Figure 14.8 The 1972 Mexico–United States Screwworm Eradication Agreement set 93°W longitude as the fly-free barrier objective to be attained. The eastern limit of sterile fly dispersal was 91°W. Primary packaging and dispersal centers were Tampico, Guadalajara, and Tuxtla Gutierrez. Tuxtla is the only center used presently. (Redrawn from Krafsur, et al., 1987, *Parasitology Today*, volume 3: 131-137, with permission)

Figure 14.9 Map of Central America showing feasible goals based on current sterile fly production levels from the plant in Tuxtla. The proposed "Darien barrier" (stipled area in eastern Panama) would require only about 50 million flies per week to maintain compared to the present 150 million flies per week. (Redrawn from Krafsur, et al., 1987, *Parasitology Today*, volume 3: 131-137, with permission)

Tropical fruit fly programs. On a worldwide basis, tropical fruit flies (Diptera: Tephritidae) are among the most destructive pests of fruits. Some of the most significant are the following: Mediterranean fruit fly (*Ceratitis capitata*), olive fruit fly (*Dacus oleae*), Mexican fruit fly (*Anastrepha ludens*), melon fly (*Dacus cucurbitae*), oriental fruit fly (*Dacus dorsalis*), and Queensland fruit fly (*Dacus tryoni*). Since the 1950s, much research has been concentrated on the use of sterile-male releases against these species, and a great deal of the basic technology has been established. Moreover, programs have been implemented against some of these species outside the United States. In these programs, isolated populations of the melon fly and oriental fruit fly have been eradicated, and nonisolated populations of the Queensland fruit fly have been suppressed.

Of the tropical and subtropical fruit flies, introductions of the Mediterranean and Mexican fruit flies have been of greatest concern in the United States. Here, sterile-male releases have been practiced to help prevent introductions and eliminate early infestations of both species. With the particularly difficult Mediterranean fruit fly problem in California, sterile-male releases have been combined with several other tactics. In a recent campaign, fruit was stripped from trees and fallen fruit picked up. This was followed by weekly aerial spraying of the infested area using bait sprays and ground spraying of pest-infested soil with diazinon. Afterward, approximately 100 sterile males for each wild male were released in strategic locations. Using these techniques and quarantine programs, the seventh eradication campaign in California was successfully concluded in 1982.

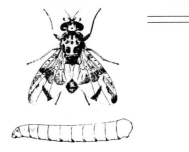

BOX 14.4 MEDITERRANEAN FRUIT FLY

SPECIES: *Ceratitis capitata* (Wiedemann) (Diptera: Tephritidae)

DISTRIBUTION: The Mediterranean fruit fly is native to central Africa and, as a result of transport by humans, it is now found in practically all the subtropical regions of the world, except North America.

IMPORTANCE: The Mediterranean fruit fly, popularly known as the Medfly, is among the most destructive pests of over 100 species of fruit, including orange, peach, grapefruit, plum, apple, and pear. In subtropical areas, larvae (maggots) feed on the pulp of the fruits, causing the development of infection courts for fruit diseases. Also, holes made by female flies during oviposition scar the fruits, which lowers quality. In 1929, the pest was found in Florida and eradicated at a cost of $7 million. In 1956, also in Florida, the species was detected and subsequently eradicated at a cost of $10 million. In 1975, the pest was discovered in southern California and eradicated. In 1980, the pest was again found in southern California, and from 1980 to 1982, an intensive, much publicized program undertaken to eliminate the pest, resulting in over $100 million in eradication costs.

APPEARANCE: Adults similar in size to house flies, have a yellow and black thorax and a yellow abdomen with two silver crossbands. The wings are transparent, with bands of yellow, brown, and black. Larvae, about 10 mm long when full-grown, are white maggots. Puparia are brown.

LIFE HISTORY: Adults or puparia overwinter in cool regions, whereas in warmer areas, the species is active year-round. Female flies, which may produce up to 600 eggs, oviposit 2 to 10 eggs in holes under the skins of fruits. Eggs hatch in 2 to 20 days, and larvae feed in the fruits for about one week. After completion of development, larvae fall to the ground and pupate in soil for approximately ten days. Generation time is variable (three weeks to three months), depending on climatic conditions.

Other insects. In addition to the screwworm and fruit fly programs, sterile-insect releases have been studied for an array of other insects. Some of these include boll weevil, tsetse flies (*Glossina* species), horn fly (*Haematobia irritans*), mosquitoes, codling moth (*Cydia pomonella*), and pink bollworm (*Pectinophora gossypiella*). Of these, the pink bollworm program stands out as being one of the most successful.

As mentioned earlier, the pink bollworm is a major pest of cotton in the southwestern United States. The pest spread from Texas into New Mexico, Arizona, southern California, and northwest Mexico during the mid-1960s and threatened to invade major cotton-growing areas of the San Joaquin valley in California. Beginning in 1967, sterile moths have been released into the valley at a rate of about 100 million per year, which has produced an estimated sterile-male to invading-male ratio exceeding 1,000 : 1 in early season. Such a ratio has resulted in an extremely low probability of successful matings and is believed responsible for preventing establishment of pink bollworm in the area.

BOX 14.5 HORN FLY

SPECIES: *Haematobia irritans* (Linnaeus) (Diptera: Muscidae)

DISTRIBUTION: The horn fly is native to Europe and was introduced into the United States (New Jersey) about 1887. Currently, the species is distributed throughout all of North America.

IMPORTANCE: The horn fly is a common cattle pest throughout its range. The flies receive their common name from their resting behavior, which is typically on the horns of cattle. Several thousand individuals can be found on a single cow, piercing the skin and feeding on the blood. Each fly takes many blood meals both day and night; indeed, it is not uncommon for females to feed up to 38 times per day, with each feeding session lasting up to four minutes. Cattle are irritated and annoyed, which causes significant reductions in both weight gain and milk production.

APPEARANCE: Adults are 4 mm long and dark gray. Larvae (maggots) are white and about 10 mm long when full-grown. Eggs are oval, about 1.2 mm long, and reddish brown. Puparia are barrel-shaped, about 3.3 mm long, and brown.

LIFE HISTORY: Pupae overwinter in soil, and adults emerge in early spring. Females leave the cattle for a short period of time to oviposit 1 to 14 eggs in fresh dung. Females can oviposit from 350 to 400 eggs. After one day, the eggs hatch and larvae feed in dung for four to five days. Pupation occurs in puparia in the soil. After five to seven days, adults emerge. Each generation is about 14 days long, and there are many generations per year.

Requirements and Limitations of Sterile-insect Programs

One of the foremost requirements for success with this approach is to have capabilities for mass-rearing the insect economically. Even with low numbers in the natural population, most programs call for millions of insects to be produced and released weekly over a period of several weeks. This capability sometimes necessitates development of a synthetic diet on which healthy and environmentally competitive insects can be reared. To understand net reproductive rates and low density points, sampling programs that will yield precise absolute estimates also are required. In addition, the approach is applicable only when releases of sterile insects do not produce major injury or cause an undue nuisance. For instance, the innundation of insects capable of transmitting arboviruses or plant pathogens in an area probably would be unacceptable.

Experience in suppressing or eradicating populations with the sterile-insect technique is still too limited to determine which species would be highly susceptible to the approach. Even for existing large-scale programs, like those for the screwworm and fruit flies, results are not yet fully understood. A great deal of ecological, production, and operational research will be required before suitability of the sterile-insect technique can be adequately predicted. However, the benefits of research in this area extend to other management tactics and so would not be wasted effort even in instances where the approach proves impractical.

BOX 14.6 PINK BOLLWORM

SPECIES: *Pectinophora gossypiella* (Saunders) (Lepidoptera: Gelechiidae)

DISTRIBUTION: The pink bollworm is native to India, and because of shipments of cotton-seed, is now found in North America. The species, moving north from Mexico, was first detected in the United States by 1917 in Texas. Currently, it is distributed in the states of California, Nevada, Arizona, New Mexico, Oklahoma, Missouri, Arkansas, and Louisiana.

IMPORTANCE: The species is a serious cotton pest throughout the world, and in some areas, it can cause total crop destruction. Early in the growing season, larvae feed in the squares, attacking developing flower structures. Usually, this damage is not severe. Later in the growing season, however, larvae feed in the bolls on lint, carpel tissues, and seeds, which causes reductions in both yield and quality. Also, larval feeding provides infection courts for pathogens like boll rot. Despite quarantines, the species continues to expand its range.

APPEARANCE: Adults have a 12 mm wingspan and are grayish brown, with inconsistent markings. Young larvae are white, with a brown head, whereas older larvae (fourth instars) display a distinctive pink coloration. Full-grown larvae are approximately 13 mm long. Eggs are oval, about 0.5 mm long, and white. Before hatching, they become red. Pupae are brown.

LIFE HISTORY: Larvae overwinter in hollowed-out cottonseeds or in plant debris in the field. In early spring, larvae pupate and in about ten days, adults emerge. Adults are active at night and feed on nectar. Females oviposit up to 200 eggs on the host plant, near the bolls. After five days, the eggs hatch, and developing larvae undergo three or four molts (10 to 14 days). Larvae then leave the boll or remain in damaged seed to pupate. Generation time is approximately 25 to 30 days, and there may be up to six generations per year.

OTHER GENETIC TACTICS

Manipulations of insect genetics for human benefit have been practiced for hundreds of years, since the domestication of the silkworm, *Bombyx mori*, and semidomestication of the honey bee, *Apis mellifera*. Superior strains, with increased production and efficiency, have been obtained through selection and hybridization, and such breeding activities are expected to show even greater progress in the future.

The manipulation of insect genetics for suppression, however, is quite a different idea. This activity, described earlier as part of genetic control, includes any kind of artificial manipulation of insect gene composition to reduce population numbers. It is genetics used in self-destruction of populations as opposed to, for instance, plant protection.

These genetic tactics for autocidal control are tentative at present; they have not been applied. However, several genetic processes are understood, and their use has been suggested. Fundamentally, the proposal is to alter genetic processes in such a way as to make insects less fecund, less vigorous, or altogether sterile; the effect would be to decrease population fitness. The objective would be suppression or complete eradication of the species in a large area.

The use of genetic principles to improve the fitness of insect natural enemies is another area that has engendered a great deal of interest recently. Much of the emphasis here is in the development of beneficial insects (and other arthropods) that are resistant to pesticides. This aspect of the genetic approach is discussed in more detail in Chapter 15.

Several properties make insects amenable to autocide through genetic manipulations. One of the foremost properties is genetic plasticity. Most species maintain many forms throughout their geographical range, each representing a different genotype; potentially, these are critical resources in selection and breeding programs. Additionally, insects have relatively short life cycles and high reproductive potentials, which enhances breeding programs and genetic experimentation. Moreover, technologies developed for rearing large numbers of some species may lead to practical implementation of programs in the field.

Research to achieve programs of genetic control has been concerned with several specific processes. The most important of these include conditional lethal mutations, inherited sterility, hybrid sterility, cytoplasmic incompatibility, chromosomal rearrangements, and meiotic drive mechanisms.

Conditional Lethal Mutations

With this process researchers have attempted to breed insects that are less fit than normal for certain kinds of environmental conditions. The approach relies on specific alleles of genes that determine fitness of individuals for factors like temperature tolerance. These alleles result in inherited traits that do not allow survival in all conditions encountered by the insect. For example, susceptibility to cold winter temperatures could be a conditionally lethal trait in southern populations of an insect species. The trait has no effect on individuals of these populations in their normal habitat. However, when individuals bearing the trait are released into the northern part of the species range, they are killed.

Of the conditionally lethal traits, temperature sensitivity has received much of the experimental attention. Theoretically, male insects with a dominant, homozygous, cold-sensitivity trait could be reared and released into the environment. There, they would mate with feral females, which, in turn, would produce progeny with the trait. The progeny then would be killed by normally low winter temperatures.

Sensitivity to low temperatures is often associated with obligatory dormancy (diapause). In insects with wide geographic ranges, both diapausing and nondiapausing genotypes may be found. Releasing nondiapausing genotypes in populations of obligatory diapausing genotypes could produce populations of progeny that would not diapause and would be killed by winter extremes. Such a tactic has been suggested for the European corn borer, *Ostrinia nubilalis*, in the United States and crickets, *Teleogryllus commodus* and *Teleogryllus oceanicus*, in Australia, but only limited studies have been conducted.

Inherited Sterility

Inherited sterility, also called delayed sterility, has been suggested as a method to substantially increase sterility ratios in populations, as compared with the conventional sterile-insect release technique. Particularly, it has been recommended for Lepidoptera because these insects usually require enormous dosages of irradiation to achieve high levels of dominant lethality, and such high levels impair competitiveness of the released insects.

In theory, if a 9:1 ratio of sterile to fertile males were established by releases in a conventional program, this would result in a 90 percent reduction of fertile crosses. If, on the other hand, the sterility produced by the release was not expressed until the F_1 generation, it would be enhanced by 9 percent. This is because a 9:1 original release would create a 9:1 ratio in both males and females of the F_1 generation. When F_1 individuals cross, the probabilities of all possible outcomes would be as follows:

$$0.9 \text{ sterile male} \times 0.9 \text{ sterile female} = 0.81 \text{ sterile matings}$$

$$0.9 \text{ sterile male} \times 0.1 \text{ fertile female} = 0.09 \text{ sterile matings}$$

$$0.1 \text{ fertile male} \times 0.9 \text{ sterile female} = 0.09 \text{ sterile matings}$$

$$0.1 \text{ fertile male} \times 0.1 \text{ fertile female} = 0.01 \text{ fertile matings}$$

Therefore, genetic death can be increased from 90 percent to as much as 99 percent by delaying it until the next generation. Such an approach has been recommended for continued investigation.

Hybrid Sterility

Hybrid sterility has been observed in laboratory studies with closely related species. Similar to horse-by-donkey crosses that result in sterile mules, this idea has been suggested as a method of obtaining sterile insects for release.

Particularly notable in this regard have been studies of hybridization between the tobacco budworm, *Heliothis virescens*, and a related species, *Heliothis subflexa*. These species cross quite readily in the laboratory, producing partially sterile F_1 offspring. Most important in this phenomenon is that the hybrid males are sterile when they mate with normal *H. virescens* and *H. subflexa* females. Although technical problems exist, this approach has been suggested for use in sterile-insect releases against populations of the tobacco budworm.

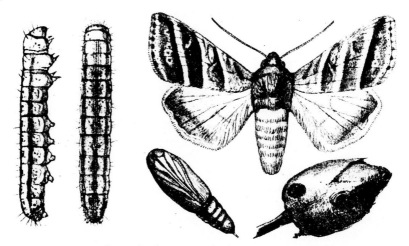

Figure 14.10 Tobacco budworm, *Heliothis virescens*, larva (side and top aspect left), adult (top center), pupa, and injured tobacco bud. (Courtesy USDA)

Cytoplasmic Incompatibility

This phenomenon occurs when individuals from different populations are crossed and reproductive potentials are reduced. The reduction is derived because of incompatibility factors in the cytoplasm causing sterility in individual eggs. Here, sterility results when a sperm enters an egg and stimulates meiosis, but it does not fuse with the egg pronucleus to form a zygote.

The principle of cytoplasmic incompatibility has been studied extensively with a mosquito, *Culex fatigans*. Programs for suppression of the mosquito would require the release of only males, because releasing both sexes simply would result in replacing one strain with another. Moreover, because females bite and transmit pathogens, their release would be unacceptable. Unfortunately, sorting out large numbers of male mosquitoes from a population for release presents many technical problems. Research into "sex-killing systems," as accomplished with house flies and other Diptera, has offered a possible solution to this problem.

Chromosomal Rearrangements

This technique selects and breeds insects with certain genetic defects or chromosome translocations or rearrangements for use in release programs. The genetically rearranged males mate with feral females and produce partially sterile progeny. The technique has also been suggested as a means to "transport" economically advantageous genes to pest populations; advantageous genes would confer traits in the pests that are helpful to humans. Some desirable traits might include insecticide susceptibility, avoidance of a crop plant, and inability to tolerate normal temperature extremes.

Chromosomal translocations have been intensively studied in a number of insect pests, particularly mosquitoes. Other pests with research emphasis include the house fly, cabbage looper, *Trichoplusia ni*, and onion maggot, *Delia antiqua*. This listing is, of course, only a partial one.

Meiotic Drive Mechanisms

Meiotic drive refers to the unequal recovery of homologous chromosomes during meiosis. Here, a conditionally lethal gene introduced into the population increases in frequency when population numbers decline. An example of the mechanism can be made with the XY system of sex determination present in many insect species. In this system XX chromosome-pairing produces females, and XY produces males. If a Y-linked mutation is induced, only Y sperm would be produced, rather than a 1:1 ratio of X:Y sperm. When mutant males, carrying only Y sperm, mate with normal females, only sons are produced. A distortion of the male-to-female sex ratio would result, favoring males, and the population would decline. As it declines the drive-mechanism effect increases, because there is an accompanying increase of mutant Y chromosomes in the population. Such a population should eventually become extinct. Although meiotic drive has been induced in laboratory house flies with substerilizing doses of gamma radiation, some researchers have raised serious questions as to its practicality.

Replacement by Innocuous Forms

Up to this point, the techniques mentioned for genetic control have emphasized establishing genetic loads on insect populations that result in numerical reduc-

BOX 14.7 CABBAGE LOOPER

SPECIES: *Trichoplusia ni* (Hübner) (Lepidoptera: Noctuidae)

DISTRIBUTION: The cabbage looper is a native North American species, common to all of Canada, Mexico, and the United States.

IMPORTANCE: The cabbage looper is a common pest of many plant species, such as cabbage, cauliflower, celery, lettuce, broccoli, tomato, tobacco, potato, bean, and turnip. Larvae defoliate host plants, causing both reduced yield and quality. Under heavy infestations, 25 to 35 larvae can infest a single cabbage head. Larvae move in a characteristic looping fashion from which the species gets its common name. Adults migrate into Canada, occasionally causing outbreaks.

APPEARANCE: Adults have brown mottled forewings and brown hindwings, with a wingspan of 40 mm. A distinctive silver marking is displayed on each forewing. Larvae, about 38 mm long when full-grown, are green, with white lines running lengthwise along the body. Eggs are greenish white and dome-shaped. Pupae are brown and enclosed in a white silken cocoon.

LIFE HISTORY: Pupae overwinter attached to leaf debris. Adults emerge in early spring, and females oviposit 275 to 350 eggs, singly on leaf surfaces. Larvae feed about 14 to 28 days, then pupate. After about ten days, adults emerge. There are three or more generations per year, depending on climatic conditions.

---

tions and, in some instances, extinction. A conceivable alternative to this autocidal approach would be replacement of the pest by forms that are not harmful. In this instance, genetically altered replacements could be mass-reared and released with the expressed purpose of lowering pest status through changes in species characteristics.

One of the most appropriate subjects of this potential approach are nonvectoring strains of insects that normally vector pathogenic microorganisms of humans. Recently, a strain of the mosquito, *Anopheles gambiae*, has been discovered that is refractory (immune) to *Plasmodium* species, malaria-causing pathogens. After ingesting infected blood, this mosquito strain is capable of encapsulating the infectious stage of the microorganism, which causes its death. Therefore, this strain is unable to transmit the infection. Production of a fully refractory strain has been achieved by selective breeding, encouraging further research on potential releases in replacement campaigns. If natural populations could be diluted or completely replaced by breeding with the released refractory strain, it is conceivable that malaria incidence could be reduced significantly. However, annoyance from biting mosquitoes would remain the same.

Although replacement by innocuous forms has not been accomplished at the present time, the approach holds promise for alleviating certain types of pest problems.

BOX 14.8 ANOPHELES MOSQUITOES

SPECIES: numerous *Anopheles* species (Diptera: Culicidae)

DISTRIBUTION: *Anopheles* mosquitoes are distributed throughout the tropic regions of the world, and some species occur in temperate regions. There are about 390 species worldwide, with 90 in the western hemisphere, and 15 in North America.

IMPORTANCE: The genus *Anopheles* vectors malaria, the most significant human parasitic disease of the tropics and subtropics. These pests have been largely responsible for deterring economic development in the tropical areas of the world. Presently, over 200 million humans are infected with malaria per year. Malaria is currently a significant disease in Africa, Asia, and Central and South America; however, in the past, the disease was distributed throughout the world. The disease is caused by a protozoan in the genus *Plasmodium*. The life cycle of malaria is complex, but basically, the mosquito ingests the sexual forms (gametocytes) of the *Plasmodium* when taking a blood-meal from a human. Then the malaria forms spores (sporozoites) in the mosquito body, which pass to the salivary glands. When the mosquito feeds again, the sporozoites are introduced into human blood. In humans, the malaria attains various forms (merozoites and trophozoites), causing fever and chill cycles. Gametocytes are then produced, which are ingested by the mosquito, and the cycle is repeated. There are several types or species of malaria that vary in severity to humans.

APPEARANCE: Adults are frail flies, with long maxillary palps and spotted wings. Males have plumose antennae. Adults rest with their bodies angled to the surface on which they are resting. Larvae (wrigglers) lie parallel to the water surface and lack a breathing tube, which is a common feature in other genera of Culicidae. Pupae (tumblers) lie at the surface of the water.

LIFE HISTORY: Where year-round activity does not occur, inseminated female adults overwinter; however, some species overwinter as larvae in standing water in tree holes. Females take a blood-meal and oviposit from 150 to 300 eggs, singly, on still water. Each female can oviposit more than 1,000 eggs. Adult males do not feed on blood, and both sexes maintain themselves on nectar and plant juices. Eggs hatch in two to six days, and larvae feed on algae for about 14 days. Larvae then pupate in the water, and in about three days, adults emerge. Typically, generation time is from 21 to 30 days.

CONCLUSIONS

Using sterile insects and genetic techniques to achieve autocide or replacement in insect populations is a highly specialized tactic. Objectives of the approach are to reduce population levels or cause complete eradication over a broad area. Because the area-wide procedures involved require a combined effort, the tactics are usually beyond the realm of decision making by individual producers. More likely, such programs will be initiated and conducted by federal and state agencies, in cooperation with various interest groups. In agriculture, economic benefits of such area-wide programs to individuals is substantial, because they represent tax-based subsidies to crop production.

Except where chemosterilants are considered, sterile-insect and genetic tactics would seem to present few personal or environmental hazards. Moreover, if eradication is achieved with the approach, continuing management inputs are reduced significantly, with only surveillance programs and occasional corrective actions being necessary. Yet other advantages of the approach may be high returns for expenditures and applicability with low-level infestations.

Unfortunately, many factors limit the practicality of the method, and progress toward development has been slow. Some of the most significant obstacles to development include the absence of economical mass-rearing technology for certain species and inability to produce fully competitive sterile insects. Many of the newer genetic tactics may allow the latter limitation to be overcome. However, much basic research is needed to make the newer approaches operational, and this research will require substantial expenditures. Therefore, unless priorities change, we would expect autocidal control to play an important but somewhat limited role in insect-pest management.

Selected References

COLLINS, F.H. , R.K. SAKAI, K.D. VERNICK, S. PASKEWITZ, D.C. SEELEY, L.H. MILLER, W.E. COLLINS, C.C. CAMBELL, and R.W. GWADZ. 1986. Genetic selection of a *Plasmodium*-refractory strain of the malaria vector *Anopheles gambiae*. *Science* 234: 607–609.

DAVIDSON, G. 1974. *Genetic Control of Insect Pests*. New York: Academic Press.

HOY, A., and J.J. MCKELVEY, JR., eds. 1979. *Genetics in Relation to Insect Management*. Rockefeller Foundation Conference, March 31 to April 5, 1978, Bellagio, Italy. New York: Rockefeller Foundation.

KNIPLING, E.F. 1979. *The Basic Principles of Insect Population Suppression and Management*. United States Department of Agriculture Handbook 512, chapter 10.

——— 1982. Present status and future trends of the SIT approach to the control of arthropod pests. In *Sterile Insect Technique and Radiation in Insect Control*. Proceedings International Symposium on the Sterile Insect Technique and Use of Radiation in Genetic Insect Control, June 29 to July 3, 1981. Vienna, Austria: International Atomic Energy Agency, pp. 3–23.

KRAFSUR, E.S., C.J. WHITTEN, and J.E. NOVY. 1987. Screwworm eradication in North and Central America. *Parasitology Today* 3:131–137.

LABRECQUE, G.C. , and C.N. SMITH, eds. 1968. *Principles of Insect Chemosterilization*. New York: Appleton-Century-Crofts.

LORIMER, N. 1981. Genetic means for controlling agricultural insects. In D. Pimentel, ed., *CRC Handbook of Pest Management in Agriculture*. Boca Raton, Fla. : CRC Press, vol. 2, pp. 299–305.

METCALF, R.L., and R.A. METCALF. 1982. Attractants, repellents, and genetic control in pest management. In R.L. METCALF, and W.H. LUCKMANN, eds., *Introduction to Insect Pest Management*, 2nd ed., New York: Wiley, pp. 299–314.

NATIONAL ACADEMY OF SCIENCES. 1969. *Insect Pest Management and Control*. *Principles of Plant and Animal Pest Control*, vol. 3. U.S. National Academy of Sciences Publication 1695, chapters 9 and 15.

PAL, R., and M.J. WHITTEN, eds. 1974. *The Use of Genetics in Insect Control*. New York: American Elsevier.

WARE, G.W. 1983. *Pesticides—Theory and Application*. San Francisco: Freeman, pp. 225–228.

WHITTEN, C.J. 1982. The sterile insect technique in the control of the screwworm. In *Sterile Insect Technique and Radiation in Insect Control*. Proceedings International Symposium on the Sterile Insect Technique and Use of Radiation in Genetic Insect Control, June 29 to July 3, 1981. Vienna, Austria: International Atomic Energy Agency, pp. 79–84.

15 ECOLOGICAL BACKLASH AND ITS MANAGEMENT

THE OBJECTIVE OF APPLYING pest management tactics is to lower pest status by reducing pest numbers and/or injury potential. Unfortunately this objective is not always achieved, and if achieved, the effect may be all too brief. Some common reasons for failure are selecting inappropriate or incompatible tactics and improper application of tactics. Even without these problems, failure to achieve lasting, economical suppression may occur because of ecological backlash to a tactic or combination of tactics.

Ecological backlash involves the counter-responses of pest populations or other biotic factors in the environment that diminish the effectiveness of management tactics. Many of these responses result from heavy mortality burdens placed on species, burdens not unlike those that nature levies. Others arise from disruption of ecological processes or changes in resource levels of the biotic community, causing readjustment in community structure. Such readjustments are counter-productive to the purposes of agriculture and medicine.

An important feature of ecological backlash is that it is usually delayed. Therefore, we may apply one or more management tactics with seemingly good results, encouraging us to continue with the same management program. But the program gradually—or perhaps abruptly—loses effectiveness and may even become totally useless.

Ecological backlash has occurred with shocking frequency in the history of pest control. Yet many growers and other agricultural specialists seem surprised and are usually caught off guard when it occurs. Agriculturists with knowledge of the phenomenon and contingency plans to deal with it may avoid significant economic penalties if it occurs. Moreover, understanding this potential problem can aid in the design of management programs that are durable, because such problems are not inevitable.

An understanding of ecological backlash rests on a knowledge of the major sources of the phenomenon. These sources can be called the three "R"s of pest management awareness. They are population resistance, resurgence, and replacement. All three are ecologically based population phenomena.

RESISTANCE OF POPULATIONS TO PEST MANAGEMENT TACTICS

Because of its pervasiveness, resistance of populations to pest management tactics is the most important of the three Rs. It is a genetic-evolutionary phenomenon most frequently associated with the use of pesticides. However, the phenomenon has been demonstrated with almost every pest management tactic and, theoretically, it is possible with any factor that causes high mortality in populations.

Resistance to a pest management tactic occurs when there is a change in the susceptibility of the population to the tactic. For example, resistance may occur after exposure of several generations to an insecticide, applied at a given rate that renders the insecticide ineffective in subsequent applications at that dose. Formally, **resistance** is the ability of certain individuals to tolerate or avoid factors that would prove lethal or reproductively degrading to the majority of individuals in a normal population.

Principles of Resistance

Resistance in a population can be explained by the same principles that explain evolution by natural selection. Briefly, every biological population possesses genotypes with differential abilities to survive and reproduce in the environment. A given set of environmental conditions favors genotypes with traits that are best adapted to the situation. Although several genotypes usually are present in a population at any one time, the best-adapted genotype assumes numerical dominance. As environmental factors change, other genotypes may be favored that subsequently displace the dominant ones. Natural selection, sometimes called survival of the fittest or Darwinian selection, is a selection of the fittest genotypes for a given set of environmental conditions. This selection has led to physiological and morphological changes in biological species and has allowed their persistence for long periods of (geological) time, despite unusual and sometimes violent changes in the environment.

Insects are some of the most adaptable animals on earth. Their adaptability has allowed them to occupy nearly every terrestrial habitat and has resulted in an unequaled diversity of species. Indeed, proof of this adaptability is seen in the species of today, which have existed for millennia and survived all sorts of natural disasters.

Just as nature selects for well-adapted genotypes, so do we when we apply pest management tactics. We do this unwittingly by imposing burdensome mortality or reproductive antagonisms on the pest population. Consequently, the population responds as it would to natural adversities; it adapts to overcome the burden and survive. This is something most insect populations do very well.

In other words, resistance is preadaptive. It is inherited from parents and never acquired through habituation during the lifetime of an individual (i.e., it is not postadaptive). For instance, it is not possible to produce resistance within a single generation by exposing insects to sublethal doses of an insecticide.

Resistance to pest management tactics is difficult to predict. As a general rule, however, the greater the population burden, the greater the rate of development. In other words, a pest management tactic that causes greatest mortalities tends to be effective for a shorter time because susceptible genotypes are eliminated quickly, and only resistant genotypes are left in the population. Once resistance is fully expressed, continued application of a pest management tactic has no economically beneficial effect on a population.

The rate of resistance development also depends on genetics of the resistance factor. In most documented cases, resistance originates with mutations occurring regularly in populations. These mutations result in new genotypes, some of which are predisposed to resist adverse factors. If the character required for resistance

(A)

(B)

(C)

Susceptible individual

Resistant individual

Figure 15.1 Schematic drawing of development of resistance in a pest population. A. Some individuals in a pest population are genetically adapted to survive aplications of a pest management tactic. B. Some of the survivors' offspring inherit resistance traits and will survive other applications. C. If the same tactic is continued, the pest population consists mostly of resistant individuals. (Redrawn from University of California Integrated Pest Management Publication 3303, *Integrated Pest Management for Citrus*)

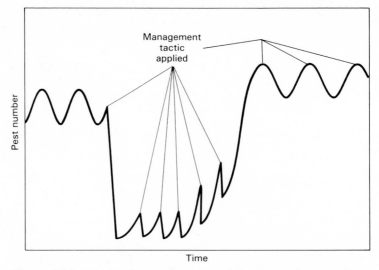

Figure 15.2 Graphical representation of insect numbers as they would respond during the development of resistance to a pest management tactic.

can be obtained through expression of a single gene (**monogenic resistance**), resistance may occur after only a few generations. For instance, monogenic resistance to some insecticides has occurred after only a few years of use (house flies, *Musca domestica*, resistant to DDT). However, if many genes are required (**polygenic resistance**), development may be much slower.

Resistance to Conventional Insecticides

Resistance to insecticides is the preeminent form of resistance in insects. The prevalence of this phenomenon reflects the widespread effectiveness of these compounds in causing pest mortality.

Magnitude of the insecticide resistance problem. The first instance of insecticide resistance was noted in 1908 in Washington State with the San Jose scale, *Quadraspidiotus perniciosus*, and lime-sulfur sprays. This case was followed by infrequently reported incidents of resistance until the development of DDT insecticide.

In 1946, house flies were discovered resistant to DDT in Sweden. Through the early 1950s, the golden period of insecticides, resistance was still rare, and most insect populations were fully susceptible to insecticides. Since then, however, the number of instances has grown exponentially. In 1986, 447 arthropod species have been proven resistant to one or more insecticides of the major groups, including DDT, cyclodienes, organophosphates, carbamates, and pyrethroids.

Of these resistant species, 59 percent are of agricultural importance, 38 percent of medical or veterinary importance, and 3 percent are natural enemies. Of the insect groups, Diptera have evolved the greatest number of resistant species (35 percent), which mainly reflects the intense insecticidal pressure on mosquitoes. Indeed, insecticide resistance is the major problem in control of arthropod-borne diseases, particularly malaria, in the world today.

Figure 15.3 Historical development of pesticide resistance worldwide. (After Georghiou, 1986, *Pesticide Resistance: Strategies and Tactics for Management*, courtesy National Academy of Sciences)

Nearly half the total resistant species resist compounds in more than one insecticide group, and at least 17 can resist chemicals in all five groups, which shows an additional dimension of the problem. Moreover, pesticide resistance is not limited to the arthropods; there are at least 100 species of plant pathogens, 55 species of weeds, 2 species of nematodes, and 5 species of rodents also resistant to one or more pesticides formerly effective in their control.

Mechanisms of resistance to insecticides. Throughout their evolutionary history, insects have had to deal with a plethora of naturally occurring environmental toxicants. To survive in the face of these natural hazards, species have evolved a variety of mechanisms to make the materials innocuous to them. Just as these mechanisms have reduced the toxicity of natural substances, they have also reduced the toxicity of human-contrived chemicals. The result is what we call insecticide resistance. As C. F. Wilkinson (1983) stated, insects were

Table 15.1. Numbers of Arthopod Species Resistant to Insecticides in 1984

| Order | \textit{Insecticide Group} | | | | | | |
	Cyclodiene	DDT	Organophosphate	Carbamate	Pyrethroid	Fumigant	Other
Diptera	108	107	62	11	10	—	1
Lepidoptera	41	41	34	14	10	—	2
Coleoptera	57	24	26	9	4	8	5
Acarina	16	18	45	13	2	—	27
Homoptera	15	14	30	13	5	3	1
Hemiptera	16	8	6	1	—	—	—
Other	23	21	9	3	1	—	2
Total	276	233	212	64	32	11	38
Percent	62	52	47	14	7	2	9

SOURCE: Data from G.P. Georghiou, University of California-Riverside. In Lafarge (1985), with permission.

"forewarned and forearmed to meet the challenges presented them" by modern synthetic organic insecticides. Insect species deal with toxic chemicals through three major mechanisms: biochemical resistance, physiological resistance, and behavioral resistance. All three forms may occur simultaneously to produce the fittest animals.

Biochemical resistance is probably the most common type of resistance in insects. With this form, an insecticide usually is attacked by one or more enzymes that detoxify it before it can reach its site of action. This metabolic process is often accomplished in two stages with high concentrations of **mixed-function oxidases** and other enzymes that produce primary products. Although these primary products may be excreted directly, most often they undergo a secondary metabolism, forming water-soluble conjugates (covalent addition of sugar, amino acids, sulfates, phosphates, and other materials) that, in turn, are excreted.

Physiological resistance is any form of resistance that reduces toxicity through changes in basic physiology. With this mechanism, the chemical is not broken

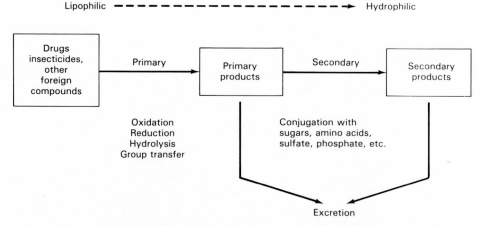

Figure 15.4 Metabolism of lipophilic foreign compounds, including insecticides, in insects. (Reprinted with permission of Plenum Press from *Pest Resistance to Pesticides* ed. by Georghiou and Saito, Copyright ©1986)

down into a less toxic form by the insect. Rather, it is accommodated by one or more physiological functions. One of the most important of these involves alterations at the site of insecticide activity. For instance, knockdown resistance in house flies to DDT and pyrethroids is believed caused by reductions in numbers of target-site receptors, making nerve sheaths less sensitive to the toxicants (a quantitative factor based on numbers of receptors). In other species, cholinesterases may be altered so that they are not affected by organophosphates and carbamates. Physiological resistance also may include decreased penetration of the insecticide through the body wall. This resistance usually is conferred through modifications in composition or structure of the cuticle (for example, additional waxy layers). Other physiological mechanisms include increased rates of excretion of the insecticide and sequestering the chemical. In the latter instance, lipophilic insecticides like DDT may be stored in body fat and are thereby prevented from reaching their site of action.

Behavioral resistance involves changes in behavior by which insects avoid insecticides. One of the best-known examples of behavioral resistance occurs with mosquitoes that vector malaria-causing plasmodia, *Anopheles gambiae* and other mosquitoes in Africa. In these species, an endophilic strain, inhabiting human structures, was susceptible to sprays of DDT applied to walls indoors. An exophilic strain not inhabiting buildings became dominant because its behavior allowed it to avoid exposure to the insecticide.

Cross-resistance. When insecticide resistance occurs in a species, pest managers are faced with the difficult problem of how to deal with it. Historically, they have turned to insecticides in other classes to which the species is still susceptible. An example of this strategy and its outcome can be seen with the Colorado potato beetle, *Leptinotarsa decemlineata*, on potatoes in Long Island, New York, where growers have used a single-factor insecticide strategy since the late 1800s. One insecticide after another produced resistance in the population beginning with the arsenicals in the 1940s, followed by the chlorinated hydrocarbons, organophosphates, and carbamates through the 1970s, and culminating in resistance to the pyrethroids in the 1980s. Today, this pest is one of the most insecticide-resistant insects in North America.

Changing to an insecticide in another class when resistance occurs is frequently necessitated by cross-resistance. **Cross-resistance** implies that an insect with resistance to one insecticide is able to resist other insecticides. Therefore, insects that have become resistant to DDT also have cross-resistance to methoxychlor, and those resistant to dieldrin have cross-resistance to heptachlor and toxaphene.

This within-class cross-resistance has led to categories of resistance according to insecticide class. Some of the major classes mentioned in the literature include the following: (1) DDT resistance, (2) dieldrin or cyclodiene resistance, (3) organophosphate resistance, (4) carbamate resistance, and (5) pyrethroid resistance.

Although most instances of cross-resistance have been observed within each of these classes, it may also exist between classes. For instance, house flies with knockdown resistance to DDT have cross-resistance to all DDT analogs, pyrethrins, and pyrethroids. Likewise, flies with resistance to organophosphates may have cross-resistance to carbamates.

Cross-resistance usually occurs because of similar toxicity modes of the chemicals involved and the inheritance of a resistance gene for that mode. Therefore from the example above, both DDT and the pyrethroids are axonic poisons (they

Table 15.2. Colorado Potato Beetle Resistance to Insecticides in Long Island, New York

Insecticide	Year Introduced	Year First Detected
Arsenicals	1880	1940s
DDT	1945	1952
Dieldrin	1954	1957
Endrin	1957	1958
Carbaryl	1959	1963
Azinphosmethyl	1959	1964
Monocrotophos	1973	1973
Phosmet	1973	1973
Phorate	1973	1974
Disulfoton	1973	1974
Carbofuran	1974	1976
Oxamyl	1978	1978
Fenvalerate	1979	1981
Permethrin	1979	1981
Fenvalerate + piperonyl butoxide	1982	1983
Rotenone + piperonyl butoxide	1984	?

SOURCE: Data of A.J. Forgash, New Jersey Agricultural Experiment Station, Cook College, Rutgers University. In Lafarge (1985), with permission.

upset impulse transmission along the nerve axon), and in house flies, cross-resistance to axonic poisons is inherited from the so-called *kdr* gene. Similarly, house flies, which also are cross-resistant to organophosphates and carbamates through altered cholinesterases, receive this trait through inheritance of the *AChE-R* gene. Other identified genes that function in resistance and cross-

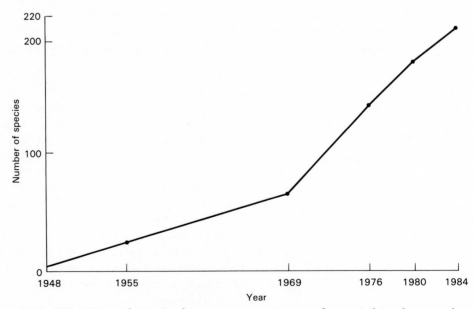

Figure 15.5 Historical growth of species resistant to more than one class of insecticides. (Based on data from Georghiou, 1986, *Pesticide Resistance: Strategies and Tactics for Management,* National Academy of Sciences)

resistance include *pen* (for decreased-penetration resistance) and *dld-r* (for dieldrin and cyclodiene resistance). All these genes except *AChE-R* are incompletely or fully recessive, but *AChE-R* and others are codominant.

Dangers and costs of the resistance phenomenon.   The problem of insecticide resistance is acute. Although the rate of growth in resistant species has declined recently, this does not mean that the problem is lessening. Rather, it means that most major pests have already become resistant to at least one insecticide group and that multiple resistance (resistance to more than one group) is increasing. In other words, an increasingly larger proportion of new records of resistance involve species that are already resistant to at least one insecticide group.

Therefore, we face a dilemma: the more often we use insecticides, the quicker they will become useless. They sort of "wear out." The responses of users to early signs of resistance are quite predictable. When an insecticide begins to lose effectiveness, a first response is to increase dosage and frequency of application. These increases enhance the rate of resistance, and when the compound is no longer effective, it is discarded for another insecticide, probably from another chemical group. The repetition of this cycle and the ever increasing need for more poisons has been called the **pesticide treadmill** (much activity with little forward motion).

Although switching to other groups has been a somewhat effective tactic in the past, such a policy cannot be sustained in the long run. One reason is that cross-resistance produces a shortened effective life of each succeeding insecticide. For example, DDT resistance provides a foundation for development of pyrethroid resistance, and the pyrethroid may be effective for a shorter time than would otherwise have been expected.

With this loss of insecticide life-span and greater costs of registration, chemical companies are increasingly reluctant to develop new insecticides. Indeed, since 1970 the introduction of new insecticides declined precipitously; only numbers of pyrethroids have increased. Pyrethroids have replaced many organophosphates and carbamates since about 1976; however, over 15 species already show resistance to these newer compounds.

Useful insecticides are a valuable resource, and monetary loss can be attributed to their demise. Chemical companies may spend between $30 million and $40 million in developing a new compound, only to see resistance diminish returns, perhaps to the point of not recouping original investments. In addition, growers spend millions of dollars for extra chemicals and additional treatments when resistance occurs.

Although this growing problem has stimulated work on new and unusual compounds (for example, insect growth regulators), most experts agree there is a limit to the number of new chemical groups that can be discovered and developed as practical alternatives to present materials. An end to the insecticide treadmill seems to be in sight, but only because it is being forced on us by resistance. The question is, will we reduce the resistance problem through proper insecticide use, or will we absorb the horrendous production losses caused by "super" pests with which we cannot deal? The least costly path is insecticide conservation through proper use.

Resistance to Other Pest Management Tactics
Although resistance to insecticides is the form most often mentioned, resistance certainly is not restricted to insecticides. Indeed, insects, having survived

so well for so long, eventually may overcome any pest management tactic we employ. Moreover, the principles of selection to explain insecticide resistance apply equally well to these other forms. As was true with insecticides, the more widespread and destructive the management tactic, the more likely these other forms of resistance will occur.

Resistance to insect growth regulators. Although originally believed refractory to resistance problems, several insect growth regulators (IGRs) recently have been added to the long list of insecticides to which insects are resistant. At first, resistance to juvenile hormones and their mimics seemed unlikely because insects would be required to overcome the effects of molecules similar to those already present in their bodies. However, we can understand resistance more readily when we recognize that insects regulate their own hormones by metabolism and that exogenous molecules are usually sensed as foreign materials subject to deactivation processes.

Cross-resistance in various insecticide-resistant strains has been a major factor in IGR resistance. Presently, at least 13 insect species, representing the Diptera, Coleoptera, Homoptera, and Lepidoptera show cross-resistance to IGRs. Included in the list of common IGR compounds are methoprene, hydroprene, kinoprene, and diflubenzuron. Although resistance mechanisms are still being investigated, it seems as if most IGR resistance results from reduced penetration and increased metabolism of the compounds.

Resistance to microbial insecticides. Insect resistance to pathogenic microorganisms is yet another concern. Although not presently widespread, several instances of such resistance have been reported. Resistance to the most widely used microbial insecticide, *Bacillus thuringiensis* (Bt), has been observed in experiments with the Indianmeal moth, *Plodia interpunctella*, and the house fly. Resistance in the Indianmeal moth, a major pest of stored grain, increased nearly 100 times over 15 generations when larvae were fed diets containing the pathogen. House flies exposed to beta-exotoxin-producing Bt, causing 50 to 90 percent mortality each generation, developed a thirteenfold resistance level in 50 generations.

Increased levels of resistance have also been observed in Lepidoptera and Hymenoptera populations that suffered viral epidemics. Pests with increased resistance to viruses include the spruce budworm, *Choristoneura fumiferana*, and sawflies, among others. From a beneficial standpoint, silkworms, *Bombyx*

Figure 15.6 Indianmeal moth, *Plodia interpunctella*, adult (top center), pupa (left) and larva. Resistance to applications of the microbial pesticide *Bacillus thuringiensis* has been observed with this stored-grain pest. (Courtesy USDA)

mori, also are capable of virus resistance, and strains have been purposefully selected for this trait.

Resistance to parasites. Insect resistance to internal parasites is possible, but has not been commonly reported. An instance of resistance to an introduced ichneumonid parasitoid, *Mesoleius tenthredinis,* has been recorded with larch sawfly, *Pristiphora erichsonii,* populations in central and eastern North America. In laboratory studies, parasite resistance has been induced in house flies to a pteromalid parasitoid, *Nasonia vitripennis,* and in an anopheline mosquito to a nematode parasite, *Diximermis peterseni.*

Virulence to resistant plants. The development of virulent biotypes that over-come the defenses of normally resistant plants was discussed previously. The occurrence of this form of resistance (resistance to resistant plants) has been of concern to plant breeders and growers for many years, although it seems most probable with parthenogenetic insect species. To date, most insect biotypes have developed among aphids, although one of the best-known examples of the phe-nomenon occurs in the Hessian fly, *Mayetiola destructor,* a dipteran.

Resistance to plant rotations. This is a very unusual form of resistance, but it exemplifies well the persistent capability of insects. As mentioned earlier, a genotype of the northern corn rootworm, *Diabrotica barberi,* has been selected that shows resistance to a two-year rotation of corn and soybeans (or small grain) in the north central United States. The resistant genotype displays a so-called extended diapause, a state of dormancy in the egg stage that lasts for two winters rather than one. The presence of the extended-diapause genotype was reported in 1965, but problematic resistance was not a factor until about 1985, when the incidence of damage to corn grown in rotation became widespread. The adaptive feature of this trait is that eggs laid in cornfields one season could survive in a nonhost habitat the alternate season and subsequently hatch in the presence of the proper host the following season. Studies conducted in the problem region showed that only 10 percent of the rootworm eggs from continuously grown cornfields underwent extended diapause, versus nearly 50 percent in cornfields from a rotation scheme.

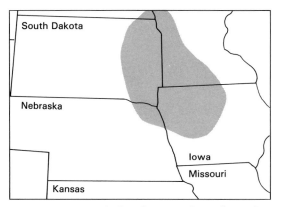

Figure 15.7 Estimated distribution of a subpopulation of the northern corn rootworm, *Diabrotica barberi,* showing extended diapause (two-year life cycle). This species normally has a one-year life cycle.

This reccurring resistance problem means that previously effective two-year rotation schemes are becoming less practical for rootworm management. Where problems exist, growers will need to consider a three-year rotation cycle (one year in corn and two years in a nonhost crop) or apply preventive soil insecticides in the management scheme (the latter alternative is not considered a good pest management practice). Two-year rotations used against this pest in other regions of the corn belt are still effective.

Resistance to sterile-male releases. Certain outbreaks of the screwworm, *Cochliomyia hominivorax*, in the southwestern United States and Mexico during the early 1970s were claimed by some workers to have been caused by replacement types not reproductively compatible with released sterile insects. R.H. Richardson and colleagues reported that at least nine screwworm types existed (five in the United States), with type intermating being low or nonexistent. According to their hypothesis, after the screwworm eradication program was begun in the Southwest, the dominant types were gradually suppressed and replaced by incompatible types. The remaining pest population in certain areas then became a mixture of types, with previously rare types being the more common. Subsequent releases of sterile males then became less successful, with performance being dependent on an adequate match of laboratory flies with the dominant type (or types) in a given area. Thus, mismatches were said to explain program failures in local areas. Richardson and coworkers suggested that future recurrences of these outbreaks could be reduced by prevention of reinfestations in the United States from Mexico and by facilitation of rapid, appropriate responses (with properly matched releases) when outbreaks occur.

Although resistance to sterile-insect releases is believed possible, the likelihood of screwworm outbreaks seems to be contradicted by the successful eradication of the pest from the United States and Mexico.

Resistance to pheromones. Insect resistance to pheromones or pheromone mimics has not yet been recorded. However, these compounds have not been used against insects on a widespread or intensive scale. Most experts agree that, if used widely and successfully for insect suppression, resistance is a distinct possibility.

Management of Resistance

Because of its pervasiveness, insecticide resistance has received the most attention with regard to management approaches, and we emphasize it here. However, it should be noted that many principles of insecticide management used to slow resistance development apply generally to other types of resistance.

Conditions that promote resistance. The development of resistance management is undergirded by an understanding of the causes of resistance. As mentioned, the basic cause is an intensive mortality (or reduced reproduction) burden placed on an insect population, resulting in an inadvertent selection for individuals capable of overcoming the burden. Resistance can be expected to occur against any tactic that imposes a significant population burden, the greatest question being how quickly it will develop. Therefore, resistance management is mostly a matter of anticipating and slowing the rate of development.

Factors that influence the rate of insecticide resistance development have been discussed by several authors and summarized by T. C. Sparks and colleagues. A partial list of the factors involved in rapid development is as follows:

Operational Factors

1. The insect has a prolonged exposure to a single insecticide, or the insecticide is used in a slow-release form.
2. Every generation of the insect is selected.
3. Insecticide selection pressure is high (mortality is great).
4. No functional refugia exist; that is, coverage by the insecticide is effectively complete so that no part of the population remains unselected.
5. A large geographical area is covered; all populations in a given area are likely to have been treated.
6. Selection occurs prior to mating.
7. The insecticide is closely related to one used earlier.
8. A low population threshold (economic threshold) is recommended for application of the insecticide.
9. The insecticide is inherently irritating and/or repellent.

Biological Factors

1. No (or little) migration occurs between populations.
2. The species has a monophagous habit.
3. The species has a relatively short generation time.
4. Numerous offspring per generation are produced.
5. The species is highly mobile, increasing the probability of exposure.

Of course, not all these factors are required to produce resistance, but the greater the number present in a situation, the greater the chance of occurrence and the more rapidly will resistance develop.

Slowing the development of resistance. The prevention of resistance to any effective pest management tactic is practically impossible in many situations. However, its rate of development can be slowed by considering the operational factors that enhance it and modifying the pest management program accordingly.

The most basic resistance management routine is use of combined tactics to achieve suppression. By integrating ecological tactics, natural enemy suppression, and resistant plants with chemical insecticides, undue reliance is not placed on any one tactic. Multiple tactics place diverse pressures on the pest population, making it more difficult for the species to overcome the effects of any one tactic. Moreover, if resistance does develop to one tactic in the integrated scheme (such as insecticide resistance), its effects will be lessened because other tactics will still contribute to suppression. Therefore, tactic diversification is the key to insect pest management.

Another possible approach to slow resistance is to employ passive tactics in the management program, when possible. Passive tactics place no known burden on pest populations. These tactics include such measures as irrigation and fertilization to produce thrifty plants, which can better tolerate insect injury. The tactics also involve developing plant varieties capable of acceptable yields despite pest injury (i.e., growing tolerant varieties). When passive tactics are used, selective pressures are lessened and resistance development is less likely. Indeed, passive tactics provide the most enduring management solutions. Unfortunately, these tactics are not always practical, particularly when high-value, blemish-free produce is desired.

Perhaps a more practical approach to slowing resistance, particularly with insecticides, is to modify use patterns. Several methods have been suggested and employed that disrupt or significantly delay resistance in pest populations. According to G.P. Georghiou (1983), these methods, which he calls "chemical strategies of resistance management," can be summarized as follows:

Management by Moderation

Objective: reduce selection pressure and conserve susceptible genes in the population.

Advantage: helps conserve environmental quality and natural enemies.

Limitation: less practical with high-value crops and medical pests.

1. Use low dosages, sparing a proportion of susceptible genotypes.
2. Use less frequent applications.
3. Use chemicals of brief environmental persistence.
4. Avoid slow-release formulations.
5. Apply selection against adults after reproduction, allowing susceptible genes to be passed on.
6. Make local, rather than areawide, applications. The use of trap crops fits here.
7. Leave some generations or population segments untreated.
8. Preserve refugia (untreated pockets for population segments). This relates to point 7 and is sometimes accomplished by treating only "hot spots" in a field or directing nozzles toward only part of a plant canopy.
9. Utilize higher economic thresholds for insecticide application. This relates to point 2 because it should result in fewer applications.

Management by Saturation

Objective: saturate insect defense mechanisms by doses that can overcome resistance.

Advantage: applicable to high-value crops like blemish-free apples and vegetables and medical pests.

Limitations: possible adverse impact on environment.

1. Render the resistance gene "functionally" recessive by applying higher dosages on target. This tactic is based on computer simulations and laboratory tests that have shown rapid resistance development when the resistance gene (R gene) is dominant. Resistance develops slowly when the R gene is recessive. Because there are few R genes in the population initially, resistance develops quickly with a high proportion of heterozygotes (hybrids with the R gene and the susceptible gene). Applying very high dosages of insecticides kills both susceptible individuals (ss) and heterozygotes (Rs), keeping the frequency of the R gene low. This tactic is not believed practical if some selection has already begun and there is a large proportion of heterozygotes already present in the population.
2. Suppress detoxification mechanisms with the use of synergists. Synergists function by inhibiting specific detoxification enzymes. For example the common synergist, piperonyl butoxide, is an oxidase inhibitor and is commonly added to insecticides to enhance action with low to moderate levels of resistance. Although synergists may eliminate the selective advantage of individuals with detoxifying enzymes, some insects like house flies have been known to develop resistance to insecticide-plus-synergist mixtures.

Management by Multiple Attack

Objective: reduce selection pressure by imposing several independently acting forces, such that pressure from any one is below the level to achieve resistance.

This is similar to using several different management tactics as discussed earlier; only here, several different insecticides are involved.

Advantage: potentially useful against pests of high-value crops and medical pests.

Limitations: imposes environmental risks, destructive of natural enemies, and risk of "super" resistance (resistance to several compounds at once).

1. Use insecticide mixtures. The idea of using mixtures to slow resistance assumes that the mechanisms of resistance are different for each member chemical, and furthermore, that all the mechanisms required to resist the mixture are not present in any one individual. In other words, insects developing resistance to one compound would succumb to another in the mixture. Success with this approach is enhanced when the insecticides mixed are synergistic and have similar decay rates and short environmental stability. Also, mixtures should be applied early before resistance occurs to any one of the components. Although not many definitive studies have been conducted with mixtures, it has been shown that when chemicals with contrasting modes of action of detoxification pathways are used, some delay in the onset of resistance occurs. An example of where a mixture delayed resistance in the laboratory involved a combination of temephos (an organophosphate), propoxur (a carbamate), and permethrin (a pyrethroid) against the mosquito *Culex quinquefasciatus*.

2. Apply insecticides in a mosaic pattern across an area. This is somewhat like a mixture, but a mixture in a spatial sense. Here, planned applications of unrelated compounds are applied in different parts of the pest population range to avoid selection of the whole population for the same resistance mechanism. With the approach, insects not killed by an insecticide treatment in one sector are killed by a different insecticide when they migrate to another sector. This approach is relatively new and has been proposed for use with mosquitoes, where opposite walls of a house would be treated with different chemicals, and with horn flies, *Haematobia irritans*, where different chemicals would be strategically applied to pasture-sized areas.

3. Apply insecticides in a rotation scheme. The idea of rotations of dissimilar insecticides to combat resistance is based on the assumption that individuals resistant to a chemical have substantially lower survival potential to natural environmental factors than susceptible individuals. Consequently, the frequency of resistant individuals would be expected to decline between the times the "resistant" chemical is used. In other words, the proportion of resistant individuals would increase during the season with a chemical but would return to its original low level by the time that chemical is used again. Insecticide rotations are presently receiving much attention as a management tool.

PEST POPULATION RESURGENCE AND REPLACEMENT

Resurgence and replacement are ecological backlash phenomena observed in a number of agricultural systems and other situations. Although evidence has shown that they are most often associated with conventional insecticides, the possibility of their occurrence exists with any management tactic. This is particularly true if the tactic is directly favorable to the physiology of the insect (for example, enhanced nutrition) or has an adverse effect on important natural enemies.

Dynamics of Resurgence and Replacement

Resurgence and replacement problems have been discussed by various authors and have been a particularly important consideration in biological control. Resurgence can be defined as a situation where a population, after having been suppressed, rebounds to numbers greater than before suppression occurred.

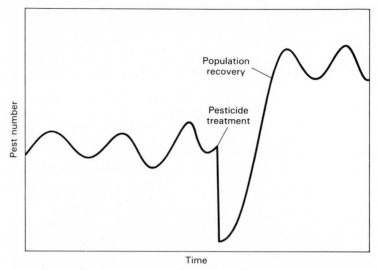

Figure 15.8 Diagrammatic representation of insect numbers as they would respond during pest resurgence. Note that numbers are higher after population recovery than before treatment.

Replacement, also frequently referred to as a secondary pest outbreak, occurs when a major pest is suppressed and continues to be suppressed by a tactic but is replaced by another pest, previously with minor status. In this instance, the primary pest is strongly affected by the tactic, but the secondary pest is not.

These so-called pest upsets and the bases for them were reviewed in detail by W.E. Ripper (1956) in his review of effects of pesticides on the balance of arthropod populations. Although dated, Ripper's review summarizes the impor-

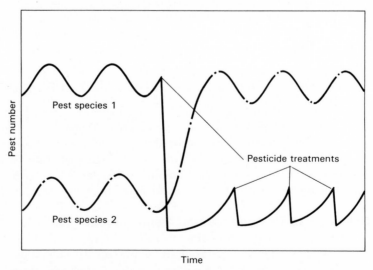

Figure 15.9 Diagrammatic representation of insect numbers of two pest species as they would respond during pest replacement. Here, species 2 was a subeconomic pest before replacing species 1.

tant principles of pest upsets that continue to be valid today. He extends three possible causes that should be examined when attempting to explain pest upsets.

1. Reduction of natural enemies by pesticides, along with the pest.
2. Direct favorable influences of pesticides on physiology and behavior of arthropods.
3. Removal of competitive species.

Upsets from Reduction of Natural Enemies

Presently, most experimental evidence supports the first cause listed. In many instances, reductions in numbers of arthropod natural enemies after insecticide applications has been believed the basis for the upset.

Susceptibility of arthropod natural enemies to insecticides. Insecticides may reduce natural enemies either by direct or indirect means. Mortality from direct exposure to an insecticide is perhaps the most obvious means of natural enemy reduction. As a group, predators and parasites have been shown to differ in their susceptibility. Records indicate that predators, particularly ladybird beetles (Coleoptera: Coccinellidae), may have greater tolerance to insecticides than parasitoids, and in some instances predators are no more susceptible than pests. Conversely, parasitoids have been found more susceptible than their hosts. From toxicological studies, it appears that insecticides detoxified oxidatively (several of the organophosphates like parathion) are much more toxic to natural enemies than to their phytophagous prey (hosts), because the former lack the detoxifying enzymes of the latter. On the other hand, some insecticide groups like the pyrethroids are believed generally less toxic to natural enemies than to pests.

Natural enemies can be suppressed by insecticides in a number of indirect ways. The more important of these for predators are (1) starvation because of host removal, (2) death from secondary poisoning after feeding on contaminated prey, and (3) repellency of the insecticide. Of these, the first is thought to be the most important.

In general, parasitoids are suppressed in some of the same indirect ways as predators. For internal parasitoids, insecticide effects are determined by host response. If the host is killed before development is completed, death by starvation is assured. If the host survives treatment, the insecticide effect may be mediated by the host's physiology. The host may detoxify the poison or possibly convert it to an even more toxic metabolite.

Paradigm for resurgence. A likely paradigm for pest resurgence could be described as follows. The pest population is fluctuating at an intolerable level for a grower. However, that level is still lower than its potential because of natural enemies. The grower sprays the crop and reduces both pest and natural enemy populations at the same time. As the effects of spraying wane, the pest population grows, but because the natural enemies must wait for an adequate food supply, their populations lag in regrowth. Unencumbered by natural enemies, pest population growth accelerates and exceeds previous levels. If left untreated, the natural enemies would finally return the pest to its original level; however, additional sprays are usually applied before this happens. Such continued activity results in a more or less permanent displacement of the natural enemy populations.

Examples of resurgence from natural enemy reduction. Most evidence for resurgence from natural-enemy destruction has been based on studies that show negative correlations of pest numbers with enemy numbers. In other words, a reduced natural-enemy-to-pest ratio is offered as proof of the cause of pest upset (this method of proof has been criticized by some workers as not really demonstrating cause and effect).

Accordingly, resurgence was studied in a springtail, the lucerne flea, *Sminthurus viridis*, in Australia, after applications of DDT in alfalfa plots. Here, pest

(A)

(B)

(C)

र्ऊ   Pest

गिर्ऊ   Natural enemy

Figure 15.10 A schematic drawing of development of pest resurgence caused by destruction of natural enemies. A. Both pests and natural enemies are killed by the pesticide application. B. Plant pests recolonize first. C. Plant pests, unrestrained by the natural enemy, increase to a level greater than in A.

reduction was observed for 12 days, followed by rapid growth of the population. On the other hand, populations of predaceous mites were reduced significantly for 23 to 45 days after treatment. This predator reduction was used to explain resurgence of the springtail population in treated plots to levels five times greater than in untreated plots.

In another study with the cyclamen mite, *Steneotarsonemus pallidus*, on strawberries in California, plots were kept free of a predator mite, *Typhlodromus reticulatus*, with treatments of a selective insecticide. Because the insecticide had little or no effect on the cyclamen mite, numbers in treated plots increased 15 to 35 times, whereas there was no significant increase in untreated plots. In greenhouse studies, hand-picking predator mites from plants gave similar results as removal by the insecticide, indicating that pest resurgence was not caused by any stimulatory effect of the compound.

Paradigm for pest replacement.   We now describe a possible paradigm for pest replacement. An agroecosystem contains a number of arthropod species that feed on the crop. Many of these species are present at low population levels and are held in check by the action of arthropod natural enemies. These so-called **secondary pests** go largely unnoticed by the grower. The grower applies insecticide to suppress the most destructive pest in the crop, the **key pest**, and continues applications to prevent further damage. In addition to killing the key pest, the pesticide is destructive of secondary-pest natural enemies but causes little mortality in secondary-pest populations. Lacking the natural enemy check, secondary pests quickly multiply and attain key pest status. In other words, one key pest is exchanged for one or more different pests, with no net gain from the insecticide program.

Examples of replacement from natural enemy reduction.   There are many examples of pest replacement in a variety of crops. One of the most infamous occurred in the mid-1960s with attempts to eradicate invading populations of the pink bollworm, *Pectinophora gossypiella*, in California cotton (Imperial Valley) with carbaryl sprays. In this program, destruction of natural enemies resulted in severe and costly outbreaks of the cotton leafperforator, *Bucculatrix thurberiella*, and to a lesser extent, certain spider mites.

Another common replacement problem has occurred in Michigan apple orchards, where spider mites (*Panonychus ulmi*, *Tetranychus urticae*, and *Aculus schlechtendali*) are under natural control by a complex of predators, including three predaceous mites (*Amblyseius fallacis*, *Agistemus fleshneri*, *Zetzellia mali*) and a ladybird beetle, *Stethorus punctum*. When frequent sprays of nonselective pesticides were made during mid- to late season against insects like the codling moth (*Cydia pomonella*) or the apple maggot (*Rhagoletis pomonella*), predators did not become established, and replacement with phytophagous spider mites occurred.

Favorable Effects of Pesticides on Arthropod Physiology and Behavior

In some instances, pest upsets cannot be explained on the basis of natural enemy destruction. Rather, they are caused by direct favorable influences on arthropod physiology and/or behavior. This phenomenon, called **hormoligosis**, occurs when

(A)

(B)

Pest 1

Pest 2

Natural enemy

Figure 15.11 A schematic drawing of development of pest replacement caused by destruction of natural enemies. A. Pesticide is applied to suppress pest 1 but also kills natural enemies that feed on pest 2. B. Released from regulation by natural enemies, pest 2 becomes a major problem. (Redrawn from University of California Integrated Pest Management Publication 3303, *Integrated Pest Management for Citrus*)

sublethal quantities of a stressful agent increase an organism's sensitivity and response to environmental factors. In several instances, this response has been increased reproduction. Examples of hormoligosis have been found in many arthropods, including the western corn rootworm (*Diabrotica virgifera*), twospotted spider mite (*Tetranychus urticae*), granary weevil (*Sitophilus granarius*), and elongate hemlock scale (*Fiorinia externa*).

Resurgence from sublethal sprays also has been observed experimentally with *Heliothis* species on cotton and the brown planthopper, *Nilaparvata lugens*, on rice. Resurgence of *Heliothis* species after applications of monocrotophos

BOX 15.1 TWOSPOTTED SPIDER MITE

SPECIES: *Tetranychus urticae* Koch (Acarina: Tetranychidae)

DISTRIBUTION: The twospotted spider mite is distributed throughout the world and is common on nearly all types of crops, such as vegetables, legumes, cotton, fruits, and ornamentals. The pest can be especially numerous in greenhouses.

IMPORTANCE: The twospotted spider mite is a problem on greenhouse crops and, under long term hot and dry environmental conditions, can cause significant damage to vegetable, field, and orchard crops. Before the advent of synthetic organic pesticides, such as DDT, the twospotted spider mite was not considered a serious pest. Pesticides elevated the pest status of this species by killing naturally occurring predators, and by eliminating the need for oil additives mixed with lead arsenate insecticides, which played a large role in keeping mite populations under control. The species damages plants by feeding on tissues on the undersides of the leaves. Chlorophyll from host tissue is removed, causing a decrease in photosynthesis and transpiration. Under light infestations, leaves become covered with pale white specks, which eventually turn necrotic. Under heavy infestations, leaves become covered with silk webbing produced by the mites. In addition, the leaves turn bronze, and then drop from the plant. Fruit, produced from the plants or trees infested by these mites, may be reduced in both size and quality. The pest is especially destructive on bean, cotton, strawberry, curcubits, and apples. Resistance to insecticides is common with these pests, particularly in greenhouse situations.

APPEARANCE: Adult mites, which are only 0.3 to 0.5 mm long, are greenish yellow with a distinctive pair of dark red spots on the back. These spots, from which the species gets its name, are actually internal food materials seen through the integument. This pest also has stages that include a six-legged larva, an eight-legged protonymph, and an eight-legged deutonymph. All these stages appear somewhat similar to that of the adult stage. The pale yellow eggs are attached to the undersides of leaves.

LIFE HISTORY: Adults overwinter as bright orange females under ground litter or in tree bark. After becoming active in early spring, females oviposit, and eggs hatch in three to ten days. The mites pass through larval, protonymph, and deutonymph stages before becoming adults. A typical life cycle takes from 7 to 21 days, and adult females may live up to nine weeks. There are numerous generations per year, with population peaks occurring in August and September.

(Azodrin®) and aldicarb (Temik®) was caused by increased ovipositional preference for treated plants. Here, it was hypothesized that the insecticides may have changed growth rate, fruiting rate, and physiology of the plants, making them more attractive. Resurgence in the brown planthopper following methyl parathion and decamethrin treatments also was believed caused by increased plant attractiveness, but reduction in length of nymphal stadia and longer adult ovipositional period also seemed to contribute to the phenomenon.

As evidence has accumulated, it appears that hormoligosis is much more important than was previously believed. Indeed, in many instances of resurgence, hormoligosis may play as great a role, if not more so, than the destruction of natural enemies. Particularly crucial to understanding the problem is more research into the effect of pesticides on the growth and phenology characteristics of plants and how these characteristics change survival and reproductive habits of arthropod pests.

Upsets from Removal of Competitors

If two or more species are competing in an area for the same requisite and one species is dominant, some workers believe that removal of the dominant will allow replacement by subordinate species, regardless of changes in natural enemy populations. Although suggested as an alternative cause when replacement occurs, experimental evidence to support this hypothesis is not very impressive. However, competition is a very difficult phenomenon to evaluate experimentally.

Managing Resurgence and Replacement

As with insecticide resistance, the basic principle in dealing with resurgence and replacement is to combine several management tactics and thereby reduce the need for insecticide application. In addition, the manner of application can be modified to take these problems into account. The fundamental objective of these modifications is to avoid hormoligosis and the destruction of natural enemies in the agroecosystem. Another approach is to make innoculative releases of natural enemies after insecticide treatments.

Avoiding hormoligosis. At present, less is known and, therefore, less has been written about avoiding hormoligosis. From a practical standpoint, a crucial step would involve eliminating sublethal doses where the problem has occurred or is expected. Sublethal doses can be minimized by applying insecticides on target and avoiding drift. In other words, following the rules of good pesticide practice, including proper chemical and formulation selection, can significantly lessen the problem. Other ways are yet to be developed.

Avoiding natural enemy destruction. Much has been written about reducing destruction of natural enemies to avoid pest upsets. Most of the approaches suggested are based on insecticide selectivity. There are many levels of selectivity, ranging from the ability of a poison to select between insects and warm-blooded animals to those that are selective between species of arthropods. It is the latter, more specific type of selectivity that is important in conserving natural enemies in ecosystems.

Insecticide selectivity can be achieved in one of two ways. One is to use selective chemicals that have a stronger depressive influence on the pest population than on its natural enemies. This form of selectivity has been termed physiological selectivity. Another way is to use otherwise nonselective chemicals in a selective manner. This way has been called ecological selectivity.

Physiological selectivity. Although much has been accomplished in medicine with the selective use of chemicals in therapy (for example, chemotherapy to cure cancer), little progress has been made with physiological selectivity in insect pest management. To date, there are few insecticides that will achieve the desired result of killing pests without also killing their natural enemies. Some of the more important examples of physiological selectivity are found with nonconventional materials like microbial insecticides and some insect growth regulators. For present conventional insecticides the list becomes very small indeed, with such selective materials as the aphicide pirimicarb (now discontinued in the United States), cryolite (Kryocide®) (a stomach poison), trichlorfon (Dylox®), and certain acaricides being noteworthy exceptions. Additionally, pyrethroids may offer a degree of selectivity with certain pest/natural-enemy complexes.

The reasons for the lack of development of selective insecticides are well known. The most important of these involves economics. Because these compounds are developed by private industry, profits become an overriding consideration. The present cost of development of a new pesticide is between $30 to $40 million, and up to $100 million or more may be needed for production facilities. These costs prevent most companies from developing and marketing a candidate compound unless potential sales are at least $40 to $50 million per year. Consequently, broad-spectrum compounds (nonselective among arthropods) are most desirable from an industry perspective, because these have market potentials necessary to cover costs of research, development, and manufacture. Without public subsidies and governmental policies to develop physiologically selective insecticides, we would not expect much future development of them.

Instead, specific selectivity will most often need to be devised through using broad-spectrum insecticides. In other words, we will need to use these materials as a dagger, rather than as a sickle.

Ecological selectivity. Several ecological approaches to achieve selectivity have been developed through the years. The major ones can be summarized as follows:

1. Monitor populations and follow recommended economic thresholds. Very often, objective determinations of treatment need will allow reductions in amount of insecticide applied, conserve natural enemies, and still provide optimal economic returns.
2. Treat with lowest rates possible. In some instances, high mortalities are not required to produce acceptable suppression. Actually, the pest needs only to be reduced to a level where surviving individuals will not cause any appreciable loss. This may call for mortality levels of 75 percent, rather than 95 percent. Leaving a residual pest population usually results in natural enemy conservation and provides hosts for population recovery. The greatest risks with low rates are weather factors such as rainfall that may reduce low-dosage treatment to unacceptable levels and sublethal doses that could cause hormoligosis. However, most often the potential benefits of reduced doses outweigh the risks.
3. Avoid treating broad areas. If trap crops can be treated or parts of a field sprayed for acceptable suppression, untreated areas can serve as refugia for natural enemies. Subsequently, these untreated reservoirs supply natural enemies to repopulate treated areas. This approach works best with relatively immobile pests like scales and highly mobile natural enemies like braconid parasitoids.
4. Time treatments to avoid heaviest natural enemy destruction. Sometimes treatments can be timed to avoid natural enemy activity periods and peaks. In the previous example with mite resurgence on apples, the most advantageous spray period was early season. In cotton, however, a delayed scheme is deemed best, with late application

resulting in better natural enemy conservation and fewer pest upsets. Understanding important natural enemies and their phenology is crucial in determining proper timing.

5. Use the most selective insecticide formulation. In some instances changing to another insecticide formulation can aid in achieving a degree of selectivity. For instance, granules applied to corn for management of second-generation European corn borer, *Ostrinia nubilalis*, concentrate in the whorl where larvae are most likely to be found. Therefore, because this formulation concentrates the insecticide at the target site, it exerts minimal effects on natural enemies on other parts of the plant. Baits are another selective formulation. These give selectivity by concentrating a lethal dose in certain areas and utilize pest activity to optimize contact.

Inoculative Releases of Natural Enemies

Another management approach that has been used to help prevent pest upsets is to repopulate an area with inoculative releases of natural enemies after insecticide application. This tactic involves applying an insecticide when pest populations cannot be restrained by natural enemies, realizing that the natural enemy population will be destroyed. After insecticide residues have subsided and pest populations are recovering, releases of insectary-reared natural enemies are made, which effect natural control once more. This approach is not unlike providing untreated refugia for recolonization, but it requires more overt management activity and is potentially more effective.

The idea of inoculative releases in biological control is not a new one. As discussed previously, natural enemy augmentation has been practiced for many years, being made most feasible by successful rearing programs that provide an inexpensive supply of common predators and parasitoids. Many of these natural enemies are available from commercial insectaries.

The newest, most unusual modification of the inoculative-release tactic is to release insecticide-resistant natural enemies. This tactic greatly improves the potential for integration of biological control with insecticides. It is based on the finding that some natural enemy populations have the ability to develop insecticide resistance with repeated exposure to the toxicant. Although the list of known resistant natural enemies is not yet long, it includes diverse groups, the most notable of which are the phytoseiid mites, *Metaseiulus species* (= *Typhlodromus*) and *Amblyseius*. Resistance has also been found in ladybird beetles, *Coleomegilla maculata*, braconid and aphelinid parasitoids, a common predator, the green lacewing *Crysoperla carnea*, and a few other species.

Figure 15.12 Ladybird beetle, *Coleomegilla maculata*, larva (left), pupa attached to leaf (center), and adult. Resistance to insecticides has been found in this predator. (Courtesy USDA)

Currently, predatory mites in the family Phytoseiidae are the only group where selection of resistance has been adequate for use in practical programs. Organophosphate-resistant genotypes of *Metaseilus occidentalis* and *Amblyseius fallacis* have been used in conjunction with certain organophosphates, like azinphosmethyl (Guthion®) and phosmet (Imidan®), for management of key pests (for example, codling moth, plum curculio, *Conotrachelus nenuphar*, apple maggot, *Ragoletis pomonella*, and others) in apples pears, plums, peaches, grapes, and hops. With these programs, substantial reductions of acaricides have been possible and the potential for pest replacements minimized. In California almonds, certain genotypes of *M. occidentalis* have developed resistance to organophosphates in the field and to carbamates and pyrethroids in the laboratory. These resistant genotypes have been released against mite pests on the crop; this has achieved biological control in the presence of sprays directed toward the navel orangeworm, *Amyelois transitella*.

OTHER FORMS OF ECOLOGICAL BACKLASH

Although the three Rs are the most obvious forms of ecological backlash, many other biological factors can respond to changes imposed by pest management tactics. Two other such factors are **enhanced microbial degradation**, which involves unusually rapid degradation of soil insecticides, and **upsets in community balance**.

Enhanced Microbial Degradation

The problem of enhanced microbial degradation applies to insecticides and herbicides placed on or in soil. Degradation of these compounds is a natural process affected by numerous chemical, physical, and biological factors that interact to modulate pesticide persistence and, therefore, efficacy against intended pests.

Various microorganisms, especially bacteria, are significant components of the degradation process. These microorganisms secrete enzymes that biochemically alter pesticides placed in contact with the soil. The altered compounds are inactive as pesticides and may serve as nutrients for growth of microbial populations. This degradation process is usually beneficial, because it prevents the long-term accumulation of pesticides in the environment. With normal microbial degradation, pesticides applied one season will not affect pests or cropping plans the following season. Indeed, relatively rapid degradation after killing action has been achieved has been a desirable characteristic in development of modern, biodegradable pesticides.

Problems arise, however, when the degradation rate is so rapid that the pesticide does not achieve its primary purpose—to kill the targeted pest population. In some instances, repeated application of the same chemical to the same soil encourages buildup of specific pesticide-degrading microorganisms to the point where succeeding applications have a progressively shorter active life. Continuing applications of the same chemical can eventually lead to complete loss of pesticide efficacy. Fundamentally, this is the description of enhanced microbial degradation, and the soils where this phenomenon occurs are called problem soils or aggressive soils.

Problem soils have been identified in the midwestern United States and elsewhere where continuous cropping and preventative pesticide treatments are common. Most evidence for enhanced microbial degradation of soil insecticides to date has accumulated on the carbamate insecticide carbofuran (Furadan®). This insecticide is applied in granular form at planting time as a preventive against the corn-rootworm complex, *Diabrotica* species, in fields continuously cropped with corn. The first failures of this previously effective insecticide began in the late 1970s, and the problem has worsened since then.

Recent laboratory and field studies on enhanced microbial degradation of other rootworm insecticides have indicated that the problem also can occur with organophosphates. Failure of isofenphos (Amaze®) to suppress the corn-rootworm complex in Iowa prompted in-depth studies into the cause. These studies showed that enhanced microbial degradation could occur quickly in some soils, and subsequently, the insecticide was withdrawn from the market.

Enhanced microbial degradation is a significant development that may become as important as insecticide resistance in some situations. Although objective guidelines for managing the problem are not developed at this time, tentative measures are similar to those for slowing the rate of insecticide resistance. Most recommendations call for rotating insecticides from one year to the next with those in different classes. In soils without degradation problems carbofuran might be applied one year, followed by terbufos (Counter®, an organophosphate) the following year, and so on. Of course, the rotation scheme may not be possible when problem soils have already developed. Here, switching to another class of compounds on a continuing basis may be the only option available. But there are no assurances that either procedure will provide a long-term solution.

The incorporation of microbial inhibitors into pesticide formulations and the use of slow-release formulations are two additional tactics employed to inhibit or delay enhanced degradation. Unfortunately, susceptibility of several available carbamate and organophosphate insecticides to enhanced microbial degradation seems to be encouraging the development of more persistent, and thus environmentally unacceptable, replacements.

Upsets in Community Balance

Basically, community balance exists when the elements in a biotic community (interacting species in an area) have reached a stable condition with regard to one another; that is, proportions of biomass (amount of living matter) for individual species have reached a status quo. Community upsets invariably occur when we change the status quo by applying management tactics. We have already discussed an example of this with the pest replacement problem.

Many times, changes in community structure may go unnoticed, particularly if there is no immediate deleterious effect on the management program. However, unplanned changes created by management tactics can sometimes produce belated problems. Such problems are exemplified by the successful eradication of the screwworm fly in Florida and Texas. Elimination of the screwworm was believed to have accounted for significant increases in white-tailed deer populations, which were partly regulated by the parasite. The white-tailed deer is also a good host for *Boophilus microplus*, a tick parasite of cattle. Some specialists believe that white-tailed deer increases because of screwworm eradication caused subsequent tick outbreaks in areas where the tick had previously been controlled.

Dealing with community upsets is very difficult, because the full consequences of community modification can rarely be predicted. But simply following good pest management practice would seem to be the most logical course.

CONCLUSIONS

The results of pest management tactics are varied and sometimes insidious. The main purpose in applying them is to reduce the status of pests to tolerable levels; even when recommendations are followed, however, results may be unexpected. These unexpected, often significant events—ecological backlash—can result from tactic-induced changes in the innate properties of the pest population and/or ecological modifications in the biotic community.

The ramifications of inept management and failure to recognize and deal with ecological backlash are costly, both from continued management and environmental quality perspectives. In general, using tactics in moderation, integrating several tactics, emphasizing selectivity, and accepting the presence of low pest numbers will reduce significant backlash. These are the fundamental principles of insect pest management. Sustainable dissolution of pest problems require their adoption.

Selected References

BRATTSTEN, L.B.; C.W. HOLYOKE, Jr., J.R. LEEPER, and K.F. RAFFA. 1986. Insecticide resistance: challenge to pest management and basic research. *Science* 231:1255–1260.

BYFORD, R.L., J.A. LOCKWOOD, and T.C. SPARKS. 1987. A novel resistance management strategy for horn flies (Diptera: Muscidae). *Journal Economic Entomology* 80:291–296.

CHELLIAH, S., and E.A. HEINRICHS. 1980. Factors affecting insecticide-induced resurgence of the brown planthopper, *Nilaparvata lugens*, on rice. *Environmental Entomology* 9: 773–777.

CROFT, B.A., and A.W.A. BROWN. 1975. Responses of arthropod natural enemies to insecticides. *Annual Review Entomology* 20:285–335.

CROFT, B.A., and K. STRICKLER. 1983. Natural enemy resistance to pesticides: documentation, characterization, theory and application. In G. P. GEORGHIOU, and T. SAITO, eds., *Pest Resistance to Pesticides*. New York: Plenum Press, pp 669–702.

FELSOT, A., J.V. MADDOX, and W. BRUCE. 1981. Enhanced microbial degradation of carbofuran in soils with histories of Furadan® use. *Bulletin Environmental Contamination and Toxicology* 26:781–788.

GEORGHIOU, G.P. 1983. Management of resistance in arthropods. In G.P. GEORGHIOU and T. SAITO, eds., *Pest Resistance to Pesticides*. New York: Plenum Press, pp. 769–792.

———1986. The magnitude of the resistance problem. In U.S. National Academy of Sciences. *Pesticide Resistance: Strategies and Tactics for Management*. Washington, D.C.: National Academy Press, pp. 14–43.

GRAFTON-CARDWELL, E.E. , and M.A. HOY. 1986. Genetic improvement of common green lacewing, *Chrysoperla carnea* (Neuroptera: Chrysopidae): selection for carbaryl resistance. *Environmental Entomology* 15:1130–1136.

KAUFMAN, D.D., and D.F. EDWARDS. 1983. Pesticide–microbe interactive effects on persistence of pesticides in soil. In J. MIYAMOTO and P.C. KEARNEY, eds., *Pesticide Chemistry: Human Welfare and the Environment* 3:177–182.

KRYSAN, J.L., D.E. FOSTER, T.F. BRANSON, K.R. OSTLIE, and W.S. CRANSHAW. 1986. Two years before the hatch: rootworms adapt to crop rotation. *Bulletin Entomological Society America* 32:250–253.

LAFARGE, A.D. 1985. The persistence of resistance. *Agrichemical Age*. 29:10–24.

MCCLURE, M.S. 1977. Resurgence of the scale, *Fiorinia externa* (Homoptera: Diaspididae) on hemlock following insecticide application. *Environmental Entomology* 6:482–483.

NEWSOM, L.D., R.F. SMITH, and W.H. WHITCOMB. 1976. Selective pesticides and selective use of pesticides. In C.B. HUFFAKER, and P.S. MESSENGER, eds., *Theory and Practice of Biological Control*. New York: Academic Press, pp. 565–591.

OSTLIE, K.R. 1987. Extended diapause. *Crops and Soils Magazine*, June-July, pp. 23–26.

PEDIGO, L.P. 1985. Integrated pest management. *McGraw-Hill Yearbook of Science and Technology 1985*. New York: McGraw-Hill, pp. 22–31.

PLAPP, F.W., Jr. 1981. The nature, modes of action, and toxicity of insecticides. In D. Pimentel, ed., *CRC Handbook of Pest Management in Agriculture*. Boca Raton, Fla. : CRC Press, vol. 3. pp. 3–16.

————1986. Genetics and biochemistry of insecticide resistance in arthropods: prospects for the future. In U.S. National Academy of Sciences, *Pesticide Resistance: Strategies and Tactics for Management*. Washington, D.C., National Academy Press, pp. 74–86.

RACKE, K.D., and J.R. COATS. 1987. Enhanced degradation of isofenphos by soil microorgansims. *Journal Agriculture and Food Chemistry*. 35:94–99.

RICHARDSON, R.H., J.R. ELLISON, and W. W. AVERHOFF. 1982. Autocidal control of screwworms in North America. *Science* 215:361–370.

RIPPER, W.E. 1956. Effect of pesticides on balance of arthropod populations. *Annual Review Entomology* 1:403–436.

SPARKS, T.C., and B.D. HAMMOCK. 1983. Insect growth regulators: resistance and the future. In G.P. GEORGHIOU and T. SAITO, eds., *Pest Resistance to Pesticides*. New York: Plenum Press, pp. 615–668.

SPARKS, T.C., S.S. QUISENBERRY, J.A. LOCKWOOD, R.L. BYFORD, and R.T. ROUSH. 1985. Insecticide resistance in the horn fly, *Haematobia irritans*. *Journal Agricultural Entomology* 2:217–233.

STAAL, G.B. 1975. Insect growth regulators with juvenile hormone activity. *Annual Review Entomology*. 20:417–460.

TABASHNIK, B.E., and B.A. CROFT. 1982. Managing pesticide resistance in crop-arthropod complexes: interactions between biological and operational factors. *Environmental Entomology* 11:1137–1144.

WATSON, T.F. 1975. Practical considerations in use of selective insecticides against major crop pests. In J.C. STREET, ed., *Pesticide Selectivity*. New York: Marcel Dekker, pp. 47–65.

WILKINSON, C.F. 1983. Role of mixed-function oxidases in insecticide resistance. In G.P. GEORGHIOU and T. SAITO, eds., *Pest Resistance to Pesticides*. New York: Plenum Press 175–205.

16 INSECT PEST MANAGEMENT PROGRAMS

IT HAS BEEN SAID that analysis is sterile without synthesis. So it is with insect pest management. Up to this point we have analyzed the mechanisms of pest management tactics; the next step is to consider synthesizing our information for application to realistic situations.

Recall that there are four fundamental strategies to consider in dealing with insect pests: do nothing, reduce numbers, reduce susceptibility of the host, or use combinations of the last two. Implementation of these strategies depends on several primary activities involving pest identification, pest population assessment, and economic evaluation. Except where no action is taken, the pest management strategies assume that economic damage is imminent and that combinations of tactics need to be employed to reduce the threat.

PEST MANAGEMENT PRACTICE

Approaches to applying these tactics can be viewed as paralleling those found in human medicine. Therefore, we can view applications as attempts to prevent problems or to cure existing problems. The first approach is **preventive practice** and the latter is **therapeutic practice**.

Preventive Practice

Preventive practice utilizes tactics to lower the environmental carrying capacity (reduce the general equilibrium position) or increase tolerance of the host to pest injury. For example, crops may be rotated and sanitation practiced, as well as receiving enhanced fertilizer and other cultural treatments that make them less susceptible to damage. Prevention involves extended-term action against a key pest or pest complex. Usually, preventive practices are employed as a matter of course, often without knowledge of pest status in a particular season. In other words, prevention is implemented because the pest has been a problem in the past, and insect population assessment is not necessarily involved in immediate decisions to initiate the activity.

For implementation, prevention relies on an intimate understanding of pest life cycle, behavior, and ecology. It involves manipulation of both the biotic and the physical environment of the insect. Manipulations make use of several tactics from natural enemy, host resistance, and ecological management categories. Although

517

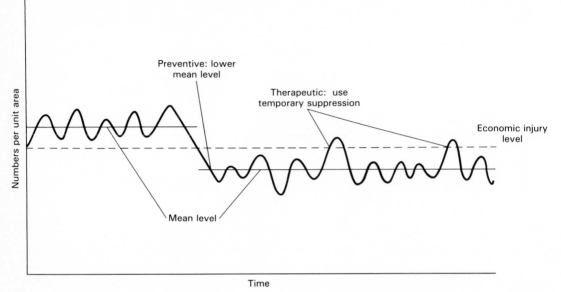

Figure 16.1 Diagrammatic representation of preventive and therapeutic practices on pest population numbers. (Redrawn from *Ecological Entomology*, C. B. Huffaker and R. L. Rabb, eds. Copyright ©1984. Reprinted by permission of John Wiley & Sons.)

they are used as such, insecticide applications for prevention of insect problems in agriculture are not usually considered a wise practice because of possible ecological backlash.

In addition to recommended preventive tactics, an important component not previously discussed in detail is plant and animal quarantine. Quarantines are legal instruments to prevent or slow the spread of pests from infested into uninfested areas by restricting movement of infested stock. The objective of quarantine is containment of the pest population. In the United States, quarantines are administered by the Animal and Plant Health Inspection Service (APHIS), which regulates both domestic quarantines (in force within the United States and its territories) and foreign quarantines, directed against pests associated with agricultural commodities of foreign countries. Quarantines are enforced by inspectors at state borders and at 85 foreign ports of entry. Quarantined goods without import or export certification may be seized by federal agents. This approach to prevention has been termed **regulatory control**.

To develop a program of prevention, several of these tactics, including quarantine when appropriate, are combined for use in a complementary manner. Optimally, the prevention program becomes the base pest management system to which therapeutic tactics may be added.

Therapeutic Practice

Therapy is applied if preventive tactics fail to keep injury below tolerable levels. Therapeutics are applied as a "correction" to the system when necessary. Moreover, unless eradication is appropriate (rare instances), therapy should aim to suppress the population minimally, reducing pest densities to levels that can be tolerated economically. Hence, the objective of therapy is to dampen population peaks, not eliminate every individual from an area.

Figure 16.2 Travelers often bring back quarantined agricultural products that are seized at international ports of entry by inspectors. Many of these items harbor exotic insects and disease organisms. (Reproduced from *Introduction to Crop Protection*, 1979, pages 342-373 by permission of the American Society of Agronomy, Inc., and Crop Science Society of America, Inc.)

Tactics used in therapy usually kill insects outright. Presently, the most important tactics involve the use of conventional insecticides, microbial insecticides, and insect growth regulators. These tactics may be used singly or in combination with preventive tactics.

Therapeutic practice is based on pest population surveillance and assessment of injury potentials. It includes sampling, estimating density, predicting population trends, and evaluating relationships to the economic-injury level. Guidelines call for corrective actions when the economic threshold is exceeded.

Selection of Tactics

Design of the insect pest management program calls for selection of a set of tactics to achieve both prevention and therapy. In making selections, the principles discussed in management of ecological backlash become paramount. To follow these, we would favor therapeutic tactics with the narrowest spectrum or those with potential for ecological selectivity. The major objective in this approach is to protect natural enemies and conserve environmental quality. In a similar vein, we would also choose the tactics that are relatively nonpersistent to accomplish our objectives.

Figure 16.3 Within the United States, quarantined items are certified pest-free before being shipped from a quarantined area. (Reproduced from *Introduction to Crop Protection*, 1979, pages 342-373 by permission of the American Society of Agronomy, Inc., and Crop Science Society of America, Inc.)

Moreover, because several tactics are to be used, they must be compatible with one another; that is, conflicts between tactics of prevention and therapeutics need to be avoided or minimized. It would do little good to apply a tactic, only to have its potential benefits reduced or nullified by another tactic.

Finally, the least expensive alternatives to achieve an objective should be chosen. Prices of tactics may vary widely, and efforts to reduce costs are an important aspect of successful pest management in a frugal agricultural production environment.

Integration of Tactics

After the proper tactics for prevention and therapy have been chosen, the next procedure is integrating them in a complementary manner. The importance of this process cannot be overestimated because it is the basis for developing stable and environmentally safe solutions to pest problems.

Integrating management tactics is not simply adding a number of these to form a program. Instead, integration considers the effect of each tactic with regard to the pest and other potential tactics. For instance, integrating insecticide use with natural enemies requires a plan to dampen the pest population adequately without decimating natural enemy populations.

By integrating several tactics, reliance is not placed on any single one. A diversity of tactics may significantly slow the development of resistance to any single tactic because genotypes with multiple resistance mechanisms occur infrequently, and selection pressure will be low.

Actual integration involves proper choice of compatible tactics and blending them so that each complements the other. For instance, host-plant resistance is usually compatible and can be easily blended with crop sanitation to obtain complementary results. Likewise, insecticide use, a therapeutic tactic, is often compatible with many preventive ones.

The most difficulty arises when attempting to blend tactics with natural enemy conservation, because important natural enemies, being arthropods themselves, succumb to the same destructive forces as pests. Unfortunately, in many instances they are even more sensitive than pests and require longer to recover from destruction. Therefore, moderation and timing are important when implementing tactic combinations where natural enemies are ecologically significant.

Although we have emphasized that several tactics are preferable to a single one, integration admittedly becomes more complex and difficult with each additional tactic. As more are considered, the chances of negative interactions increase. Therefore, there are limits to the number of tactics and combinations that are practical for any one situation.

Levels of integration and the accompanying complexity can be seen in a hypothetical scheme presented by R. L. Rabb and coauthors (1984). Using a field-crop pest example, these authors suggest that integration might proceed through seven different levels and that both positive and negative interactions among tactics may occur at any level. In their scheme, level 1 integration involves inserting one or more tactics in a management system for a pest from a single taxonomic group (taxon). At this level, the focus is on augmenting natural control forces like weather and natural enemies and avoiding resistance and resurgence. Level 2 integration occurs when two or more pests in the same taxon are managed by combined tactics. The challenge here is to blend tactics to allow suppression of all pests in the complex without upsetting any one, to avoid replacements. Integrations in levels 1 and 2 are of greatest concern in our discussion of insect pest management.

As higher levels of integration are sought, our perspective becomes interdisciplinary and involves considerations of plant-parasitic nematodes, pathogenic microorganisms, and weeds (also rodents and other pests). At level 3, we move from insect pest management to integrated pest management and upward through levels of crop and farm management.

Although not indicated by Rabb et al., we would suppose that the chances of negative interactions would tend to increase with each level of integration. Witnessing the incompatibilities that often arise in insect pest management alone, we can imagine that even more formidable conflicts may occur as we ascend in the hierarchy, particularly at crop producer levels (levels 4 and 5). To make a point, compromises are frequently necessary in almost any insect pest management program we devise because of other pests and crop production priorities.

Insect Pest Management and Crop Values

The ratio of management costs to commodity prices (market values) is an important index for making pest management decisions in agriculture. This ratio forms a primary constraint on choice, integration, and employment of useful tactics. Furthermore, because costs of tactics probably are less variable between crops than are commodity prices, it is the latter that mostly limits our alternatives.

For discussion, it is meaningful to locate crops from low to high along a commodity-value continuum. The continuum represents potential value per

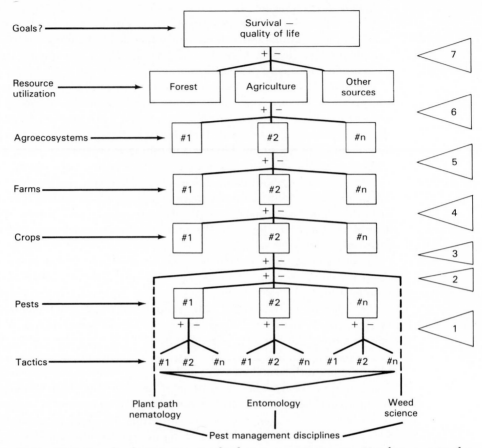

Figure 16.4 Levels of integration involved in pest management. Numbers in triangles refer to seven levels of integration, and plus and minus symbols indicate both positive and negative interactions among tactics (see text for explanation). (From *Ecological Entomology*, C. B. Huffaker and R. L. Rabb, eds. Copyright ©1984. Reprinted by permission of John Wiley & Sons, Inc.)

area, where value has little relationship to importance in terms of human consumption.

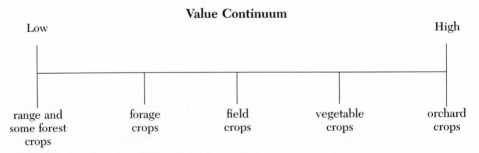

Along the continuum, low-value crops are represented by forests and pastures, moderate value by forage crops, field crops, and vegetables, and high value by orchards. Low-value crops require large land areas for production and have lower economic values per area than other types. Substantial injury of these crops usually can be tolerated before economic damage occurs because each unit injured has a relatively low price. Therefore, expensive management practices, especially frequent use of therapeutic tactics, usually are not practical. Rather, preventive tactics like sanitation, harvest practices, and controlled burning are the more likely tactics to be integrated into pest management programs.

Moderate-value crops have greater value per land area, and therefore injury is less tolerable than with low-value crops. Higher values tend to create lower economic-injury levels, and as a result, pest populations exceed these levels more often. Therefore, whereas preventive tactics are useful and important, there is a greater need for and use of therapeutic tactics.

High-value crops present some of the most difficult pest management challenges of all, because high prices and quality considerations usually place economic-injury levels just above or even below a pest's general equilibrium position. Consequently, the need for pest population corrections is frequent and continuing. For these crops, preventive practices are useful, but the need for therapeutics becomes prevalent. Therefore, we might anticipate changes in reliance on prevention and therapeutics as value per land area increases and economic-injury levels fall. This trend can be noticed in the examples that follow.

INSECT PEST MANAGEMENT IN A LOW-VALUE PRODUCTION SYSTEM

An example of insect pest management in a low-value production system can be seen with the spruce bark beetle, *Ips typographus*, on Norway spruce in Europe. This insect is credited with being the most destructive bark beetle of coniferous forests on the continent. Management approaches for this pest have been analyzed by A.A. Berryman (1986).

Pest life cycle and biology. Spruce bark beetles overwinter as adults in forest litter or in the bark of spruce trees. Activity begins mid- to late May when males attack spruce trees or logs. These colonizing males release aggregating pheromones that attract other males and females, which subsequently attack trees or logs. Females bore into trees where they construct individual egg galleries in phloem tissue and lay eggs in notches cut into walls of the gallery. If the host site is crowded, some adults may leave, and one or two more broods are produced at other sites.

Following egg hatch, larvae feed on the phloem tissue and pupate at the tip of tunnels formed during their feeding. Adults emerge, feed for a short time beneath the bark, and bore out of their tunnel either to continue another generation or to overwinter. In northern Europe, only one generation occurs each year, whereas two or three may develop in southern Europe.

Injury and interactions with the tree. As with many bark beetles, spruce bark beetles transmit several fungi pathogenic to the tree as they bore beneath the bark. One of these fungi, *Ceratocystis polonica*, is particularly destructive and causes considerable tree mortality, especially in unhealthy trees.

Healthy trees can resist damage to a degree. In these trees, resin flow from severed ducts isolates and kills some beetles and repels others. Wound response to the fungus involves development of dead resin-impregnated tissue that lacks nutrients necessary for fungal growth and development. These responses in healthy trees are usually sufficient to prevent significant damage. If beetle attack

Figure 16.5 *Ips typographus* life cycle in southern Norway. A. Adults overwinter in litter and under bark. B. Adults emerge and attack trees, logs, and windthrown trees in early spring. C. Parent beetles may reemerge in summer and attack new trees or logs. D. Offspring emerge in fall and enter overwintering habitats. (Reprinted with permission of Plenum Press from *Forest Insects* by A. A. Berryman. Copyright ©1986)

is intense, however, this resistance can be overwhelmed, and the tree will succumb. Trees stressed by other factors have little or no resistance to beetle attack.

Beetle outbreaks.   Outbreaks of the spruce bark beetle have occurred irregularly through the years and in various regions. In many instances, outbreaks are caused by unusual storms that result in windthrown trees. These trees, when knocked down by high winds, often continue to survive for a time (because some roots remain in the soil) but do so in a weakened state. Such windthrown trees become excellent sources for beetle reproduction and survival. Other factors implicated in outbreaks include poor management practices and trees growing on unfavorable soils or on steep north and east-facing mountain slopes.

Pest management program for spruce bark beetle. The suggested pest management program for this pest proceeds through a number of steps that Berryman calls diagnosis, prognosis, prescription, and conducting management operations (implementation). **Diagnosis** involves recognizing the presence and extent of an infestation and determining the causal agent. **Prognosis** includes anticipating problems before they occur and predicting their outcome, if unattended. **Prescription** is recommending a strategy or course of action. Conducting the actual program is self-explanatory.

Spruce bark beetle outbreaks are diagnosed on the basis of symptoms expressed by injured trees. These symptoms include dead and discolored (orange) trees, which drop their needles a few weeks to several months after infestation. The damaged trees are usually found in clumps of several to many individuals. Diagnosis involves inspections that may be conducted in intensive programs by ground personnel or extensive programs by airplane surveys. The intensive program is desirable for accessible, high-value stands, whereas extensive programs are used with inaccessible or low-value stands.

Based on these surveys, prognoses of risk to bark beetle outbreaks are made. Subsequently, individual stands may be grouped into treatment or no treatment categories on the basis of this risk.

In the prescription phase, recommendations are made that attempt to prevent eruptive outbreaks in individual stands or groups of stands. These recommendations are based on stand prognosis and depend on the course of an outbreak in an area. They include the following: (1) thinning the stand before outbreaks occur; (2) clear-cutting (removing all trees in one cut) or shelterwood harvesting (two or more cuttings at different times) if the stand is old or stagnating; here, slash from logging operations must be handled by burning or other means to eliminate reproductive habitats for the pest; (3) selective logging of spruce and encouragement of resistant species like Scots pine, where growing conditions for spruce are poor; and (4) sanitation-salvage cuts to remove damaged trees. These silvicultural practices are mainly preventive approaches aimed at avoiding eruptive outbreaks. If preventive measures are impractical because of terrain or threat of the current outbreak, use of pheromone traps is suggested. These traps (pipe traps) contain pheromone mixtures impregnated in plastic strips or in plastic bags that efficiently capture large numbers of flying beetles. They are usually established in open areas to reduce attacks on adjacent trees.

Finally, scheduling of pest management programs is done by forest managers, who allocate personnel, equipment, and material to accomplish the planned activities. In scheduling and allocation, contingency plans are also in place to accommodate unexpected windthrows or other disturbances that produce new outbreak potentials.

INSECT PEST MANAGEMENT IN MODERATE-VALUE PRODUCTION SYSTEMS

Moderate-value crops include some of the most important with regard to feeding and clothing the world's population. Although there are many examples of insect pest management in this category, we discuss two, which involve cotton in Texas and corn in the north-central United States.

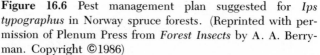

MANAGEMENT OPTION
(one or both)

Intensive Program
(every 10 years)

Extensive Program
(annual)

DIAGNOSIS

Damage Prognosis ← forecast — Damage Appraisal
(risk assessment) (aerial survey)

PRESCRIPTION

Sanitation
Thinning
Harvest
Conversion

Trap trees
Pheromone traps
Clearcut

OPERATIONS
Priorities for treatment
Scheduling treatments
Allocating men and machines
Contingency planning

Figure 16.6 Pest management plan suggested for *Ips typographus* in Norway spruce forests. (Reprinted with permission of Plenum Press from *Forest Insects* by A. A. Berryman. Copyright ©1986)

Insect Pest Management in Cotton

Insects are one of the major limiting factors of cotton production in the United States. Yield losses of over 8 percent per annum can be attributed to these pests. Cotton is attacked by insects from planting until harvest, with nearly 100 species being recorded as pests. All parts of the plant may be attacked; however, injury to reproductive structures, including squares (flower buds), blooms, and bolls, accounts for the greatest damage (estimated at 80 percent).

The key insect pests of cotton in the West include the pink bollworm, *Pectinophora gossypiella*, and plant bugs, *Lygus* species, whereas the boll weevil, *Anthonomus grandis*, and certain plant bugs (Hemiptera: Miridae) dominate in the Southwest, Midsouth, and Southeast. In most areas, repeated insecticide applications have caused replacement of key pests with other species, particularly the corn earworm, *Heliothis zea*, and tobacco budworm, *Heliothis virescens*, which have assumed major pest status in most areas.

To understand the development and current status of insect pest management in cotton, a brief survey of important historical events is helpful.

Historical background of cotton-insect control. The historical development of pest control in cotton has been shaped to a great degree by the boll weevil. This insect invaded the United States from Mexico in 1892 and spread rapidly through the southern high plains of Texas to the Atlantic coast. Invasion by this pest resulted in yield reductions of up to one-third to one-half in some areas. Because much of the South's economic welfare at the time depended on cotton, bankruptcies of growers, merchants, and bankers followed in the weevil's aftermath.

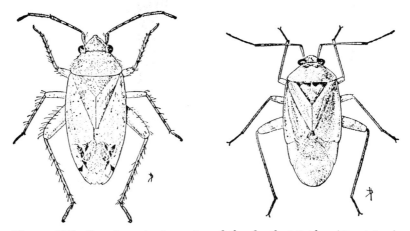

Figure 16.7 Two important species of the family Miridae (Hemiptera), attacking cotton. Cotton fleahopper, *Pseudatomoscelis seriatus*, and *Lygus* species. (Courtesy USDA)

The boll weevil does most of its damage when squares begin to form. At this time, females gouge out small cavities with their snouts and lay eggs in the feeding cavities (usually one per square). The combined injury causes bracts to open or "flare" and squares to shed. Feeding and egg laying continue in bolls (the fruit) as the plant matures, but bolls do not fall from the plant. Larvae develop inside the square or boll and pupate in about 7 to 12 days. Pupae remain in the square or boll for three to six days, after which the adult weevil chews its way out. From two to seven generations occur annually, depending on latitude, and the fall generation migrates to overwintering sites under ground litter in cotton fields, at edges of woods, in fence rows, and in other protected places. Winter is spent in dormancy, and most adults become active and begin feeding again in June.

Early pest control programs relied mainly on preventive practices for reducing damage, including early planting of rapidly fruiting cotton varieties and stalk destruction after harvest to eliminate overwintering sites. Before World War II, these procedures were supplemented by low-key applications of inorganic insecticides (mostly calcium arsenate), but these insecticides had limited effectiveness.

Following the discovery of the insecticidal properties of DDT and its development for agriculture, use of insecticides became widespread. Growers began to apply vast quantities of DDT, HCH, aldrin, dieldrin, endrin, and toxaphene on a regularly scheduled and frequent basis. With the great successes of this approach, shifts were made from short-season cotton varieties to long-season, higher-yielding varieties. These long-season varieties called for even more insecticide treatment, and for about five to seven years, yields with this system were unparalleled. By the mid-1950s, boll weevils developed resistance to many organochlorine compounds in Louisiana, and this resistance spread quickly through the South and Southwest.

At this time, growers turned to other insecticide groups, especially organophosphates and carbamates, for solutions. Compounds from these groups included methyl parathion, azinphosmethyl, ethyl parathion, malathion, EPN, and carbaryl, and resistant weevil populations were effectively suppressed once more.

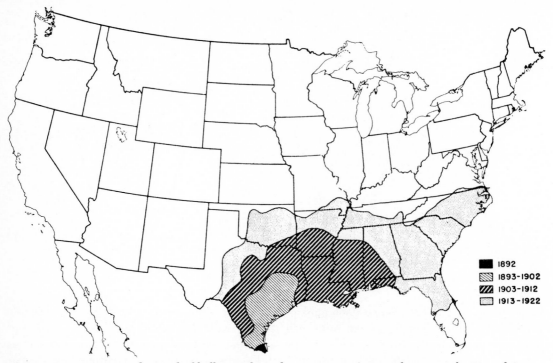

Figure 16.8 Invasion and spread of boll weevil; *Anthonomus grandis,* populations in the United States from 1892 through 1922. (Courtesy USDA)

Although low rates of these compounds would control the boll weevil, they were not effective for corn earworms and tobacco budworms (*Heliothis* complex). The *Heliothis* complex had been only an occasional problem before widespread insecticide use, having been kept mostly in check by insect natural enemies and effectively suppressed by the organochlorines. When growers switched to organophosphates and carbamates and there was no adequate natural enemy check, *Heliothis* populations erupted (a classic example of the replacement phenomenon).

To protect the long-season cotton, growers used higher doses of organophosphates mixed with chlorinated hydrocarbons. Although boll weevil and *Heliothis* suppression was achieved for a time, by the early 1960s resistance of *Heliothis* to the mixtures began to develop in Texas. Other cotton-producing areas were affected by the early 1970s. This cross-resistance problem was addressed with increased insecticide rates and more frequent applications. The more intense insecticide program was costly and resulted in reduced profits and a search for alternative practices.

Insect pest management in Texas cotton. Declining profits from programs of heavy insecticide use contributed greatly to changes in cotton pest control strategy. Although high yields could be maintained in long-season cotton, greater costs of production reduced profits significantly.

An example of present strategies can be made with pest management recommendations for cotton production in southern Texas. The foundation of the program begins with preventive tactics aimed at boll weevils. The basic tactic in

Cotton plant showing *a*, punctured squares on ground; *b*, square showing egg puncture; *c*, larva in square; *d*, pupa in square; *e*, adult emerging from square; *f*, larva and pupa in boll; *g*, adult. (*a*, one-fourth natural size; *b*, *c*, *d*, *e*, and *f*, natural size; *g*, six times natural size.)

Figure 16.9 The boll weevil and injury to the cotton plant. (Courtesy USDA)

BOX 16.1 CORN EARWORM

SPECIES: *Heliothis zea* (Boddie) (Lepidoptera: Nocturidae)

DISTRIBUTION: The corn earworm is native to North America and is found wherever corn is grown. The pest causes the most significant damage in the southern United States, where it also damages tobacco, cotton, tomato, legumes, and other crops.

IMPORTANCE: The corn earworm is one of the most serious sweet corn and field corn pests in the United States; it causes an estimated $75 million to $140 million damage annually. The larvae feed on corn silks, interfering with pollination, which results in the formation of "nubbins." Developing kernels also are fed on, providing entry wounds for secondary pests like sap beetles and molds. Heavy growths of mold on the ears result in unsafe feed for livestock. Damaged sections of sweet corn ears must be cut out, which reduces market value and quality. In field corn, larvae burrow into germ tissue under kernels, causing the kernels to drop from the ear. Besides field and sweet corn, larvae are also significant tomato and cotton pests. Known to growers as the tomato fruitworm, corn earworm larvae burrow into the tomato fruits, causing fruit drop. The larvae, also known as cotton bollworms, damage cotton by burrowing into the squares, thus reducing yield.

APPEARANCE: Adult moths are large, with a wingspan of about 38 mm. Male moths are yellow-brown and females are orange-brown. The wings are banded black, and each forewing displays a dark spot. Larvae vary from green to black, with alternating light and dark stripes running lengthwise along the body. The larvae, having yellow heads and black legs, may be 50 mm long. Eggs are yellow, hemispherical, and oviposited singly. Pupae are about 20 mm long and turn from green to brown during development.

LIFE HISTORY: Pupae overwinter in soil south of 40° North latitude. Moths begin migration to the northern United States and Canada in early spring from the southern overwintering areas. Female moths oviposit from 350 to 3,000 eggs singly on host plants. Eggs hatch in two to eight days, and the larvae initially feed on their empty eggshells. After five molts (about 14 to 30 days) they cease feeding, drop to the ground, and pupate in the soil. After 14 to 21 days, adults emerge and feed on nectar. There are from two to seven generations per year, depending on the climate and on the type of host plant.

prevention is a return to early planting of short-season cotton cultivars, moderate fertilizer use, and well-timed irrigation. In addition, plant thinning is delayed or not implemented, which suppresses vegetative growth and stimulates early fruiting. These practices shorten the production season and the period of time cotton is vulnerable to insect attack.

Early harvesting and stalk destruction are the other major preventive tactics of the program. Early harvest is facilitated by defoliants in some operations, after which stalks are cut and shredded. These late-season tactics prevent further weevil reproduction and weaken or starve weevils going into hibernation. Consequently, hibernating populations suffer higher-than-normal mortalities.

a, Eggs; *b,* egg (15 times natural size); *c,* young larva on square; *d,* damaged square; *e,* full-grown larvae showing color differences; *f,* pupa in soil; *g,* adult. (*a, c, d, e, f,* and *g,* about natural size.)

Figure 16.10 The corn earworm, *Heliothis zea,* also known as the cotton bollworm, and injury to the cotton plant. (Courtesy USDA)

Such preventive tactics are supplemented by pest surveillance (scouting) and therapeutic treatments of organophosphates against boll weevil and cotton flea-hopper, *Pseudatomoscelis seriatus*. Additionally, pyrethroids (among other insecticides) are applied when *Heliothis* outbreaks occur. In making applications, natural enemy conservation is practiced by closely following economic thresholds, using lowest possible rates, and, when possible, treating only "hot spots" in a field.

Such programs have been effective and more profitable than traditional full-season management programs. In one cost/benefit study, profits from growing short-season cotton and practicing insect pest management were from 56 to 130 percent greater than those in a high-input, full-season program.

Insect Pest Management in Corn

Corn is the major cash grain crop grown in the United States; it is produced for silage, grain, seed, and fresh market produce. Of these products, corn grown for feed grain (field corn) occupies the majority of the total acreage (over 65 million acres), and nearly 75 percent of this is grown in the north-central United States (Illinois, Iowa, Indiana, Minnesota, Nebraska, Ohio, and Wisconsin).

A large complex of insects attacks corn, with over 30 species or species groups causing economic damage. Some of the most significant problems occur in field corn when it is grown continuously on the same land. In the north-central states (corn belt), this practice gives rise to problems from a key beetle complex comprising the northern corn rootworm, *Diabrotica barberi*, and the western corn rootworm, *Diabrotica virgifera*. A second key pest is the European corn borer, *Ostrinia nubilalis*, which causes damage to field corn grown continuously or in rotation with other crops. Together, these species account for most insecticide use in field corn in the corn belt, and they contribute to this crop's top ranking in the United States for acreage treated.

Life history and injury from key pests. Both northern and western rootworm species have similar life histories and behavior patterns. In these species, eggs from the previous season hatch in May and June, and larvae move in the soil until a corn root is encountered. Northern and western corn rootworms are monophagous; that is, they feed mostly on one species, corn. However, western corn rootworms are also known to survive on foxtail and a limited number of other grasses. Corn is injured by larvae as they chew off and tunnel inside roots, including main roots and brace roots touching the soil. Root pruning is believed to reduce the flow of water and nutrients to the plant, which disrupts plant physiology and causes yield loss. An additional, sometimes more serious loss occurs when root pruning allows lodging during rain and wind storms. Lodged plants lose yield because of modified architecture (goose necking), and further losses occur from harvest difficulties.

Upon reaching full growth in July, larvae pupate in earthen cells. Adults emerge in late July and August, crawl to the soil surface, and feed on corn pollen and silks or pollen of other plants. If adult densities are high, corn silk clipping can reduce kernel set and grain yield by interfering with pollination. After mating, females lay clusters of eggs in the soil from late July to September. These eggs hibernate until the following spring, when the next corn crop is planted.

The European corn borer, a species introduced to the United States over 70 years ago, is probably the most important pest across all corn-growing regions. Different genetic strains or so-called ecotypes produce various numbers of gen-

BOX 16.2 EUROPEAN CORN BORER

SPECIES: *Ostrinia nubilalis* (Hübner) (Lepidoptera: Pyralidae)

DISTRIBUTION: The European corn borer entered the United States at Boston harbor prior to 1917 and has successfully moved across the continent to the Rocky Mountains. The species has evolved into ecotypes based on the number of generations per year. Southern populations may have four generations per year, Eastern and lower Midwest populations have two generations, and Northern populations typically have only one generation per year. The European corn borer is also distributed over most of Europe and parts of Asia.

IMPORTANCE: The European corn borer is one of the most serious corn pests in the United States. In addition to feeding on corn, the species is capable of surviving in many herbaceous plants with stems large enough for the larvae to enter. Extensive larval boring in plant stalks or stems results in premature breakage or lodging that leads to complications in harvesting and ultimately reduced yield. Later in the season, larvae tunnel in ear shanks, causing ears to drop. Several control strategies have been investigated to limit losses to this pest. Host-plant resistance and cultural manipulation have proven effective on many occasions. The use of insecticides, when timed on the basis of egg mass and larval scouting, is a major tactic for the management of the second generation.

APPEARANCE: Adult female moths are pale yellowish brown, with irregular darker bands running in wavy lines across the wings. Male moths are distinctly darker, having olive brown wings. Eggs are in masses of 5 to 50. They are initially white but turn yellow with continued development. Young larvae are yellow, with several rows of brown spots. Full-grown larvae are about 25 mm long, gray to light brown or pink, and faintly spotted on the dorsal surface.

LIFE CYCLE: Full-grown larvae overwinter inside their tunnels in stubble, stalks, ears, or weeds. Prior to pupation, larvae cut escape holes at the base of the plants to allow the moth to exit following pupation. Moths emerge in spring and mate within 24 hours, usually at dusk. Three days after mating, adults begin to oviposit egg masses on the lower surfaces of host leaves. Eggs hatch in 3 to 12 days, and larvae feed on the surface of the leaves initially before entering the whorl, where they develop. The borers have five or six larval stadia and become full-grown in 25 to 35 days. Larvae of second and third generations readily bore in the plant or feed on ears. Diapause in full-grown larvae is induced by temperature and photoperiod.

erations, with only one generation in Canada and northern Minnesota and four in Georgia and other southern states. Two or three generations are the most common in the corn belt.

The life cycle of the two-generation ecotype can be described as beginning with full-grown larvae, which hibernate through the winter mostly in corn stalks and cobs. These mature larvae pupate in May, and adults emerge in late May and June. Adults mate in grassy areas near cornfields and fly back into the cornfield for egg laying. Eggs are laid in masses on the undersides of leaves. After egg

Figure 16.11 Uninjured (left) and corn rootworm, *Diabrotica* species, injured corn roots. (Courtesy J. Tollefson and Iowa State University)

hatch, surviving larvae feed in the leaf whorl for a few days and subsequently burrow into the stalk. Once in the stalk, they tunnel in conductive tissues and cause physiological stress that ultimately reduces yields. First-generation larvae pupate in July and produce adults that lay eggs for the second generation. Second-generation larvae tunnel in the stalk as well as in ear shanks, the latter injury causing ears to drop and harvest losses. When second-generation larvae become full-grown, they hibernate to pass the following winter.

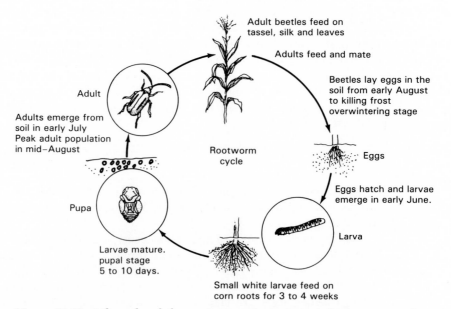

Figure 16.12 Life cycle of the western corn rootworm, *Diabrotica virgifera*. (Courtesy J. Tollefson and Iowa State Univesity)

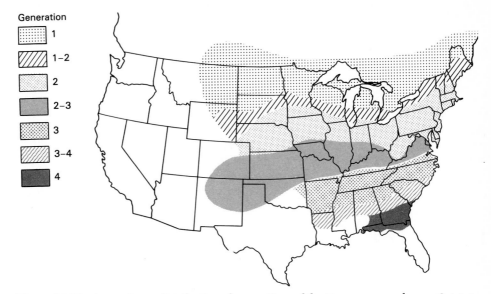

Figure 16.13 Approximate distribution of generations of the European corn borer, *Ostrinia nubilalis*, in the United States and Canada. (After Showers et al., 1983, North Central Regional Publication Number 22)

Historical aspects of pest control. Cultural approaches to suppress corn borers have included stalk shredding and deep plowing to increase mortality of overwintering larvae. However, many of these practices have been abandoned with reduced tillage and the need for soil conservation. Early planting to avoid heaviest losses from second-generation corn borers and early harvest to reduce yield losses from dropped ears and lodged plants remain as attractive preventive practices.

Resistant hybrids are also practical for use against the corn borer. Today,

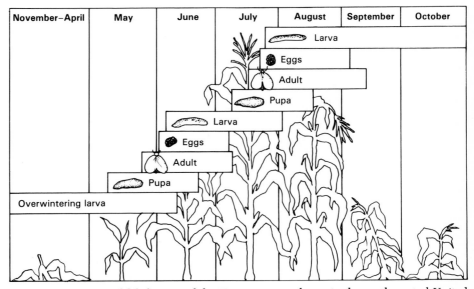

Figure 16.14 Typical life history of the European corn borer in the north-central United States. (After Showers et al., 1983, North Central Regional Publication Number 22)

most seed companies have hybrids available with intermediate resistance to leaf feeding by small, first-generation larvae. The resistance is lost before the hybrids reach the pretassel stage; this makes them susceptible to late first-generation and second-generation larvae. Unfortunately, since the mid-1970s, there has been a trend away from planting resistant hybrids in favor of the higher-yielding but susceptible B73 × Mo17. This shift has resulted in greater insecticide use against first-generation corn borers.

Through the years, insecticides have been a mainstay in dealing with both corn rootworms and the European corn borer. Insecticides are used as a preventive measure against rootworms (applied regularly at planting time) in continuous corn and are sometimes applied as a therapeutic treatment for corn borers. In the latter instance, prediction of phenology and proper timing of treatments is critical, because larvae are vulnerable only a short time before they enter the plant. The development of degree-day models, scouting procedures, and accurate economic-injury levels have contributed significantly to effective therapeutics for corn borer problems.

Historically, insecticide programs have not resulted in major pest upsets with corn in the north-central United States, as has occurred with cotton in the South.

Figure 16.15 Leaf injury to corn from young European corn borer larvae. Whorl feeding of the first generation (left) and leaf-axil feeding of the second generation. (After Showers et al., 1983, North Central Regional Publication Number 22)

A likely reason is that rootworms have few natural enemies, and it does not seem that the natural enemy complex strongly regulates corn borer populations. But rootworms have become resistant to a number of insecticides in each of the major groups, and this nagging problem is accompanied by the growing threat of enhanced microbial degradation and resistance to rotation in some areas. Such events argue for the adoption of a durable insect pest management program for corn.

Insect pest management program for corn in the north-central United States. The key element of management design for corn rootworms and the European corn borer is prevention. The prevention program begins in most regions with a plan of crop rotation to eliminate losses from the corn-rootworm complex. This plan would include alternate year rotations with soybean or other nonhost crops like oats or involve a three-year rotation scheme. The three-year plan would be necessary in areas of extended rootworm diapause.

In the preventive program, a hybrid like A619 $\times$ A632 or A619 $\times$ H99, with intermediate resistance to corn borer, would be chosen and planted as early as possible. This preventive program could be rounded out by early harvest and grain drying, if on-farm facilities are available, or by selling the grain at a moisture-discounted price if they are not.

During many years, such a preventive program would eliminate the need for insecticides in corn completely. However, occasional therapeutic treatments would be required for second-generation corn borers and pests like black cut-worms, *Agrotis ipsilon*, when outbreaks occur. The therapeutic program relies on scouting and early detection of these problems and is particularly refined for second-generation corn borers.

Conventional guidelines with second-generation corn borers call for insecticide treatment when 50 percent of the plants in pre- to posttasseling stages of development have an egg mass on them. A more advanced procedure is to use scouting information with a computer model to arrive at treatment decisions. Utilizing computer software (for use with a personal computer), growers can input current insecticide costs and estimate benefits in making a decision. If the benefit gained from the insecticide exceeds the cost of suppression, then applying the tactic is cost-effective. To use the model, egg-mass density is entered into the program to predict population trend before all eggs are laid, and the resulting output presents information for timely therapeutics based on a straightforward cost/benefit analysis. If therapeutics are required, aerial applications with a granular insecticide can be made to minimize drift and reduce deleterious effects on natural enemies and honey bees.

This example of insect pest management in corn shows very well how several tactics and pests can be integrated into a farm-level management system. Moreover, it shows that both prevention and therapy are crucial in achieving program objectives.

INSECT PEST MANAGEMENT IN HIGH-VALUE PRODUCTION SYSTEMS

Insect pest management for high-value crops is both tedious and complex. These crops usually require significant economic inputs for establishment and continued production, enhancing risks of insect losses and leaving little margin for error in decision making. High values of produce, accompanied by consumer demand for cosmetically "perfect" or blemish-free products raise the stakes for pest management and create a system where therapeutic corrections are prevalent. One of the best examples of pest management in this type of production system is that for apples.

Insect Pest Management in Apples

Insect pests have traditionally received much attention from apple producers because these pests have had a significant impact on apple crops. Over 500 species feed on trees and fruit on a worldwide basis and cause millions of dollars in losses. In the United States, 43 species have been designated as economic pests, but fewer than 10 are considered serious pests.

Most apple pest management focuses on integrating chemical pesticides with natural enemy control. In particular, the major emphasis has been to suppress key pests without causing replacements from secondary pests. Preventive practices are implemented with these programs to some extent, but they would not be expected to eliminate the need for therapeutics as the growing season progresses.

For brevity, the management examples to be mentioned here are taken from production areas of the eastern and midwestern United States. Although key pests may vary in other regions, pest types, relationships, and management approaches are similar.

Key pests and injury.　There are four key insect pests of apple in the midwestern and eastern United States: codling moth (*Cydia pomonella*), plum curculio (*Conotrachelus nenuphar*), apple maggot (*Rhagoletis pomonella*), and redbanded leafroller (*Argyrotaenia velutinana*). In addition to these insects, two diseases, apple scab and powdery mildew, are considered key to management operations in commercial orchards.

All these pest problems involve direct injury of fruit, and combined pest injury cannot exceed 1 percent of the harvest. This low economic-injury level reflects the high market value of the produce and gives a perspective of how little injury can be tolerated economically. Because of the importance of these key pests, most pest management programs are directed primarily toward them.

Codling moth larvae injure fruit when they tunnel and feed on seeds. There are usually two or more generations per season. Plum curculios make crescentlike punctures on fruit surfaces by laying eggs, and emerging adults later make additional holes as they feed. The apple maggot may produce the heaviest infestations of all, with larvae feeding, often undetected, inside the fruit. Both apple maggot and plum curculio produce one generation per year. Redbanded leafroller larvae feed on leaves and the surface of fruit and produce two to four generations per season.

Secondary pests and injury.　Because of serious problems with insecticides and pest replacements, secondary pests are a major consideration in designing pest management programs for apple production. These secondary pests include various species of scales, aphids, leafhoppers, and mites. Of these, phytophagous mites especially have been a problem.

The major mite pests of concern are the European red mite (*Panonychus ulmi*), twospotted spider mite (*Tetranychus urticae*), and the apple rust mite (*Aculus schlechtendali*). These species injure foliage by sucking the contents of individual cells, eventually producing a yellow-green (spider mites) or silvery

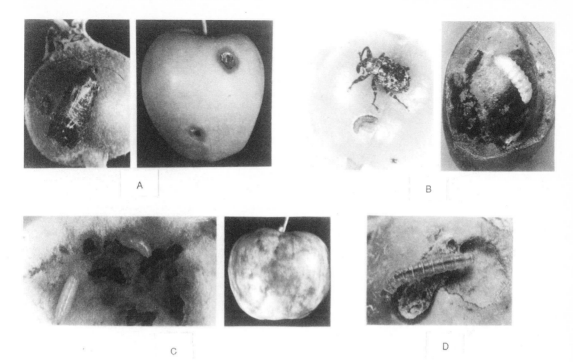

Figure 16.16 Key insect pests of apples in the midwestern and eastern United States. A. Codling moth, *Cydia pomonella*, (left) and injury. B. Plum curculio, *Conotrachelus nenuphar*, on cherry (left) and larva with injury. C. Apple maggot, *Rhagoletis pomonella*, and injury (left) and blotched and streaked apple infested with larvae. D. Redbanded leafroller, *Argyrotaenia velutinana*, larva and injury to apple. (Courtesy USDA)

(rust mites) look to leaves. A bronze appearance of foliage may occur with very heavy infestations. A result of this indirect injury can be reduced fruit size, poor fruit color, and excessive fruit drop in the year of initial infestation. Reduced blossoms and fruit set and poor tree vigor may occur in following years.

Studies of unmanaged orchards have shown that European red mite and twospotted spider mite populations are strongly regulated by predatory mites and insects. One of the major predators is *Amblyseius fallacis*, a pear-shaped mite that can virtually eliminate phytophagous mites from apple trees. Very often the phytophagous mites have developed resistance to many of the broad-spectrum insecticides used against key pests, but predators may succumb after applications. As mentioned earlier, phytophagous mite populations allowed to reproduce without pressure from their natural enemies can quickly become major pests.

Insect pest management in commercial apple orchards. Because natural control of the key pests is unacceptable, much of the insect pest management program in apples is focused on proper therapeutics. Preventive measures are frequently (but not exclusively) directed toward secondary mite pests and reducing the threat of sporadic outbreaks from occasional pests. According to B. A. Croft and W. M. Bode (1983), these measures include pruning water sprouts, sucker growth, or foliage preferred by aphids, leafhoppers, and others; destruction of cull or dropped fruits and other sites used by key pests; elimination of gnarled wood formations and other sites preferred by scale insects; fertilizer management to

BOX 16.3 CODLING MOTH

SPECIES: *Cydia pomonella* (Linnaeus) (Lepidoptera: Olethreutidae)

DISTRIBUTION: The codling moth is found throughout the world except in Japan and western Australia. This pest was introduced into North America from southeastern Europe and now occurs wherever apples are grown. The species is also found on pear, English walnut, and quince.

IMPORTANCE: The codling moth is one of the most destructive apple pests in the world. Damage is caused by larvae that tunnel into the fruit and feed to the core, sometimes on the seeds. Larvae also initially chew into insecticide-treated apples, leaving blemish marks called "stings." Feeding injury also provides infection courts for diseases that further damage the fruit. Injury caused by larvae lowers the quality of the fruit and therefore lowers commercial value. Because of tunneling, larvae are very difficult to suppress. As a result, infestations affect 20 to 95 percent of the apples in a typical orchard.

APPEARANCE: Adult moths have a wingspan of about 18 mm. The wings are grayish brown and crossed with bands of light gray; they display a dark brown spot at the tip of each forewing. Larvae, which are from 12 to 20 mm long when full-grown, are pinkish white, with a brown head. Pupae are approximately 10 mm long and vary from yellow to brown. The minute white eggs are flattened and shaped like pancakes.

LIFE CYCLE: Full-grown larvae overwinter in silken cocoons concealed on or under the bark of apple trees. Pupation occurs in early spring, and adult moths emerge when apple trees are in bloom. The adults, living just 14 to 21 days, mate and females oviposit from 30 to 40 eggs singly on leaves, twigs, and developing fruits. After 5 to 15 days, the eggs hatch, and the larvae bore into fruits. Larvae feed for three to five weeks, then burrow out of the apple and spin silken cocoons under loose bark on the tree or under plant debris on the ground. In warm climates, adult moths emerge from the cocoon in 14 to 30 days, and the life cycle is repeated. The number of generations per year ranges from one to three, depending on climatic conditions.

---

reduce excessive plant growth that fosters piercing-sucking insects; and removal of alternate hosts and habitats in areas adjacent to the orchard.

Preventive measures also may be achieved through ground cover management. In particular, where *Amblyseius fallacis* is an important regulator of European red mite, maintenance of ground cover with some broadleaf weeds is beneficial, because the broadleaves serve as hosts for twospotted spider mites on which *Amblyseius fallacis* can feed. After populations of these predators increase in the ground cover, they invade the tree habitat and subdue populations of the European red mite. Consequently, management recommendations emphasize proper timing of mowing, herbicide applications, and cultivation to reduce deleterious effects on predator-and-prey interactions.

As with other pest management programs, the basis for therapeutics is surveillance of pest populations. Therefore, weekly scouting and inspections are conducted in an orchard-by-orchard program to alert producers of impending problems and to serve as a foundation for decision making. According to J. R. Leeper and J. P. Tette (1981), scouting should be started in the preseason by searching for the San Jose scale, *Quadraspidiotus perniciosus*, where it occurs, and for eggs of the European red mite. Population densities of these pests above economic thresholds may require oil applications during the prebloom period. After trees break dormancy, scouts check for developing aphid populations and, where red-banded leafroller has been a problem, establish pheromone traps to continuously monitor populations.

Beginning with the tight-cluster to pink stage, inspections are made for young European red mites on interior leaves of the tree, along with predator mites. Other pests that might be anticipated at this time include the tarnished plant bug, *Lygus lineolaris*, and the green fruitworm, *Lithophane antennata*. Most orchards would require an insecticide spray for aphids and mites during this stage.

During the bloom stage, codling moth and obliquebanded leafroller, *Choristoneura rosaceana*, traps are placed in the orchard for surveillance. Checks are also continued on mites, aphids, leafhoppers, and other pests, and petal-fall management strategies are developed from projections of pest status.

In the petal-fall period, most orchards would receive a selective insecticide spray to reduce populations of insect pests that have built up by this time. This spray may be delayed for up to ten days, depending on the status of mites and other pests, to coincide with activity of the plum curculio.

Figure 16.17 A predator mite, *Amblyseius fallacis* (top), about to feed on a plant-feeding mite, the twospotted spider mite, *Tetranychus urticae*. (After Michigan State University Extension Bulletin E-825)

During the post-petal-fall period, aphids, mites, and apple maggot may require corrective action, as well as other species as indicated by pheromone traps. If pesticides are used effectively in the earlier part of the season and discriminately later on, most arthropod problems should be kept under control without mite upsets. This is because early-season applications have little effect on *Amblyseius fallacis*, because the species inhabits ground vegetation during heaviest spray periods.

Checks of phytophagous mites are made through the season, however, and a close watch is kept on pest-to-predator ratios. As spider mites increase to densities of one to five per leaf, pest status is determined from sample counts of pest and predator. Subsequent action is taken based on pest status and specific decision categories.

Computer software designed to analyze data inputs for pest status and to conduct cost/benefit analyses of spider mite situations has been developed by several states.

This example of insect pest management is important for showing the significance of therapeutics in dealing with high-value crops. Previously, scheduled sprays to offer protection have been most common because of pest omnipresence. However, by following the spray-as-needed program advocated by pest management, one can reduce production costs and conserve environmental quality.

BOX 16.4 SAN JOSE SCALE

SPECIES: *Quadraspidiotus perniciosus* (Comstock) (Homoptera: Diaspididae)

DISTRIBUTION: The San Jose scale is native to China but was introduced from Japan into San Jose, California, about 1880. Currently, the species is distributed throughout southern Canada and the United States.

IMPORTANCE: The species is a serious fruit tree pest throughout its range. Both adults and nymphs suck sap from the wood and leaves of trees, reducing vigor and subsequently crop yield. Developing fruits are also attacked, leaving gray, mottled blemishes that reduce quality. Under high infestations, trees may be killed. The scales also attack ornamental trees and shrubs. Humans have aided in transport and dispersal of this pest.

APPEARANCE: Adult females are round, about 2 mm in diameter, and covered with waxy scales secreted by the body. Adult males are oval and about 1 mm long. Both males and females are brown to black and have a raised nipple at the top of the scale cover. Nymphal scales are light but turn dark with age. Young nymphs that have not produced scales are commonly called "crawlers," which are yellow and resemble mites.

LIFE HISTORY: Partially grown nymphs (sooty-black stage) overwinter on bark. In spring, nymphs resume feeding and are full-grown during fruit blossom. Two-winged males appear and mate with scale-covered females, which produce active offspring. The crawlers begin feeding on the hosts and soon after molt into the recognizable scale form. There are from two to six overlapping generations per year.

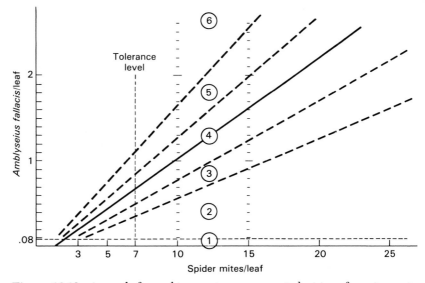

Figure 16.18 A graph for making pest management decisions for mite pests in apples. See Table 16.1 for the meaning of circled-number codes. (After Michigan State University Extension Bulletin E-825)

Table 16.1. Management Recommendations for Mite–Pest to Predator Ratios from Decision Regions in Figure 16.18

Region	Suggested Recommendation	Probability of Natural Control
1	As bronzing appears, spray Plictran® acaricide 50 WP at full rate (4-6 oz./ 100 gal. or 16–20 oz./acre)	Very low
2	If bronzing appears, spray Plictran 50 WP at 2 oz./100 gal. or 8 oz./acre.	Equal to or less than 10%
3	If bronzing appears, spray Omite® 30 WP at 1.25 lb./100 gal. or 5 lb./acre or Plictran 50 WP at 1.5 oz./100 gal. or 6 oz./acre.	Greater than 10% but less than 50%
4	Wait one week; natural control should occur soon; if not, spray Omite at 1.25 oz./100 gal.	About 50%
5	Same as 4.	Greater than 50% but less than 90%
6	Wait one week; natural control is almost certain.	Greater than 90%

SOURCE: From Michigan State Cooperative Extension Service Bulletin E825 (1975).

CONCLUSIONS

All insect pest management programs have elements of both prevention and therapeutics. Prevention's practice is based on a past history of pest problems and relies on a detailed understanding of insect life history and ecology. Preventive tactics are often less expensive to employ and become routine in producing crops.

Therapeutic tactics are integrated into the pest management program when preventive practices fail to maintain a pest within tolerable economic bounds. Most often, therapeutics involves killing insects outright with conventional broad-spectrum insecticides. Infrequently, selective materials like microbial insecticides and insect growth regulators are practical alternatives. As crop values increase, economic-injury levels fall; consequently, there is a greater need for therapeutic tactics.

The key to durable insect pest management is to understand pest life cycles and ecology, then to integrate several preventive tactics with minimal therapeutics. Integration may occur at different levels, but several tactics applied to one pest is the most basic. As more pests are added to the system and greater areas and time periods are encompassed, integration becomes more difficult. Difficulty arises from incompatible tactics and countereffects in the system. Much greater understanding of the entire agroecosystem and total crop management will be required to resolve these conflicts in the future and to allow development of truly integrated pest management systems.

Selected References

BERRYMAN, A.A. 1986. *Forest Insects—Principles and Practice of Population Management*. New York: Plenum Press, pp. 145–245.

CROFT, B.A. 1982. Apple pest management. In R.L. METCALF, and W.H. LUCKMANN, eds., *Introduction to Insect Pest Management*, 2nd ed., New York: Wiley, pp. 465–498.

————.1975. Integrated control of apple mites. Michigan State University Extension Bulletin E825.

CROFT, B.A. , and W.M. BODE. 1983. Tactics for deciduous fruit IPM. In B.A. CROFT, and S.C. HOYT, eds., *Integrated Management of Insect Pests of Pome and Stone Fruits*. New York: Wiley, pp. 219–270.

CROFT, B.A., and L.A. HULL. 1983. The orchard as an ecosystem. In B.A. CROFT, and S.C. HOYT, eds., *Integrated Management of Insect Pests of Pome and Stone Fruits*. New York: Wiley, pp. 19–42.

FRISBIE, R.E., and P.L. ADKISSON, eds. 1985. *Integrated Pest Management on Major Agricultural Systems*. College Station, Tex.: Texas Agricultural Experiment Station MP-1616.

FRISBIE, R.E., and J.K. WALKER. 1981. Pest management systems for cotton insects. In D. PIMENTEL, ed., *CRC Handbook of Pest Management in Agriculture*. Boca Raton, Fla: CRC Press, vol. 3. pp. 187–202.

LEEPER, J.R., and J.P. TETTE. 1981. Pest management systems for apple insects. In D. PIMENTEL, ed., *CRC Handbook of Pest Management in Agriculture*. Boca Raton, Fla.: CRC Press, vol. 3, pp. 243–255.

LUCKMANN, W.H. 1982. Integrating the cropping system for corn insect pest management. In R.L. METCALF, and W.H. LUCKMANN, eds., *Introduction to Insect Pest Management*, 2nd ed. New York: Wiley, pp. 499–519.

PFADT, R.E. 1985. Insect pests of cotton. In R.E. PFADT, ed., *Fundamentals of Applied Entomology*, 4th ed. New York: Macmillan, 339–370.

RABB, R.L., G.K. DEFOLIART, and G.G. KENNEDY. 1984. An ecological approach to managing insect populations. In C.B. HUFFAKER, and R.L. RABB, eds., *Ecological Entomology*. New York: Wiley, pp. 697–728.

REYNOLDS, H.T., P.L. ADKISSON, R.F. SMITH, and R.E. FRISBIE. 1982. Cotton insect pest management. In R.L. METCALF, and W.H. LUCKMANN, eds., *Introduction to Insect Pest Management*, 2nd ed. New York: Wiley, pp. 375–441.

SHOWERS, W.B., J.F. WITKOWSKI, C.E. MASON, F.L. POSTON, S.M. WELCH, A.J. KEASTER, W.D. GUTHRIE, and H.C. CHIANG. 1983. Management of the European corn borer. North Central Regional Publication No. 22. Ames, Iowa: Iowa State University.

TANIGOSHI, L.K., S.C. HOYT, and B.A. CROFT. 1983. Basic biology and management components for insect IPM. In B.A. CROFT, and S.C. HOYT, eds., *Integrated Management of Insect Pests of Pome and Stone Fruits*. New York: Wiley, pp. 153–202.

Epilogue: Toward Integrated Pest Management

With increasing demands for food and fiber, today's world is witnessing an exploitation of the land as never seen before. This exploitation, based on a fossil fuel resource, has allowed unprecedented standards of living in many parts of the world. Yet with the human population growing at exponential rates, the need for improved agriculture has never been greater.

This is not to say that food and fiber production can increase infinitely to meet population growth. Rather, it suggests that agriculture can help alleviate human suffering in the short run, while solutions to the social problem of population growth are being devised.

There are many impediments to significant advancements in agriculture. Some of the more important socioeconomic factors are inflation, trade barriers, diversion of land and water for urban interests, embargoes, and lack of grower cooperation. Some of the most important technical problems are resource management vis-à-vis dwindling fossil fuel supplies, efficiency of domesticated plants and animals based on current genetics, adverse weather patterns, and pests. Of these problems, reducing losses from pests would seem to be the most tractable.

A Pest-limited Agriculture

Significant improvements in pest control could result in vast gains in human welfare through limiting pest impact. The potential for gain in agriculture can be envisioned by considering present losses involved. Currently, world losses to all pest types are estimated at about 35 percent of harvest potential, with most of these caused by weeds, microorganisms, and arthropods. Added to this figure are postharvest losses of about 10 percent, mostly from insects, microorganisms, and rodents.

Fortunately, significant pest control improvements are possible, even without new technology. Indeed, significant gains could be made simply by implementing pest management programs already developed. Currently, some of the most practical programs are those for key insect pests in major crops.

Challenges to Continued Development of Pest Management

Although pest science has come far, there is still much left to do. Today, pest management programs still lag or are wanting for most weeds and plant pathogens and for many insects. Future development will depend on overcoming an array of obstacles, not the least of which is funding for research and development of new pest management programs.

Even with increased funding, unexpected and unpredicted events in the cropping environment will continue to present a series of new challenges. Chief among these are (1) introductions of new crops where they have not been grown previously, (2) introductions of new pests or expanded ranges of old ones, and (3) loss of tactics, particularly pesticides. The latter problem is becoming especially acute because of resistance, not only in insects, but in other pest types as well.

Another recent challenge to pest science in the United States is groundwater contamination from pesticides and other contaminants. Groundwater exists below the level where pore spaces in the soil or rock are saturated with water. Groundwater contamination is of growing national concern because about 50 percent of the total population (85 percent of the rural population) depends on groundwater for human consumption. Concern arises when pesticides and other chemicals (principally nitrates from fertilizers) leach through soils to contaminate groundwater aquifers. Presently, at least 17 pesticides have been found in groundwater in 23 states. Principal among these are the insecticides aldicarb, carbofuran, oxamyl, and dinoseb; the nematicides ethylene dibromide (EDB), D-D, and DBCP; and the herbicides alachlor, atrazine, and simazine. We can expect greater federal regulation of these and other agricultural chemicals because of the groundwater problem, as well as more dependence on integrated pest management as a partial solution to the problem.

Needs and Future Outlook for Pest Management

Pest management has been one of the most important developments in pest control since the advent of modern organic pesticides. Pest management continues to be an important step in our ability to deal with pests in a manner compatible both with the environment and with production economics. Further advances, however, are needed before ultimate pest management objectives can be realized.

One of the foremost needs is to develop truly integrated pest management programs that span the pest disciplines. To date, most programs relate to separate pest disciplines. For example, we have an abundance of separate insect pest management programs, and several weed and plant-pathogen management programs. But integrated pest management should be more than a haphazard juxtaposition of separate programs. True integration, involving modifications within programs to make diverse tactics compatible and reduce countereffects, is the essence of the approach. With this requirement of effective integration in mind, it is probably safe to say that presently integrated pest management is more of an ideal than a reality.

How can integrated pest management become a reality? Progress will be very slow if we rely only on empirical (trial-and-error) approaches. Instead, we need a better theoretical base that can be addressed by priority experimentation.

Development of an adequate theoretical framework for integrated pest management will depend on an improved understanding of agroecosystems and their characteristics with regard to the production system. The agroecosystem is biological, and the production system is socioeconomic. Changes in one influence changes in the other. Researchers must clearly delineate both the changes that can be made in production systems and the consequences as well as the factors beyond control. This delineation requires the perception of integrated pest management as an integral part of the production system; it argues for an interdisciplinary approach to research and development. Without a strong commitment to interdisciplinary research, it is doubtful that integrated pest management will become a reality.

Probably most crucial in an interdisciplinary approach is an understanding of the crops (plants or animals) being produced and the reaction of these to pest attack or interference. Trying to understand host-plant or animal reaction to pests is the impetus for bringing together many specialists in the same experiment. Such interdisciplinary experiments should attempt to explain interactions among insects, weeds, plant pathogens, and other pests, and crop reactions to these multiple stressors in practical situations. Such information would allow more detailed models of agroecosystems, giving the theoretical framework for system manipulations. The integration of pest management tactics then could move from a mode of trial and error to one of analytical optimization based on sound theory.

Finally, before complete implementation can occur, many public attitudes must change. The idea that "the only good bug is a dead bug" must give way to a greater appreciation of the ecological role of pest species and acceptance of their presence. This is particularly true of commodities that most consumers feel should be blemish-free. This attitude, termed the "cosmetic effect," is responsible for application of large pesticide quantities annually, with little or no gain in yield or nutritional quality. The same is true of aesthetic pests (those that are unsightly) in the home, on ornamentals, and in yards. An economic-injury level for many of these is purely psychological. Problems of attitude ultimately must be rectified through education at all levels.

As we have seen, entomology has made substantial contributions to improving human welfare through better management of pests. Clearly, new problems and changing situations threaten that progress, but through the ecologically based principles of pest management and through more research and education, we have the means to surmount these challenges.

Selected References

Council for Agricultural Science and Technology. 1985. Agriculture and groundwater quality. Ames, Iowa: Council for Agricultural Science and Technology, Report No. 103.

PEDIGO, L.P. 1984. Integrated pest management. In *McGraw-Hill Yearbook of Science & Technology 1985*, New York: McGraw-Hill, pp. 22–31.

PIMENTEL, D. 1981. Introduction. In D. PIMENTEL, ed., *CRC Handbook of Pest Management in Agriculture*. Boca Raton, Fla.: CRC Press, vol. 1, pp. 3–11.

Appendix 1:
Key To the Orders of Insecta

The following key is slightly modified from that of Steyskal, Murphy and Hoover (1986, Agricultural Research Service, Miscellaneous Publication 1443) and is useful in identifying most insect specimens. For uncommon and aberrant forms that do not fit the couplets, a more comprehensive key such as that found in *An Introduction to the Study of Insects*, fifth edition, by Borrer, De Long, and Triplehorn (1981, New York: Saunders College Publishing) should be used. In certain instances suborder and superfamily names are listed, which indicates that some specialists believe that the group should have ordinal status.

1. Wings present and well developed 2
   Wings absent or unsuitable for flight (wingless adults and immature stages) ... 34
2. Forewings (on mesothorax) wholly or partly horny, leathery, or otherwise strongly
       differing from wholly membranous hindwings; hindwings sometimes
       lacking 3
   Forewings wholly membranous, or so at least at base 12
3. Forewings (wing covers, elytra) uniformly horny, without apparent veins;
       hindwings, if present, folded both lengthwise and crosswise, hidden under
       forewings when at rest; mouth with mandibles........................ 4
   Forewings (hemelytra or tegmina) with veins; hindwings not folded crosswise . 5
4. End of abdomen with heavy, forcepslike cerci; wings short, leaving most of
       abdomen exposed; hindwings very delicate, almost circular, radially folded
       .... DERMAPTERA
   Without such cerci; wings usually covering most of abdomen, or if forewings
       short, then hindwings elongate, but sometimes absent ..... COLEOPTERA
5. Mouthparts fitted for sucking, forming jointed beak...................... 6
   Mouthparts fitted for chewing, with mandibles moving sideways 7
6. Forewings hardened or leathery basally, with membranous apical part, usually
       lying flat on abdomen with apical part overlapping; beak arising from front part
       of head .... HEMIPTERA
   Forewings usually wholly membranous or very nearly so, held more or less
       rooflike over abdomen; beak arising from hindpart of head, projecting
       downward between forelegs .... HOMOPTERA
7. Hindwings not folded, similar to forewings; both wings with thickened, very short
       basal part separated from rest of wing by suture, so that most of wing can easily
       be broken off; social insects living in colonies .... ISOPTERA
   Hindwings folded fanwise, broader than forewings; wings without breaking suture   8

8. Minute insects, usually less than 6 mm long; forewings small, clublike; antennae short, with few segments; parasites of other insects .. STREPSIPTERA males
Usually large or moderately large insects; forewings usually flat and long; antennae usually lengthened and slender, many-segmented 9

9. Hindfemora enlarged, modified for jumping .... ORTHOPTERA
Hindfemora not enlarged, similar to other legs.......................... 10

10. Cerci short, unsegmented; body usually elongate and slender (stick-like)..................... ORTHOPTERA, suborder PHASMATODEA
Cerci long or short but segmented; body usually not sticklike.............. 11

11. Shape oval; all legs similar, adapted to walking..........................
.... ORTHOPTERA, superfamily BLATTOIDEA
Shape elongate; forelegs raptorial (with spines and modified for grasping prey).
.... ORTHOPTERA, superfamily MANTOIDEA

12. With but 2 well-developed wings, forepair functional, hindpair not winglike and sometimes small and clublike 13
With 4 wings, hindpair sometimes small but flat or straplike, not clublike..... 15

13. Mouthparts forming a sucking or lapping proboscis, rarely rudimentary or virtually lacking; hindwings replaced by clublike halteres; abdomen without tail filaments... DIPTERA
Mouthparts not functional; hindwings not formed into clublike halteres; abdomen with tail filaments.. 14

14. Without halteres; antennae inconspicuous, with small scape and pedicel and bristlelike flagellum; forewings with numerous crossveins (few mayflies) .... EPHEMEROPTERA
With hindwings reduced to halterlike structures; antennae evident, not bristlelike; venation of forewings apparently reduced to 1 forked vein (male scale insects)...................................... HOMOPTERA

15. Wings long, narrow, almost veinless, with long marginal fringes; tarsi 1- or 2-segmented, with swollen tip; mouthparts conical, fitted for piercing and sucking plant tissues (minute insects)........................ THYSANOPTERA
Wings broader; if fringed, fringe not longer than width of wing; veins usually conspicuous and at least 1 crossvein present; tarsi with more than 2 segments and tip not swollen 16

16. Wings, legs, and body at least in part with elongate, flattened scales and often also with hairs; wings hyaline (glasslike) under color pattern formed by scales; mouthparts consisting of tongue (rarely rudimentary) formed of helically coiled tube; mandibles (fitted for chewing) present only in a few families of small moths with wingspread not over 12 mm .... LEPIDOPTERA
Wings, legs, and body not covered with scales, although a few scales sometimes present; color pattern of wing involving wing membrane and/or hair 17

17. Hindwings with broad anal area, plaited when wings folded, usually larger than forewings; antennae prominent...................................... 18
Hindwings without plaited anal area, not larger than forewings; antennae often inconspicuous, bristlelike.. 20

18. Tarsi 3-segmented; cerci well developed, usually long and many-segmented
.... PLECOPTERA
Tarsi 5-segmented; cerci not prominent................................ 19

19. Wings with several subcostal crossveins, surface without hairs or scales..................... NEUROPTERA, suborder MEGALOPTERA
Wings without subcostal crossveins, surface with hairs or scales ....
.... TRICHOPTERA

20. Antennae short, bristlelike; wings with numerous crossveins forming overall network; mouthparts with mandibles close to eyes...................... 21
Antennae larger or wings with few crossveins or mouthparts at end of beak ... 22

21. Hindwings much smaller than forewings; abdomen with long tail filaments
    . . . . EPHEMEROPTERA
    Hindwings very similar to forewings; abdomen without long tail filaments
    . . . . ODONATA

22. Head extended into beak with mandibles at end; hindwings not folded; wings
    usually with color pattern and numerous crossveins; male genitalia usually
    swollen, turned forward, and with strong pair of forceps . . . . . MECOPTERA
    Head not extended into beak; male genitalia without conspicuous forceps   23

23. Mouthparts (sometimes lacking) consisting of proboscis without chewing
    mandibles; cerci lacking; wings with few crossveins   24
    Mouthparts including mandibles fitted for chewing   26

24. Wings covered with scales forming color pattern; antennae many-segmented;
    mouthparts (when present) consisting of helically coiled haustellum (tongue)
    . . . . LEPIDOPTERA
    Wings not covered with scales; antennae with few segments; mouthparts
    consisting of segmented piercing beak   25

25. Beak arising from anterior part of head . . . . HEMIPTERA
    Beak arising from posterior part of head, extended downward between forelegs
    . . . . HOMOPTERA

26. Body and wings covered with whitish powder; wings bordered anteriorly by
    very narrow cell without row of crossveins; insects less than 5 mm long
    . . . . NEUROPTERA, suborder PLANIPENNIA (Coniopterygidae)
    Body and wings not covered with whitish powder; otherwise differing   27

27. Tarsi 5-segmented   28
    Tarsi with 4 or fewer segments.   31

28. Prothorax fused with mesothorax; hindwings smaller than forewings, latter with
    no more than 20 cells; abdomen often constricted at base . . HYMENOPTERA
    Prothorax more or less free, sometimes long; forewings and hindwings
    approximately equal in size, with more than 20 cells   29

29. Prothorax much longer than head, cylindrical; forelegs similar to others, not
    enlarged . . . . NEUROPTERA, suborder RAPHIDIODEA
    Prothorax not longer than head; if longer, then forelegs enlarged for grasping
    prey.   30

30. Costal cell with many crossveins . . NEUROPTERA, suborder PLANIPENNIA
    Costal cell without series of crossveins . . . . MECOPTERA

31. Wings equal in size or rarely hindwings larger; tarsi 3- or 4-segmented   32
    Hindwings smaller than forewings; tarsi 2- or 3-segmented.   33

32. Tarsi apparently 4-segmented; forebasitarsi unswollen; wings dehiscent (see also
    couplet 7) . . . . ISOPTERA
    Tarsi 3-segmented, forebasitarsi swollen . . . . EMBIOPTERA

33. Cerci absent; wings remaining attached; antennae slender, with 13 or more
    segments. . . . . PSOCOPTERA
    Cerci evident, although short, ending in bristle; wings shed eventually; antennae
    with 9 beadlike segments; seldom-encountered small insects . . ZORAPTERA

34. Body with more or less distinct head, thorax, and abdomen; with jointed legs and
    ability to move about.   35
    Without distinctly separate body parts, or without legs, or not able to move about   78

35. Parasites of warm-blooded animals.   36
    Not parasites of warm-blooded animals   40

36. Body strongly flattened sideways; mouth a sharp, downturned beak (jumping
    insects) . . . . SIPHONAPTERA
    Body flattened dorsoventrally or maggots of more or less cylindrical form.   37

37. Mouthparts with mandibles for chewing, directed forward; generally
    oval insects with more or less triangular head; parasites of birds and

mammals .... MALLOPHAGA
Mouthparts in form of beak for piercing and sucking 38
38. Antennae inserted in pits, not visible from above (also maggot-shaped larvae without antennae)............................ DIPTERA (a few families)
Antennae present, although short, not in pits 39
39. Beak not jointed; tarsi forming hook for grasping hairs of host; parasites remaining on host .... ANOPLURA
Beak jointed; tarsi not hooked; parasites not remaining on host (bedbugs and related insects) .... HEMIPTERA
40. Aquatic, usually breathing by gills; larval and some pupal forms............. 41
Terrestrial, breathing by spiracles or rarely without breathing organs......... 49
41. Mouth forming a strong, pointed, downcurved beak .. Immature HEMIPTERA
Mouth with mandibles.. 42
42. Mandibles extending straight forward, united with maxillae to form piercing jaws .... Some larval NEUROPTERA
Mandibles moving sideways, forming biting jaws 43
43. Living in case formed of sand, pebbles, leaves, twigs, etc.; usually with external tracheae serving as gills................. Larvae of some TRICHOPTERA
Not living in case.. 44
44. Abdomen with lateral organs serving as gills (a few larval Trichoptera and Coleoptera key out here also) 45
Abdomen without external gills (some larval Trichoptera will key out here also). 46
45. Abdomen with 2 or 3 long tail filaments .... naiads of EPHEMEROPTERA
Abdomen with short end processes (larvae of some Trichoptera will run to this point)............ Larvae of NEUROPTERA, suborder MEGALOPTERA
46. Lower lip (labium) folded backward, extensible, and furnished with pair of jawlike hooks... naiads of ODONATA
Labium not so constructed 47
47. Abdomen with nonjointed false legs (prolegs) in pairs on several segments ....
.... Few larvae of LEPIDOPTERA
Abdomen without prolegs 48
48. Thorax in 3 loosely united divisions; antennae and tail filaments long and slender
.... naiads of PLECOPTERA
Thoracic divisions without constrictions; antennae and tail filaments short (larvae of some aquatic Diptera and Trichoptera also run to this point) ....
.... Larvae of COLEOPTERA
49. Mouthparts retracted into head and hardly or not at all visible; underside of abdomen with appendages; very delicate, small, or minute insects, sometimes without antennae 50
External mouthparts conspicuous; antennae always present; underside of abdomen rarely with appendages 52
50. Head pear shaped, without antennae; abdomen without long cerci, pincers, jumping apparatus, or basal ventral sucker .... PROTURA
Head usually not pear shaped, antennae conspicuous; abdomen with long cerci, pincers, or basal ventral sucker.................................... 51
51. Abdomen consisting of 6 or fewer segments, with forked sucker at base below and usually with conspicuous jumping apparatus near end, but without conspicuous long cerci or pincers .... COLLEMBOLA
Abdomen with more than 8 evident segments and ending in long, many-jointed cerci or strong pincers; eyes and ocelli lacking .... DIPLURA
52. Mouthparts with mandibles fitted for chewing 53
Mouthparts in form of proboscis fitted for sucking 74
53. Body usually covered with scales; abdomen with 3 prominent tail filaments and at least 2 pairs of ventral appendages (styles)............................

.... THYSANURA and MICROCORYPHIA

Body never covered with scales; abdomen never with 3 tail filaments nor ventral styles... 54

54. Abdomen bearing pairs of false legs (prolegs) beneath, not jointed, and differing from true legs on thorax, which is not distinctly separated from abdomen; body caterpillarlike; larval forms 55

Underside of abdomen without legs or prolegs........................... 57

55. Prolegs 5 pairs or less, none on 1st, 2d, or 7th segments; prolegs tipped with many tiny hooklets and rarely present on 2d and 7th segments ....

.... Larvae of most LEPIDOPTERA

Prolegs 6 to 10 pairs, not tipped with tiny hooks; 1 pair of prolegs on 2d segment 56

56. Head with single ocellus on each side ...... Larvae of some HYMENOPTERA

Head with several ocelli on each side .... Larvae of MECOPTERA

57. Antennae long and distinct.. 58

Antennae short; larval forms.. 71

58. Abdomen ending in strong pincerlike forceps; prothorax free . . DERMAPTERA

Abdomen not ending in forceps 59

59. Abdomen strongly constricted at base; prothorax fused with mesothorax

.... HYMENOPTERA

Abdomen not strongly constricted at base, broadly joined to thorax.......... 60

60. Head produced into beak with mandibles at end .... MECOPTERA

Head not produced into beak.. 61

61. Very small insects with soft body; tarsi 2- or 3-segmented.................. 62

Usually much larger insects; tarsi usually with more than 5 segments, or body hard and cerci absent 63

62. Cerci absent... PSOCOPTERA

Cerci of single segment, prominent........................ ZORAPTERA

63. Hindlegs fitted for jumping, femora enlarged; wing pads of immatures inverted, hindpads overlapping forepads .... ORTHOPTERA

Hindlegs not enlarged for jumping; wing pads, if present, in normal position . . 64

64. Prothorax much longer than mesothorax; front legs modified for grasping prey (raptorial) .... ORTHOPTERA, superfamily MANTOIDEA

Prothorax not greatly lengthened...................................... 65

65. Without cerci; body often hard shelled; antennae usually with 11 segments

.... COLEOPTERA

Cerci present; antennae usually with more than 15 segments.............. 66

66. Cerci with more than 3 segments 67

Cerci with 1-3 segments 69

67. Body flattened, oval; head turned down and backward ....

.... ORTHOPTERA, superfamily BLATTOIDEA

Body elongate; head nearly horizontal.................................. 68

68. Cerci long; ovipositor evident, hardened; tarsi 5-segmented ....

.... ORTHOPTERA, suborder GRYLLOBLATTODEA

Cerci short; ovipositor lacking; tarsi 4-segmented................ ISOPTERA

69. Tarsi 5-segmented (3-segmented only in *Timema*, in Pacific Coast States, most antennal segments several times as long as wide); body usually slender and long...................... ORTHOPTERA, suborder PHASMATODEA

Tarsi 2- or 3-segmented; antennal segments beadlike...................... 70

70. Front tarsi with 1st segment swollen, containing silk-spinning gland for producing web in which insects live; cerci conspicuous............. EMBIOPTERA

Front tarsi not so, not producing silk; cerci inconspicuous .... ISOPTERA

71. Body cylindrical, caterpillarlike 72

Body more or less depressed, not caterpillarlike 73

72. Head with 6 ocelli on each side; antennae inserted in membranous area at base

of mandibles......................... Some larvae of LEPIDOPTERA

Head with more than 6 ocelli on each side; 3d pair of legs distinctly larger than 1st pair .... MECOPTERA (larvae of Boreidae)

73. Mandibles united with maxillae to form sucking jaws......................
.... Larvae of NEUROPTERA, suborder PLANIPENNIA

Mandibles nearly always separate from maxillae.........................
Larvae of some COLEOPTERA; NEUROPTERA, suborder RAPHIDIODEA; STREPSIPTERA; DIPTERA

74. Body densely clothed with scales and hairs; proboscis, if present, coiled under head .... LEPIDOPTERA

Body bare, with scattered hairs or waxy coating 75

75. Last tarsal segment bladderlike, without claws; mouth a triangular unsegmented beak; very small insects............................ THYSANOPTERA

Tarsi not swollen at tip, with distinct claws 76

76. Prothorax small, hidden when viewed from above .... DIPTERA

Prothorax evident when viewed from above 77

77. Beak arising from front part of head....................... HEMIPTERA

Beak arising from lower back part of head.................. HOMOPTERA

78. Legless grubs or maggots, moving about by squirming, larvae of DIPTERA HYMENOPTERA; LEPIDOPTERA; COLEOPTERA; SIPHONAPTERA; STREPSIPTERA (in body of wasps or bees with flattened head exposed)

Forms legless or with a single claw 79

79. Small forms with little resemblance to most insects, with filamentous mouthparts inserted in plant tissue; usually covered with waxy scale, powder, or cottony tufts... HOMOPTERA

Body unable to move or only able to bend from side to side, enclosed in tight skin, sometimes wholly covering body or sometimes with appendages free, but rarely movable; sometimes enclosed in cocoon (pupae) 80

80. Legs, wings, etc., more or less free from body; biting mouthparts visible 81

Skin enclosing body holding appendages tightly against body; mouthparts evident as proboscis, without mandibles 83

81. Prothorax small, fused with mesothorax; sometimes enclosed in thin cocoon...
.... Pupae of HYMENOPTERA

Prothorax larger and not fused with mesothorax........................ 82

82. Wing cases with few or no veins .... Pupae of COLEOPTERA

Wing cases with several branched veins .... Pupae of NEUROPTERA

83. Proboscis usually long, rarely absent; wing cases 4; often in cocoon ....
.... Pupae of LEPIDOPTERA

Proboscis usually short; wing cases 2; rarely in cocoon, but often tightly enclosed in hardened last larval skin .... Pupae of DIPTERA

Appendix 2:
List of Some Insects
and Related Species
Alphabetized by Common Name

A

abbreviated wireworm	*Hypolithus abbreviatus* (Say)	COLEOPTERA: Elateridae
acacia psyllid	*Psylla uncatoides* (Ferris & Klyver)	HOMOPTERA: Psyllidae
achemon sphinx	*Eumorpha achemon* (Drury)	LEPIDOPTERA: Sphingidae
acuminate scale	*Kilifia acuminata* (Signoret)	HOMOPTERA: Coccidae
acute-angled fungus beetle	*Cryptophagus acutangulus* (Gyllenhal)	COLEOPTERA: Cryptophagidae
aerial yellowjacket	*Dolichovespula arenaria* (Fabricius)	HYMENOPTERA: Vespidae
African mole cricket	*Gryllotalpa africana* Palisot de Beauvois	ORTHOPTERA: Gryllotalpidae
ailanthus webworm	*Atteva punctella* (Cramer)	LEPIDOPTERA: Yponomeutidae
alder bark beetle	*Alniphagus aspericollis* (LeConte)	COLEOPTERA: Scolytidae
alder flea beetle	*Altica ambiens* LeConte	COLEOPTERA: Chrysomelidae
alder spittlebug	*Clastoptera obtusa* (Say)	HOMOPTERA: Cercopidae
alfalfa blotch leafminer	*Agromyza frontella* (Rondani)	DIPTERA: Agromyzidae
alfalfa caterpillar	*Colias eurytheme* Boisduval	LEPIDOPTERA: Pieridae
alfalfa gall midge	*Asphondylia websteri* Felt	DIPTERA: Cecidomyiidae
alfalfa leafcutting bee	*Megachile rotundata* (Fabricius)	HYMENOPTERA: Megachilidae
alfalfa leaftier	*Dichomeris ianthes* (Meyrick)	LEPIDOPTERA: Gelechiidae
alfalfa looper	*Autographa californica* (Speyer)	LEPIDOPTERA: Noctuidae
alfalfa plant bug	*Adelphocoris lineolatus* (Goeze)	HEMIPTERA: Miridae
alfalfa seed chalcid	*Bruchophagus roddi* (Gussakovsky)	HYMENOPTERA: Eurytomidae
alfalfa snout beetle	*Otiorhynchus ligustici* (Linnaeus)	COLEOPTERA: Curculionidae
alfalfa webworm	*Loxostege commixtalis* (Walker)	LEPIDOPTERA: Pyralidae
alfalfa weevil	*Hypera postica* (Gyllenhal)	COLEOPTERA: Curculionidae
alkali bee	*Nomia melanderi* Cockerell	HYMENOPTERA: Halictidae
Allegheny mound ant	*Formica exsectoides* Forel	HYMENOPTERA: Formicidae
Allegheny spruce beetle	*Dendroctonus punctatus* LeConte	COLEOPTERA: Scolytidae
almond moth	*Cadra cautella* (Walker)	LEPIDOPTERA: Pyralidae
aloe mite	*Eriophyes aloinis* Keifer	ACARI: Eriophyidae
American aspen beetle	*Gonioctena americana* (Schaeffer)	COLEOPTERA: Chrysomelidae
American black flour beetle	*Tribolium audax* Halstead	COLEOPTERA: Tenebrionidae
American cockroach	*Periplaneta americana* (Linnaeus)	ORTHOPTERA: Blattidae
American dagger moth	*Acronicta americana* (Harris)	LEPIDOPTERA: Noctuidae
American dog tick	*Dermacentor variabilis* (Say)	ACARI: Ixodidae
American grasshopper	*Schistocerca americana* (Drury)	ORTHOPTERA: Acrididae
American hornet moth	*Sesia tibialis* (Harris)	LEPIDOPTERA: Sesiidae
American house dust mite	*Dermatophagoides farinae* Hughes	ACARI: Epidermoptidae
American plum borer	*Euzophera semifuneralis* (Walker)	LEPIDOPTERA: Pyralidae
American spider beetle	*Mezium americanum* (Laporte)	COLEOPTERA: Ptinidae
Angora goat biting louse	*Bovicola crassipes* (Rudow)	MALLOPHAGA: Trichodectidae
Angoumois grain moth	*Sitotroga cerealella* (Olivier)	LEPIDOPTERA: Gelechiidae
angularwinged katydid	*Microcentrum retinerve* (Burmeister)	ORTHOPTERA: Tettigoniidae
angulate leafhopper	*Acinopterus angulatus* Lawson	HOMOPTERA: Cicadellidae
apple aphid	*Aphis pomi* De Geer	HOMOPTERA: Aphididae
apple bark borer	*Synanthedon pyri* (Harris)	LEPIDOPTERA: Sesiidae
apple barkminer	*Marmara elotella* (Busck)	LEPIDOPTERA: Gracillariidae
apple blotch leafminer	*Phyllonorycter crataegella* (Clemens)	LEPIDOPTERA: Gracillariidae
apple curculio	*Tachypterellus quadrigibbus* (Say)	COLEOPTERA: Curculionidae
apple flea weevil	*Phynchaenus pallicornis* (Say)	COLEOPTERA: Curculionidae
apple fruit moth	*Argyresthia conjugella* Zeller	LEPIDOPTERA: Argyresthiidae
apple fruitminer	*Marmara pomonella* Busck	LEPIDOPTERA: Gracillariidae
apple grain aphid	*Rhopalosiphum fitchii* (Sanderson)	HOMOPTERA: Aphididae
apple leafhopper	*Empoasca maligna* (Walsh)	HOMOPTERA: Cicadellidae
apple maggot	*Rhagoletis pomonella* (Walsh)	DIPTERA: Tephritidae
apple mealybug	*Phenacoccus aceris* (Signoret)	HOMOPTERA: Pseudococcidae
apple red bug	*Lygidea mendax* Reuter	HEMIPTERA: Miridae
apple rust mite	*Aculus schlechtendali* (Nalepa)	ACARI: Eriophyidae
apple seed chalcid	*Torymus varians* (Walker)	HYMENOPTERA: Torymidae
apple sucker	*Psylla mali* (Schmidberger)	HOMOPTERA: Psyllidae
apple twig beetle	*Hypothenemus obscurus* (Fabricius)	COLEOPTERA: Scolytidae

apple twig borer	*Amphicerus bicaudatus* (Say)	COLEOPTERA: Bostrichidae
apple-and-thorn skeletonizer	*Choreutis pariana* (Clerck)	LEPIDOPTERA: Choreutidae
appleleaf skeletonizer	*Psorosina hammondi* (Riley)	LEPIDOPTERA: Pyralidae
appleleaf trumpet miner	*Tischeria malifoliella* Clemens	LEPIDOPTERA: Tischeriidae
araucaria aphid	*Neophyllaphis araucariae* Takahashi	HOMOPTERA: Aphididae
arborvitae leafminer	*Argyresthia thuiella* (Packard)	LEPIDOPTERA: Argyresthiidae
arborvitae weevil	*Phyllobius intrusus* Kono	COLEOPTERA: Curculionidae
Argentine ant	*Iridomyrmex humilis* (Mayr)	HYMENOPTERA: Formicidae
argus tortoise beetle	*Chelymorpha cassidea* (Fabricius)	COLEOPTERA: Chrysomelidae
army cutworm	*Euxoa auxiliaris* (Grote)	LEPIDOPTERA: Noctuidae
armyworm	*Pseudaletia unipuncta* (Haworth)	LEPIDOPTERA: Noctuidae
artichoke plume moth	*Platyptilia carduidactyla* (Riley)	LEPIDOPTERA: Pterophoridae
ash plant bug	*Tropidosteptes amoenus* Reuter	HEMIPTERA: Miridae
ashgray blister beetle	*Epicauta fabricii* (LeConte)	COLEOPTERA: Meloidae
Asiatic garden beetle	*Maladera castanea* (Arrow)	COLEOPTERA: Scarabaeidae
Asiatic oak weevil	*Cyrtepistomus castaneus* (Roelofs)	COLEOPTERA: Curculionidae
Asiatic rice borer	*Chilo suppressalis* (Walker)	LEPIDOPTERA: Pyralidae
Asiatic rose scale	*Aulacaspis rosarum* Borchsenius	HOMOPTERA: Diaspididae
asparagus beetle	*Crioceris asparagi* (Linnaeus)	COLEOPTERA: Chrysomelidae
asparagus miner	*Ophiomyia simplex* (Loew)	DIPTERA: Agromyzidae
asparagus spider mite	*Schizotetranychus asparagi* (Oudemans)	ACARI: Tetranychidae
aspen blotchminer	*Phyllonorycter tremuloidiella* (Braun)	LEPIDOPTERA: Gracillariidae
aspen leaf beetle	*Chrysomela crotchi* Brown	COLEOPTERA: Chrysomelidae
aster leafhopper	*Macrosteles fascifrons* (Stål)	HOMOPTERA: Cicadellidae
aster leafminer	*Calycomyza humeralis* (Roser)	DIPTERA: Agromyzidae
Australian cockroach	*Periplaneta australasiae* (Fabricius)	ORTHOPTERA: Blattidae
Australian fern weevil	*Syagrius fulvitarsis* Pascoe	COLEOPTERA: Curculionidae
Australian mantid	*Tenodera australasiae* (Leach)	ORTHOPTERA: Mantidae
Australian rat flea	*Xenopsylla vexabilis* Jordan	SIPHONAPTERA: Pulicidae
Australian spider beetle	*Ptinus ocellus* Brown	COLEOPTERA: Ptinidae
Australianpine borer	*Chrysobothris tranquebarica* (Gmelin)	COLEOPTERA: Buprestidae
avocado brown mite	*Oligonychus punicae* (Hirst)	ACARI: Tetranychidae
avocado red mite	*Oligonychus yothersi* (McGregor)	ACARI: Tetranychidae
avocado whitefly	*Trialeurodes floridensis* (Quaintance)	HOMOPTERA: Aleyrodidae
azalea bark scale	*Eriococcus azaleae* Comstock	HOMOPTERA: Eriococcidae
azalea lace bug	*Stephanitis pyrioides* (Scott)	HEMIPTERA: Tingidae
azalea leafminer	*Caloptilia azaleella* (Brants)	LEPIDOPTERA: Gracillariidae
azalea plant bug	*Rhinocapsus vanduzeei* Uhler	HEMIPTERA: Miridae
azalea whitefly	*Pealius azaleae* (Baker & Moles)	HOMOPTERA: Aleyrodidae

B

bagworm	*Thyridopteryx ephemeraeformis* (Haworth)	LEPIDOPTERA: Psychidae
Bahaman swallowtail	*Papilio andraemon bonhotei* Sharpe	LEPIDOPTERA: Papilionidae
baldcypress coneworm	*Dioryctria pygmaeella* Ragonot	LEPIDOPTERA: Pyralidae
baldfaced hornet	*Dolichovespula maculata* (Linnaeus)	HYMENOPTERA: Vespidae
balsam fir sawfly	*Neodiprion abietis* (Harris)	HYMENOPTERA: Diprionidae
balsam fir sawyer	*Monochamus marmorator* Kirby	COLEOPTERA: Cerambycidae
balsam gall midge	*Paradiplosis tumifex* Gagné	DIPTERA: Cecidomyiidae
balsam shootboring sawfly	*Pleroneura brunneicornis* Rohwer	HYMENOPTERA: Xyelidae
balsam twig aphid	*Mindarus abietinus* Koch	HOMOPTERA: Aphididae
balsam woolly adelgid	*Adelges piceae* (Ratzeburg)	HOMOPTERA: Adelgidae
bamboo borer	*Chlorophorus annularis* (Fabricius)	COLEOPTERA: Cerambycidae
bamboo mealybug	*Chaetococcus bambusae* (Maskell)	HOMOPTERA: Pseudococcidae
bamboo powderpost beetle	*Dinoderus minutus* (Fabricius)	COLEOPTERA: Bostrichidae
bamboo spider mite	*Schizotetranychus celarius* (Banks)	ACARI: Tetranychidae
banana aphid	*Pentalonia nigronervosa* Coquerel	HOMOPTERA: Aphididae
banana root borer	*Cosmopolites sordidus* (Germar)	COLEOPTERA: Curculionidae
banana skipper	*Pelopidas thrax* (Linnaeus)	LEPIDOPTERA: Hesperiidae

banded alder borer	*Rosalia funebris* Motschulsky	COLEOPTERA: Cerambycidae
banded ash clearwing	*Podosesia aureocincta* Purrington & Nielsen	LEPIDOPTERA: Sesiidae
banded cucumber beetle	*Diabrotica balteata* LeConte	COLEOPTERA: Chrysomelidae
banded greenhouse thrips	*Hercinothrips femoralis* (O. M. Reuter)	THYSANOPTERA: Thripidae
banded hickory borer	*Knulliana cincta* (Drury)	COLEOPTERA: Cerambycidae
banded sunflower moth	*Cochylis hospes* Walsingham	LEPIDOPTERA: Cochylidae
banded woollybear	*Pyrrharctia isabella* (J. E. Smith)	LEPIDOPTERA: Arctiidae
bandedwinged whitefly	*Trialeurodes abutilonea* (Haldeman)	HOMOPTERA: Aleyrodidae
Banks grass mite	*Oligonychus pratensis* (Banks)	ACARI: Tetranychidae
banyan aphid	*Thoracaphis fici* (Takahashi)	HOMOPTERA: Aphididae
Barber brown lacewing	*Sympherobius barberi* (Banks)	NEUROPTERA: Hemerobiidae
barberpole caterpillar	*Mimoschinia rufofascialis* (Stephens)	LEPIDOPTERA: Pyralidae
barley jointworm	*Tetramesa hordei* (Harris)	HYMENOPTERA: Eurytomidae
barnacle scale	*Ceroplastes cirripediformis* Comstock	HOMOPTERA: Coccidae
basswood lace bug	*Gargaphia tiliae* (Walsh)	HEMIPTERA: Tingidae
basswood leafminer	*Baliosus ruber* (Weber)	COLEOPTERA: Chrysomelidae
basswood leafroller	*Pantographa limata* Grote & Robinson	LEPIDOPTERA: Pyralidae
beachgrass scale	*Eriococcus carolinae* Williams	HOMOPTERA: Eriococcidae
bean aphid	*Aphis fabae* Scopoli	HOMOPTERA: Aphididae
bean butterfly	*Lampides boeticus* (Linnaeus)	LEPIDOPTERA: Lycaenidae
bean capsid	*Pycnoderes quadrimaculatus* Guérin-Méneville	HEMIPTERA: Miridae
bean fly	*Ophiomyia phaseoli* (Tryon)	DIPTERA: Agromyzidae
bean leaf beetle	*Cerotoma trifurcata* (Forster)	COLEOPTERA: Chrysomelidae
bean leafroller	*Urbanus proteus* (Linnaeus)	LEPIDOPTERA: Hesperiidae
bean leafskeletonizer	*Autoplusia egena* (Guenée)	LEPIDOPTERA: Noctuidae
bean pod borer	*Maruca testulalis* (Geyer)	LEPIDOPTERA: Pyralidae
bean stalk weevil	*Sternechus paludatus* (Casey)	COLEOPTERA: Curculionidae
bean thrips	*Caliothrips fasciatus* (Pergande)	THYSANOPTERA: Thripidae
bean weevil	*Acanthoscelides obtectus* (Say)	COLEOPTERA: Bruchidae
Beardsley leafhopper	*Balclutha beardsleyi* Namba	HOMOPTERA: Cicadellidae
bed bug	*Cimex lectularius* Linnaeus	HEMIPTERA: Cimicidae
beech blight aphid	*Fagiphagus imbricator* (Fitch)	HOMOPTERA: Aphididae
beech scale	*Cryptococcus fagisuga* Lindinger	HOMOPTERA: Eriococcidae
beet armyworm	*Spodoptera exigua* (Hübner)	LEPIDOPTERA: Noctuidae
beet leaf beetle	*Erynephala puncticollis* (Say)	COLEOPTERA: Chrysomelidae
beet leafhopper	*Circulifer tenellus* (Baker)	HOMOPTERA: Cicadellidae
beet leafminer	*Pegomya hyoscyami* (Panzer)	DIPTERA: Anthomyiidae
beet webworm	*Loxostege sticticalis* (Linnaeus)	LEPIDOPTERA: Pyralidae
bella moth	*Utetheisa bella* (Linnaeus)	LEPIDOPTERA: Arctiidae
Bermudagrass mite	*Eriophyes cynodoniensis* Sayed	ACARI: Eriophyidae
bertha armyworm	*Mamestra configurata* Walker	LEPIDOPTERA: Noctuidae
bidens borer	*Epiblema otiosana* (Clemens)	LEPIDOPTERA: Tortricidae
bigheaded ant	*Pheidole megacephala* (Fabricius)	HYMENOPTERA: Formicidae
bigheaded grasshopper	*Aulocara elliotti* (Thomas)	ORTHOPTERA: Acrididae
birch bark beetle	*Dryocoetes betulae* Hopkins	COLEOPTERA: Scolytidae
birch casebearer	*Coleophora serratella* (Linnaeus)	LEPIDOPTERA: Coleophoridae
birch leafminer	*Fenusa pusilla* (Lepeletier)	HYMENOPTERA: Tenthredinidae
birch sawfly	*Arge pectoralis* (Leach)	HYMENOPTERA: Argidae
birch skeletonizer	*Bucculatrix canadensisella* Chambers	LEPIDOPTERA: Lyonetiidae
birch tubemaker	*Acrobasis betulella* Hulst	LEPIDOPTERA: Pyralidae
bird tick	*Haemaphysalis chordeilis* (Packard)	ACARI: Ixodidae
black army cutworm	*Actebia fennica* (Tauscher)	LEPIDOPTERA: Noctuidae
black blister beetle	*Epicauta pennsylvanica* (De Geer)	COLEOPTERA: Meloidae
black blow fly	*Phormia regina* (Meigen)	DIPTERA: Calliphoridae
black carpenter ant	*Camponotus pennsylvanicus* (De Geer)	HYMENOPTERA: Formicidae
black carpet beetle	*Attagenus megatoma* (Fabricius)	COLEOPTERA: Dermestidae
black cherry aphid	*Myzus cerasi* (Fabricius)	HOMOPTERA: Aphididae
black cherry fruit fly	*Rhagoletis fausta* (Osten Sacken)	DIPTERA: Tephritidae
black citrus aphid	*Toxoptera aurantii* (Fonscolombe)	HOMOPTERA: Aphididae
black cockroach wasp	*Dolichurus stantoni* (Ashmead)	HYMENOPTERA: Ampulicidae

black cutworm	*Agrotis ipsilon* (Hufnagel)	LEPIDOPTERA: Noctuidae
black dung beetle	*Copris incertus prociduus* (Say)	COLEOPTERA: Scarabaeidae
black earwig	*Chelisoches morio* (Fabricius)	DERMAPTERA: Chelisochidae
black elm bark weevil	*Magdalis barbita* (Say)	COLEOPTERA: Curculionidae
black flower thrips	*Haplothrips gowdeyi* (Franklin)	THYSANOPTERA: Phlaeothripidae
black fungus beetle	*Alphitobius laevigatus* (Fabricius)	COLEOPTERA: Tenebrionidae
black grain stem sawfly	*Trachelus tabidus* (Fabricius)	HYMENOPTERA: Cephidae
black horse fly	*Tabanus atratus* Fabricius	DIPTERA: Tabanidae
black hunter thrips	*Leptothrips mali* (Fitch)	THYSANOPTERA: Phlaeothripidae
black imported fire ant	*Solenopsis richteri* Forel	HYMENOPTERA: Formicidae
black ladybird beetle	*Rhizobius ventralis* (Erichson)	COLEOPTERA: Coccinellidae
black larder beetle	*Dermestes ater* De Geer	COLEOPTERA: Dermestidae
black peach aphid	*Brachycaudus persicae* (Passerini)	HOMOPTERA: Aphididae
black pecan aphid	*Melanocallis caryaefoliae* (Davis)	HOMOPTERA: Aphididae
black pineleaf scale	*Nuculaspis californica* (Coleman)	HOMOPTERA: Diaspididae
black potter wasp	*Delta pyriformis philippinensis* (Bequaert)	HYMENOPTERA: Vespidae
black scale	*Saissetia oleae* (Olivier)	HOMOPTERA: Coccidae
black soldier fly	*Hermetia illucens* (Linnaeus)	DIPTERA: Stratiomyidae
black stink bug	*Coptosoma xanthogramma* (White)	HEMIPTERA: Plataspidae
black swallowtail	*Papilio polyxenes asterius* Stoll	LEPIDOPTERA: Papilionidae
black thread scale	*Ischnaspis longirostris* (Signoret)	HOMOPTERA: Diaspididae
black turfgrass ataenius	*Ataenius spretulus* (Haldeman)	COLEOPTERA: Scarabaeidae
blank turpentine beetle	*Dendroctonus terebrans* (Olivier)	COLEOPTERA: Scolytidae
black twig borer	*Xylosandrus compactus* (Eichhoff)	COLEOPTERA: Scolytidae
black vine weevil	*Otiorhynchus sulcatus* (Fabricius)	COLEOPTERA: Curculionidae
black walnut curculio	*Conotrachelus retentus* (Say)	COLEOPTERA: Curculionidae
black widow spider	*Latrodectus mactans* (Fabricius)	ARANEAE: Theridiidae
black witch	*Ascalapha odorata* (Linnaeus)	LEPIDOPTERA: Noctuidae
blackbellied clerid	*Enoclerus lecontei* (Wolcott)	COLEOPTERA: Cleridae
blackberry skeletonizer	*Schreckensteinia festaliella* (Hübner)	LEPIDOPTERA: Heliodinidae
Blackburn butterfly	*Vaga blackburni* (Tuely)	LEPIDOPTERA: Lycaenidae
Blackburn damsel bug	*Nabis blackburni* (White)	HEMIPTERA: Nabidae
Blackburn dragonfly	*Nesogonia blackburni* (McLachlan)	ODONATA: Libellulidae
blackfaced leafhopper	*Graminella nigrifrons* (Forbes)	HOMOPTERA: Cicadellidae
blackheaded ash sawfly	*Tethida barba* (Say)	HYMENOPTERA: Tenthredinidae
blackheaded fireworm	*Rhopobota unipunctana* (Haworth)	LEPIDOPTERA: Tortricidae
blackheaded pine sawfly	*Neodiprion excitans* Rohwer	HYMENOPTERA: Diprionidae
blackhorned pine borer	*Callidium antennatum hesperum* Casey	COLEOPTERA: Cerambycidae
blackhorned tree cricket	*Oecanthus nigricornis* Walker	ORTHOPTERA: Gryllidae
blackjacket	*Vespula consobrina* (Saussure)	HYMENOPTERA: Vespidae
blacklegged tick	*Ixodes scapularis* Say	ACARI: Ixodidae
blacklegged tortoise beetle	*Jonthonota nigripes* (Olivier)	COLEOPTERA: Chrysomelidae
blackmargined aphid	*Monellia caryella* (Fitch)	HOMOPTERA: Aphididae
blister coneworm	*Dioryctria clarioralis* (Walker)	LEPIDOPTERA: Pyralidae
bloodsucking conenose	*Triatoma sanguisuga* (LeConte)	HEMIPTERA: Reduviidae
blue alfalfa aphid	*Acyrthosiphon kondi* Shinji	HOMOPTERA: Aphididae
blue cactus borer	*Melitara dentata* (Grote)	LEPIDOPTERA: Pyralidae
blue horntail	*Sirex cyaneus* Fabricius	HYMENOPTERA: Siricidae
blue soldier fly	*Neoexaireta spinigera* (Wiedemann)	DIPTERA: Stratiomyidae
blueberry bud mite	*Acalitus vaccinii* (Keifer)	ACARI: Eriophyidae
blueberry case beetle	*Neochlamisus cribripennis* (Leconte)	COLEOPTERA: Chrysomelidae
blueberry flea beetle	*Altica sylvia* Malloch	COLEOPTERA: Chrysomelidae
blueberry maggot	*Rhagoletis mendax* (Curran)	DIPTERA: Tephritidae
blueberry thrips	*Frankliniella vaccinii* Morgan	THYSANOPTERA: Thripidae
blueberry tip midge	*Contarinia vaccinii* (Felt)	DIPTERA: Cecidomyiidae
bluegrass billbug	*Sphenophorus parvulus* Gyllenhal	COLEOPTERA: Curculionidae
bluegrass webworm	*Parapediasia teterrella* (Zincken)	LEPIDOPTERA: Pyralidae
bluntnosed cranberry leafhopper	*Scleroracus vaccinii* (Van Duzee)	HOMOPTERA: Cicadellidae
body louse	*Pediculus humanus humanus* Linnaeus	ANOPLURA: Pediculidae
Boisduval scale	*Diaspis boisduvalli* Signoret	HOMOPTERA: Diaspididae

boll weevil	*Anthonomus grandis grandis* Boheman	COLEOPTERA: Curculionidae
bollworm	*Heliothis zea* (Boddie)	LEPIDOPTERA: Noctuidae
booklouse	*Liposcelis corrodens* Heymons	PSOCOPTERA: Liposcelidae
boxelder aphid	*Periphyllus negundinis* (Thomas)	HOMOPTERA: Aphididae
boxelder bug	*Leptocoris trivittatus* (Say)	HEMIPTERA: Rhopalidae
boxelder leafroller	*Caloptilia negundella* (Chambers)	LEPIDOPTERA: Gracillariidae
boxelder psyllid	*Psylla negundinis* Mally	HOMOPTERA: Psyllidae
boxelder twig borer	*Proteoteras willingana* (Kearfott)	LEPIDOPTERA: Tortricidae
boxwood leafminer	*Monarthropalpus buxi* (Laboulbène)	DIPTERA: Cecidomyiidae
boxwood psyllid	*Psylla buxi* (Linnaeus)	HOMOPTERA: Psyllidae
bramble leafhopper	*Ribautiana tenerrima* (Herrich-Schäffer)	HOMOPTERA: Cicadellidae
Brasilian leafhopper	*Protalebrella brasiliensis* (Baker)	HOMOPTERA: Cicadellidae
bristly cutworm	*Lacinipolia renigera* (Stephens)	LEPIDOPTERA: Noctuidae
bristly roseslug	*Cladius difformis* (Panzer)	HYMENOPTERA: Tenthredinidae
broad mite	*Polyphagotarsonemus latus* (Banks)	ACARI: Tarsonemidae
broadbean weevil	*Bruchus rufimanus* Boheman	COLEOPTERA: Bruchidae
broadhorned flour beetle	*Gnathocerus cornutus* (Fabricius)	COLEOPTERA: Tenebrionidae
broadnecked root borer	*Prionus laticollis* (Drury)	COLEOPTERA: Cerambycidae
broadnosed grain weevil	*Caulophilus oryzae* (Gyllenhal)	COLEOPTERA: Curculionidae
broadwinged katydid	*Microcentrum rhombifolium* (Saussure)	ORTHOPTERA: Tettigoniidae
bromegrass seed midge	*Contarinia bromicola* (Marikovskij & Agafonova)	DIPTERA: Cecidomyiidae
bronze appletree weevil	*Magdalis aenescens* LeConte	COLEOPTERA: Curculionidae
bronze birch borer	*Agrilus anxius* Gory	COLEOPTERA: Buprestidae
bronze leaf beetle	*Diachus auratus* (Fabricius)	COLEOPTERA: Chrysomelidae
bronze poplar borer	*Agrilus liragus* Barter & Brown	COLEOPTERA: Buprestidae
bronzed cutworm	*Nephelodes minians* (Guenée)	LEPIDOPTERA: Noctuidae
brown chicken louse	*Goniodes dissimilis* Denny	MALLOPHAGA: Philopteridae
brown citrus aphid	*Toxoptera citricida* (Kirkaldy)	HOMOPTERA: Aphididae
brown cockroach	*Periplaneta brunnea* Burmeister	ORTHOPTERA: Blattidae
brown cotton leafworm	*Acontia dacia* Druce	LEPIDOPTERA: Noctuidae
brown dog tick	*Rhipicephalus sanguineus* (Latreille)	ACARI: Ixodidae
brown dung beetle	*Onthophagus gazella* Fabricius	COLEOPTERA: Scarabaeidae
brown flour mite	*Gohieria fusca* (Oudemans)	ACARI: Glycyphagidae
brown house moth	*Hofmannophila pseudospretella* (Stainton)	LEPIDOPTERA: Oecophoridae
brown mite	*Bryobia rubrioculus* (Scheuten)	ACARI: Tetranychidae
brown pineapple scale	*Melanaspis bromeliae* (Leonardi)	HOMOPTERA: Diaspididae
brown recluse spider	*Loxosceles reclusa* Gertsch & Mulaik	ARANEAE: Loxoscelidae
brown saltmarsh mosquito	*Aedes cantator* (Coquillett)	DIPTERA: Culicidae
brown soft scale	*Coccus hesperidum* Linnaeus	HOMOPTERA: Coccidae
brown spider beetle	*Ptinus clavipes* Panzer	COLEOPTERA: Ptinidae
brown stink bug	*Euschistus servus* (Say)	HEMIPTERA: Pentatomidae
brown wheat mite	*Petrobia latens* (Müller)	ACARI: Tetranychidae
brown widow spider	*Latrodectus geometricus* (Fabricius)	ARANEAE: Theridiidae
brownbanded cockroach	*Supella longipalpa* (Fabricius)	ORTHOPTERA: Blattellidae
brownheaded ash sawfly	*Tomostethus multicinctus* (Rohwer)	HYMENOPTERA: Tenthredinidae
brownheaded jack pine sawfly	*Neodiprion dubiosus* Schedl	HYMENOPTERA: Diprionidae
browlegged grain mite	*Aleuroglyphus ovatus* (Troupeau)	ACARI: Acaridae
browntail moth	*Euproctis chrysorrhoea* (Linnaeus)	LEPIDOPTERA: Lymantriidae
Bruce spanworm	*Operophtera bruceata* (Hulst)	LEPIDOPTERA: Geometridae
buck moth	*Hemileuca maia* (Drury)	LEPIDOPTERA: Saturniidae
buckthorn aphid	*Aphis nasturtii* Kaltenbach	HOMOPTERA: Aphididae
buffalo treehopper	*Stictocephala bisonia* Kopp & Yonke	HOMOPTERA: Membracidae
buffalograss webworm	*Surattha indentella* Kearfott	LEPIDOPTERA: Pyralidae
bulb mite	*Rhizoglyphus echinopus* (Fumouze & Robin)	ACARI: Acaridae
bulb scale mite	*Steneotarsonemus laticeps* (Halbert)	ACARI: Tarsonemidae
bumble flower beetle	*Euphoria inda* (Linnaeus)	COLEOPTERA: Scarabaeidae
bumelia fruit fly	*Pseudodacus pallens* (Coquillett)	DIPTERA: Tephritidae
burdock borer	*Papaipema cataphracta* (Grote)	LEPIDOPTERA: Noctuidae
Burmeister mantid	*Orthodera burmeisteri* Wood-Mason	ORTHOPTERA: Mantidae
butternut curculio	*Conotrachelus juglandis* LeConte	COLEOPTERA: Curculionidae

C

cabbage aphid	*Brevicoryne brassicae* (Linnaeus)	HOMOPTERA: Aphididae
cabbage curculio	*Ceutorhynchus rapae* Gyllenhal	COLEOPTERA: Curculionidae
cabbage looper	*Trichoplusia ni* (Hübner)	LEPIDOPTERA: Noctuidae
cabbage maggot	*Delia radicum* (Linnaeus)	DIPTERA: Anthomyiidae
cabbage seedpod weevil	*Ceutorhynchus assimilis* (Paykull)	COLEOPTERA: Curculionidae
cabbage seedstalk curculio	*Ceutorhynchus quadridens* (Panzer)	COLEOPTERA: Curculionidae
cabbage webworm	*Hellula rogatalis* (Hulst)	LEPIDOPTERA: Pyralidae
cactus moth	*Cactoblastis cactorum* (Berg)	LEPIDOPTERA: Pyralidae
cactus scale	*Diaspis echinocacti* (Bouché)	HOMOPTERA: Diaspididae
cadelle	*Tenebroides mauritanicus* (Linnaeus)	COLEOPTERA: Trogositidae
Caledonia seed bug	*Nysius caledoniae* Distant	HEMIPTERA: Lygaeidae
calico scale	*Eulecanium cerasorum* (Cockerell)	HOMOPTERA: Coccidae
California fivespined ips	*Ips paraconfusus* Lanier	COLEOPTERA: Scolytidae
California flatheaded borer	*Melanophila californica* Van Dyke	COLEOPTERA: Buprestidae
California harvester ant	*Pogonomyrmex californicus* (Buckley)	HYMENOPTERA: Formicidae
California oakworm	*Phryganidia californica* Packard	LEPIDOPTERA: Dioptidae
California pear sawfly	*Pristiphora abbreviata* (Hartig)	HYMENOPTERA: Tenthredinidae
California prionus	*Prionus californicus* Motschulsky	COLEOPTERA: Cerambycidae
California red scale	*Aonidiella aurantii* (Maskell)	HOMOPTERA: Diaspididae
California saltmarsh mosquito	*Aedes squamiger* (Coquillett)	DIPTERA: Culicidae
California tortoiseshell	*Nymphalis californica* (Boisduval)	LEPIDOPTERA: Nymphalidae
camellia scale	*Lepidosaphes camelliae* Hoke	HOMOPTERA: Diaspididae
camphor scale	*Pseudaonidia duplex* (Cockerell)	HOMOPTERA: Diaspididae
camphor thrips	*Liothrips floridensis* (Watson)	THYSANOPTERA: Phlaeothripidae
caragana aphid	*Acyrthosiphon caraganae* (Cholodkovsky)	HOMOPTERA: Aphididae
caragana blister beetle	*Epicauta subglabra* (Fall)	COLEOPTERA: Meloidae
caragana plant bug	*Lopidea dakota* Knight	HEMIPTERA: Miridae
Caribbean black scale	*Saissetia neglecta* De Lotto	HOMOPTERA: Coccidae
Caribbean pod borer	*Fundella pellucens* Zeller	LEPIDOPTERA: Pyralidae
carmine spider mite	*Tetranychus cinnabarinus* (Boisduval)	ACARI: Tetranychidae
carnation maggot	*Delia brunnescens* (Zetterstedt)	DIPTERA: Anthomyiidae
carnation tip maggot	*Delia echinata* (Séguy)	DIPTERA: Anthomyiidae
Carolina conifer aphid	*Cinara atlantica* (Wilson)	HOMOPTERA: Aphididae
Carolina grasshopper	*Dissosteira carolina* (Linnaeus)	ORTHOPTERA: Acrididae
Carolina mantid	*Stagmomantis carolina* (Johannson)	ORTHOPTERA: Mantidae
carpenter bee	*Xylocopa virginica* (Linnaeus)	HYMENOPTERA: Xylocopidae
carpenterworm	*Prionoxystus robiniae* (Peck)	LEPIDOPTERA: Cossidae
carpet beetle	*Anthrenus scrophulariae* (Linnaeus)	COLEOPTERA: Dermestidae
carpet moth	*Trichophaga tapetzella* (Linnaeus)	LEPIDOPTERA: Tineidae
carrot beetle	*Bothynus gibbosus* (De Geer)	COLEOPTERA: Scarabaeidae
carrot rust fly	*Psila rosae* (Fabricius)	DIPTERA: Psilidae
carrot weevil	*Listronotus oregonensis* (LeConte)	COLEOPTERA: Curculionidae
casemaking clothes moth	*Tinea pellionella* Linnaeus	LEPIDOPTERA: Tineidae
cat flea	*Ctenocephalides felis* (Bouché)	SIPHONAPTERA: Pulicidae
cat follicle mite	*Demodex cati* Mégnin	ACARI: Demodicidae
cat louse	*Felicola subrostratus* (Burmeister)	MALLOPHAGA: Trichodectidae
catalpa midge	*Contarinia catalpae* (Comstock)	DIPTERA: Cecidomyiidae
catalpa sphinx	*Ceratomia catalpae* (Boisduval)	LEPIDOPTERA: Sphingidae
cattle biting louse	*Bovicola bovis* (Linnaeus)	MALLOPHAGA: Trichodectidae
cattle follicle mite	*Demodex bovis* Stiles	ACARI: Demodicidae
cattle itch mite	*Sarcoptes bovis* Robin	ACARI: Sarcoptidae
cattle tail louse	*Haematopinus quadripertusus* Fahrenholz	ANOPLURA: Haematopinidae
cattle tick	*Boophilus annulatus* (Say)	ACARI: Ixodidae
Cayenne tick	*Amblyomma cajennense* (Fabricius)	ACARI: Ixodidae
ceanothus silk moth	*Hyalophora euryalus* (Boisduval)	LEPIDOPTERA: Saturniidae
cecropia moth	*Hyalophora cecropia* (Linnaeus)	LEPIDOPTERA: Saturniidae
cedartree borer	*Semanotus ligneus* (Fabricius)	COLEOPTERA: Cerambycidae
celery aphid	*Brachycolus heraclei* Takahashi	HOMOPTERA: Aphididae

celery leaftier	*Udea rubigalis* (Guenée)	LEPIDOPTERA: Pyralidae
celery looper	*Syngrapha falcitera* (Kirby)	LEPIDOPTERA: Noctuidae
cereal leaf beetle	*Oulema melanopus* (Linnaeus)	COLEOPTERA: Chrysomelidae
chaff scale	*Parlatoria pergandii* Comstock	HOMOPTERA: Diaspididae
chainspotted geometer	*Cingilia catenaria* (Drury)	LEPIDOPTERA: Geometridae
changa	*Scapteriscus vicinus* Scudder	ORTHOPTERA: Gryllotalpidae
charcoal beetle	*Melanophila consputa* LeConte	COLEOPTERA: Buprestidae
cheese mite	*Tyrolichus casei* Oudemans	ACARI: Acaridae
cheese skipper	*Piophila casei* (Linnaeus)	DIPTERA: Piophilidae
cherry casebearer	*Coleophora pruniella* Clemens	LEPIDOPTERA: Coleophoridae
cherry fruit fly	*Rhagoletis cingulata* (Loew)	DIPTERA: Tephritidae
cherry fruit sawfly	*Hoplocampa cookei* (Clarke)	HYMENOPTERA: Tenthredinidae
cherry fruitworm	*Grapholita packardi* Zeller	LEPIDOPTERA: Tortricidae
cherry leaf beetle	*Pyrrhalta cavicollis* (LeConte)	COLEOPTERA: Chrysomelidae
cherry maggot	*Rhagoletis cingulata* (Loew)	DIPTERA: Tephritidae
chestnut timberworm	*Melittomma sericeum* (Harris)	COLEOPTERA: Lymexylonidae
chicken body louse	*Menacanthus stramineus* (Nitzsch)	MALLOPHAGA: Menoponidae
chicken dung fly	*Fannia pusio* (Wiedemann)	DIPTERA: Muscidae
chicken head louse	*Cuclotogaster heterographus* (Nitzsch)	MALLOPHAGA: Philopteridae
chicken mite	*Dermanyssus gallinae* (De Geer)	ACARI: Dermanyssidae
chigoe	*Tunga penetrans* Linnaeus	SIPHONAPTERA: Tungidae
chinch bug	*Blissus leucopterus leucopterus* (Say)	HEMIPTERA: Lygaeidae
Chinese dryinid	*Pseudogonatopus hospes* Perkins	HYMENOPTERA: Dryinidae
Chinese mantid	*Tenodera aridifolia sinensis* Saussure	ORTHOPTERA: Mantidae
Chinese obscure scale	*Parlatoreopsis chinensis* (Marlatt)	HOMOPTERA: Diaspididae
Chinese rose beetle	*Adoretus sinicus* Burmeister	COLEOPTERA: Scarabaeidae
Christmas berry webworm	*Cryptoblabes gnidiella* (Millière)	LEPIDOPTERA: Pyralidae
chrysanthemum aphid	*Macrosiphoniella sanborni* (Gillette)	HOMOPTERA: Aphididae
chrysanthemum flower borer	*Lorita abornana* Busck	LEPIDOPTERA: Cochylidae
chrysanthemum gall midge	*Rhopalomyia chrysanthemi* (Ahlberg)	DIPTERA: Cecidomyiidae
chrysanthemum lace bug	*Corythucha marmorata* (Uhler)	HEMIPTERA: Tingidae
chrysanthemum leafminer	*Phytomyza syngenesiae* (Hardy)	DIPTERA: Agromyzidae
chrysanthemum thrips	*Thrips nigropilosus* Uzel	THYSANOPTERA: Thripidae
cicada killer	*Sphecius speciosus* (Drury)	HYMENOPTERA: Sphecidae
cigar casebearer	*Coleophora serratella* (Linnaeus)	LEPIDOPTERA: Coleophoridae
cigarette beetle	*Lasioderma serricorne* (Fabricius)	COLEOPTERA: Anobiidae
cinereous cockroach	*Nauphoeta cinerea* (Olivier)	ORTHOPTERA: Blaberidae
cinnabar moth	*Tyria jacobaeae* (Linnaeus)	LEPIDOPTERA: Arctiidae
citricola scale	*Coccus pseudomagnoliarum* (Kuwana)	HOMOPTERA: Coccidae
citrophilus mealybug	*Pseudococcus calceolariae* (Maskell)	HOMOPTERA: Pseudococcidae
citrus blackfly	*Aleurocanthus woglumi* Ashby	HOMOPTERA: Aleyrodidae
citrus bud mite	*Eriophyes sheldoni* Ewing	ACARI: Eriophyidae
citrus flat mite	*Brevipalpus lewisi* McGregor	ACARI: Tenuipalpidae
citrus mealybug	*Planococcus citri* (Risso)	HOMOPTERA: Pseudococcidae
citrus red mite	*Panonychus citri* (McGregor)	ACARI: Tetranychidae
citrus root weevil	*Pachnaeus litus* (Germar)	COLEOPTERA: Curculionidae
citrus rust mite	*Phyllocoptruta oleivora* (Ashmead)	ACARI: Eriophyidae
citrus snow scale	*Unaspis citri* (Comstock)	HOMOPTERA: Diaspididae
citrus swallowtail	*Papilio xuthus* Linnaeus	LEPIDOPTERA: Papilionidae
citrus thrips	*Scirtothrips citri* (Moulton)	THYSANOPTERA: Thripidae
citrus whitefly	*Dialeurodes citri* (Ashmead)	HOMOPTERA: Aleyrodidae
claybacked cutworm	*Agrotis gladiaria* Morrison	LEPIDOPTERA: Noctuidae
claycolored billbug	*Sphenophorus aequalis aequalis* Gyllenhal	COLEOPTERA: Curculionidae
claycolored leaf beetle	*Anomoea laticlavia* (Forster)	COLEOPTERA: Chrysomelidae
Clear Lake gnat	*Chaoborus astictopus* Dyar & Shannon	DIPTERA: Chaoboridae
clearwinged grasshopper	*Camnula pellucida* (Scudder)	ORTHOPTERA: Acrididae
clematis blister beetle	*Epicauta cinerea* (Forster)	COLEOPTERA: Meloidae
clidemia leafroller	*Blepharomastix ebulealis* (Guenée)	LEPIDOPTERA: Pyralidae
clidemia thrips	*Liothrips urichi* Karny	THYSANOPTERA: Phlaeothripidae

clouded plant bug	*Neurocolpus nubilus* (Say)	HEMIPTERA: Miridae
clouded sulfur	*Colias philodice* Godart	LEPIDOPTERA: Pieridae
cloudywinged whitefly	*Dialeurodes citrifolii* (Morgan)	HOMOPTERA: Aleyrodidae
clover aphid	*Nearctaphis bakeri* (Cowen)	HOMOPTERA: Aphididae
clover aphid parasite	*Aphelinus lapisligni* Howard	HYMENOPTERA: Encyrtidae
clover cutworm	*Scotogramma trifolii* (Hufnagel)	LEPIDOPTERA: Noctuidae
clover hayworm	*Hypsopygia costalis* (Fabricius)	LEPIDOPTERA: Pyralidae
clover head caterpillar	*Grapholita interstinctana* (Clemens)	LEPIDOPTERA: Tortricidae
clover head weevil	*Hypera meles* (Fabricius)	COLEOPTERA: Curculionidae
clover leaf midge	*Dasineura trifolii* (Loew)	DIPTERA: Cecidomyiidae
clover leaf weevil	*Hypera punctata* (Fabricius)	COLEOPTERA: Curculionidae
clover leafhopper	*Aceratagallia sanguinolenta* (Provancher)	HOMOPTERA: Cicadellidae
clover looper	*Caenurgina crassiuscula* (Haworth)	LEPIDOPTERA: Noctuidae
clover mite	*Bryobia praetiosa* Koch	ACARI: Tetranychidae
clover root borer	*Hylastinus obscurus* (Marsham)	COLEOPTERA: Scolytidae
clover root curculio	*Sitona hispidulus* (Fabricius)	COLEOPTERA: Curculionidae
clover seed chalcid	*Bruchophagus platypterus* (Walker)	HYMENOPTERA: Eurytomidae
clover seed midge	*Dasineura leguminicola* (Lintner)	DIPTERA: Cecidomyiidae
clover seed weevil	*Tychius picirostris* (Fabricius)	COLEOPTERA: Curculionidae
clover stem borer	*Languria mozardi* Latreille	COLEOPTERA: Languriidae
cluster fly	*Pollenia rudis* (Fabricius)	DIPTERA: Calliphoridae
cochineal insect	*Dactylopius coccus* Costa	HOMOPTERA: Dactylopiidae
cocklebur weevil	*Rhodobaenus tredecimpunctatus* (Illiger)	COLEOPTERA: Curculionidae
coconut leafminer	*Agonoxena argaula* Meyrick	LEPIDOPTERA: Agonoxenidae
coconut leafroller	*Hedylepta blackburni* (Butler)	LEPIDOPTERA: Pyralidae
coconut mealybug	*Nipaecoccus nipae* (Maskell)	HOMOPTERA: Pseudococcidae
coconut scale	*Aspidiotus destructor* Signoret	HOMOPTERA: Diaspididae
codling moth	*Cydia pomonella* (Linnaeus)	LEPIDOPTERA: Tortricidae
coffee bean weevil	*Araecerus fasciculatus* (De Geer)	COLEOPTERA: Anthribidae
Colorado potato beetle	*Leptinotarsa decemlineata* (Say)	COLEOPTERA: Chrysomelidae
Columbia Basin wireworm	*Limonius subauratus* LeConte	COLEOPTERA: Elateridae
Columbian timber beetle	*Corthylus columbianus* Hopkins	COLEOPTERA: Scolytidae
columbine borer	*Papaipema purpurifascia* (Grote & Robinson)	LEPIDOPTERA: Noctuidae
Comanche lacewing	*Chrysopa comanche* Banks	NEUROPTERA: Chrysopidae
common Australian lady beetle	*Coelophora inaequalis* (Fabricius)	COLEOPTERA: Coccinellidae
common cattle grub	*Hypoderma lineatum* (Villers)	DIPTERA: Oestridae
common damsel bug	*Reduviolus americoferus* (Carayon)	HEMIPTERA: Nabidae
common green darner	*Anax junius* (Drury)	ODONATA: Aeschnidae
common green lacewing	*Chrysoperla carnea* Stephens	NEUROPTERA: Chrysopidae
common malaria mosquito	*Ahopheles quadrimaculatus* Say	DIPTERA: Culicidae
composite thrips	*Microcephalothrips abdominalis* (D. L. Crawford)	THYSANOPTERA: Thripidae
Comstock mealybug	*Pseudococcus comstocki* (Kuwana)	HOMOPTERA: Pseudococcidae
conchuela	*Chlorochroa ligata* (Say)	HEMIPTERA: Pentatomidae
confused flour beetle	*Tribolium confusum* Jacquelin du Val	COLEOPTERA: Tenebrionidae
conifer spider mite	*Oligonychus coniferarum* (McGregor)	ACARI: Tetranychidae
convergent ladybird beetle	*Hippodamia convergens* Guérin-Méneville	COLEOPTERA: Coccinellidae
Cooley spruce gall adelgid	*Adelges cooleyi* (Gillette)	HOMOPTERA: Adelgidae
corn delphacid	*Peregrinus maidis* (Ashmead)	HOMOPTERA: Delphacidae
corn earworm	*Heliothis zea* (Boddie)	LEPIDOPTERA: Noctuidae
corn flea beetle	*Chaetocnema pulicaria* Melsheimer	COLEOPTERA: Chrysomelidae
corn leaf aphid	*Rhopalosiphum maidis* (Fitch)	HOMOPTERA: Aphididae
corn root aphid	*Anuraphis maidiradicis* (Forbes)	HOMOPTERA: Aphididae
corn root webworm	*Crambus caliginosellus* Clemens	LEPIDOPTERA: Pyralidae
corn sap beetle	*Carpophilus dimidiatus* (Fabricius)	COLEOPTERA: Nitidulidae
corn silk beetle	*Calomicrus brunneus* (Crotch)	COLEOPTERA: Chrysomelidae
cornfield ant	*Lasius alienus* (Foerster)	HYMENOPTERA: Formicidae
cosmopolitan grain psocid	*Lachesilla pedicularia* (Linnaeus)	PSOCOPTERA: Lachesillidae
cotton aphid	*Aphis gossypii* Glover	HOMOPTERA: Aphididae

cotton blister mite	*Acalitus gossypii* (Banks)	ACARI: Eriophyidae
cotton fleahopper	*Pseudatomoscelis seriatus* (Reuter)	HEMIPTERA: Miridae
cotton lace bug	*Corythucha gossypii* (Fabricius)	HEMIPTERA: Tingidae
cotton leafminer	*Stigmella gossypii* (Forbes & Leonard)	LEPIDOPTERA: Nepticulidae
cotton leafperforator	*Bucculatrix thurberiella* Busck	LEPIDOPTERA: Lyonetiidae
cotton leafworm	*Alabama argillacea* (Hübner)	LEPIDOPTERA: Noctuidae
cotton square borer	*Strymon melinus* Hübner	LEPIDOPTERA: Lycaenidae
cotton stainer	*Dysdercus suturellus* (Herrich-Schäffer)	HEMIPTERA: Pyrrhocoridae
cotton stem moth	*Platyedra subcinerea* (Haworth)	LEPIDOPTERA: Gelechiidae
cottonwood borer	*Plectrodera scalator* (Fabricius)	COLEOPTERA: Cerambycidae
cottonwood dagger moth	*Acronicta lepusculina* Guenée	LEPIDOPTERA: Noctuidae
cottonwood leaf beetle	*Chrysomela scripta* Fabricius	COLEOPTERA: Chrysomelidae
cottonwood twig borer	*Gypsonoma haimbachiana* (Kearfott)	LEPIDOPTERA: Tortricidae
cottony maple scale	*Pulvinaria innumerabilis* (Rathvon)	HOMOPTERA: Coccidae
cottony peach scale	*Pulvinaria amygdali* Cockerell	HOMOPTERA: Coccidae
cottonycushion scale	*Icerya purchasi* (Maskell)	HOMOPTERA: Margarodidae
coulee cricket	*Peranabrus scabricollis* (Thomas)	ORTHOPTERA: Tettigoniidae
cowpea aphid	*Aphis craccivora* Koch	HOMOPTERA: Aphididae
cowpea curculio	*Chalcodermus aeneus* Boheman	COLEOPTERA: Curculionidae
cowpea weevil	*Callosobruchus maculatus* (Fabricius)	COLEOPTERA: Bruchidae
crab louse	*Pthirus pubis* (Linnaeus)	ANOPLURA: Pediculidae
crabhole mosquito	*Deinocerites cancer* Theobald	DIPTERA: Culicidae
cranberry fruitworm	*Acrobasis vaccinii* Riley	LEPIDOPTERA: Pyralidae
cranberry girdler	*Chrysoteuchia topiaria* (Zeller)	LEPIDOPTERA: Pyralidae
cranberry rootworm	*Rhabdopterus picipes* (Olivier)	COLEOPTERA: Chrysomelidae
cranberry weevil	*Anthonomus musculus* Say	COLEOPTERA: Curculionidae
crapemyrtle aphid	*Tinocallis kahawaluokalani* (Kirkaldy)	HOMOPTERA: Aphididae
crazy ant	*Paratrechina longicornis* (Latreille)	HYMENOPTERA: Formicidae
crescentmarked lily aphid	*Neomyzus circumflexus* (Buckton)	HOMOPTERA: Aphididae
cribrate weevil	*Otiorhynchus cribricollis* Gyllenhal	COLEOPTERA: Curcualionidae
crinkled flannel moth	*Lagoa crispata* (Packard)	LEPIDOPTERA: Megalopygidae
cross-striped cabbageworm	*Evergestis rimosalis* (Guenée)	LEPIDOPTERA: Pyralidae
croton caterpillar	*Achaea janata* (Linnaeus)	LEPIDOPTERA: Noctuidae
croton mussel scale	*Lepidosaphes tokionis* (Kuwana)	HOMOPTERA: Diaspididae
Cuban cockroach	*Panchlora nivea* (Linnaeus)	ORTHOPTERA: Blaberidae
Cuban laurel thrips	*Gynaikothrips ficorum* (Marchal)	THYSANOPTERA: Phlaeothripidae
cucurbit longicorn	*Apomecyna saltator* (Fabricius)	COLEOPTERA: Cerambycidae
cucurbit midge	*Prodiplosis citrulli* (Felt)	DIPTERA: Cecidomyiidae
curled rose sawfly	*Allantus cinctus* (Linnaeus)	HYMENOPTERA: Tenthredinidae
currant aphid	*Cryptomyzus ribis* (Linnaeus)	HOMOPTERA: Aphididae
currant borer	*Synanthedon tipuliformis* (Clerck)	LEPIDOPTERA: Sesiidae
currant bud mite	*Cecidophyopsis ribis* (Westwood)	ACARI: Eriophyidae
currant fruit fly	*Epochra canadensis* (Loew)	DIPTERA: Tephritidae
currant fruit weevil	*Pseudanthonomus validus* Dietz	COLEOPTERA: Curculionidae
currant spanworm	*Itame ribearia* (Fitch)	LEPIDOPTERA: Geometridae
currant stem girdler	*Janus integer* (Norton)	HYMENOPTERA: Cephidae
cyclamen mite	*Steneotarsonemus pallidus* (Banks)	ACARI: Tarsonemidae
cynthia moth	*Samia cynthia* (Drury)	LEPIDOPTERA: Saturniidae

D

dandelion gall wasp	*Phanacis taraxaci* (Ashmead)	HYMENOPTERA: Cynipidae
dark mealworm	*Tenebrio obscurus* Fabricius	COLEOPTERA: Tenebrionidae
darksided cutworm	*Euxoa messoria* (Harris)	LEPIDOPTERA: Noctuidae
datebug	*Asarcopus palmarum* Horvath	HOMOPTERA: Issidae
deodar weevil	*Pissodes nemorensis* Germar	COLEOPTERA: Curculionidae
depluming mite	*Knemidokoptes gallinae* (Railliet)	ACARI: Sarcoptidae
depressed flour beetle	*Palorus subdepressus* (Wollaston)	COLEOPTERA: Tenebrionidae
desert corn flea beetle	*Chaetocnema ectypa* Horn	COLEOPTERA: Chrysomelidae

desert spider mite	*Tetranychus desertorum* Banks	ACARI: Tetranychidae
devastating grasshopper	*Melanoplus devastator* Scudder	ORTHOPTERA: Acrididae
diamondback moth	*Plutella xylostella* (Linnaeus)	LEPIDOPTERA: Plutellidae
diamondbacked spittlebug	*Lepyronia quadrangularis* (Say)	HOMOPTERA: Cercopidae
dictyospermum scale	*Chrysomphalus dictyospermi* (Morgan)	HOMOPTERA: Diaspididae
differential grasshopper	*Melanoplus differentialis* (Thomas)	ORTHOPTERA: Acrididae
dingy cutworm	*Feltia ducens* Walker	LEPIDOPTERA: Noctuidae
dobsonfly	*Corydalus cornutus* (Linnaeus)	NEUROPTERA: Corydalidae
dock sawfly	*Ametastegia glabrata* (Fallén)	HYMENOPTERA: Tenthredinidae
dodder gall weevil	*Smicronyx sculpticollis* Casey	COLEOPTERA: Curculionidae
dog biting louse	*Trichodectes canis* (De Geer)	MALLOPHAGA: Trichodectidae
dog flea	*Ctenocephalides canis* (Curtis)	SIPHONAPTERA: Pulicidae
dog follicle mite	*Demodex canis* Leydig	ACARI: Demodicidae
dog sucking louse	*Linognathus setosus* (Olfers)	ANOPLURA: Linognathidae
dogwood borer	*Synanthedon scitula* (Harris)	LEPIDOPTERA: Sesiidae
dogwood clubgall midge	*Resseliella clavula* (Beutenmüller)	DIPTERA: Cecidomyiidae
dogwood scale	*Chionaspis corni* Cooley	HOMOPTERA: Diaspididae
dogwood spittlebug	*Clastoptera proteus* Fitch	HOMOPTERA: Cercopidae
dogwood twig borer	*Oberea tripunctata* (Swederus)	COLEOPTERA: Cerambycidae
Douglas-fir beetle	*Dendroctonus pseudotsugae* Hopkins	COLEOPTERA: Scolytidae
Douglas-fir cone moth	*Barbara colfaxiana* (Kearfott)	LEPIDOPTERA: Tortricidae
Douglas-fir engraver	*Scolytus unispinosus* LeConte	COLEOPTERA: Scolytidae
Douglas-fir pitch moth	*Synanthedon novaroensis* (Edwards)	LEPIDOPTERA: Sesiidae
Douglas-fir tussock moth	*Orgyia pseudotsugata* (McDunnough)	LEPIDOPTERA: Lymantriidae
Douglas-fir twig weevil	*Cylindrocopturus furnissi* Buchanan	COLEOPTERA: Curculionidae
driedfruit beetle	*Carpophilus hemipterus* (Linnaeus)	COLEOPTERA: Nitidulidae
driedfruit mite	*Carpoglyphus lactis* (Linnaeus)	ACARI: Carpoglyphidae
driedfruit moth	*Vitula edmandsae serratilineella* Ragonot	LEPIDOPTERA: Pyralidae
drone fly	*Eristalis tenax* (Linnaus)	DIPTERA: Syrphidae
drugstore beetle	*Stegobium paniceum* (Linnaeus)	COLEOPTERA: Anobiidae
dryberry mite	*Phyllocoptes gracilis* (Nalepa)	ACARI: Eriophyidae
dryland wireworm	*Ctenicera glauca* (Germar)	COLEOPTERA: Elateridae
dusky birch sawfly	*Croesus latitarsus* Norton	HYMENOPTERA: Tenthredinidae
dusky sap beetle	*Carpophilus lugubris* Murray	COLEOPTERA: Nitidulidae
dusky stink bug	*Euschistus tristigmus* (Say)	HEMIPTERA: Pentatomidae

E

ear tick	*Otobius megnini* (Dugès)	ACARI: Argasidae
eastern blackheaded budworm	*Acleris variana* (Fernald)	LEPIDOPTERA: Tortricidae
eastern field wireworm	*Limonius agonus* (Say)	COLEOPTERA: Elateridae
eastern Hercules beetle	*Dynastes tityus* (Linnaeus)	COLEOPTERA: Scarabaeidae
eastern larch beetle	*Dendroctonus simplex* LeConte	COLEOPTERA: Scolytidae
eastern lubber grasshopper	*Romalea guttata* (Latreille)	ORTHOPTERA: Acrididae
eastern pine seedworm	*Laspeyresia toreuta* (Grote)	LEPIDOPTERA: Tortricidae
eastern pine shoot borer	*Eucosma gloriola* Heinrich	LEPIDOPTERA: Tortricidae
eastern raspberry fruitworm	*Byturus rubi* Barber	COLEOPTERA: Byturidae
eastern spruce gall adelgid	*Adelges abietis* (Linnaeus)	HOMOPTERA: Adelgidae
eastern subterranean termite	*Reticulitermes flavipes* (Kollar)	ISOPTERA: Rhinotermitidae
eastern tent caterpillar	*Malacosoma americanum* (Fabricius)	LEPIDOPTERA: Lasiocampidae
eastern yellowjacket	*Vespula maculifrons* (Buysson)	HYMENOPTERA: Vespidae
eggplant flea beetle	*Epitrix fuscula* Crotch	COLEOPTERA: Chrysomelidae
eggplant lace bug	*Gargaphia solani* Heidemann	HEMIPTERA: Tingidae
eggplant leafminer	*Tildenia inconspicuella* (Murtfeldt)	LEPIDOPTERA: Gelechiidae
Egyptian alfalfa weevil	*Hypera brunneipennis* (Boheman)	COLEOPTERA: Curculionidae
eightspotted forester	*Alypia octomaculata* (Fabricius)	LEPIDOPTERA: Noctuidae
El Segundo blue	*Euphilotes battoides allyni* (Shields)	LEPIDOPTERA: Lycaenidae
elder shoot borer	*Achatodes zeae* (Harris)	LEPIDOPTERA: Noctuidae
elm borer	*Saperda tridentata* Olivier	COLEOPTERA: Cerambycidae

elm calligrapha	*Calligrapha scalaris* (LeConte)	COLEOPTERA: Chrysomelidae
elm casebearer	*Coleophora ulmifoliella* McDunnough	LEPIDOPTERA: Coleophoridae
elm cockscombgall aphid	*Colopha ulmicola* (Fitch)	HOMOPTERA: Aphididae
elm flea beetle	*Altica carinata* Germar	COLEOPTERA: Chrysomelidae
elm lace bug	*Corythucha ulmi* Osborn & Drake	HEMIPTERA: Tingidae
elm leaf aphid	*Tinocallis ulmifolii* (Monell)	HOMOPTERA: Aphididae
elm leaf beetle	*Pyrrhalta luteola* (Müller)	COLEOPTERA: Chrysomelidae
elm leafminer	*Fenusa ulmi* Sundevall	HYMENOPTERA: Tenthredinidae
elm sawfly	*Cimbex americana* Leach	HYMENOPTERA: Cimbicidae
elm scurfy scale	*Chionaspis americana* Johnson	HOMOPTERA: Diaspididae
elm spanworm	*Ennomos subsignarius* (Hübner)	LEPIDOPTERA: Geometridae
elm sphinx	*Ceratomia amyntor* (Geyer)	LEPIDOPTERA: Sphingidae
elongate flea beetle	*Systena elongata* (Fabricius)	COLEOPTERA: Chrysomelidae
elongate hemlock scale	*Fiorinia externa* (Ferris)	HOMOPTERA: Diaspididae
emerald cockroach wasp	*Ampulex compressa* (Fabricius)	HYMENOPTERA: Ampulicidae
Engelmann spruce weevil	*Pissodes strobi* (Peck)	COLEOPTERA: Curculionidae
English grain aphid	*Macrosiphum avenae* (Fabricius)	HOMOPTERA: Aphididae
erigeron root aphid	*Aphis middletonii* (Thomas)	HOMOPTERA: Aphididae
ermine moth	*Yponomeuta padella* (Linnaeus)	LEPIDOPTERA: Yponomeutidae
eugenia caterpillar	*Phlegetonia delatrix* Guenée	LEPIDOPTERA: Noctuidae
euonymus scale	*Unaspis euonymi* (Comstock)	HOMOPTERA: Diaspididae
eupatorium gall fly	*Procecidochares utilis* Stone	DIPTERA: Tephritidae
Eurasian pine adelgid	*Pineus pini* (Macquart)	HOMOPTERA: Adelgidae
European alder leafminer	*Fenusa dohrnii* (Tischbein)	HYMENOPTERA: Tenthredinidae
European apple sawfly	*Hoplocampa testudinea* (Klug)	HYMENOPTERA: Tenthredinidae
European chafer	*Rhizotrogus majalis* (Razoumowsky)	COLEOPTERA: Scarabaeidae
European chicken flea	*Ceratophyllus gallinae* (Schrank)	SIPHONAPTERA: Ceratophyllidae
European corn borer	*Ostrinia nubilalis* (Hübner)	LEPIDOPTERA: Pyralidae
European crane fly	*Tipula paludosa* Meigen	DIPTERA: Tipulidae
European earwig	*Forficula auricularia* Linnaeus	DERMAPTERA: Forficulidae
European elm scale	*Gossyparia spuria* (Modeer)	HOMOPTERA: Eriococcidae
European fruit lecanium	*Parthenolecanium corni* (Bouché)	HOMOPTERA: Coccidae
European fruit scale	*Quadraspidiotus ostreaeformis* (Curtis)	HOMOPTERA: Diaspididae
European grain moth	*Nemapogon granella* (Linnaeus)	LEPIDOPTERA: Tineidae
European honeysuckle leafroller	*Ypsolophus dentella* (Fabricius)	LEPIDOPTERA: Plutellidae
European hornet	*Vespa crabro germana* Christ	HYMENOPTERA: Vespidae
European house dust mite	*Dermatophagoides pteronyssinus* (Trouessart)	ACARI: Epidermoptidae
European mantid	*Mantis religiosa* Linnaeus	ORTHOPTERA: Mantidae
European mouse flea	*Leptopsylla segnis* (Schönherr)	SIPHONAPTERA: Leptopsyllidae
European peach scale	*Parthenolecanium persicae* (Fabricius)	HOMOPTERA: Coccidae
European pine sawfly	*Neodiprion sertifer* (Geoffroy)	HYMENOPTERA: Diprionidae
European pine shoot moth	*Rhyacionia buoliana* (Denis & Schiffermüller)	LEPIDOPTERA: Tortricidae
European red mite	*Panonychus ulmi* (Koch)	ACARI: Tetranychidae
European spruce beetle	*Dendroctonus micans* (Kugelann)	COLEOPTERA: Scolytidae
European spruce sawfly	*Gilpinia hercyniae* (Hartig)	HYMENOPTERA: Diprionidae
European wheat stem sawfly	*Cephus pygmaeus* (Linnaeus)	HYMENOPTERA: Cephidae
eyed click beetle	*Alaus oculatus* (Linnaeus)	COLEOPTERA: Elateridae
eyespotted bud moth	*Spilonota ocellana* (Denis & Schiffermüller)	LEPIDOPTERA: Tortricidae

F

face fly	*Musca autumnalis* De Geer	DIPTERA: Muscidae
fall armyworm	*Spodoptera frugiperda* (J. E. Smith)	LEPIDOPTERA: Noctuidae
fall cankerworm	*Alsophila pometaria* (Harris)	LEPIDOPTERA: Geometridae
fall webworm	*Hyphantria cunea* (Drury)	LEPIDOPTERA: Arctiidae
false celery leaftier	*Udea profundalis* (Packard)	LEPIDOPTERA: Pyralidae
false chinch bug	*Nysius raphanus* Howard	HEMIPTERA: Lygaeidae
false German cockroach	*Blattella lituricollis* (Walker)	ORTHOPTERA: Blattellidae
false hemlock looper	*Nepytia canosaria* (Walker)	LEPIDOPTERA: Geometridae

false potato beetle	*Leptinotarsa juncta* (Germar)	COLEOPTERA: Chrysomelidae
false stable fly	*Muscina stabulans* (Fallén)	DIPTERA: Muscidae
feather mite	*Megninia cubitalis* (Mégnin)	ACARI: Analgidae
fern aphid	*Idiopterus nephrelepidis* Davis	HOMOPTERA: Aphididae
fern caterpillar	*Callopistria spp.*	LEPIDOPTERA: Noctuidae
fern scale	*Pinnaspis aspidistrae* (Signoret)	HOMOPTERA: Diaspididae
field crickets	*Gryllus spp.*	ORTHOPTERA: Gryllidae
fiery hunter	*Calosoma calidum* (Fabricius)	COLEOPTERA: Carabidae
fiery skipper	*Hylephila phyleus* (Drury)	LEPIDOPTERA: Hesperiidae
fig mite	*Eriophyes ficus* Cotte	ACARI: Eriophyidae
fig scale	*Lepidosaphes conchiformis* (Gmelin)	HOMOPTERA: Diaspididae
fig wasp	*Blastophaga psenes* (Linnaeus)	HYMENOPTERA: Agaonidae
Fijian ginger weevil	*Elytroteinus subtruncatus* (Fairmaire)	COLEOPTERA: Curculionidae
filament bearer	*Nematocampa filamentaria* Guenée	LEPIDOPTERA: Geometridae
filbert aphid	*Myzocallis coryli* (Goetze)	HOMOPTERA: Aphididae
filbert bud mite	*Phytocoptella avellanae* (Nalepa)	ACARI: Nalepellidae
filbert weevil	*Curculio occidentis* (Casey)	COLEOPTERA: Curculionidae
filbertworm	*Melissopus latiferreanus* (Walsingham)	LEPIDOPTERA: Tortricidae
fir cone looper	*Eupithecia spermaphaga* (Dyar)	LEPIDOPTERA: Geometridae
fir engraver	*Scolytus ventralis* LeConte	COLEOPTERA: Scolytidae
fir seed moth	*Laspeyresia bracteatana* (Fernald)	LEPIDOPTERA: Tortricidae
fire ant	*Solenopsis geminata* (Fabricius)	HYMENOPTERA: Formicidae
firebrat	*Thermobia domestica* (Packard)	THYSANURA: Lepismatidae
firtree borer	*Semanotus litigiosus* (Casey)	COLEOPTERA: Cerambycidae
flat grain beetle	*Cryptolestes pusillus* (Schönherr)	COLEOPTERA: Cucujidae
flatheaded appletree borer	*Chrysobothris femorata* (Olivier)	COLEOPTERA: Buprestidae
flatheaded cone borer	*Chrysophana placida conicola* Van Dyke	COLEOPTERA: Buprestidae
flatheaded fir borer	*Melanophila drummondi* (Kirby)	COLEOPTERA: Buprestidae
flax bollworm	*Heliothis ononis* (Denis & Schiffermüller)	LEPIDOPTERA: Noctuidae
Fletcher scale	*Parthenolecanium fletcheri* (Cockerell)	HOMOPTERA: Coccidae
floodwater mosquito	*Aedes sticticus* (Meigen)	DIPTERA: Culicidae
Florida carpenter ant	*Camponotus abdominalis* (Fabricius)	HYMENOPTERA: Formicidae
Florida fern caterpillar	*Callopistria floridensis* (Guenée)	LEPIDOPTERA: Noctuidae
Florida harvester ant	*Pogonomyrmex badius* (Latreille)	HYMENOPTERA: Formicidae
Florida red scale	*Chrysomphalus aonidum* (Linnaeus)	HOMOPTERA: Diaspididae
Florida wax scale	*Ceroplastes floridensis* Comstock	HOMOPTERA: Coccidae
flower thrips	*Frankliniella tritici* (Fitch)	THYSANOPTERA: Thripidae
fluff louse	*Goniocotes gallinae* (De Geer)	MALLOPHAGA: Philopteridae
follicle mite	*Demodex folliculorum* (Simon)	ACARI: Demodicidae
forage looper	*Caenurgina erechtea* (Cramer)	LEPIDOPTERA: Noctuidae
Forbes scale	*Quadraspidiotus forbesi* (Johnson)	HOMOPTERA: Diaspididae
foreign grain beetle	*Ahasverus advena* (Waltl)	COLEOPTERA: Cucujidae
forest day mosquito	*Aedes albopictus* (Skuse)	DIPTERA: Culicidae
forest tent caterpillar	*Malacosoma disstria* Hübner	LEPIDOPTERA: Lasiocampidae
forest tree termite	*Neotermes connexus* Snyder	ISOPTERA: Kalotermitidae
forktailed bush katydid	*Scudderia furcata* Brunner von Wattenwyl	ORTHOPTERA: Tettigoniidae
Formosan subterranean termite	*Coptotermes formosanus* Shiraki	ISOPTERA: Rhinotermitidae
fourlined plant bug	*Poecilocapsus lineatus* (Fabricius)	HEMIPTERA: Miridae
fourspotted spider mite	*Tetranychus canadensis* (McGregor)	ACARI: Tetranychidae
fourspotted tree cricket	*Oecanthus quadripunctatus* Beutenmüller	ORTHOPTERA: Gryllidae
fowl tick	*Argas persicus* (Oken)	ACARI: Argasidae
foxglove aphid	*Acyrthosiphon solani* (Kaltenbach)	HOMOPTERA: Aphididae
frigate bird fly	*Olfersia spinifera* (Leach)	DIPTERA: Hippoboscidae
fringed orchid aphid	*Cerataphis orchidearum* (Westwood)	HOMOPTERA: Aphididae
frit fly	*Oscinella frit* (Linnaeus)	DIPTERA: Chloropidae
fruittree leafroller	*Archips argyrospila* (Walker)	LEPIDOPTERA: Tortricidae
Fuller rose beetle	*Pantomorus cervinus* (Boheman)	COLEOPTERA: Curculionidae
furniture beetle	*Anobium punctatum* (De Geer)	COLEOPTERA: Anobiidae
furniture carpet beetle	*Anthrenus flavipes* LeConte	COLEOPTERA: Dermestidae

G

gallmaking maple borer	*Xylotrechus aceris* Fisher	COLEOPTERA: Cerambycidae
garden fleahopper	*Halticus bractatus* (Say)	HEMIPTERA: Miridae
garden millipede	*Oxidus gracilis* Koch	POLYDESMIDA: Paradoxosomatidae
garden springtail	*Bourletiella hortensis* (Fitch)	COLLEMBOLA: Sminthuridae
garden symphylan	*Scutigerella immaculata* (Newport)	SYMPHYLA: Scutigerellidae
garden webworm	*Achyra rantalis* (Guenée)	LEPIDOPTERA: Pyralidae
gardenia bud mite	*Colomerus gardeniella* (Keifer)	ACARI: Eriophyidae
genista caterpillar	*Uresiphita reversalis* (Guenée)	LEPIDOPTERA: Pyralidae
German cockroach	*Blattella germanica* (Linnaeus)	ORTHOPTERA: Blattellidae
giant African snail	*Achatina fulica* Bowdich	STYLOMMATOPHORA: Achatinidae
giant bark aphid	*Longistigma caryae* (Harris)	HOMOPTERA: Aphididae
giant Hawaiian dragonfly	*Anax strenuus* Hagen	ODONATA: Aeschnidae
giant stag beetle	*Lucanus elaphus* Fabricius	COLEOPTERA: Lucanidae
giant water bug	*Lethocerus americanus* (Leidy)	HEMIPTERA: Belostomatidae
Giffard whitefly	*Bemisia giffardi* (Kotinsky)	HOMOPTERA: Aleyrodidae
ginger maggot	*Eumerus figurans* Walker	DIPTERA: Syrphidae
gladiolus thrips	*Thrips simplex* (Morison)	THYSANOPTERA: Thripidae
glassy cutworm	*Crymodes devastator* (Brace)	LEPIDOPTERA: Noctuidae
globose scale	*Sphaerolecanium prunastri* (Fonscolombe)	HOMOPTERA: Coccidae
globular spider beetle	*Trigonogenius globulum* Solier	COLEOPTERA: Ptinidae
gloomy scale	*Melanaspis tenebricosa* (Comstock)	HOMOPTERA: Diaspididae
Glover scale	*Lepidosaphes gloveri* (Packard)	HOMOPTERA: Diaspididae
goat biting louse	*Bovicola caprae* Gurlt	MALLOPHAGA: Trichodectidae
goat follicle mite	*Demodex caprae* Railliet	ACARI: Demodicidae
goat sucking louse	*Linognathus stenopsis* (Burmeister)	ANOPLURA: Linognathidae
golden buprestid	*Buprestis aurulenta* Linnaeus	COLEOPTERA: Buprestidae
golden cricket wasp	*Liris aurulenta* (Fabricius)	HYMENOPTERA: Sphecidae
golden oak scale	*Asterolecanium variolosum* (Ratzeburg)	HOMOPTERA: Asterolecaniidae
golden paper wasp	*Polistes fuscatus aurifer* Saussure	HYMENOPTERA: Vespidae
golden spider beetle	*Niptus hololeucus* (Faldermann)	COLEOPTERA: Ptinidae
golden tortoise beetle	*Metriona bicolor* (Fabricius)	COLEOPTERA: Chrysomelidae
goldeneye lacewing	*Chrysoperla oculata* Say	NEUROPTERA: Chrysopidae
goldenglow aphid	*Dactynotus rudbeckiae* (Fitch)	HOMOPTERA: Aphididae
goose body louse	*Trinoton anserinum* (Fabricius)	MALLOPHAGA: Menoponidae
gooseberry fruitworm	*Zophodia convolutella* (Hübner)	LEPIDOPTERA: Pyralidae
gooseberry witchbroom aphid	*Kakimia houghtonensis* (Troop)	HOMOPTERA: Aphididae
gophertortoise tick	*Amblyomma tuberculatum* Marx	ACARI: Ixodidae
gorse seed weevil	*Apion ulicis* (Forster)	COLEOPTERA: Curculionidae
grain mite	*Acarus siro* Linnaeus	ACARI: Acaridae
grain rust mite	*Abacarus hystrix* (Nalepa)	ACARI: Eriophyidae
grain thrips	*Limothrips cerealium* (Haliday)	THYSANOPTERA: Thripidae
granary weevil	*Sitophilus granarius* (Linnaeus)	COLEOPTERA: Curculionidae
granulate cutworm	*Feltia subterranea* (Fabricius)	LEPIDOPTERA: Noctuidae
grape berry moth	*Endopiza viteana* Clemens	LEPIDOPTERA: Tortricidae
grape blossom midge	*Contarinia johnsoni* Felt	DIPTERA: Cecidomyiidae
grape cane gallmaker	*Ampeloglypter sesostris* (LeConte)	COLEOPTERA: Curculionidae
grape colaspis	*Colaspsis brunnea* (Fabricius)	COLEOPTERA: Chrysomelidae
grape curculio	*Craponius inaequalis* (Say)	COLEOPTERA: Curculionidae
grape erineum mite	*Colomerus vitis* (Pagenstecher)	ACARI: Eriophyidae
grape flea beetle	*Altica chalybea* Illiger	COLEOPTERA: Chrysomelidae
grape leaffolder	*Desmia funeralis* (Hübner)	LEPIDOPTERA: Pyralidae
grape mealybug	*Pseudococcus maritimus* (Ehrhorn)	HOMOPTERA: Pseudococcidae
grape phylloxera	*Daktalosphaira vitifoliae* (Fitch)	HOMOPTERA: Phylloxeridae
grape plume moth	*Pterophorus periscelidactylus* Fitch	LEPIDOPTERA: Pterophoridae
grape root borer	*Vitacea polistiformis* (Harris)	LEPIDOPTERA: Sesiidae
grape rootworm	*Fidia viticida* Walsh	COLEOPTERA: Chrysomelidae
grape sawfly	*Erythraspides vitis* (Harris)	HYMENOPTERA: Tenthredinidae

grape scale	*Diaspidiotus uvae* (Comstock)	HOMOPTERA: Diaspididae
grape seed chalcid	*Evoxysoma vitis* (Saunders)	HYMENOPTERA: Eurytomidae
grape trunk borer	*Clytoleptus albofasciatus* (Laporte & Gory)	COLEOPTERA: Cerambycidae
grape whitefly	*Trialeurodes vittata* (Quaintance)	HOMOPTERA: Aleyrodidae
grapeleaf skeletonizer	*Harrisina americana* (Guérin)	LEPIDOPTERA: Zygaenidae
grapevine aphid	*Aphis illinoisensis* Shimer	HOMOPTERA: Aphididae
grapevine looper	*Eulythis diversilineata* (Hübner)	LEPIDOPTERA: Geometridae
grass fleahopper	*Halticus chrysolepis* Kirkaldy	HEMIPTERA: Miridae
grass mite	*Siteroptes graminum* (Reuter)	ACARI: Siteroptidae
grass sawfly	*Pachynematus extensicornis* (Norton)	HYMENOPTERA: Tenthredinidae
grass scolytid	*Hypothenemus pubescens* Hopkins	COLEOPTERA: Scolytidae
grass sharpshooter	*Draeculacephala minerva* Ball	HOMOPTERA: Cicadellidae
grass thrips	*Anaphothrips obscurus* (Müller)	THYSANOPTERA: Thripidae
grass webworm	*Herpetogramma licarsisalis* (Walker)	LEPIDOPTERA: Pyralidae
grasshopper bee fly	*Systoechus vulgaris* Loew	DIPTERA: Bombyliidae
grasshopper maggots	*Blaesoxipha* spp.	DIPTERA: Sarcophagidae
gray lawn leafhopper	*Exitianus exitiosus* (Uhler)	HOMOPTERA: Cicadellidae
gray pineapple mealybug	*Dysmicoccus neobrevipes* Beardsley	HOMOPTERA: Pseudococcidae
gray sugarcane mealybug	*Dysmicoccus boninsis* (Kuwana)	HOMOPTERA: Pseudococcidae
gray willow leaf beetle	*Pyrrhalta decora decora* (Say)	COLEOPTERA: Chrysomelidae
graybanded leafroller	*Argyrotaenia mariana* (Fernald)	LEPIDOPTERA: Tortricidae
great ash sphinx	*Sphinx chersis* (Hübner)	LEPIDOPTERA: Sphingidae
Great Basin wireworm	*Ctenicera pruinina* (Horn)	COLEOPTERA: Elateridae
greater wax moth	*Galleria mellonella* (Linnaeus)	LEPIDOPTERA: Pyralidae
greedy scale	*Hemiberlesia rapax* (Comstock)	HOMOPTERA: Diaspididae
green budworm	*Hedya nubiferana* (Haworth)	LEPIDOPTERA: Tortricidae
green cloverworm	*Plathypena scabra* (Fabricius)	LEPIDOPTERA: Noctuidae
green fruitworm	*Lithophane antennata* (Walker)	LEPIDOPTERA: Noctuidae
green garden looper	*Chrysodeixis chalcites* (Esper)	LEPIDOPTERA: Noctuidae
green June beetle	*Cotinis nitida* (Linnaeus)	COLEOPTERA: Scarabaeidae
green peach aphid	*Myzus persicae* (Sulzer)	HOMOPTERA: Aphididae
green rose chafer	*Dichelonyx backi* (Kirby)	COLEOPTERA: Scarabaeidae
green scale	*Coccus viridis* (Green)	HOMOPTERA: Coccidae
green shield scale	*Pulvinaria psidii* Maskell	HOMOPTERA: Coccidae
green sphinx	*Tinostoma smaragditis* (Meyrick)	LEPIDOPTERA: Sphingidae
green spruce aphid	*Cinara fornacula* Hottes	HOMOPTERA: Aphididae
green stink bug	*Acrosternum hilare* (Say)	HEMIPTERA: Pentatomidae
greenbug	*Schizaphis graminum* (Rondani)	HOMOPTERA: Aphididae
greenheaded spruce sawfly	*Pikonema dimmockii* (Cresson)	HYMENOPTERA: Tenthredinidae
greenhouse leaftier	*Udea rubigalis* (Guenée)	LEPIDOPTERA: Pyralidae
greenhouse orthezii	*Orthezia insignis* Browne	HOMOPTERA: Ortheziidae
greenhouse stone cricket	*Tachycines asynamorus* Adelung	ORTHOPTERA: Gryllacrididae
greenhouse thrips	*Heliothrips haemorrhoidalis* (Bouché)	THYSANOPTERA: Thripidae
greenhouse whitefly	*Trialeurodes vaporariorum* (Westwood)	HOMOPTERA: Aleyrodidae
greenstriped grasshopper	*Chortophaga viridifasciata* (De Geer)	ORTHOPTERA: Acrididae
greenstriped mapleworm	*Dryocampa rubicunda* (Fabricius)	LEPIDOPTERA: Saturniidae
gregarious oak leafminer	*Cameraria cincinnatiella* (Chambers)	LEPIDOPTERA: Gracillariidae
ground mealybug	*Rhizoecus falcifer* Kunckel d'Herculais	HOMOPTERA: Pseudococcidae
Guinea ant	*Tetramorium guineense* (Fabricius)	HYMENOPTERA: Formicidae
guinea feather louse	*Goniodes numidae* Mjöberg	MALLOPHAGA: Philopteridae
Gulf Coast tick	*Amblyomma maculatum* Koch	ACARI: Ixodidae
Gulf wireworm	*Conoderus amplicollis* (Gyllenhal)	COLEOPTERA: Elateridae
gypsy moth	*Lymantria dispar* (Linnaeus)	LEPIDOPTERA: Lymantriidae

H

hackberry engraver	*Scolytus muticus* Say	COLEOPTERA: Scolytidae
hackberry lace bug	*Corythucha celtidis* Osborn & Drake	HEMIPTERA: Tingidae

hackberry nipplegall maker	*Pachypsylla celtidismamma* (Riley)	HOMOPTERA: Psyllidae
hag moth	*Phobetron pithecium* (J. E. Smith)	LEPIDOPTERA: Limacodidae
hairy chinch bug	*Blissus leucopterus hirtus* Montandon	HEMIPTERA: Lygaeidae
hairy fungus beetle	*Typhaea stercorea* (Linnaeus)	COLEOPTERA: Mycetophagidae
hairy maggot blow fly	*Chrysomya rufifacies* (Macquart)	DIPTERA: Calliphoridae
hairy rove beetle	*Creophilus maxillosus* (Linnaeus)	COLEOPTERA: Staphylinidae
hairy spider beetle	*Ptinis villiger* (Reitter)	COLEOPTERA: Ptinidae
Hall scale	*Nilotaspis halli* (Green)	HOMOPTERA: Diaspididae
hard maple budminer	*Obrussa ochrefasciella* (Chambers)	LEPIDOPTERA: Nepticulidae
harlequin bug	*Murgantia histrionica* (Hahn)	HEMIPTERA: Pentatomidae
harlequin cockroach	*Neostylopyga rhombifolia* (Stoll)	ORTHOPTERA: Blattidae
hau leafminer	*Philodoria hauicola* (Swezey)	LEPIDOPTERA: Gracillariidae
Hawaiian antlion	*Eidoleon wilsoni* (McLachlan)	NEUROPTERA: Myrmeleontidae
Hawaiian beet webworm	*Spoladea recurvalis* (Fabricius)	LEPIDOPTERA: Pyralidae
Hawaiian bud moth	*Heliothis hawaiiensis* (Quaintance & Brues)	LEPIDOPTERA: Noctuidae
Hawaiian carpenter ant	*Camponotus variegatus* (F. Smith)	HYMENOPTERA: Formicidae
Hawaiian flower thrips	*Thrips hawaiiensis* (Morgan)	THYSANOPTERA: Thripidae
Hawaiian grass thrips	*Anaphothrips swezeyi* Moulton	THYSANOPTERA: Thripidae
Hawaiian pelagic water strider	*Halobates hawaiiensis* Usinger	HEMIPTERA: Gerridae
Hawaiian sphinx	*Hyles calida* (Butler)	LEPIDOPTERA: Sphingidae
hawthorn lace bug	*Corythucha cydoniae* (Fitch)	HEMIPTERA: Tingidae
hazelnut weevil	*Curculio neocorylus* Gibson	COLEOPTERA: Curculionidae
head louse	*Pediculus humanus capitis* De Geer	ANOPLURA: Pediculidae
heath spittlebug	*Clastoptera saintcyri* Provancher	HOMOPTERA: Cercopidae
hellgrammite	*Corydalus cornutus* (Linnaeus)	NEUROPTERA: Corydalidae
hemispherical scale	*Saissetia coffeae* (Walker)	HOMOPTERA: Coccidae
hemlock borer	*Melanophila fulvoguttata* (Harris)	COLEOPTERA: Buprestidae
hemlock looper	*Lambdina fiscellaria fiscellaria* (Guenée)	LEPIDOPTERA: Geometridae
hemlock sawfly	*Neodiprion tsugae* Middleton	HYMENOPTERA: Diprionidae
hemlock scale	*Abgrallaspis ithacae* (Ferris)	HOMOPTERA: Diaspididae
Hessian fly	*Mayetiola destructor* (Say)	DIPTERA: Cecidomyiidae
hibiscus leafminer	*Philodoria hibiscella* (Swezey)	LEPIDOPTERA: Gracillariidae
hibiscus mealybug	*Nipaecoccus vastator* (Maskell)	HOMOPTERA: Pseudococcidae
hibiscus whitefly	*Pealius hibisci* (Kotinsky)	HOMOPTERA: Aleyrodidae
hickory bark beetle	*Scolytus quadrispinosus* Say	COLEOPTERA: Scolytidae
hickory horned devil	*Citheronia regalis* (Fabricius)	LEPIDOPTERA: Saturniidae
hickory leafroller	*Argyrotaenia juglandana* (Fernald)	LEPIDOPTERA: Tortricidae
hickory plant bug	*Lygocoris caryae* (Knight)	HEMIPTERA: Miridae
hickory shuckworm	*Laspeyresia caryana* (Fitch)	LEPIDOPTERA: Tortricidae
hickory tussock moth	*Lophocampa caryae* (Harris)	LEPIDOPTERA: Arctiidae
hide beetle	*Dermestes maculatus* De Geer	COLEOPTERA: Dermestidae
High Plains grasshopper	*Dissosteira longipennis* (Thomas)	ORTHOPTERA: Acrididae
hog follicle mite	*Demodex phylloides* Csokor	ACARI: Demodicidae
hog louse	*Haematopinus suis* (Linnaeus)	ANOPLURA: Haematopinidae
holly leafminer	*Phytomyza ilicis* Curtis	DIPTERA: Agromyzidae
holly scale	*Dynaspidiotus britannicus* (Newstead)	HOMOPTERA: Diaspididae
hollyhock plant bug	*Brooksetta althaeae* (Hussey)	HEMIPTERA: Miridae
hollyhock weevil	*Apion longirostre* Olivier	COLEOPTERA: Curculionidae
honey bee	*Apis mellifera* Linnaeus	HYMENOPTERA: Apidae
honey bee mite	*Acarapis woodi* (Rennie)	ACARI: Tarsonemidae
honeylocust plant bug	*Diaphnocoris chlorionis* (Say)	HEMIPTERA: Miridae
honeysuckle leafminer	*Swezeyula lonicerae* Zimmerman & Bradley	LEPIDOPTERA: Elachistidae
honeysuckle sawfly	*Zaraea inflata* Norton	HYMENOPTERA: Cimbicidae
hop aphid	*Phorodon humuli* (Schrank)	HOMOPTERA: Aphididae
hop flea beetle	*Psylliodes punctulata* Melsheimer	COLEOPTERA: Chrysomelidae
hop looper	*Hypena humuli* (Harris)	LEPIDOPTERA: Noctuidae
hop plant bug	*Taedia hawleyi* (Knight)	HEMIPTERA: Miridae
hop vine borer	*Hydraecia immanis* Guenée	LEPIDOPTERA: Noctuidae
horn fly	*Haematobia irritans* (Linnaeus)	DIPTERA: Muscidae
horned passalus	*Odontotaenius disjunctus* (Illiger)	COLEOPTERA: Passalidae

horned squash bug	*Anasa armigera* (Say)	HEMIPTERA: Coreidae
hornet moth	*Sesia apiformis* (Clerck)	LEPIDOPTERA: Sesiidae
hornets	*Vespidae spp.*	HYMENOPTERA: Vespidae
horse biting louse	*Bovicola equi* (Denny)	MALLOPHAGA: Trichodectidae
horse bot fly	*Gasterophilus intestinalis* (De Geer)	DIPTERA: Gasterophilidae
horse follicle mite	*Demodex equi* Railliet	ACARI: Demodicidae
horse sucking louse	*Haematopinus asini* (Linnaeus)	ANOPLURA: Haematopinidae
horseradish flea beetle	*Phyllotreta armoraciae* (Koch)	COLEOPTERA: Chrysomelidae
house centipede	*Scutigera coleoptrata* (Linnaeus)	SCUTIGEROMORPHA: Scutigeridae
house cricket	*Acheta domesticus* (Linnaeus)	ORTHOPTERA: Gryllidae
house fly	*Musca domestica* Linnaeus	DIPTERA: Muscidae
house mite	*Glycyphagus domesticus* (De Geer)	ACARI: Glycyphagidae
house mouse mite	*Lyponyssoides sanguineus* (Hirst)	ACARI: Macronyssidae
household casebearer	*Phereoeca uterella* Walsingham	LEPIDOPTERA: Tineidae
human flea	*Pulex irritans* Linnaeus	SIPHONAPTERA: Pulicidae
hunting billbug	*Sphenophorus venatus vestitus* Chittenden	COLEOPTERA: Curculionidae
hyaline grass bug	*Liorhyssus hyalinus* (Fabricius)	HEMIPTERA: Rhopalidae

I

ilima leafminer	*Philodoria marginestrigata* (Walsingham)	LEPIDOPTERA: Gracillariidae
ilima moth	*Amyna natalis* (Walker)	LEPIDOPTERA: Noctuidae
imbricated snout beetle	*Epicaerus imbricatus* (Say)	COLEOPTERA: Curculionidae
immigrant acacia weevil	*Orthorhinus klugi* Boheman	COLEOPTERA: Curculionidae
imperial moth	*Eacles imperialis* (Drury)	LEPIDOPTERA: Saturniidae
imported cabbageworm	*Artogeia rapae* (Linnaeus)	LEPIDOPTERA: Pieridae
imported crucifer weevil	*Baris lepidii* Germar	COLEOPTERA: Curculionidae
imported currantworm	*Nematus ribesii* (Scopoli)	HYMENOPTERA: Tenthredinidae
imported longhorned weevil	*Calomycterus setarius* Roelofs	COLEOPTERA: Curculionidae
imported willow leaf beetle	*Plagiodera versicolora* (Laicharting)	COLEOPTERA: Chrysomelidae
incense-cedar wasp	*Syntexis libocedrii* Rohwer	HYMENOPTERA: Anaxyelidae
Indianmeal moth	*Plodia interpunctella* (Hübner)	LEPIDOPTERA: Pyralidae
inornate scale	*Aonidiella inornata* McKenzie	HOMOPTERA: Diaspididae
introduced pine sawfly	*Diprion similis* (Hartig)	HYMENOPTERA: Diprionidae
io moth	*Automeris io* (Fabricius)	LEPIDOPTERA: Saturniidae
iris borer	*Macronoctua onusta* Grote	LEPIDOPTERA: Noctuidae
iris thrips	*Frankliniella iridis* (Watson)	THYSANOPTERA: Thripidae
iris weevil	*Mononychus vulpeculus* (Fabricius)	COLEOPTERA: Curculionidae
Italian pear scale	*Epidiaspis leperii* (Signoret)	HOMOPTERA: Diaspididae
itch mite	*Sarcoptes scabiei* (De Geer)	ACARI: Sarcoptidae
ivy aphid	*Aphis hederae* Kaltenbach	HOMOPTERA: Aphididae

J

jack pine budworm	*Choristoneura pinus* Freeman	LEPIDOPTERA: Tortricidae
jack pine sawfly	*Neodiprion pratti banksianae* Rohwer	HYMENOPTERA: Diprionidae
jack pine tip beetle	*Conophthorus banksianae* McPherson	COLEOPTERA: Scolytidae
Japanese beetle	*Popillia japonica* Newman	COLEOPTERA: Scarabaeidae
Japanese broadwinged katydid	*Holochlora japonica* Brunner von Wattenwyl	ORTHOPTERA: Tettigoniidae
Japanese grasshopper	*Oxya japonica* (Thunberg)	ORTHOPTERA: Acrididae
Jeffrey pine beetle	*Dendroctonus jeffreyi* Hopkins	COLEOPTERA: Scolytidae
Jerusalem cricket	*Stenopelmatus fuscus* Haldeman	ORTHOPTERA: Stenopelmatidae
juniper midge	*Contarinia juniperina* Felt	DIPTERA: Cecidomyiidae
juniper scale	*Carulaspis juniperi* (Bouché)	HOMOPTERA: Diaspididae
juniper tip midge	*Oligotrophus betheli* Felt	DIPTERA: Cecidomyiidae
juniper webworm	*Dichomeris marginella* (Fabricius)	LEPIDOPTERA: Gelechiidae

K

Kamehameha butterfly	*Vanessa tameamea* Eschscholtz	LEPIDOPTERA: Nymphalidae
keyhole wasp	*Pachodynerus nasidens* (Latreille)	HYMENOPTERA: Vespidae
khapra beetle	*Trogoderma granarium* Everts	COLEOPTERA: Dermestidae
kiawe bean weevil	*Algarobius bottimeri* Kingsolver	COLEOPTERA: Bruchidae
kiawe flower moth	*Ithome concolorella* (Chambers)	LEPIDOPTERA: Cosmopterigidae
kiawe roundheaded borer	*Placosternus cyclene crinicornis* (Chevrolat)	COLEOPTERA: Cerambycidae
kiawe scolytid	*Hypothenemus birmanus* (Eichhoff)	COLEOPTERA: Scolytidae
Kirkaldy whitefly	*Dialeurodes kirkaldyi* (Kotinsky)	HOMOPTERA: Aleyrodidae
Klamathweed beetle	*Chrysolina quadrigemina* (Suffrian)	COLEOPTERA: Chrysomelidae
koa bug	*Coleotichus blackburniae* White	HEMIPTERA: Pentatomidae
koa haole seed weevil	*Araecerus levipennis* Jordan	COLEOPTERA: Anthribidae
koa moth	*Scotorythra paludicola* (Butler)	LEPIDOPTERA: Geometridae
koa seedworm	*Cryptophlebia illepida* (Butler)	LEPIDOPTERA: Tortricidae
kou leafworm	*Ethmia nigroapicella* (Saalmüller)	LEPIDOPTERA: Oecophoridae

L

Lange metalmark	*Apodemia mormo langei* Comstock	LEPIDOPTERA: Riodinidae
lantana cerambycid	*Plagiohammus spinipennis* Thomson	COLEOPTERA: Cerambycidae
lantana defoliator caterpillar	*Hypena strigata* (Fabricius)	LEPIDOPTERA: Noctuidae
lantana gall fly	*Eutreta xanthochaeta* Aldrich	DIPTERA: Tephritidae
lantana hispid	*Uroplata girardi* Pic	COLEOPTERA: Chrysomelidae
lantana lace bug	*Teleonemia scrupulosa* Stål	HEMIPTERA: Tingidae
lantana leaf beetle	*Octotoma scabripennis* Guérin-Méneville	COLEOPTERA: Chrysomelidae
lantana leafminer	*Cremastobombycia lantanella* (Schrank)	LEPIDOPTERA: Gracillariidae
lantana leaftier	*Salbia haemorrhoidalis* Guenée	LEPIDOPTERA: Pyralidae
lantana plume moth	*Lantanophaga pusillodactyla* (Walker)	LEPIDOPTERA: Pterophoridae
lantana seed fly	*Ophiomyia lantanae* (Froggatt)	DIPTERA: Agromyzidae
lantana stick caterpillar	*Neogalea esula* (Druce)	LEPIDOPTERA: Noctuidae
lappet moth	*Phyllodesma americana* (Harris)	LEPIDOPTERA: Lasiocampidae
larch aphid	*Cinara laricis* (Walker)	HOMOPTERA: Aphididae
larch casebearer	*Coleophora laricella* (Hübner)	LEPIDOPTERA: Coleophoridae
larch sawfly	*Pristiphora erichsonii* (Hartig)	HYMENOPTERA: Tenthredinidae
larder beetle	*Dermestes lardarius* Linnaeus	COLEOPTERA: Dermestidae
large aspen tortrix	*Choristoneura conflictana* (Walker)	LEPIDOPTERA: Tortricidae
large bigeyed bug	*Geocoris bullatus* (Say)	HEMIPTERA: Lygaeidae
large brown spider	*Heteropoda venatoria* (Linnaeus)	ARANEAE: Sparassidae
large chestnut weevil	*Curculio caryatrypes* (Boheman)	COLEOPTERA: Curculionidae
large chicken louse	*Goniodes gigas* (Taschenberg)	MALLOPHAGA: Philopteridae
large cottony scale	*Pulvinaria mammeae* Maskell	HOMOPTERA: Coccidae
large duck louse	*Trinoton querquedulae* (Linnaeus)	MALLOPHAGA: Menoponidae
large kissing bug	*Triatoma rubrofasciata* (De Geer)	HEMIPTERA: Reduviidae
large milkweed bug	*Oncopeltus fasciatus* (Dallas)	HEMIPTERA: Lygaeidae
large turkey louse	*Chelopistes meleagridis* (Linnaeus)	MALLOPHAGA: Philopteridae
larger black flour beetle	*Cynaeus angustus* (LeConte)	COLEOPTERA: Tenebrionidae
larger canna leafroller	*Calpodes ethlius* (Stoll)	LEPIDOPTERA: Hesperiidae
larger elm leaf beetle	*Monocesta coryli* (Say)	COLEOPTERA: Chrysomelidae
larger grain borer	*Prostephanus truncatus* (Horn)	COLEOPTERA: Bostrichidae
larger Hawaiian cutworm	*Agrotis crinigera* (Butler)	LEPIDOPTERA: Noctuidae
larger lantana butterfly	*Strymon echion* (Linnaeus)	LEPIDOPTERA: Lycaenidae
larger pale trogiil	*Trogium pulsatorium* (Linnaeus)	PSOCOPTERA: Trogiidae
larger shothole borer	*Scolytus mali* (Bechstein)	COLEOPTERA: Scolytidae
larger yellow ant	*Acanthomyops interjectus* (Mayr)	HYMENOPTERA: Formicidae
latania scale	*Hemiberlesia lataniae* (Signoret)	HOMOPTERA: Diaspididae
latrine fly	*Fannia scalaris* (Fabricius)	DIPTERA: Muscidae
lawn armyworm	*Spodoptera mauritia* (Boisduval)	LEPIDOPTERA: Noctuidae
lawn leafhopper	*Recilia hospes* (Kirkaldy)	HOMOPTERA: Cicadellidae

leadcable borer	*Scobicia declivis* (LeConte)	COLEOPTERA: Bostrichidae
leaf crumpler	*Acrobasis indigenella* (Zeller)	LEPIDOPTERA: Pyralidae
leaffooted bug	*Leptoglossus phyllopus* (Linnaeus)	HEMIPTERA: Coreidae
leaffooted pine seed bug	*Leptoglossus corculus* (Say)	HEMIPTERA: Coreidae
leafhopper assassin bug	*Zelus renardii* Kolenati	HEMIPTERA: Reduviidae
leek moth	*Acrolepiopsis assectella* (Zeller)	LEPIDOPTERA: Acrolepiidae
leopard moth	*Zeuzera pyrina* (Linnaeus)	LEPIDOPTERA: Cossidae
lespedeza webworm	*Tetralopha scortealis* (Lederer)	LEPIDOPTERA: Pyralidae
lesser appleworm	*Grapholita prunivora* (Walsh)	LEPIDOPTERA: Tortricidae
lesser bud moth	*Recurvaria nanella* (Denis & Schiffermüller)	LEPIDOPTERA: Gelechiidae
lesser bulb fly	*Eumerus tuberculatus* Rondani	DIPTERA: Syrphidae
lesser canna leafroller	*Geshna cannalis* (Quaintance)	LEPIDOPTERA: Pyralidae
lesser clover leaf weevil	*Hypera nigrirostris* (Fabricius)	COLEOPTERA: Curculionidae
lesser cornstalk borer	*Elasmopalpus lignosellus* (Zeller)	LEPIDOPTERA: Pyralidae
lesser ensign wasp	*Szepligetella sericia* (Cameron)	HYMENOPTERA: Evaniidae
lesser follicle mite	*Demodex brevis* Bulanova	ACARI: Demodicidae
lesser grain borer	*Rhyzopertha dominica* (Fabricius)	COLEOPTERA: Bostrichidae
lesser lawn leafhopper	*Graminella sonorus* (Ball)	HOMOPTERA: Cicadellidae
lesser mealworm	*Alphitobius diaperinus* (Panzer)	COLEOPTERA: Tenebrionidae
lesser orchid weevil	*Orchidophilus peregrinator* Buchanan	COLEOPTERA: Curculionidae
lesser peachtree borer	*Synanthedon pictipes* (Grote & Robinson)	LEPIDOPTERA: Sesiidae
lesser wax moth	*Achroia grisella* (Fabricius)	LEPIDOPTERA: Pyralidae
lettuce root aphid	*Pemphigus bursarius* (Linnaeus)	HOMOPTERA: Aphididae
light brown apple moth	*Austrotortrix postvittana* (Walker)	LEPIDOPTERA: Tortricidae
lilac borer	*Podosesia syringae* (Harris)	LEPIDOPTERA: Sesiidae
lilac leafminer	*Caloptilia syringella* (Fabricius)	LEPIDOPTERA: Gracillariidae
lily bulb thrips	*Liothrips vaneeckei* Priesner	THYSANOPTERA: Phlaeothripidae
lily weevil	*Agasphaerops nigra* Horn	COLEOPTERA: Curculionidae
limabean pod borer	*Etiella zinckenella* (Treitschke)	LEPIDOPTERA: Pyralidae
limabean vine borer	*Monoptilota pergratialis* (Hulst)	LEPIDOPTERA: Pyralidae
linden borer	*Saperda vestita* Say	COLEOPTERA: Cerambycidae
linden looper	*Erannis tiliaria* (Harris)	LEPIDOPTERA: Geometridae
lined click beetle	*Agriotes lineatus* (Linnaeus)	COLEOPTERA: Elateridae
lined spittlebug	*Neophilaenus lineatus* (Linnaeus)	HOMOPTERA: Cercopidae
lined stalk borer	*Oligia fractilinea* (Grote)	LEPIDOPTERA: Noctuidae
lion beetle	*Ulochaetes leoninus* LeConte	COLEOPTERA: Cerambycidae
litchi fruit moth	*Cryptophlebia ombrodelta* (Lower)	LEPIDOPTERA: Tortricidae
litchi mite	*Eriophyes litchii* Keifer	ACARI: Eriophyidae
little black ant	*Monomorium minimum* (Buckley)	HYMENOPTERA: Formicidae
little carpenterworm	*Prionoxystus macmurtrei* (Guérin)	LEPIDOPTERA: Cossidae
little fire ant	*Ochetomyrmex auropunctatus* (Roger)	HYMENOPTERA: Formicidae
little green leafhopper	*Balclutha hospes* (Kirkaldy)	HOMOPTERA: Cicadellidae
little house fly	*Fannia canicularis* (Linnaeus)	DIPTERA: Muscidae
little yellow ant	*Plagiolepis alluaudi* Emery	HYMENOPTERA: Formicidae
loblolly pine sawfly	*Neodiprion taedae linearis* Ross	HYMENOPTERA: Diprionidae
locust borer	*Megacyllene robiniae* (Forster)	COLEOPTERA: Cerambycidae
locust leafminer	*Odontota dorsalis* (Thunberg)	COLEOPTERA:Chrysomelidae
locust leafroller	*Nephopterix subcaesiella* (Clemens)	LEPIDOPTERA: Pyralidae
locust twig borer	*Ecdytolopha insiticiana* Zeller	LEPIDOPTERA: Tortricidae
lodgepole cone beetle	*Conophthorus contortae* Hopkins	COLEOPTERA: Scolytidae
lodgepole needleminer	*Coleotechnites milleri* (Busck)	LEPIDOPTERA: Gelechiidae
lodgepole pine beetle	*Dendroctonus murrayanae* Hopkins	COLEOPTERA: Scolytidae
lodgepole sawfly	*Neodiprion burkei* Middleton	HYMENOPTERA: Diprionidae
lodgepole terminal weevil	*Pissodes terminalis* Hopping	COLEOPTERA: Curculionidae
lone star tick	*Amblyomma americanum* (Linnaeus)	ACARI: Ixodidae
long brown scale	*Coccus longulus* (Douglas)	HOMOPTERA: Coccidae
longheaded flour beetle	*Latheticus oryzae* Waterhouse	COLEOPTERA: Tenebrionidae
longleaf pine seedworm	*Laspeyresia ingens* Heinrich	LEPIDOPTERA: Tortricidae
longlegged ant	*Anoplolepis longipes* (Jerdon)	HYMENOPTERA: Formicidae

longnosed cattle louse	*Linognathus vituli* (Linnaeus)	ANOPLURA: Linognathidae
longtailed fruit fly parasite	*Opius longicaudatus* Ashmead	HYMENOPTERA: Braconidae
longtailed mealybug	*Pseudococcus longispinus* (Targioni-Tozzetti)	HOMOPTERA: Pseudococcidae
lotis blue	*Lycaeides argyrognomon lotis* (Lintner)	LEPIDOPTERA: Lycaenidae
lowland tree termite	*Incisitermes immigrans* (Snyder)	ISOPTERA: Kalotermitidae
lubber grasshopper	*Brachystola magna* (Girard)	ORTHOPTERA: Acrididae
luna moth	*Actias luna* (Linnaeus)	LEPIDOPTERA: Saturniidae

M

Macao paper wasp	*Polistes macaensis* (Fabricius)	HYMENOPTERA: Vespidae
Madeira cockroach	*Leucophaea maderae* (Fabricius)	ORTHOPTERA: Blaberidae
magnolia scale	*Neolecanium cornuparvum* (Thro)	HOMOPTERA: Coccidae
maize billbug	*Sphenophorus maidis* Chittenden	COLEOPTERA: Curculionidae
maize weevil	*Sitophilus zeamais* Motschulsky	COLEOPTERA: Curculionidae
mango bark beetle	*Hypocryphalus mangiferae* (Stebbing)	COLEOPTERA: Scolytidae
mango bud mite	*Eriophyes mangiferae* (Sayed)	ACARI: Eriophyidae
mango flower beetle	*Protaetia fusca* (Herbst)	COLEOPTERA: Scarabaeidae
mango shoot caterpillar	*Bombotelia jocosatrix* Guenée	LEPIDOPTERA: Noctuidae
mango spider mite	*Oligonychus mangiferus* (Rahman & Sapra)	ACARI: Tetranychidae
mango weevil	*Cryptorhynchus mangiferae* (Fabricius)	COLEOPTERA: Curculionidae
maple bladdergall mite	*Vasates quadripedes* Shimer	ACARI: Eriophyidae
maple callus borer	*Synanthedon acerni* (Clemens)	LEPIDOPTERA: Sesiidae
maple leafcutter	*Paraclemensia acerifoliella* (Fitch)	LEPIDOPTERA: Incurvariidae
maple petiole borer	*Caulocampus acericaulis* (MacGillivray)	HYMENOPTERA: Tenthredinidae
maple trumpet skeletonizer	*Epinotia aceriella* (Clemens)	LEPIDOPTERA: Tortricidae
margined blister beetle	*Epicauta pestifera* Werner	COLEOPTERA: Meloidae
masked hunter	*Reduvius personatus* (Linnaeus)	HEMIPTERA: Reduviidae
mauna loa bean beetle	*Araeocorynus cumingi* Jekel	COLEOPTERA: Anthribidae
McDaniel spider mite	*Tetranychus mcdanieli* McGregor	ACARI: Tetranychidae
meadow plant bug	*Leptopterna dolabrata* (Linnaeus)	HEMIPTERA: Miridae
meadow spittlebug	*Philaenus spumarius* (Linnaeus)	HOMOPTERA: Cercopidae
meal moth	*Pyralis farinalis* Linnaeus	LEPIDOPTERA: Pyralidae
mealy plum aphid	*Hyalopterus pruni* (Geoffroy)	HOMOPTERA: Aphididae
mealybug destroyer	*Cryptolaemus montrouzieri* Mulsant	COLEOPTERA: Coccinellidae
Mediterranean flour moth	*Anagasta kuehniella* (Zeller)	LEPIDOPTERA: Pyralidae
Mediterranean fruit fly	*Ceratitis capitata* (Wiedemann)	DIPTERA: Tephritidae
melastoma borer	*Selca brunella* (Hampson)	LEPIDOPTERA: Noctuidae
melon aphid	*Aphis gossypii* Glover	HOMOPTERA: Aphididae
melon fly	*Dacus cucurbitae* Coquillett	DIPTERA: Tephritidae
melonworm	*Diaphania hyalinata* (Linnaeus)	LEPIDOPTERA: Pyralidae
merchant grain beetle	*Oryzaephilus mercator* (Fauvel)	COLEOPTERA: Cucujidae
Mexican bean beetle	*Epilachna varivestis* Mulsant	COLEOPTERA: Coccinellidae
Mexican bean weevil	*Zabrotes subfasciatus* (Boheman)	COLEOPTERA: Bruchidae
Mexican black scale	*Saissetia miranda* (Cockerell & Parrott)	HOMOPTERA: Coccidae
Mexican corn rootworm	*Diabrotica virgifera zeae* Krysan & Smith	COLEOPTERA: Chrysomelidae
Mexican fruit fly	*Anastrepha ludens* (Loew)	DIPTERA: Tephritidae
Mexican leafroller	*Amorbia emigratella* Busck	LEPIDOPTERA: Tortricidae
Mexican mealybug	*Phenacoccus gossypii* Townsend & Cockerell	HOMOPTERA: Pseudococcidae
Mexican pine beetle	*Dendroctonus approximatus* Dietz	COLEOPTERA: Scolytidae
migratory grasshopper	*Melanoplus sanguinipes* (Fabricius)	ORTHOPTERA: Acrididae
mimosa webworm	*Homadaula anisocentra* Meyrick	LEPIDOPTERA: Plutellidae
mining scale	*Howardia biclavis* (Comstock)	HOMOPTERA: Diaspididae
mint aphid	*Ovatus crataegarius* (Walker)	HOMOPTERA: Aphididae
minute egg parasite	*Trichogramma minutum* Riley	HYMENOPTERA: Trichogrammatidae
minute pirate bug	*Orius tristicolor* (White)	HEMIPTERA: Anthocoridae
mission blue	*Icaricia icarioides missionensis* (Hovanitz)	LEPIDOPTERA: Lycaenidae
mold mite	*Tyrophagus putrescentiae* (Schrank)	ACARI: Acaridae

monarch butterfly	*Danaus plexippus* (Linnaeus)	LEPIDOPTERA: Danaidae
monkeypod moth	*Polydesma umbricola* Boisduval	LEPIDOPTERA: Noctuidae
monkeypod roundheaded borer	*Xystrocera globosa* (Olivier)	COLEOPTERA: Cerambycidae
Monterey pine cone beetle	*Conophthorus radiatae* Hopkins	COLEOPTERA: Scolytidae
Monterey pine resin midge	*Cecidomyia resinicoloides* Williams	DIPTERA: Cecidomyiidae
Monterey pine weevil	*Pissodes radiatae* Hopkins	COLEOPTERA: Curculionidae
Mormon cricket	*Anabrus simplex* Haldeman	ORTHOPTERA: Tettigoniidae
morningglory leafminer	*Bedellia somnulentella* (Zeller)	LEPIDOPTERA: Lyonetiidae
Morrill lace bug	*Corythucha morrilli* Osborn & Drake	HEMIPTERA: Tingidae
mossyrose gall wasp	*Diplolepis rosae* (Linnaeus)	HYMENOPTERA: Cynipidae
mottled tortoise bettle	*Deloyala guttata* (Olivier)	COLEOPTERA: Chrysomelidae
mountain leafhopper	*Colladonus montanus* (Van Duzee)	HOMOPTERA: Cicadellidae
mountain pine beetle	*Dendroctonus ponderosae* Hopkins	COLEOPTERA: Scolytidae
mountain pine coneworm	*Dioryctria yatesi* Mutuura & Munroe	LEPIDOPTERA: Pyralidae
mountain-ash sawfly	*Pristiphora geniculata* (Hartig)	HYMENOPTERA: Tenthredinidae
mourningcloak butterfly	*Nymphalis antiopa* (Linnaeus)	LEPIDOPTERA: Nymphalidae
mulberry whitefly	*Tetraleurodes mori* (Quaintance)	HOMOPTERA: Aleyrodidae
mullein thrips	*Haplothrips verbasci* (Osborn)	THYSANOPTERA: Phlaeothripidae

N

Nantucket pine tip moth	*Rhyacionia frustrana* (Comstock)	LEPIDOPTERA: Tortricidae
narcissus bulb fly	*Merodon equestris* (Fabricius)	DIPTERA: Syrphidae
narrownecked grain beetle	*Anthicus floralis* (Linnaeus)	COLEOPTERA: Anthicidae
narrowwinged mantid	*Tenodera augustipennis* Saussure	ORTHOPTERA: Mantidae
native elm bark beetle	*Hylurgopinus opaculus* (LeConte)	COLEOPTERA: Scolytidae
native holly leafminer	*Phytomyza ilicicola* Loew	DIPTERA: Agromyzidae
navel orangeworm	*Amyelois transitella* (Walker)	LEPIDOPTERA: Pyralidae
negro bug	*Corimelaena pulicaria* (Germar)	HEMIPTERA: Thyreocoridae
Nevada sage grasshopper	*Melanoplus rugglesi* Gurney	ORTHOPTERA: Acrididae
New Guinea sugarcane weevil	*Rhabdoscelus obscurus* (Boisduval)	COLEOPTERA: Curculionidae
new house borer	*Arhopalus productus* (LeConte)	COLEOPTERA: Cerambycidae
New York weevil	*Ithycerus noveboracensis* (Forster)	COLEOPTERA: Curculionidae
nigra scale	*Parasaissetia nigra* (Nietner)	HOMOPTERA: Coccidae
northeastern sawyer	*Monochamus notatus* (Drury)	COLEOPTERA: Cerambycidae
northern cattle grub	*Hypoderma bovis* (Linnaeus)	DIPTERA: Oestridae
northern corn rootworm	*Diabrotica barberi* Smith & Lawrence	COLEOPTERA: Chrysomelidae
northern fowl mite	*Ornithonyssus sylviarum* (Canestrini & Fanzago)	ACARI: Macronyssidae
northern house mosquito	*Culex pipiens* Linnaeus	DIPTERA: Culicidae
northern masked chafer	*Cyclocephala borealis* Arrow	COLEOPTERA: Scarabaeidae
northern mole cricket	*Neocurtilla hexadactyla* (Perty)	ORTHOPTERA: Gryllotalpidae
northern pine weevil	*Pissodes approximatus* Hopkins	COLEOPTERA: Curculionidae
northern pitch twig moth	*Petrova albicapitana* (Busck)	LEPIDOPTERA: Tortricidae
northern rat flea	*Nosopsyllus fasciatus* (Bosc)	SIPHONAPTERA: Ceratophyllidae
northwest coast mosquito	*Aedes aboriginis* Dyar	DIPTERA: Culicidae
Norway maple aphid	*Periphyllus lyropictus* (Kessler)	HOMOPTERA: Aphididae
nose bot fly	*Gasterophilus haemorrhoidalis* (Linnaeus)	DIPTERA: Gasterophilidae
nutgrass armyworm	*Spodoptera exempta* (Walker)	LEPIDOPTERA: Noctuidae
nutgrass billbug	*Sphenophorus cariosus* (Olivier)	COLEOPTERA: Curculionidae
nutgrass borer moth	*Bactra venosoma* (Zeller)	LEPIDOPTERA: Tortricidae
nutgrass weevil	*Athesapeuta cyperi* Marshall	COLEOPTERA: Curculionidae
Nuttall blister beetle	*Lytta nuttalli* (Say)	COLEOPTERA: Meloidae

O

oak clearwing moth	*Paranthrene asilipennis* (Boisduval)	LEPIDOPTERA: Sesiidae
oak lace bug	*Corythucha arcuata* (Say)	HEMIPTERA: Tingidae
oak leafroller	*Archips semiferana* (Walker)	LEPIDOPTERA: Tortricidae
oak leaftier	*Croesia semipurpurana* (Kearfott)	LEPIDOPTERA: Tortricidae

oak lecanium	*Parthenolecanium quercifex* (Fitch)	HOMOPTERA: Coccidae
oak sapling borer	*Goes tesselatus* (Haldeman)	COLEOPTERA: Cerambycidae
oak skeletonizer	*Bucculatrix ainsliella* Murtfeldt	LEPIDOPTERA: Lyonetiidae
oak timberworm	*Arrhenodes minutus* (Drury)	COLEOPTERA: Brentidae
oak webworm	*Archips fervidana* (Clemens)	LEPIDOPTERA: Tortricidae
oatbird-cherry aphid	*Rhopalosiphum padi* (Linnaeus)	HOMOPTERA: Aphididae
obliquebanded leafroller	*Choristoneura rosaceana* (Harris)	LEPIDOPTERA: Tortricidae
obscure mealybug	*Pseudococcus obscurus* Essig	HOMOPTERA: Pseudococcidae
obscure root weevil	*Sciopithes obscurus* Horn	COLEOPTERA: Curculionidae
obscure scale	*Melanaspis obscura* (Comstock)	HOMOPTERA: Diaspididae
oceanic burrower bug	*Geotomus pygmaeus* (Dallas)	HEMIPTERA: Cydnidae
oceanic embiid	*Oligotoma oceanica* Ross	EMBIOPTERA: Oligotomidae
oceanic field cricket	*Teleogryllus oceanicus* (Le Guillou)	ORTHOPTERA: Gryllidae
odd beetle	*Thylodrias contractus* Motschulsky	COLEOPTERA: Dermestidae
odorous house ant	*Tapinoma sessile* (Say)	HYMENOPTERA: Formicidae
old house borer	*Hylotrupes bajulus* (Linnaeus)	COLEOPTERA: Cerambycidae
oleander aphid	*Aphis nerii* Fonscolombe	HOMOPTERA: Aphididae
oleander hawk moth	*Daphnis nerii* (Linnaeus)	LEPIDOPTERA: Sphingidae
oleander pit scale	*Asterolecanium pustulans* (Cockerell)	HOMOPTERA: Asterolecaniidae
oleander scale	*Aspidiotus nerii* Bouché	HOMOPTERA: Diaspididae
olive fruit fly	*Dacus oleae* (Gmelin)	DIPTERA: Tephritidae
olive scale	*Parlatoria oleae* (Colvée)	HOMOPTERA: Diaspididae
omnivorous leaftier	*Chephasia longana* (Haworth)	LEPIDOPTERA: Tortricidae
omnivorous looper	*Sabulodes caberata* Guenée	LEPIDOPTERA: Geometridae
onespotted stink bug	*Euschistus variolarius* (Palisot de Beauvois)	HEMIPTERA: Pentatomidae
onion aphid	*Keotoxoptera formosana* (Takahashi)	HOMOPTERA: Aphididae
onion bulb fly	*Kumerus strigatus* (Fallén)	DIPTERA: Syrphidae
onion maggot	*Delia antiqua* (Meigen)	DIPTERA: Anthomyiidae
onion plant bug	*Labopidicola allii* (Knight)	HEMIPTERA: Miridae
onion thrips	*Thrips tabaci* Lindeman	THYSANOPTERA: Thripidae
orange spiny whitefly	*Aleurocanthus spiniferus* (Quaintance)	HOMOPTERA: Aleyrodidae
orange tortrix	*Argyrotaenia citrana* (Fernald)	LEPIDOPTERA: Tortricidae
orangedog	*Papilio cresphontes* Cramer	LEPIDOPTERA: Papilionidae
orangehumped mapleworm	*Symmerista leucitys* Franclemont	LEPIDOPTERA: Notodontidae
orangestriped oakworm	*Anisota senatoria* (J. E. Smith)	LEPIDOPTERA: Saturniidae
orangetailed potter wasp	*Delta latreillei petiolaris* (Schulz)	HYMENOPTERA: Vespidae
orchid aphid	*Macrosiphum luteum* (Buckton)	HOMOPTERA: Aphididae
orchidfly	*Eurytoma orchidearum* (Westwood)	HYMENOPTERA: Eurytomidae
Oregon fir sawyer	*Monochamus oregonensis* (LeConte)	COLEOPTERA: Cerambycidae
Oregon wireworm	*Melanotus longulus oregonensis* (LeConte)	COLEOPTERA: Elateridae
oriental beetle	*Anomala orientalis* Waterhouse	COLEOPTERA: Scarabaeidae
oriental cockroach	*Blatta orientalis* Linnaeus	ORTHOPTERA: Blattidae
oriental fruit fly	*Dacus dorsalis* Hendel	DIPTERA: Tephritidae
oriental fruit moth	*Grapholita molesta* (Busck)	LEPIDOPTERA: Tortricidae
oriental house fly	*Musca domestica vicina* Macquart	DIPTERA: Muscidae
oriental moth	*Cnidocampa flavescens* (Walker)	LEPIDOPTERA: Limacodidae
oriental rat flea	*Xenopsylla cheopis* (Rothschild)	SIPHONAPTERA: Pulicidae
oriental stink bug	*Plautia stali* Scott	HEMIPTERA: Pentatomidae
ornate aphid	*Myzus ornatus* Laing	HOMOPTERA: Aphididae
orthezia ladybird beetle	*Hyperaspis jocosa* (Mulsant)	COLEOPTERA: Coccinellidae
oval guineapig louse	*Gyropus ovalis* Burmeister	MALLOPHAGA: Gyropidae
oxalis spider mite	*Tetranychina harti* (Ewing)	ACARI: Tetranychidae
oxalis whitefly	*Aleyrodes shizuokensis* Kuwana	HOMOPTERA: Aleyrodidae
oystershell scale	*Lepidosaphes ulmi* (Linnaeus)	HOMOPTERA: Diaspididae

P

Pacific beetle cockroach	*Diploptera punctata* (Eschscholtz)	ORTHOPTERA: Blaberidae
Pacific Coast tick	*Dermacentor occidentalis* Marx	ACARI: Ixodidae

Pacific Coast wireworm	*Limonius canus* LeConte	COLEOPTERA: Elateridae
Pacific cockroach	*Euthyrrhapha pacifica* (Coquebert)	ORTHOPTERA: Polyphagidae
Pacific dampwood termite	*Zootermopsis angusticollis* (Hagen)	ISOPTERA: Hodotermitidae
Pacific flatheaded borer	*Chrysobothris mali* Horn	COLEOPTERA: Buprestidae
Pacific kissing bug	*Oncocephalus pacificus* (Kirkaldy)	HEMIPTERA: Reduviidae
Pacific pelagic water strider	*Halobates sericeus* Eschscholtz	HEMIPTERA: Gerridae
Pacific spider mite	*Tetranychus pacificus* McGregor	ACARI: Tetranychidae
Pacific tent caterpillar	*Malacosoma constrictum* (Henry Edwards)	LEPIDOPTERA: Lasiocampidae
Pacific willow leaf beetle	*Pyrrhalta decora carbo* (LeConte)	COLEOPTERA: Chrysomelidae
Packard grasshopper	*Melanoplus packardii* Scudder	ORTHOPTERA: Acrididae
painted beauty	*Vanessa virginiensis* (Drury)	LEPIDOPTERA: Nymphalidae
painted hickory borer	*Megacyllene caryae* (Gahan)	COLEOPTERA: Cerambycidae
painted lady	*Vanessa cardui* (Linnaeus)	LEPIDOPTERA: Nymphalidae
painted leafhopper	*Endria inimica* (Say)	HOMOPTERA: Cicadellidae
painted maple aphid	*Drepanaphis acerifoliae* (Thomas)	HOMOPTERA: Aphididae
pale damsel bug	*Tropiconabis capsiformis* (Germar)	HEMIPTERA: Nabidae
pale juniper webworm	*Aethes rutilana* (Hübner)	LEPIDOPTERA: Cochylidae
pale leafcutting bee	*Megachile concinna* Smith	HYMENOPTERA: Megachilidae
pale legume bug	*Lygus elisus* Van Duzee	HEMIPTERA: Miridae
pale tussock moth	*Halysidota tessellaris* (J. E. Smith)	LEPIDOPTERA: Arctiidae
pale western cutworm	*Agrotis orthogonia* Morrison	LEPIDOPTERA: Noctuidae
pales weevil	*Hylobius pales* (Herbst)	COLEOPTERA: Curculionidae
palesided cutworm	*Agrotis malefida* Guenée	LEPIDOPTERA: Noctuidae
palestriped flea beetle	*Systena blanda* Melsheimer	COLEOPTERA: Chrysomelidae
palm leafskeletonizer	*Homaledra sabalella* (Chambers)	LEPIDOPTERA: Coleophoridae
palm mealybug	*Palmicultor palmarum* (Ehrhorn)	HOMOPTERA: Pseudococcidae
palmerworm	*Dichomeris ligulella* Hübner	LEPIDOPTERA: Gelechiidae
pandanus mealybug	*Laminicoccus pandani* (Cockerell)	HOMOPTERA: Pseudococcidae
pandora moth	*Coloradia pandora* Blake	LEPIDOPTERA: Saturniidae
papaya fruit fly	*Toxotrypana curvicauda* Gerstaecker	DIPTERA: Tephritidae
paper wasps	*Polistes* spp.	HYMENOPTERA: Vespidae
parasitic grain wasp	*Cephalonomia waterstoni* Gahan	HYMENOPTERA: Bethylidae
parlatoria date scale	*Parlatoria blanchardi* (Targioni-Tozzetti)	HOMOPTERA: Diaspididae
parsleyworm	*Papilio polyxenes asterius* Stoll	LEPIDOPTERA: Papilionidae
parsnip webworm	*Depressaria pastinacella* (Duponchel)	LEPIDOPTERA: Oecophoridae
pavement ant	*Tetramorium caespitum* (Linnaeus)	HYMENOPTERA: Formicidae
pea aphid	*Acyrthosiphon pisum* (Harris)	HOMOPTERA: Aphididae
pea leaf weevil	*Sitona lineatus* (Linnaeus)	COLEOPTERA: Curculionidae
pea leafminer	*Liriomyza huidobrensis* (Blanchard)	DIPTERA: Agromyzidae
pea moth	*Cydia nigricana* (Fabricius)	LEPIDOPTERA: Tortricidae
pea weevil	*Bruchus pisorum* (Linnaeus)	COLEOPTERA: Bruchidae
peach bark beetle	*Phloeotribus liminaris* (Harris)	COLEOPTERA: Scolytidae
peach silver mite	*Aculus cornutus* (Banks)	ACARI: Eriophyidae
peach twig borer	*Anarsia lineatella* Zeller	LEPIDOPTERA: Gelechiidae
peachtree borer	*Synanthedon exitiosa* (Say)	LEPIDOPTERA: Sesiidae
pear midge	*Contarinia pyrivora* (Riley)	DIPTERA: Cecidomyiidae
pear plant bug	*Lygocoris communis* (Knight)	HEMIPTERA: Miridae
pear psylla	*Psylla pyricola* Foerster	HOMOPTERA: Psyllidae
pear rust mite	*Epitrimerus pyri* (Nalepa)	ACARI: Eriophyidae
pear sawfly	*Caliroa cerasi* (Linnaeus)	HYMENOPTERA: Tenthredinidae
pear thrips	*Taeniothrips inconsequens* (Uzel)	THYSANOPTERA: Thripidae
pearleaf blister mite	*Phytoptus pyri* Pagenstecher	ACARI: Eriophyidae
pecan bud moth	*Gretchena bolliana* (Slingerland)	LEPIDOPTERA: Tortricidae
pecan carpenterworm	*Cossula magnifica* (Strecker)	LEPIDOPTERA: Cossidae
pecan cigar casebearer	*Coleophora laticornella* Clemens	LEPIDOPTERA: Coleophoridae
pecan leaf casebearer	*Acrobasis juglandis* (LeBaron)	LEPIDOPTERA: Pyralidae
pecan leaf phylloxera	*Phylloxera notabilis* Pergande	HOMOPTERA: Phylloxeridae
pecan leaf scorch mite	*Eotetranychus hicoriae* (McGregor)	ACARI: Tetranychidae
pecan leafroll mite	*Eriophyes caryae* Keifer	ACARI: Eriophyidae
pecan nut casebearer	*Acrobasis nuxvorella* Neunzig	LEPIDOPTERA: Pyralidae
pecan phylloxera	*Phylloxera devastatrix* Pergande	HOMOPTERA: Phylloxeridae

pecan serpentine leafminer	*Stigmella juglandifoliella* (Clemens)	LEPIDOPTERA: Nepticulidae
pecan spittlebug	*Clastoptera achatina* Germar	HOMOPTERA: Cercopidae
pecan weevil	*Curculio caryae* (Horn)	COLEOPTERA: Curculionidae
pepper maggot	*Zonosemata electa* (Say)	DIPTERA: Tephritidae
pepper weevil	*Anthonomus eugenii* Cano	COLEOPTERA: Curculionidae
pepper-and-salt moth	*Biston betularia cognataria* (Guenée)	LEPIDOPTERA: Geometridae
peppergrass beetle	*Galeruca browni* Blake	COLEOPTERA: Chrysomelidae
periodical cicada	*Magicicada septendecim* (Linnaeus)	HOMOPTERA: Cicadidae
persimmon borer	*Sannina uroceriformis* Walker	LEPIDOPTERA: Sesiidae
persimmon psylla	*Trioza diospyri* (Ashmead)	HOMOPTERA: Psyllidae
phantom hemlock looper	*Nepytia phantasmaria* (Strecker)	LEPIDOPTERA: Geometridae
Pharaoh ant	*Monomorium pharaonis* (Linnaeus)	HYMENOPTERA: Formicidae
Philippine katydid	*Phaneroptera furcifera* Stål	ORTHOPTERA: Tettigoniidae
phlox plant bug	*Lopidea davisi* Knight	HEMIPTERA: Miridae
pickleworm	*Diaphania nitidalis* (Stoll)	LEPIDOPTERA: Pyralidae
pigeon fly	*Pseudolynchia canariensis* (Macquart)	DIPTERA: Hippoboscidae
pigeon tremex	*Tremex columba* (Linnaeus)	HYMENOPTERA: Siricidae
pine bark adelgid	*Pineus strobi* (Hartig)	HOMOPTERA: Phylloxeridae
pine bud mite	*Trisetacus pini* (Nalepa)	ACARI: Nalepellidae
pine butterfly	*Neophasia menapia* (Felder & Felder)	LEPIDOPTERA: Pieridae
pine candle moth	*Exoteleia nepheos* Freeman	LEPIDOPTERA: Gelechiidae
pine chafer	*Anomala oblivia* Horn	COLEOPTERA: Scarabaeidae
pine colaspis	*Colaspis pini* Barber	COLEOPTERA: Chrysomelidae
pine conelet looper	*Nepytia semiclusaria* (Walker)	LEPIDOPTERA: Geometridae
pine engraver	*Ips pini* (Say)	COLEOPTERA: Scolytidae
pine false webworm	*Acantholyda erythrocephala* (Linnaeus)	HYMENOPTERA: Pamphiliidae
pine gall weevil	*Podapion gallicola* Riley	COLEOPTERA: Curculionidae
pine leaf adelgid	*Pineus pinifoliae* (Fitch)	HOMOPTERA: Adelgidae
pine needle scale	*Chionaspis pinifoliae* (Fitch)	HOMOPTERA: Diaspididae
pine needle sheathminer	*Zelleria haimbachi* Busck	LEPIDOPTERA: Yponomeutidae
pine needleminer	*Exoteleia pinifoliella* (Chambers)	LEPIDOPTERA: Gelechiidae
pine root collar weevil	*Hylobius radicis* Buchanan	COLEOPTERA: Curculionidae
pine root tip weevil	*Hylobius rhizophagus* Millers, Benjamin, & Warner	COLEOPTERA: Curculionidae
pine rosette mite	*Trisetacus gemmavitians* Styer	ACARI: Nalepellidae
pine spittlebug	*Aphrophora parallela* (Say)	HOMOPTERA: Cercopidae
pine tortoise scale	*Toumeyella parvicornis* (Cockerell)	HOMOPTERA: Coccidae
pine tube moth	*Argyrotaenia pinatubana* (Kearfott)	LEPIDOPTERA: Tortricidae
pine tussock moth	*Dasychira pinicola* (Dyar)	LEPIDOPTERA: Lymantriidae
pine webworm	*Tetralopha robustella* Zeller	LEPIDOPTERA: Pyralidae
pineapple false spider mite	*Dolichotetranychus floridanus* (Banks)	ACARI: Tenuipalpidae
pineapple mealybug	*Dysmicoccus brevipes* (Cockerell)	HOMOPTERA: Pseudococcidae
pineapple scale	*Diaspis bromeliae* (Kerner)	HOMOPTERA: Diaspididae
pineapple tarsonemid	*Steneotarsonemus ananas* (Tryon)	ACARI: Tarsonemidae
pineapple weevil	*Metamasius ritchiei* Marshall	COLEOPTERA: Curculionidae
pink bollworm	*Pectinophora gossypiella* (Saunders)	LEPIDOPTERA: Gelechiidae
pink scavenger caterpillar	*Pyroderces rileyi* (Walsingham)	LEPIDOPTERA: Cosmopterigidae
pink sugarcane mealybug	*Saccharicoccus sacchari* (Cockerell)	HOMOPTERA: Pseudococcidae
pinkstriped oakworm	*Anisota virginiensis* (Drury)	LEPIDOPTERA: Saturniidae
pinkwinged grasshopper	*Atractomorpha sinensis* Bolivar	ORTHOPTERA: Pyrgomorphidae
pinon cone beetle	*Conophthorus edulis* Hopkins	COLEOPTERA: Scolytidae
pipevine swallowtail	*Battus philenor* (Linnaeus)	LEPIDOPTERA: Papilionidae
pistol casebearer	*Coleophora malivorella* Riley	LEPIDOPTERA: Coleophoridae
pitch mass borer	*Synanthedon pini* (Kellicott)	LEPIDOPTERA: Sesiidae
pitch pine tip moth	*Rhyacionia rigidana* (Fernald)	LEPIDOPTERA: Tortricidae
pitch twig moth	*Petrova comstockiana* (Fernald)	LEPIDOPTERA: Tortricidae
pitch-eating weevil	*Pachylobius picivorus* (Germar)	COLEOPTERA: Curculionidae
pitcherplant mosquito	*Wyeomyia smithii* (Coquillett)	DIPTERA: Culicidae
plains false wireworm	*Eleodes opacus* (Say)	COLEOPTERA: Tenebrionidae
plaster beetle	*Cartodere constricta* (Gyllenhal)	COLEOPTERA: Lathridiidae
plum curculio	*Conotrachelus nenuphar* (Herbst)	COLEOPTERA: Curculionidae

plum gouger	*Coccotorus scutellaris* (LeConte)	COLEOPTERA: Curculionidae
plum leafhopper	*Macropsis trimaculata* (Fitch)	HOMOPTERA: Cicadellidae
plum rust mite	*Aculus fockeui* (Nalepa & Trouessart)	ACARI: Eriophyidae
plum webspinning sawfly	*Neurotoma inconspicua* (Norton)	HYMENOPTERA: Pamphiliidae
plumeria borer	*Lagocheirus undatus* (Voet)	COLEOPTERA: Cerambycidae
plumeria whitefly	*Paraleyrodes perseae* (Quaintance)	HOMOPTERA: Aleyrodidae
poinciana looper	*Pericyma cruegeri* (Butler)	LEPIDOPTERA: Noctuidae
polyphemus moth	*Antheraea polyphemus* (Cramer)	LEPIDOPTERA: Saturniidae
ponderosa pine bark borer	*Canonura princeps* (Walker)	COLEOPTERA: Cerambycidae
ponderosa pine cone beetle	*Conophthorus ponderosae* Hopkins	COLEOPTERA: Scolytidae
poplar borer	*Saperda calcarata* Say	COLEOPTERA: Cerambycidae
poplar leaffolding sawfly	*Phyllocolpa bozemani* (Cooley)	HYMENOPTERA: Tenthredinidae
poplar petiolegall aphid	*Pemphigus populitransversus* Riley	HOMOPTERA: Aphididae
poplar tentmaker	*Ichthyura inclusa* (Hübner)	LEPIDOPTERA: Notodontidae
poplar twig gall aphid	*Pemphigus populiramulorum* Riley	HOMOPTERA: Aphididae
poplar vagabond aphid	*Mordvilkoja vagabunda* (Walsh)	HOMOPTERA: Aphididae
poplar-and-willow borer	*Cryptorhynchus lapathi* (Linnaeus)	COLEOPTERA: Curculionidae
portulaca leafmining weevil	*Hypurus bertrandi* Perris	COLEOPTERA: Curculionidae
potato aphid	*Macrosiphum euphorbiae* (Thomas)	HOMOPTERA: Aphididae
potato flea beetle	*Epitrix cucumeris* (Harris)	COLEOPTERA: Chrysomelidae
potato leafhopper	*Empoasca fabae* (Harris)	HOMOPTERA: Cicadellidae
potato psyllid	*Paratrioza cockerelli* (Sulc)	HOMOPTERA: Psyllidae
potato scab gnat	*Pnyxia scabiei* (Hopkins)	DIPTERA: Sciaridae
potato stalk borer	*Trichobaris trinotata* (Say)	COLEOPTERA: Curculionidae
potato stem borer	*Hydraecia micacea* (Esper)	LEPIDOPTERA: Noctuidae
potato tuberworm	*Phthorimaea operculella* (Zeller)	LEPIDOPTERA: Gelechiidae
poultry bug	*Haematosiphon inodorus* (Dugès)	HEMIPTERA: Cimicidae
poultry house moth	*Niditinea fuscipunctella* (Haworth)	LEPIDOPTERA: Tineidae
powderpost bostrichid	*Amphicerus cornutus* (Pallas)	COLEOPTERA: Bostrichidae
prairie flea beetle	*Altica canadensis* Gentner	COLEOPTERA: Chrysomelidae
prairie grain wireworm	*Ctenicera aeripennis destructor* (Brown)	COLEOPTERA: Elateridae
privet aphid	*Myzus ligustri* (Mosley)	HOMOPTERA: Aphididae
privet leafminer	*Caloptilia cuculipennella* (Hübner)	LEPIDOPTERA: Gracillariidae
privet mite	*Brevipalpus obovatus* Donnadieu	ACARI: Tenuipalpidae
privet thrips	*Dendrothrips ornatus* (Jablonowski)	THYSANOPTERA: Thripidae
promethea moth	*Callosamia promethea* (Drury)	LEPIDOPTERA: Saturniidae
pruinose bean weevil	*Stator pruininus* (Horn)	COLEOPTERA: Bruchidae
prune leafhopper	*Edwardsiana prunicola* (Edwards)	HOMOPTERA: Cicadellidae
Puget Sound wireworm	*Ctenicera aeripennis aeripennis* (Kirby)	COLEOPTERA: Elateridae
puncturevine seed weevil	*Microlarinus lareynii* (Jacquelin du Val)	COLEOPTERA: Curculionidae
puncturevine stem weevil	*Microlarinus lypriformis* (Wollaston)	COLEOPTERA: Curculionidae
purple scale	*Lepidosaphes beckii* (Newman)	HOMOPTERA: Diaspididae
purplebacked cabbageworm	*Evergestis pallidata* (Hufnagel)	LEPIDOPTERA: Pyralidae
purplespotted lily aphid	*Macrosiphum lilii* (Monell)	HOMOPTERA: Aphididae
puss caterpillar	*Megalopyge opercularis* (J. E. Smith)	LEPIDOPTERA: Megalopygidae
Putnam scale	*Diaspidiotus ancylus* (Putnam)	HOMOPTERA: Diaspididae
pyramid ant	*Conomyrma insana* (Buckley)	HYMENOPTERA: Formicidae
pyriform scale	*Protopulvinaria pyriformis* (Cockerell)	HOMOPTERA: Coccidae

Q

quince curculio	*Conotrachelus crataegi* Walsh	COLEOPTERA: Curculionidae
quince treehopper	*Glossonotus crataegi* (Fitch)	HOMOPTERA: Membracidae

R

rabbit louse	*Haemodipsus ventricosus* (Denny)	ANOPLURA: Hoplopleuridae
rabbit tick	*Haemaphysalis leporispalustris* (Packard)	ACARI: Ixodidae
ragweed borer	*Epiblema strenuana* (Walker)	LEPIDOPTERA: Tortricidae

ragweed plant bug	*Chlamydatus associatus* (Uhler)	HEMIPTERA: Miridae
raisin moth	*Cadra figulilella* (Gregson)	LEPIDOPTERA: Pyralidae
range caterpillar	*Hemileuca oliviae* Cockerell	LEPIDOPTERA: Saturniidae
range crane fly	*Tipula simplex* Doane	DIPTERA: Tipulidae
rapid plant bug	*Adelphocoris rapidus* (Say)	HEMIPTERA: Miridae
raspberry bud moth	*Lampronia rubiella* (Bjerkander)	LEPIDOPTERA: Incurvariidae
raspberry cane borer	*Oberea bimaculata* (Olivier)	COLEOPTERA: Cerambycidae
raspberry cane maggot	*Pegomya rubivora* (Coquillett)	DIPTERA: Anthomyiidae
raspberry crown borer	*Pennisetia marginata* (Harris)	LEPIDOPTERA: Sesiidae
raspberry leafroller	*Olethreutes permundana* (Clemens)	LEPIDOPTERA: Tortricidae
raspberry sawfly	*Monophadnoides geniculatus* (Hartig)	HYMENOPTERA: Tenthredinidae
red admiral	*Vanessa atalanta rubria* (Fruhstorfer)	LEPIDOPTERA: Nymphalidae
red and black flat mite	*Brevipalpus phoenicis* (Geijskes)	ACARI: Tenuipalpidae
red assassin bug	*Haematoloecha rubescens* Distant	HEMIPTERA: Reduviidae
red carpenter ant	*Camponotus ferrugineus* (Fabricius)	HYMENOPTERA: Formicidae
red clover seed weevil	*Tychius stephensi* (Schönherr)	COLEOPTERA: Curculionidae
red date scale	*Phoenicococcus marlatti* Cockerell	HOMOPTERA: Phoenicococcidae
red elm bark weevil	*Magdalis armicollis* (Say)	COLEOPTERA: Curculionidae
red flour beetle	*Tribolium castaneum* (Herbst)	COLEOPTERA: Tenebrionidae
red grasshopper mite	*Eutrombidium trigonum* (Hermann)	ACARI: Trombidiidae
red harvester ant	*Pogonomyrmex barbatus* (F. Smith)	HYMENOPTERA: Formicidae
red imported fire ant	*Solenopsis invicta* Buren	HYMENOPTERA: Formicidae
red milkweed beetle	*Tetraopes tetrophthalmus* (Forster)	COLEOPTERA: Cerambycidae
red oak borer	*Enaphalodes rufulus* (Haldeman)	COLEOPTERA: Cerambycidae
red orchid scale	*Furcaspis biformis* (Cockerell)	HOMOPTERA: Diaspididae
red pine cone beetle	*Conophthorus resinosae* Hopkins	COLEOPTERA: Scolytidae
red pine sawfly	*Neodiprion nanulus nanulus* Schedl	HYMENOPTERA: Diprionidae
red pine scale	*Matsucoccus resinosae* Bean & Godwin	HOMOPTERA: Margarodidae
red turnip beetle	*Entomoscelis americana* Brown	COLEOPTERA: Chrysomelidae
red turpentine beetle	*Dendroctonus valens* LeConte	COLEOPTERA: Scolytidae
red wax scale	*Ceroplastes rubens* Maskell	HOMOPTERA: Coccidae
redbacked cutworm	*Euxoa ochrogaster* (Guenée)	LEPIDOPTERA: Noctuidae
redbanded leafroller	*Argyrotaenia velutinana* (Walker)	LEPIDOPTERA: Tortricidae
redbanded thrips	*Selenothrips rubrocinctus* (Giard)	THYSANOPTERA: Thripidae
redberry mite	*Acalitus essigi* (Hassan)	ACARI: Eriophyidae
redblack oedemerid	*Eobia bicolor* (Fairmaire)	COLEOPTERA: Oedemeridae
redbud leaffolder	*Fascista cercerisella* (Chambers)	LEPIDOPTERA: Gelechiidae
redheaded ash borer	*Neoclytus acuminatus* (Fabricius)	COLEOPTERA: Cerambycidae
redheaded jack pine sawfly	*Neodiprion rugifrons* Middleton	HYMENOPTERA: Diprionidae
redheaded pine sawfly	*Neodiprion lecontei* (Fitch)	HYMENOPTERA: Diprionidae
redhumped caterpillar	*Schizura concinna* (J. E. Smith)	LEPIDOPTERA: Notodontidae
redlegged flea beetle	*Derocrepis erythropus* (Melsheimer)	COLEOPTERA: Chrysomelidae
redlegged grasshopper	*Melanoplus femurrubrum* (De Geer)	ORTHOPTERA: Acrididae
redlegged ham beetle	*Necrobia rufipes* (De Geer)	COLEOPTERA: Cleridae
redmargined assassin bug	*Scadra rufidens* Stål	HEMIPTERA: Reduviidae
rednecked cane borer	*Agrilus ruficollis* (Fabricius)	COLEOPTERA: Buprestidae
rednecked peanutworm	*Stegasta bosqueella* (Chambers)	LEPIDOPTERA: Gelechiidae
redshouldered ham beetle	*Necrobia ruficollis* (Fabricius)	COLEOPTERA: Cleridae
redshouldered stink bug	*Thyanta accerra* McAtee	HEMIPTERA: Pentatomidae
redtailed spider wasp	*Tachypompilus analis* (Fabricius)	HYMENOPTERA: Pompilidae
redtailed tachina	*Winthemia quadripustulata* (Fabricius)	DIPTERA: Tachinidae
regal moth	*Citheronia regalis* (Fabricius)	LEPIDOPTERA: Saturniidae
relapsing fever tick	*Ornithodoros turicata* (Dugès)	ACARI: Argasidae
resplendent shield bearer	*Coptodisca splendoriferella* (Clemens)	LEPIDOPTERA: Heliozelidae
reticulate mite	*Lorryia reticulata* (Oudemans)	ACARI: Tydeidae
reticulatewinged trogiid	*Lepinotus reticulatus* Enderlein	PSOCOPTERA: Trogiidae
rhinoceros beetle	*Xyloryctes jamaicensis* (Drury)	COLEOPTERA: Scarabaeidae
Rhodesgrass mealybug	*Antonina graminis* (Maskell)	HOMOPTERA: Pseudococcidae
rhododendron borer	*Synanthedon rhododendri* Beutenmüller	LEPIDOPTERA: Sesiidae
rhododendron lace bug	*Stephanitis rhododendri* Horvath	HEMIPTERA: Tingidae

rhododendron whitefly	*Dialeurodes chittendeni* Laing	HOMOPTERA: Aleyrodidae
rhubarb curculio	*Lixus concavus* Say	COLEOPTERA: Curculionidae
rice delphacid	*Sogatodes oryzicola* (Muir)	HOMOPTERA: Delphacidae
rice leaffolder	*Lerodea eufala* (Edwards)	LEPIDOPTERA: Hesperiidae
rice leafhopper	*Nephotettix nigropictus* (Stål)	HOMOPTERA: Cicadellidae
rice root aphid	*Rhopalosiphum rufiabdominalis* (Sasaki)	HOMOPTERA: Aphididae
rice stalk borer	*Chilo plejadellus* Zincken	LEPIDOPTERA: Pyralidae
rice stink bug	*Oebalus pugnax* (Fabricius)	HEMIPTERA: Pentatomidae
rice water weevil	*Lissorhoptrus oryzophilus* Kuschel	COLEOPTERA: Curculionidae
rice weevil	*Sitophilus oryzae* (Linnaeus)	COLEOPTERA: Curculionidae
ridgewinged fungus beetle	*Thes bergrothi* (Reitter)	COLEOPTERA: Lathridiidae
ringlegged earwig	*Euborellia annulipes* (Lucas)	DERMAPTERA: Labiduridae
robust leafhopper	*Stragania robusta* (Uhler)	HOMOPTERA: Cicadellidae
Rocky Mountain grasshopper	*Melanoplus spretus* (Walsh)	ORTHOPTERA: Acrididae
Rocky Mountain wood tick	*Dermacentor andersoni* Stiles	ACARI: Ixodidae
rose aphid	*Macrosiphum rosae* (Linnaeus)	HOMOPTERA: Aphididae
rose chafer	*Macrodactylus subspinosus* (Fabricius)	COLEOPTERA: Scarabaeidae
rose curculio	*Merhynchites bicolor* (Fabricius)	COLEOPTERA: Curculionidae
rose leaf beetle	*Nodonota puncticollis* (Say)	COLEOPTERA: Chrysomelidae
rose leafhopper	*Edwardsiana rosae* (Linnaeus)	HOMOPTERA: Cicadellidae
rose midge	*Dasineura rhodophaga* (Coquillett)	DIPTERA: Cecidomyiidae
rose scale	*Aulacaspis rosae* (Bouché)	HOMOPTERA: Diaspididae
rose stem girdler	*Agrilus aurichalceus* Redtenbacher	COLEOPTERA: Buprestidae
roseroot gall wasp	*Diplolepis radicum* (Osten Sacken)	HYMENOPTERA: Cynipidae
roseslug	*Endelomyia aethiops* (Fabricius)	HYMENOPTERA: Tenthredinidae
rosy apple aphid	*Dysaphis plantaginea* (Passerini)	HOMOPTERA: Aphididae
rotund tick	*Ixodes kingi* Bishopp	ACARI: Ixodidae
rough stink bug	*Brochymena quadripustulata* (Fabricius)	HEMIPTERA: Pentatomidae
roughskinned cutworm	*Proxenus mindara* Barnes & McDunnough	LEPIDOPTERA: Noctuidae
roundheaded appletree borer	*Saperda candida* Fabricius	COLEOPTERA: Cerambycidae
roundheaded cone borer	*Paratimia conicola* Fisher	COLEOPTERA: Cerambycidae
roundheaded fir borer	*Tetropium abietis* Fall	COLEOPTERA: Cerambycidae
roundheaded pine beetle	*Dendroctonus adjunctus* Blandford	COLEOPTERA: Scolytidae
Russian wheat aphid	*Diuraphis noxiu* (Mordvilko)	HOMOPTERA: Aphididae
rustic borer	*Xylotrechus colonus* (Fabricius)	COLEOPTERA: Cerambycidae
rusty banded aphid	*Dysaphis apiifolia* (Theobald)	HOMOPTERA: Aphididae
rusty grain beetle	*Cryptolestes ferrugineus* (Stephens)	COLEOPTERA: Cucujidae
rusty plum aphid	*Hysteroneura setariae* (Thomas)	HOMOPTERA: Aphididae
rusty tussock moth	*Orgyia antiqua* (Linnaeus)	LEPIDOPTERA: Lymantriidae

S

saddleback caterpillar	*Sibine stimulea* (Clemens)	LEPIDOPTERA: Limacodidae
saddled leafhopper	*Colladonus clitellarius* (Say)	HOMOPTERA: Cicadellidae
saddled prominent	*Heterocampa guttivitta* (Walker)	LEPIDOPTERA: Notodontidae
sagebrush defoliator	*Aroga websteri* Clarke	LEPIDOPTERA: Gelechiidae
saltmarsh caterpillar	*Estigmene acrea* (Drury)	LEPIDOPTERA: Arctiidae
saltmarsh mosquito	*Aedes sollicitans* (Walker)	DIPTERA: Culicidae
San Bruno elfin	*Incisalia fotis bayensis* (R. M. Brown)	LEPIDOPTERA: Lycaenidae
San Jose scale	*Quadraspidiotus perniciosus* (Comstock)	HOMOPTERA: Diaspididae
sand wireworm	*Horistonotus uhlerii* Horn	COLEOPTERA: Elateridae
sandcherry weevil	*Coccotorus hirsutus* Bruner	COLEOPTERA: Curculionidae
sapwood timberworm	*Hylecoetus lugubris* Say	COLEOPTERA: Lymexylonidae
Saratoga spittlebug	*Aphrophora saratogensis* (Fitch)	HOMOPTERA: Cercopidae
saskatoon borer	*Saperda bipunctata* R. Hopping	COLEOPTERA: Cerambycidae
satin moth	*Leucoma salicis* (Linnaeus)	LEPIDOPTERA: Lymantriidae
Saunders embiid	*Oligotoma saundersii* (Westwood)	EMBIOPTERA: Oligotomidae
sawtoothed grain beetle	*Oryzaephilus surinamensis* (Linnaeus)	COLEOPTERA: Cucujidae
Say blister beetle	*Lytta sayi* LeConte	COLEOPTERA: Meloidae

Say stink bug	*Chlorochroa sayi* Stål	HEMIPTERA: Pentatomidae
scab mite	*Psoroptes equi* (Raspail)	ACARI: Psoroptidae
scaly grain mite	*Suidasia nesbitti* Hughes	ACARI: Acaridae
scalyleg mite	*Knemidokoptes mutans* (Robin & Lanquetin)	ACARI: Sarcoptidae
scarlet oak sawfly	*Caliroa quercuscoccineae* (Dyar)	HYMENOPTERA: Tenthredinidae
Schaus swallowtail	*Papilio aristodemus ponceanus* Schaus	LEPIDOPTERA: Papilionidae
Schoene spider mite	*Tetranychus schoenei* McGregor	ACARI: Tetranychidae
screwworm	*Cochliomyia hominivorax* (Coquerel)	DIPTERA: Calliphoridae
sculptured pine borer	*Chalcophora angulicollis* (LeConte)	COLEOPTERA: Buprestidae
scurfy scale	*Chionaspis furfura* (Fitch)	HOMOPTERA: Diaspididae
secondary screwworm	*Cochliomyia macellaria* (Fabricius)	DIPTERA: Calliphoridae
seed bugs	*Nysius spp.*	HEMIPTERA: Lygaeidae
seedcorn beetle	*Stenolophus lecontei* (Chaudoir)	COLEOPTERA: Carabidae
seedcorn maggot	*Delia platura* (Meigen)	DIPTERA: Anthomyiidae
sequoia pitch moth	*Synanthedon sequoiae* (Hy. Edwards)	LEPIDOPTERA: Sesiidae
shaft louse	*Menopon gallinae* (Linnaeus)	MALLOPHAGA: Menoponidae
shallot aphid	*Myzus ascalonicus* Doncaster	HOMOPTERA: Aphididae
sheep biting louse	*Bovicola ovis* (Schrank)	MALLOPHAGA: Trichodectidae
sheep bot fly	*Oestrus ovis* Linnaeus	DIPTERA: Oestridae
sheep follicle mite	*Demodex ovis* Railliet	ACARI: Demodicidae
sheep ked	*Melophagus ovinus* (Linnaeus)	DIPTERA: Hippoboscidae
sheep scab mite	*Psoroptes ovis* (Hering)	ACARI: Psoroptidae
shieldbacked pine seed bug	*Tetyra bipunctata* (Herrich-Schäffer)	HEMIPTERA: Pentatomidae
shortleaf pine cone borer	*Eucosma cocana* Kearfott	LEPIDOPTERA: Tortricidae
shortnosed cattle louse	*Haematopinus eurysternus* (Nitzsch)	ANOPLURA: Haematopinidae
shothole borer	*Scolytus rugulosus* (Müller)	COLEOPTERA: Scolytidae
sigmoid fungus beetle	*Cryptophagus varus* Woodroffe & Coombs	COLEOPTERA: Cryptophagidae
silkworm	*Bombyx mori* (Linnaeus)	LEPIDOPTERA: Bombycidae
silky ant	*Formica fusca* Linnaeus	HYMENOPTERA: Formicidae
silky cane weevil	*Metamasius hemipterus sericeus* (Olivier)	COLEOPTERA: Curculionidae
silverfish	*Lepisma saccharina* Linnaeus	THYSANURA: Lepismatidae
silverspotted skipper	*Epargyreus clarus* (Cramer)	LEPIDOPTERA: Hesperiidae
silverspotted tiger moth	*Lophocampa argentata* (Packard)	LEPIDOPTERA: Arctiidae
sinuate ladybird beetle	*Hippodamia sinuata* Mulsant	COLEOPTERA: Coccinellidae
sinuate peartree borer	*Agrilus sinuatus* (Olivier)	COLEOPTERA: Buprestidae
Sitka spruce weevil	*Pissodes strobi* (Peck)	COLEOPTERA: Curculionidae
sixspotted mite	*Eotetranychus sexmaculatus* (Riley)	ACARI: Tetranychidae
sixspotted thrips	*Scolothrips sexmaculatus* (Pergande)	THYSANOPTERA: Thripidae
slash pine flower thrips	*Gnophothrips fuscus* (Morgan)	THYSANOPTERA: Phlaeothripidae
slash pine sawfly	*Neodiprion merkeli* Ross	HYMENOPTERA: Diprionidae
slash pine seedworm	*Laspeyresia anaranjada* Miller	LEPIDOPTERA: Tortricidae
slender duck louse	*Anaticola crassicornis* (Scopoli)	MALLOPHAGA: Philopteridae
slender goose louse	*Anaticola anseris* (Linnaeus)	MALLOPHAGA: Philopteridae
slender guinea louse	*Lipeurus numidae* (Denny)	MALLOPHAGA: Philopteridae
slender guineapig louse	*Gliricola porcelli* (Schrank)	MALLOPHAGA: Gyropidae
slender pigeon louse	*Columbicola columbae* (Linnaeus)	MALLOPHAGA: Philopteridae
slender seedcorn beetle	*Clivina impressifrons* LeConte	COLEOPTERA: Carabidae
slender turkey louse	*Oxylipeurus polytrapezius* (Burmeister)	MALLOPHAGA: Philopteridae
slenderhorned flour beetle	*Gnathocerus maxillosus* (Fabricius)	COLEOPTERA: Tenebrionidae
small chestnut weevil	*Curculio sayi* (Gyllenhal)	COLEOPTERA: Curculionidae
small milkweed bug	*Lygaeus kalmii* Stål	HEMIPTERA: Lygaeidae
small pigeon louse	*Campanulotes bidentatus compar* (Burmeister)	MALLOPHAGA: Philopteridae
small southern pine engraver	*Ips avulsus* (Eichhoff)	COLEOPTERA: Scolytidae
smaller European elm bark beetle	*Scolytus multistriatus* (Marsham)	COLEOPTERA: Scolytidae
smaller Hawaiian cutworm	*Agrotis dislocata* (Walker)	LEPIDOPTERA: Noctuidae
smaller lantana butterfly	*Strymon bazochii gundlachianus* (Bates)	LEPIDOPTERA: Lycaenidae
smaller yellow ant	*Acanthomyops claviger* (Roger)	HYMENOPTERA: Formicidae
smalleyed flour beetle	*Palorus ratzeburgi* (Wissmann)	COLEOPTERA: Tenebrionidae
smartweed borer	*Ostrinia obumbratalis* (Lederer)	LEPIDOPTERA: Pyralidae

smeared dagger moth	*Acronicta oblinita* (J. E. Smith)	LEPIDOPTERA: Noctuidae
Smith blue	*Euphilotes enoptes smithi* (Mattoni)	LEPIDOPTERA: Lycaenidae
smokybrown cockroach	*Periplaneta fuliginosa* (Serville)	ORTHOPTERA: Blattidae
smut beetle	*Phalacrus politus* Melsheimer	COLEOPTERA: Phalacridae
snowball aphid	*Neoceruraphis viburnicola* (Gillette)	HOMOPTERA: Aphididae
snowy tree cricket	*Oecanthus fultoni* Walker	ORTHOPTERA: Gryllidae
solanaceous treehopper	*Antianthe expansa* (Germar)	HOMOPTERA: Membracidae
solitary oak leafminer	*Cameraria hamadryadella* (Clemens)	LEPIDOPTERA: Gracillariidae
sonchus fly	*Ensina sonchi* (Linnaeus)	DIPTERA: Tephritidae
Sonoran tent caterpillar	*Malacosoma tigris* (Dyar)	LEPIDOPTERA: Lasiocampidae
sorghum midge	*Contarinia sorghicola* (Coquillett)	DIPTERA: Cecidomyiidae
sorghum webworm	*Celama sorghiella* (Riley)	LEPIDOPTERA: Noctuidae
sourbush seed fly	*Acinia picturata* (Snow)	DIPTERA: Tephritidae
South African emex weevil	*Apion antiquum* Gyllenhal	COLEOPTERA: Curculionidae
south coastal coneworm	*Dioryctria ebeli* Mutuura & Munroe	LEPIDOPTERA: Pyralidae
southern armyworm	*Spodoptera eridania* (Cramer)	LEPIDOPTERA: Noctuidae
southern beet webworm	*Herpetogramma bipunctalis* (Fabricius)	LEPIDOPTERA: Pyralidae
southern buffalo gnat	*Cnephia pecuarum* (Riley)	DIPTERA: Simuliidae
southern cabbageworm	*Pontia protodice* (Boisduval & LeConte)	LEPIDOPTERA: Pieridae
southern cattle tick	*Boophilus microplus* (Canestrini)	ACARI: Ixodidae
southern chinch bug	*Blissus insularis* Barber	HEMIPTERA: Lygaeidae
southern corn billbug	*Sphenophorus callosus* (Olivier)	COLEOPTERA: Curculionidae
southern corn rootworm	*Diabrotica undecimpunctata howardi* Barber	COLEOPTERA: Chrysomelidae
southern cornstalk borer	*Diatraea crambidoides* (Grote)	LEPIDOPTERA: Pyralidae
southern fire ant	*Solenopsis xyloni* McCook	HYMENOPTERA: Formicidae
southern garden leafhopper	*Empoasca solana* DeLong	HOMOPTERA: Cicadellidae
southern green stink bug	*Nezara viridula* (Linnaeus)	HEMIPTERA: Pentatomidae
southern house mosquito	*Culex quinquefasciatus* Say	DIPTERA: Culicidae
southern lyctus beetle	*Lyctus planicollis* LeConte	COLEOPTERA: Lyctidae
southern masked chafer	*Cyclocephala immaculata* (Olivier)	COLEOPTERA: Scarabaeidae
southern mole cricket	*Scapteriscus acletus* Rehn & Hebard	ORTHOPTERA: Gryllotalpidae
southern pine beetle	*Dendroctonus frontalis* Zimmermann	COLEOPTERA: Scolytidae
southern pine coneworm	*Dioryctria amatella* (Hulst)	LEPIDOPTERA: Pyralidae
southern pine root weevil	*Hylobius aliradicis* Warner	COLEOPTERA: Curculionidae
southern pine sawyer	*Monochamus titillator* (Fabricius)	COLEOPTERA: Cerambycidae
southern potato wireworm	*Conoderus falli* Lane	COLEOPTERA: Elateridae
southern red mite	*Oligonychus ilicis* (McGregor)	ACARI: Tetranychidae
southwestern corn borer	*Diatraea grandiosella* (Dyar)	LEPIDOPTERA: Pyralidae
southwestern Hercules beetle	*Dynastes granti* Horn	COLEOPTERA: Scarabaeidae
southwestern pine tip moth	*Rhyacionia neomexicana* (Dyar)	LEPIDOPTERA: Tortricidae
southwestern squash vine borer	*Melittia calabaza* Duckworth & Eichlin	LEPIDOPTERA: Sesiidae
southwestern tent caterpillar	*Malacosoma incurvum* (Henry Edwards)	LEPIDOPTERA: Lasiocampidae
sow thistle aphid	*Nasonovia lactucae* (Linnaeus)	HOMOPTERA: Aphididae
soybean leafminer	*Odontota horni* Smith	COLEOPTERA: Chrysomelidae
soybean looper	*Pseudoplusia includens* (Walker)	LEPIDOPTERA: Noctuidae
soybean thrips	*Sericothrips variabilis* (Beach)	THYSANOPTERA: Thripidae
Spanishfly	*Lytta vesicatoria* (Linnaeus)	COLEOPTERA: Meloidae
spicebush swallowtail	*Papilio troilus* Linnaeus	LEPIDOPTERA: Papilionidae
spider mite destroyer	*Stethorus picipes* Casey	COLEOPTERA: Coccinellidae
spinach flea beetle	*Disonycha xanthomelas* (Dalman)	COLEOPTERA: Chrysomelidae
spinach leafminer	*Pegomya hyoscyami* (Panzer)	DIPTERA: Anthomyiidae
spined assassin bug	*Sinea diadema* (Fabricius)	HEMIPTERA: Reduviidae
spined rat louse	*Polyplax spinulosa* (Burmeister)	ANOPLURA: Hoplopleuridae
spined soldier bug	*Podisus maculiventris* (Say)	HEMIPTERA: Pentatomidae
spined stilt bug	*Jalysus spinosus* (Say)	HEMIPTERA: Berytidae
spiny assassin bug	*Polididus armatissimus* Stål	HEMIPTERA: Reduviidae
spiny oakworm	*Anisota stigma* (Fabricius)	LEPIDOPTERA: Saturniidae
spirea aphid	*Aphis citricola* Van der Goot	HOMOPTERA: Aphididae
spotted alfalfa aphid	*Therioaphis maculata* (Buckton)	HOMOPTERA: Aphididae
spotted asparagus beetle	*Crioceris duodecimpunctata* (Linnaeus)	COLEOPTERA: Chrysomelidae

spotted beet webworm	*Hymenia perspectalis* (Hübner)	LEPIDOPTERA: Pyralidae
spotted blister beetle	*Epicauta maculata* (Say)	COLEOPTERA: Meloidae
spotted cucumber beetle	*Diabrotica undecimpunctata howardi* Barber	COLEOPTERA: Chrysomelidae
spotted cutworm	*Amathes c-nigrum* (Linnaeus)	LEPIDOPTERA: Noctuidae
spotted hairy fungus beetle	*Mycetophagus quadriguttatus* Muller	COLEOPTERA: Mycetophagidae
spotted Mediterranean cockroach	*Ectobius pallidus* (Olivier)	ORTHOPTERA: Blattellidae
spotted pine sawyer	*Monochamus maculosus* Haldeman	COLEOPTERA: Cerambycidae
spotted tentiform leafminer	*Phyllonorycter blancardella* (Fabricius)	LEPIDOPTERA: Gracillariidae
spotted tussock moth	*Lophocampa maculata* Harris	LEPIDOPTERA: Arctiidae
spottedwinged antlion	*Dendroleon obsoletus* (Say)	NEUROPTERA: Myrmeleontidae
spring cankerworm	*Paleacrita vernata* (Peck)	LEPIDOPTERA: Geometridae
spruce aphid	*Elatobium abietinum* (Walker)	HOMOPTERA: Aphididae
spruce beetle	*Dendroctonus rufipennis* (Kirby)	COLEOPTERA: Scolytidae
spruce bud midge	*Rhabdophaga swainei* Felt	DIPTERA: Cecidomyiidae
spruce bud moth	*Zeiraphera canadensis* Mutuura & Freeman	LEPIDOPTERA: Tortricidae
spruce bud scale	*Physokermes piceae* (Schrank)	HOMOPTERA: Coccidae
spruce budworm	*Choristoneura fumiferana* (Clemens)	LEPIDOPTERA: Tortricidae
spruce coneworm	*Dioryctria reniculelloides* Mutuura & Munroe	LEPIDOPTERA: Pyralidae
spruce mealybug	*Puto sandini* Washburn	HOMOPTERA: Pseudococcidae
spruce needleminer	*Endothenia albolineana* (Kearfott)	LEPIDOPTERA: Tortricidae
spruce seed moth	*Laspeyresia youngana* (Kearfott)	LEPIDOPTERA: Tortricidae
spruce spider mite	*Oligonychus ununguis* (Jacobi)	ACARI: Tetranychidae
squarenecked grain beetle	*Cathartus quadricollis* (Guérin-Méneville)	COLEOPTERA: Cucujidae
squarenosed fungus beetle	*Lathridius minutus* (Linnaeus)	COLEOPTERA: Lathridiidae
squash beetle	*Epilachna borealis* (Fabricius)	COLEOPTERA: Coccinellidae
squash bug	*Anasa tristis* (De Geer)	HEMIPTERA: Coreidae
squash vine borer	*Melittia cucurbitae* (Harris)	LEPIDOPTERA: Sesiidae
stable fly	*Stomoxys calcitrans* (Linnaeus)	DIPTERA: Muscidae
stalk borer	*Papaipema nebris* (Guenée)	LEPIDOPTERA: Noctuidae
star jasmine thrips	*Thrips orientalis* (Bagnall)	THYSANOPTERA: Thripidae
steelblue ladybird beetle	*Orcus chalybeus* (Boisduval)	COLEOPTERA: Coccinellidae
Stevens leafhopper	*Empoasca stevensi* Young	HOMOPTERA: Cicadellidae
sticktight flea	*Echidnophaga gallinacea* (Westwood)	SIPHONAPTERA: Pulicidae
stinging rose caterpillar	*Parasa indetermina* (Boisduval)	LEPIDOPTERA: Limacodidae
stink beetle	*Nomius pygmaeus* (Dejean)	COLEOPTERA: Carabidae
stored nut moth	*Aphomia gularis* (Zeller)	LEPIDOPTERA: Pyralidae
straw itch mite	*Pyemotes tritici* (Lagrèze-Fossat & Montané)	ACARI: Pyemotidae
strawberry aphid	*Chaetosiphon fragaefolii* (Cockerell)	HOMOPTERA: Aphididae
strawberry bud weevil	*Anthonomus signatus* Say	COLEOPTERA: Curculionidae
strawberry crown borer	*Tyloderma fragariae* (Riley)	COLEOPTERA: Curculionidae
strawberry crown moth	*Synanthedon bibionipennis* (Boisduval)	LEPIDOPTERA: Sesiidae
strawberry crownminer	*Aristotelia fragariae* Busck	LEPIDOPTERA: Gelechiidae
strawberry leafroller	*Ancylis comptana* (Froelich)	LEPIDOPTERA: Tortricidae
strawberry root aphid	*Aphis forbesi* Weed	HOMOPTERA: Aphididae
strawberry root weevil	*Otiorhynchus ovatus* (Linnaeus)	COLEOPTERA: Curculionidae
strawberry rootworm	*Paria fragariae* Wilcox	COLEOPTERA: Chrysomelidae
strawberry sap beetle	*Stelidota geminata* (Say)	COLEOPTERA: Nitidulidae
strawberry spider mite	*Tetranychus turkestani* Ugarov & Nikolski	ACARI: Tetranychidae
strawberry whitefly	*Trialeurodes packardi* (Morrill)	HOMOPTERA: Aleyrodidae
striped alder sawfly	*Hemichroa crocea* (Geoffroy)	HYMENOPTERA: Tenthredinidae
striped ambrosia beetle	*Trypodendron lineatum* (Olivier)	COLEOPTERA: Scolytidae
striped blister beetle	*Epicauta vittata* (Fabricius)	COLEOPTERA: Meloidae
striped cucumber beetle	*Acalymma vittatum* (Fabricius)	COLEOPTERA: Chrysomelidae
striped cutworm	*Euxoa tessellata* (Harris)	LEPIDOPTERA: Noctuidae
striped earwig	*Labidura riparia* Pallas	DERMAPTERA: Labiduridae
striped flea beetle	*Phyllotreta striolata* (Fabricius)	COLEOPTERA: Chrysomelidae
striped garden caterpillar	*Lacanobia legitima* (Grote)	LEPIDOPTERA: Noctuidae
striped horse fly	*Tabanus lineola* Fabricius	DIPTERA: Tabanidae
striped mealybug	*Ferrisia virgata* (Cockerell)	HOMOPTERA: Pseudococcidae

subtropical pine tip moth	*Rhyacionia subtropica* Miller	LEPIDOPTERA: Tortricidae
suckfly	*Cyrtopeltis notatus* (Distant)	HEMIPTERA: Miridae
sugar maple borer	*Glycobius speciosus* (Say)	COLEOPTERA: Cerambycidae
sugar pine cone beetle	*Conophthorus lambertianae* Hopkins	COLEOPTERA: Scolytidae
sugarbeet crown borer	*Hulstia undulatella* (Clemens)	LEPIDOPTERA: Pyralidae
sugarbeet root aphid	*Pemphigus populivenae* Fitch	HOMOPTERA: Aphididae
sugarbeet root maggot	*Tetanops myopaeformis* (Röder)	DIPTERA: Otitidae
sugarbeet wireworm	*Limonius californicus* (Mannerheim)	COLEOPTERA: Elateridae
sugarcane aphid	*Melanaphis sacchari* (Zehntner)	HOMOPTERA: Aphididae
sugarcane beetle	*Euetheola humilis rugiceps* (LeConte)	COLEOPTERA: Scarabaeidae
sugarcane borer	*Diatraea saccharalis* (Fabricius)	LEPIDOPTERA: Pyralidae
sugarcane bud moth	*Neodecadarachis flavistriata* (Walsingham)	LEPIDOPTERA: Tineidae
sugarcane delphacid	*Perkinsiella saccharicida* Kirkaldy	HOMOPTERA: Delphacidae
sugarcane leaf mite	*Oligonychus indicus* (Hirst)	ACARI: Tetranychidae
sugarcane leafroller	*Hedylepta accepta* (Butler)	LEPIDOPTERA: Pyralidae
sugarcane stalk mite	*Steneotarsonemus bancrofti* (Michael)	ACARI: Tarsonemidae
sugarcane thrips	*Baliothrips minutus* (van Deventer)	THYSANOPTERA: Thripidae
sunflower beetle	*Zygogramma exclamationis* (Fabricius)	COLEOPTERA: Chrysomelidae
sunflower bud moth	*Suleima helianthana* (Riley)	LEPIDOPTERA: Tortricidae
sunflower maggot	*Strauzia longipennis* (Wiedemann)	DIPTERA: Tephritidae
sunflower moth	*Homoeosoma electellum* (Hulst)	LEPIDOPTERA: Pyralidae
sunflower seed midge	*Neolasioptera murtfeldtiana* (Felt)	DIPTERA: Cecidomyiidae
sunflower spittlebug	*Clastoptera xanthocephala* Germar	HOMOPTERA: Cercopidae
superb plant bug	*Adelphocoris superbus* (Uhler)	HEMIPTERA: Miridae
Surinam cockroach	*Pycnoscelus surinamensis* (Linnaeus)	ORTHOPTERA: Blaberidae
Swaine jack pine sawfly	*Neodiprion swainei* Middleton	HYMENOPTERA: Diprionidae
swallow bug	*Oeciacus vicarius* Horvath	HEMIPTERA: Cimicidae
sweetclover aphid	*Therioaphis riehmi* (Börner)	HOMOPTERA: Aphididae
sweetclover root borer	*Walshia miscecolorella* (Chambers)	LEPIDOPTERA: Cosmopterigidae
sweetclover weevil	*Sitona cylindricollis* Fåhraeus	COLEOPTERA: Curculionidae
sweetfern leaf casebearer	*Acrobasis comptoniella* (Hulst)	LEPIDOPTERA: Pyralidae
sweetpotato flea beetle	*Chaetocnema confinis* Crotch	COLEOPTERA: Chrysomelidae
sweetpotato hornworm	*Agrius cingulatus* (Fabricius)	LEPIDOPTERA: Sphingidae
sweetpotato leaf beetle	*Typophorus nigritus viridicyaneus* (Crotch)	COLEOPTERA: Chrysomelidae
sweetpotato leafminer	*Bedellia orchilella* Walsingham	LEPIDOPTERA: Lyonetiidae
sweetpotato leafroller	*Pilocrocis tripunctata* (Fabricius)	LEPIDOPTERA: Pyralidae
sweetpotato vine borer	*Omphisa anastomosalis* (Guenée)	LEPIDOPTERA: Pyralidae
sweetpotato weevil	*Cylas formicarius elegantulus* (Summers)	COLEOPTERA: Curculionidae
sweetpotato whitefly	*Bemisia tabaci* (Gennadius)	HOMOPTERA: Aleyrodidae
sycamore lace bug	*Corythucha ciliata* (Say)	HEMIPTERA: Tingidae
sycamore tussock moth	*Halysidota harrisii* Walsh	LEPIDOPTERA: Arctiidae

T

Tahitian coconut weevil	*Diocalandra taitensis* (Guérin-Menéville)	COLEOPTERA: Curculionidae
tamarind weevil	*Sitophilus linearis* (Herbst)	COLEOPTERA: Curculionidae
tamarix leafhopper	*Opsius stactogalus* Fieber	HOMOPTERA: Cicadellidae
tarnished plant bug	*Lygus lineolaris* (Palisot de Beauvois)	HEMIPTERA: Miridae
tea scale	*Fiorinia theae* Green	HOMOPTERA: Diaspididae
tenlined June beetle	*Polyphylla decemlineata* (Say)	COLEOPTERA: Scarabaeidae
tenspotted ladybird beetle	*Coelophora pupillata* (Swartz)	COLEOPTERA: Coccinellidae
terrapin scale	*Mesolecanium nigrofasciatum* (Pergande)	HOMOPTERA: Coccidae
tessellated scale	*Eucalymnatus tessellatus* (Signoret)	HOMOPTERA: Coccidae
Texas citrus mite	*Eutetranychus banksi* (McGregor)	ACARI: Tetranychidae
Texas leafcutting ant	*Atta texana* (Buckley)	HYMENOPTERA: Formicidae
thief ant	*Solenopsis molesta* (Say)	HYMENOPTERA: Formicidae
thirteenspotted ladybird beetle	*Hippodamia tredecimpunctata tibialis* (Say)	COLEOPTERA: Coccinellidae
thistle aphid	*Brachycaudus cardui* (Linnaeus)	HOMOPTERA: Aphididae
thread bug	*Empicoris rubromaculatus* (Blackburn)	HEMIPTERA: Reduviidae

threebanded leafhopper	*Erythroneura tricincta* Fitch	HOMOPTERA: Cicadellidae
threecornered alfalfa hopper	*Spissistilus festinus* (Say)	HOMOPTERA: Membracidae
threelined leafroller	*Pandemis limitata* (Robinson)	LEPIDOPTERA: Tortricidae
threelined potato beetle	*Lema trilineata* (Olivier)	COLEOPTERA: Chrysomelidae
threespotted flea beetle	*Disonycha triangularis* (Say)	COLEOPTERA: Chrysomelidae
threestriped blister beetle	*Epicauta lemniscata* (Fabricius)	COLEOPTERA: Meloidae
threestriped ladybird beetle	*Brumoides suturalis* (Fabricius)	COLEOPTERA: Coccinellidae
throat bot fly	*Gasterophilus nasalis* (Linnaeus)	DIPTERA: Gasterophilidae
thurberia weevil	*Anthonomus grandis thurberiae* Pierce	COLEOPTERA: Curculionidae
tiger swallowtail	*Papilio glaucus* Linnaeus	LEPIDOPTERA: Papilionidae
tilehorned prionus	*Prionus imbricornis* (Linnaeus)	COLEOPTERA: Cerambycidae
tipdwarf mite	*Calepiterimerus thujae* (Garman)	ACARI: Eriophyidae
toad bug	*Gelastocoris oculatus* (Fabricius)	HEMIPTERA: Gelastocoridae
tobacco budworm	*Heliothis virescens* (Fabricius)	LEPIDOPTERA: Noctuidae
tobacco flea beetle	*Epitrix hirtipennis* (Melsheimer)	COLEOPTERA: Chrysomelidae
tobacco hornworm	*Manduca sexta* (Linnaeus)	LEPIDOPTERA: Sphingidae
tobacco moth	*Ephestia elutella* (Hübner)	LEPIDOPTERA: Pyralidae
tobacco stalk borer	*Trichobaris mucorea* (LeConte)	COLEOPTERA: Curculionidae
tobacco thrips	*Frankliniella fusca* (Hinds)	THYSANOPTERA: Thripidae
tobacco wireworm	*Conoderus vespertinus* (Fabricius)	COLEOPTERA: Elateridae
tomato bug	*Cyrtopeltis modestus* (Distant)	HEMIPTERA: Miridae
tomato fruitworm	*Heliothis zea* (Boddie)	LEPIDOPTERA: Noctuidae
tomato hornworm	*Manduca quinquemaculata* (Haworth)	LEPIDOPTERA: Sphingidae
tomato pinworm	*Keiferia lycopersicella* (Walsingham)	LEPIDOPTERA: Gelechiidae
tomato psyllid	*Paratrioza cockerelli* (Sulc)	HOMOPTERA: Psyllidae
tomato russet mite	*Aculops lycopersici* (Massee)	ACARI: Eriophyidae
toothed flea beetle	*Chaetocnema denticulata* (Illiger)	COLEOPTERA: Chrysomelidae
torsalo	*Dermatobia hominis* (Linnaeus, Jr.)	DIPTERA: Cuterebridae
transparentwinged plant bug	*Hyalopeplus pellucidus* (Stål)	HEMIPTERA: Miridae
transverse ladybird beetle	*Coccinella transversoguttata richardsoni* Brown	COLEOPTERA: Coccinellidae
trefoil seed chalcid	*Bruchophagus kolobovae* Fedoseeva	HYMENOPTERA: Eurytomidae
tropical fowl mite	*Ornithonyssus bursa* (Berlese)	ACARI: Macronyssidae
tropical horse tick	*Anocentor nitens* (Neumann)	ACARI: Ixodidae
tropical rat louse	*Hoplopleura pacifica* Ewing	ANOPLURA: Hoplopleuridae
tropical rat mite	*Ornithonyssus bacoti* (Hirst)	ACARI: Macronyssidae
tropical sod webworm	*Herpetogramma phaeopteralis* Guenée	LEPIDOPTERA: Pyralidae
tuber flea beetle	*Epitrix tuberis* Gentner	COLEOPTERA: Chrysomelidae
tule beetle	*Agonum maculicolle* Dejean	COLEOPTERA: Carabidae
tulip bulb aphid	*Dysaphis tulipae* (Fonscolombe)	HOMOPTERA: Aphididae
tuliptree aphid	*Macrosiphum liriodendri* (Monell)	HOMOPTERA: Aphididae
tuliptree scale	*Toumeyella liriodendri* (Gmelin)	HOMOPTERA: Coccidae
tumid spider mite	*Tetranychus tumidus* Banks	ACARI: Tetranychidae
tupelo leafminer	*Antispila nysaefoliella* Clemens	LEPIDOPTERA: Heliozelidae
turkey chigger	*Neoschoengastia americana* (Hirst)	ACARI: Trombiculidae
turkey gnat	*Simulium meridionale* Riley	DIPTERA: Simuliidae
turnip aphid	*Hyadaphis erysimi* (Kaltenbach)	HOMOPTERA: Aphididae
turnip maggot	*Delia floralis* (Fallén)	DIPTERA: Anthomyiidae
turpentine borer	*Buprestis apricans* Herbst	COLEOPTERA: Buprestidae
twicestabbed ladybird beetle	*Chilocorus stigma* (Say)	COLEOPTERA: Coccinellidae
twig girdler	*Oncideres cingulata* (Say)	COLEOPTERA: Cerambycidae
twig pruner	*Elaphidionoides villosus* (Fabricius)	COLEOPTERA: Cerambycidae
twobanded fungus beetle	*Alphitophagus bifasciatus* (Say)	COLEOPTERA: Tenebrionidae
twobanded Japanese weevil	*Callirhopalus bifasciatus* (Roelofs)	COLEOPTERA: Curculionidae
twolined chestnut borer	*Agrilus bilineatus* (Weber)	COLEOPTERA: Buprestidae
twolined spittlebug	*Prosapia bicincta* (Say)	HOMOPTERA: Cercopidae
twomarked treehopper	*Enchenopa binotata* (Say)	HOMOPTERA: Membracidae
twospotted ladybird beetle	*Adalia bipunctata* (Linnaeus)	COLEOPTERA: Coccinellidae
twospotted spider mite	*Tetranychus urticae* Koch	ACARI: Tetranychidae
twospotted stink bug	*Perillus bioculatus* (Fabricius)	HEMIPTERA: Pentatomidae

twostriped grasshopper	*Melanoplus bivittatus* (Say)	ORTHOPTERA: Acrididae
twostriped walkingstick	*Anisomorpha buprestoides* (Stoll)	ORTHOPTERA: Phasmatidae

U

uglynest caterpillar	*Archips cerasivorana* (Fitch)	LEPIDOPTERA: Tortricidae
unicorn caterpillar	*Schizura unicornis* (J. E. Smith)	LEPIDOPTERA: Notodontidae

V

vagabond crambus	*Agriphila vulgivagella* (Clemens)	LEPIDOPTERA: Pyralidae
vagrant grasshopper	*Schistocerca nitens nitens* (Thunberg)	ORTHOPTERA: Acrididae
Van Duzee treehopper	*Vanduzea segmentata* (Fowler)	HOMOPTERA: Membracidae
vanda thrips	*Dichromothrips corbetti* (Priesner)	THYSANOPTERA: Thripidae
variable oakleaf caterpillar	*Heterocampa manteo* (Doubleday)	LEPIDOPTERA: Notodontidae
varied carpet beetle	*Anthrenus verbasci* (Linnaeus)	COLEOPTERA: Dermestidae
variegated cutworm	*Peridroma saucia* (Hübner)	LEPIDOPTERA: Noctuidae
vedalia	*Rodolia cardinalis* (Mulsant)	COLEOPTERA: Coccinellidae
vegetable leafminer	*Liriomyza sativae* Blanchard	DIPTERA: Agromyzidae
vegetable weevil	*Listroderes costirostris obliquus* (Klug)	COLEOPTERA: Curculionidae
velvetbean caterpillar	*Anticarsia gemmatalis* (Hübner)	LEPIDOPTERA: Noctuidae
verbena bud moth	*Endothenia hebesana* (Walker)	LEPIDOPTERA: Tortricidae
vespiform thrips	*Franklinothrips vespiformis* (Crawford)	THYSANOPTERA: Aeolothripidae
vetch bruchid	*Bruchus brachialis* Fåhraeus	COLEOPTERA: Bruchidae
vexans mosquito	*Aedes vexans* (Meigen)	DIPTERA: Culicidae
viburnum aphid	*Aphis viburniphila* Patch	HOMOPTERA: Aphididae
viceroy	*Basilarchia archippus* (Cramer)	LEPIDOPTERA: Nymphalidae
violet aphid	*Micromyzus violae* (Pergande)	HOMOPTERA: Aphididae
violet sawfly	*Ametastegia pallipes* (Spinola)	HYMENOPTERA: Tenthredinidae
Virginia pine sawfly	*Neodiprion pratti pratti* (Dyar)	HYMENOPTERA: Diprionidae
Virginiacreeper leafhopper	*Erythroneura ziczac* Walsh	HOMOPTERA: Cicadellidae
Virginiacreeper sphinx	*Darapsa myron* (Cramer)	LEPIDOPTERA: Sphingidae

W

w-marked cutworm	*Spaelotis clandestina* (Harris)	LEPIDOPTERA: Noctuidae
walkingstick	*Diapheromera femorata* (Say)	ORTHOPTERA: Phasmatidae
walnut aphid	*Chromaphis juglandicola* (Kaltenbach)	HOMOPTERA: Aphididae
walnut blister mite	*Eriophyes erinea* (Nalepa)	ACARI: Eriophyidae
walnut caterpillar	*Datana integerrima* Grote & Robinson	LEPIDOPTERA: Notodontidae
walnut husk fly	*Rhagoletis completa* Cresson	DIPTERA: Tephritidae
walnut scale	*Quadraspidiotus juglansregiae* (Comstock)	HOMOPTERA: Diaspididae
walnut shoot moth	*Acrobasis demotella* Grote	LEPIDOPTERA: Pyralidae
walnut sphinx	*Cressonia juglandis* (J. E. Smith)	LEPIDOPTERA: Sphingidae
wardrobe beetle	*Attagenus fasciatus* (Thunberg)	COLEOPTERA: Dermestidae
warehouse beetle	*Trogoderma variabile* Ballion	COLEOPTERA: Dermestidae
warty grain mite	*Aeroglyphus robustus* (Banks)	ACARI: Glycyphagidae
watercress leaf beetle	*Phaedon viridus* (Melsheimer)	COLEOPTERA: Chrysomelidae
watercress sharpshooter	*Draeculacephala mollipes* (Say)	HOMOPTERA: Cicadellidae
waterlily aphid	*Rhopalosiphum nymphaeae* (Linnaeus)	HOMOPTERA: Aphididae
waterlily leaf beetle	*Pyrrhalta nymphaeae* (Linnaeus)	COLEOPTERA: Chrysomelidae
waterlily leafcutter	*Synclita obliteralis* (Walker)	LEPIDOPTERA: Pyralidae
webbing clothes moth	*Tineola bisselliella* (Hummel)	LEPIDOPTERA: Tineidae
webbing coneworm	*Dioryctria disclusa* Heinrich	LEPIDOPTERA: Pyralidae
West Indian cane weevil	*Metamasius hemipterus hemipterus* (Linnaeus)	COLEOPTERA: Curculionidae

West Indian flatid	*Melormensis antillarium* (Kirkaldy)	HOMOPTERA: Flatidae
West Indian fruit fly	*Anastrepha obliqua* (Macquart)	DIPTERA: Tephritidae
West Indian sweetpotato weevil	*Euscepes postfasciatus* (Fairmaire)	COLEOPTERA: Curculionidae
western balsam bark beetle	*Dryocoetes confusus* Swaine	COLEOPTERA: Scolytidae
western bean cutworm	*Loxagrotis albicosta* Smith	LEPIDOPTERA: Noctuidae
western bigeyed bug	*Geocoris pallens* Stål	HEMIPTERA: Lygaeidae
western black flea beetle	*Phyllotreta pusilla* Horn	COLEOPTERA: Chrysomelidae
western blackheaded budworm	*Acleris gloverana* (Walsingham)	LEPIDOPTERA: Tortricidae
western bloodsucking conenose	*Triatoma protracta* (Uhler)	HEMIPTERA: Reduviidae
western boxelder bug	*Leptocoris rubrolineatus* Barber	HEMIPTERA: Rhopalidae
western brown stink bug	*Euschistus impictiventris* Stål	HEMIPTERA: Pentatomidae
western cedar bark beetle	*Phloeosinus punctatus* LeConte	COLEOPTERA: Scolytidae
western cedar borer	*Trachykele blondeli* Marseul	COLEOPTERA: Buprestidae
western cherry fruit fly	*Rhagoletis indifferens* Curran	DIPTERA: Tephritidae
western chicken flea	*Ceratophyllus niger* Fox	SIPHONAPTERA: Ceratophyllidae
western chinch bug	*Blissus occiduus* Barber	HEMIPTERA: Lygaeidae
western corn rootworm	*Diabrotica virgifera virgifera* LeConte	COLEOPTERA: Chrysomelidae
western damsel bug	*Reduviolus alternatus* (Parshley)	HEMIPTERA: Nabidae
western drywood termite	*Incisitermes minor* (Hagen)	ISOPTERA: Kalotermitidae
western field wireworm	*Limonius infuscatus* Motschulsky	COLEOPTERA: Elateridae
western flower thrips	*Frankliniella occidentalis* (Pergande)	THYSANOPTERA: Thripidae
western grape rootworm	*Bromius obscurus* (Linnaeus)	COLEOPTERA: Chrysomelidae
western grapeleaf skeletonizer	*Harrisina brillians* Barnes & McDunnough	LEPIDOPTERA: Zygaenidae
western harvester ant	*Pogonomyrmex occidentalis* (Cresson)	HYMENOPTERA: Formicidae
western hemlock looper	*Lambdina fiscellaria lugubrosa* (Hulst)	LEPIDOPTERA: Geometridae
western lawn moth	*Tehama bonifatella* (Hulst)	LEPIDOPTERA: Pyralidae
western lily aphid	*Macrosiphum scoliopi* Essig	HOMOPTERA: Aphididae
western lygus bug	*Lygus hesperis* Knight	HEMIPTERA: Miridae
western oak looper	*Lambdina fiscellaria somniaria* (Hulst)	LEPIDOPTERA: Geometridae
western pine beetle	*Dendroctonus brevicomis* LeConte	COLEOPTERA: Scolytidae
western pine shoot borer	*Eucosma sonomana* Kearfott	LEPIDOPTERA: Tortricidae
western plant bug	*Rhinacloa forticornis* Reuter	HEMIPTERA: Miridae
western poplar clearwing	*Paranthrene robiniae* (H. Edwards)	LEPIDOPTERA: Sesiidae
western potato flea beetle	*Epitrix subcrinita* LeConte	COLEOPTERA: Chrysomelidae
western potato leafhopper	*Empoasca abrupta* DeLong	HOMOPTERA: Cicadellidae
western predatory mite	*Galandromus occidentalis* (Nesbitt)	ACARI: Phytoseiidae
western raspberry fruitworm	*Byturus bakeri* Barber	COLEOPTERA: Byturidae
western spotted cucumber beetle	*Diabrotica undecimpunctata undecimpunctata* Mannerheim	COLEOPTERA: Chrysomelidae
western spruce budworm	*Choristoneura occidentalis* Freeman	LEPIDOPTERA: Tortricidae
western striped cucumber beetle	*Acalymma trivittatum* (Mannerheim)	COLEOPTERA: Chrysomelidae
western striped flea beetle	*Phyllotreta ramosa* (Crotch)	COLEOPTERA: Chrysomelidae
western subterranean termite	*Reticulitermes hesperus* Banks	ISOPTERA: Rhinotermitidae
western tent caterpillar	*Malacosoma californicum* (Packard)	LEPIDOPTERA: Lasiocampidae
western thatching ant	*Formica obscuripes* Forel	HYMENOPTERA: Formicidae
western treehole mosquito	*Aedes sierrensis* (Ludlow)	DIPTERA: Culicidae
western tussock moth	*Orgyia vetusta* (Boisduval)	LEPIDOPTERA: Lymantriidae
western w-marked cutworm	*Spaelotis havilae* (Grote)	LEPIDOPTERA: Noctuidae
western wheat aphid	*Brachycolus tritici* (Gillette)	HOMOPTERA: Aphididae
western yellowjacket	*Vespula pensylvanica* (Saussure)	HYMENOPTERA: Vespidae
western yellowstriped armyworm	*Spodoptera praefica* (Grote)	LEPIDOPTERA: Noctuidae
wharf borer	*Nacerdes melanura* (Linnaeus)	COLEOPTERA: Oedemeridae
wheat curl mite	*Eriophyes tulipae* Keifer	ACARI: Eriophyidae
wheat head armyworm	*Faronta diffusa* (Walker)	LEPIDOPTERA: Noctuidae
wheat jointworm	*Tetramesa tritici* (Fitch)	HYMENOPTERA: Eurytomidae
wheat midge	*Sitodiplosis mosellana* (Géhin)	DIPTERA: Cecidomyiidae
wheat stem maggot	*Meromyza americana* Fitch	DIPTERA: Chloropidae
wheat stem sawfly	*Cephus cinctus* Norton	HYMENOPTERA: Cephidae
wheat strawworm	*Tetramesa grandis* (Riley)	HYMENOPTERA: Eurytomidae

wheat wireworm	*Agriotes mancus* (Say)	COLEOPTERA: Elateridae
wheel bug	*Arilus cristatus* (Linnaeus)	HEMIPTERA: Reduviidae
white apple leafhopper	*Typhlocyba pomaria* McAtee	HOMOPTERA: Cicadellidae
white cutworm	*Euxoa scandens* (Riley)	LEPIDOPTERA: Noctuidae
white fir needleminer	*Epinotia meritana* Heinrich	LEPIDOPTERA: Tortricidae
white oak borer	*Goes tigrinus* (De Geer)	COLEOPTERA: Cerambycidae
white peach scale	*Pseudaulacaspis pentagona* (Targioni-Tozzetti)	HOMOPTERA: Diaspididae
white pine aphid	*Cinara strobi* (Fitch)	HOMOPTERA: Aphididae
white pine cone beetle	*Conophthorus coniperda* (Schwarz)	COLEOPTERA: Scolytidae
white pine cone borer	*Eucosma tocullionana* Heinrich	LEPIDOPTERA: Tortricidae
white pine sawfly	*Neodiprion pinetum* (Norton)	HYMENOPTERA: Diprionidae
white pine weevil	*Pissodes strobi* (Peck)	COLEOPTERA: Curculionidae
whitebanded elm leafhopper	*Scaphoideus luteolus* (Van Duzee)	HOMOPTERA: Cicadellidae
whitecrossed seed bug	*Neacoryphus bicrucis* (Say)	HEMIPTERA: Lygaeidae
whitefringed beetles	*Graphognathus* spp.	COLEOPTERA: Curculionidae
whitelined sphinx	*Hyles lineata* (Fabricius)	LEPIDOPTERA: Sphingidae
whitemargined cockroach	*Melanozosteria soror* (Brunner)	ORTHOPTERA: Blattidae
whitemarked fleahopper	*Spanagonicus albofasciatus* (Reuter)	HEMIPTERA: Miridae
whitemarked spider beetle	*Ptinus fur* (Linnaeus)	COLEOPTERA: Ptinidae
whitemarked treehopper	*Tricentrus albomaculatus* Distant	HOMOPTERA: Membracidae
whitemarked tussock moth	*Orgyia leucostigma* (J. E. Smith)	LEPIDOPTERA: Lymantriidae
whiteshouldered house moth	*Endrosis sarcitrella* (Linnaeus)	LEPIDOPTERA: Oecophoridae
whitespotted sawyer	*Monochamus scutellatus* (Say)	COLEOPTERA: Cerambycidae
willow beaked-gall midge	*Mayetiola rigidae* (Osten Sacken)	DIPTERA: Cecidomyiidae
willow flea weevil	*Rhynchaenus rufipes* (LeConte)	COLEOPTERA: Curculionidae
willow redgall sawfly	*Pontania promixa* (Lepeletier)	HYMENOPTERA: Tenthredinidae
willow sawfly	*Nematus ventralis* Say	HYMENOPTERA: Tenthredinidae
willow shoot sawfly	*Janus abbreviatus* (Say)	HYMENOPTERA: Cephidae
Wilson sphinx	*Hyles wilsoni* (Rothschild)	LEPIDOPTERA: Sphingidae
wing louse	*Lipeurus caponis* (Linnaeus)	MALLOPHAGA: Philopteridae
winter grain mite	*Penthaleus major* (Dugès)	ACARI: Eupodidae
winter moth	*Operophtera brumata* (Linnaeus)	LEPIDOPTERA: Geometridae
winter tick	*Dermacentor albipictus* (Packard)	ACARI: Ixodidae
wood cockroaches	*Parcoblatta* spp.	ORTHOPTERA: Blattellidae
woodrose bug	*Graptostethus manillensis* (Stål)	HEMIPTERA: Lygaeidae
woods weevil	*Nemocestes incomptus* (Horn)	COLEOPTERA: Curculionidae
woolly alder aphid	*Paraprociphilus tessellatus* (Fitch)	HOMOPTERA: Aphididae
woolly apple aphid	*Eriosoma lanigerum* (Hausmann)	HOMOPTERA: Aphididae
woolly elm aphid	*Eriosoma americanum* (Riley)	HOMOPTERA: Aphididae
woolly pear aphid	*Eriosoma pyricola* Baker & Davidson	HOMOPTERA: Aphididae
woolly whitefly	*Aleurothrixus floccosus* (Maskell)	HOMOPTERA: Aleyrodidae

Y

yellow and black potter wasp	*Delta campaniformis campaniformis* (Fabricius)	HYMENOPTERA: Vespidae
yellow clover aphid	*Therioaphis trifolii* (Monell)	HOMOPTERA: Aphididae
yellow garden spider	*Argiope aurantia* Lucas	ARANEAE:Araneidae
yellow mealworm	*Tenebrio molitor* Linnaeus	COLEOPTERA: Tenebrionidae
yellow rose aphid	*Acyrthosiphon porosum* (Sanderson)	HOMOPTERA: Aphididae
yellow scale	*Aonidiella citrina* (Coquillett)	HOMOPTERA: Diaspididae
yellow spider mite	*Eotetranychus carpini borealis* (Ewing)	ACARI: Tetranychidae
yellow sugarcane aphid	*Sipha flava* (Forbes)	HOMOPTERA: Aphididae
yellow woollybear	*Spilosoma virginica* (Fabricius)	LEPIDOPTERA: Arctiidae
yellowfaced leafhopper	*Scaphytopius loricatus* (Van Duzee)	HOMOPTERA: Cicadellidae
yellowfever mosquito	*Aedes aegypti* (Linnaeus)	DIPTERA: Culicidae
yellowheaded cutworm	*Apamea amputatrix* (Fitch)	LEPIDOPTERA: Noctuidae
yellowheaded fireworm	*Acleris minuta* (Robinson)	LEPIDOPTERA: Tortricidae
yellowheaded leafhopper	*Carneocephala flaviceps* (Riley)	HOMOPTERA: Cicadellidae
yellowheaded spruce sawfly	*Pikonema alaskensis* (Rohwer)	HYMENOPTERA: Tenthredinidae

yellowjackets	*Dolichovespula spp. & Vespula spp.*	HYMENOPTERA: Vespidae
yellowmargined leaf beetle	*Microtheca ochroloma* Stål	COLEOPTERA: Chrysomelidae
yellownecked caterpillar	*Datana ministra* (Drury)	LEPIDOPTERA: Notodontidae
yellowshouldered ladybird beetle	*Scymnodes lividigaster* (Mulsant)	COLEOPTERA: Coccinellidae
yellowstriped armyworm	*Spodoptera ornithogalli* (Guenée)	LEPIDOPTERA: Noctuidae
Yosemite bark weevil	*Pissodes schwarzi* (Hopkins)	COLEOPTERA: Curculionidae
yucca moth	*Tegeticula yuccasella* (Riley)	LEPIDOPTERA: Incurvariidae
yucca plant bug	*Halticotoma valida* Townsend	HEMIPTERA: Miridae
Yuma spider mite	*Eotetranychus yumensis* (McGregor)	ACARI: Tetranychidae

Z

zebra caterpillar	*Melanchra picta* (Harris)	LEPIDOPTERA: Noctuidae
Zimmerman pine moth	*Dioryctria zimmermani* (Grote)	LEPIDOPTERA: Pyralidae

Appendix 3: List of Some Insecticides by Common and Registered Trade Name

Insecticides may be identified either by their common name or by their registered trade name. The common name is never capitalized (for example, carbaryl). It may have originated with the Entomological Society of America, the American National Standards Institute, or the International Organization for Standardization.

Most trade names are registered trademarks belonging to the product manufacturers. The symbol ® is used following trade names to indicate that the compounds have been registered. Trade names are always capitalized (for example, Sevin®).

The first part of this appendix presents an alphabetical list of insecticide common names. Also indicated is the class, a cross-reference to the registered trade name, and the oral and dermal LD_{50}. LD_{50} is a standard toxicological term that indicates the number of milligrams (mg) of pesticide per kilogram (kg) of test-animal body weight required to kill 50 percent of a test-animal population. The LD_{50} data have been obtained from several reliable sources; in instances where a source gave a range for the LD_{50} or different values were given for each test-animal sex, the LD_{50} value listed is the lowest presented. When toxicological data were not available, a dash appears in the LD_{50} column.

Fumigant toxicity is listed as VA, the acute vapor toxicity in parts per million (ppm). Acute vapor toxicity is the maximum dose that is not lethal to a human when exposed to the chemical for 60 minutes.

The second part of this appendix is a list of insecticides alphabetized by registered trade name. It also includes a cross-reference to the common name and indicates the basic manufacturer.

The information listed is from publication Pm–833, *List of Insecticides by Common and Registered Trade Name* by W. K. Wintersteen, D. R. Lewis, D. E. Foster, and K. Holscher, Cooperative Extension Service, Iowa State University (compiled March 1986). Basic manufacturers were changed where appropriate.

List of Insecticides by Common Names

Common Name	Class*	Trade Name	LD$_{50}$ Oral	LD$_{50}$ Dermal
acephate	2	Orthene®	866	>2,000
aldicarb	3	Temik®	1	5
aldrin	1	Aldrite®	39	98
allethrin	4	Pynamin®	680	11,200
aluminum phosphide	6	Phostoxin®	—	—
amitraz	8	BAAM®	600	1,600
azinphosethyl	2	Ethyl Guthion®	7	80
azinphosmethyl	2	Guthion®	11	220
Bacillus popilliae	5	Doom®, Japidemic®	—	—
Bacillus thuringiensis	5	Biotrol®, Dipel®, Thuricide® Bactur®, Top-Side Dipel®	—	—
bendiocarb	3	Ficam®	179	>1,000
benzene hexachloride	1	BHC (canceled)	600	Variable
binapacryl	8	Morocide®	421	—
boric acid	8	Roach-Prufe®	—	—
carbaryl	3	Sevin®	850	>4,000
carbofuran	3	Furadan®	8	>10,200
carbon disulfide	6	(canceled)	VA 200	—
carbon tetrachloride	6	(canceled)	VA 300	—
carbophenothion	2	Trithion®	6	22
chlordane	1	Belt®	283	580
chlordecone	1	Kepone® (discontinued)	95	345
chlordimeform	8	Fundal®, Galecron®	170	225
chlorfenvinphos	2	Supona®, Birlane® (discontinued)	10	3,200
chlorobenzilate	1	Acaraben®	1,040	>5,000
chlorpicrin	6	Picfume®, Larvicide®	VA 20	—
chlorpyrifos	2	Dursban®, Lorsban®	97	2,000
coumaphos	2	Co-Ral®	16	860
crotoxyphos	2	Ciodrin®	74	202
crufomate	2	Ruelene® (discontinued)	770	>2,000
cyhexatin	8	Plictran®	540	>2,000
cyromazine	5	Larvadex®	5,000	2,000
deet	8	OFF®, Diethyltoluamide®	2,000	—
demeton	2	Systox®	2.5	8.2
diazinon	2	Spectracide®, Diazinon®	250	455
dibromochloropropane	6	Nemagon®, Fumazone® (discontinued)	VA 103	—
dichlofenthion	2	V-C 13®	270	6,000
dichlorvos	2	DDVP, Vapona®	56	75
dicofol	1	Kelthane®	575	1,000
dicrotofos	2	Bidrin®	22	225
dieldrin	1	Dieldrex®, Dieldrite®	46	60
dienochlor	1	Pentac®	>3,160	>3,160
diflubenzuron	8	Dimilin®	>4,640	>2,000
dimethoate	2	Cygon®, De-Fend®	215	610
dinocap	8	Karathane®	980	4,700
dioxathion	2	Delnav®	23	63
disulfoton	2	Di-Syston®	2.3	6
d-Phenothrin	2	Sumithrin®	>10,000	>5,000
endosulfan	1	Thiodan®, Tiovel®	18	74
endrin	1	—	7.5	15
(EPN)	2	EPN	7	22
ethion	2	Nialate®	27	62
ethoprop	2	Mocap®	61	26
ethylan	1	Perthane® (discontinued)	8,170	—

Common Name	Class*	Trade Name	LD$_{50}$	
			Oral	Dermal
ethylene dibromide	6	EDB, Dowfume®W-85 (canceled)	VA 200	—
ethylene dichloride	6	EDC (canceled)	VA 1000	—
ethylene oxide	6	ETO	VA 72	—
etrimfos	2	Ekamet®	1,800	500
famphur	2	Warbex®	35	1,460
fenamiphos	2	Nemacur®	8	73
fenbutatin-oxide	8	Vendex®	2,631	>2,000
fensulfothion	2	Dasanit®	2	3
fenthion	2	Baytex®, Entex®, Spotton®, Tiguvon®, Lysoff®	215	330
fenvalerate	4	Pydrin®, Ectrin®, Pyrid®	451	2,500
flucythrinate	4	Pay-off®, Guardian®	67	1,000
fluvalinate	4	Mavrik®	260	>20,000
fonofos	2	Dyfonate®	8	147
formetanate hydrochloride	3	Carzol®, Dicarzol®	20	10,200
fosthietane	2	Nem-A-Tak®	4	20
heptachlor	1	—	100	195
hydroprene	5	Gencor®	34,000	4,550
isofenphos	2	Oftanol®	28	>1,000
kinoprene	5	Enstar®	4,900	9,000
(Landrin—not a common name)	3	Landrin® (discontinued)	178	2,500
leptophos	2	Phosvel® (discontinued)	43	10,000
lindane	1	Gamma BHC	76	500
malathion	2	Cythion®	1,000	>4,444
metalkamate	3	Bux® (discontinued)	87	400
metam-sodium	6	Vapam®	820	800
methamidophos	2	Monitor®	13	110
methidathion	2	Supracide®	215	330
methiocarb	3	Mesurol®	130	>2,000
methomyl	3	Lannate®, Nudrin®	17	>1,000
methoprene	5	Altosid®, Precor®	34,600	>3,500
methoxychlor	1	Marlate®	5,000	>6,000
methyl bromide	6	Dowfume®	VA 200	—
methyl parathion (encapsulated)	2	Penncap-M®	270	>5,400
methyl parathion	2	Metacide® and many others	9	63
mevinphos	2	Phosdrin®	3	3
mexacarbate	3	Zectran® (discontinued)	15	>500
mirex	1	Dechlorane® (discontinued)	235	800
monocrotophos	2	Azodrin®	21	354
naled	2	Dibrom®	250	800
nicotine sulfate	4	Black Leaf 40®	50	140
nuclear polyhedrosis virus of *Heliothis* species	5	Eclar®	—	—
oxamyl	3	Vydate®	5.4	2,960
oxydemeton-methyl	2	Metasystox-R®	47	158
oxythioquinox	8	Morestan®	2,500	>2,000
para-dichlorobenzene	6	Paracide® (discontinued)	500	—
parathion (ethyl)	2	Niran®	3.6	6.8
permethrin	4	Ambush®, Pounce®, Ectiban®, Atroban®, Permectrin®, Torpedo®, Dragnet®	4,000	4,000
petroleum oils	8	Superior oils	—	—
phorate	2	Thimet®	1	2

List of Insecticides by Common Names *(continued)*

Common Name	Class*	Trade Name	LD$_{50}$ Oral	LD$_{50}$ Dermal
phosalone	2	Zolone®	125	1,500
phosmet	2	Imidan®, Prolate®	113	1,550
phosphamidon	2	Dimecron®	17	374
piperonyl butoxide	2	Butacide®	7,500	7,500
pirimicarb	3	Pirimor®	147	500
pirimiphos methyl	2	Actellic®	2,000	2,000
propargite	7	Omite®	2,200	—
propetamphos	2	Safrotin®	82	2,300
propoxur	3	Baygon®	95	>1,000
propyl thiopyrophosphate	2	Aspon®	1,224	—
resmethrin	4	Scourge®, Synthrin®	4,240	2,500
ronnel	2	Korlan®, Rid-Ezy® (discontinued)	1,250	>5,000
rotenone	5	Noxfish®	132	940
ryania	5	(discontinued)	1,200	—
silica	8	Silica Aerogel®	—	—
stirofos	2	Rabon®	1,125	>4,000
sulfotepp	6	Bladafum®	7	—
sulfur dioxide	6	—	VA 40	—
sulfuryl fluoride	6	Vikane®	VA 400	—
TEPP	2	Tetron®	1	—
temephos	2	Abate®	1,000	>4,000
terbufos	2	Counter®	4	25
tetradifon	7	Tedion®	5,000	1,000
tetramethrin	4	Neo-Pynamin®, Multicide®	>20,000	15,000
toxaphene	1	—	80	780
trichlorfon	2	Dipterex®, Dylox® Neguvon®, Proxol®	560	>2,000
trimethacarb	3	Broot®	566	2,000

*Classes: 1. organochlorines or chlorinated hydrocarbons
2. organophosphates
3. carbamates
4. pyrethroids
5. rotenoids, biologicals, growth regulators
6. fumigants
7. sulfonates, sulfides, sulfones
8. miscellaneous

List of Insecticides by Trade Names

Trade Name	Common Name	Basic Manufacturer
Abate®	temephos	American Cyanamid
Acaraben®	chlorobenzilate	Ciba-Geigy
Actellic®	pirimiphos methyl	ICI
Aldrite®	aldrin	DuPont
Alfatox®	mixture—diazinon & methoxychlor	Ciba-Geigy
Altosid®	methoprene	Sandoz
Amaze®	isofenphos	Mobay (discontinued)
Ambush®	permethrin	ICI
Aspon®	propyl thiopyrophosphate	ICI
Atroban®	permethrin	Burroughs-Wellcome

Trade Name	Common Name	Basic Manufacturer
Azodrin®	monocrotophos	DuPont
BAAM®	amitraz	Upjohn
Bactur®	*Bacillus thuringiensis*	Thompson-Hayward
Barricade®	methyl parathion and EPN	Velsicol
Baygon®	propoxur	Mobay
Baytex®	fenthion	Mobay
Belt®	chlordane	Velsicol
BHC	benzene hexachloride	Hooker (canceled)
Bidrin®	dicrotofos	DuPont
Biotrol®	*Bacillus thuringiensis*	Thompson-Hayward
Black Leaf 40®	nicotine sulfate	Center-chem
Bladafum®	sulfotepp	Bayer
Bromofume®	ethylene dibromide	Dow (canceled)
Broot®	trimethacarb	Union Carbide
Butacide®	piperonyl butoxide	Fairfield American (discontinued)
Bux®	metalkamate	Chevron
Carzol®	formetanate hydrochloride	Nor-Am
Ciodrin®	crotoxyphos	SDS Brotech
Ciovap®	mixture—Ciodrin & Vapona	SDS Brotech
Co-Ral®	coumaphos	Mobay
Counter®	terbufos	American Cyanamid
Cygon®	dimethoate	American Cyanamid
Cythion®	malathion	American Cyanamid
Dasanit®	fensulfothion	Mobay
DDVP	dichlorvos	DuPont
Dechlorane®	mirex	Allied (discontinued)
De-Fend®	dimethoate	Thompson-Hayward
Delnav®	dioxathion	Nor-Am
Diazinon®	diazinon	Ciba-Geigy
Dibrom®	naled	Chevron
Dicarzol®	formethanate hydrochloride	Nor-Am
Diethyltolua-mide®	deet	MGK
Dimecron®	phosphamidon	Chevron
Dimilin®	diflubenzuron	Thompson-Hayward
Dipel®	*Bacillus thuringiensis*	Abbott
Dipterex®	trichlorfon	Mobay
D-Syston®	disulfoton	Mobay
Doom®	*Bacillus popilliae*	Fairfax Biol. Lab.
Dowfume®	methyl bromide	Dow
Dowfume®W-85	Ethylene dibromide	Dow (canceled)
Dragnet®	permethrin	FMC
Dursban®	chlorpyrifos	Dow
Dyfonate®	fonofos	ICI
Dylox®	trichlorfon	Mobay
Ectiban®	permethrin	Burroughs-Wellcome
Ectrin®	fenvalerate	SDS Biotech
EDB	ethylene dibromide	Dow (canceled)
EDC	ethylene dichloride	Diamond, Dow (canceled)

List of Insecticides by Trade Names (*continued*)

Trade Name	Common Name	Basic Manufacturer
Ekamet®	etrimfos	Sandoz
Elcar®	nuclear polyhedrosis virus of *Heliothis* species	Sandoz
Enstar®	kinoprene	Sandoz
Entex®	fenthion	Mobay
EPN	EPN	DuPont, Velsicol
Ethyl Guthion®	azinphosethyl	Mobay
ETO	ethylene oxide	Dow
Famix®	famphur	American Cyanamid (discontinued)
Famophos®	famphur	American Cyanamid (discontinued)
Ficam®	bendiocarb	Nor-Am
Fumazone®	dibromochloropropane	Dow (discontinued)
Fundal®	chlordimeform	Nor-Am
Furadan®	carbofuran	FMC
Galecron®	chlordimeform	Ciba-Geigy
Gamma BHC	lindane	Hooker
Gencor®	hydroprene	Sandoz
Guardian®	flucythrinate	American Cyanamid
Guthion®	azinphosmethyl	Mobay
Imidan®	phosmet	ICI
Japidemic®	*Bacillus popilliae*	Fairfax Biol. Lab.
Karathane®	dinocap	Rohm & Haas
Kelthane®	dicofol	Rohm & Haas
Kepone®	chlordecone	Allied (discontinued)
Korlan®	ronnel	Dow (discontinued)
Lannate®	methomyl	DuPont
Larvicide®	chlorpicrin	Soweco Inc.
Larvadex®	cyromazine	Ciba-Geigy
Lorsban®	chlorpyrifos	Dow
Lysoff®	fenthion	Mobay
Marlate®	methoxychlor	DuPont
Mavrik®	fluvalinate	Sandoz
Mesurol®	methiocarb	Mobay
Metacide®	methyl parathion	Bayer
Metasystox-R®	oxydemetonmethyl	Mobay
Mocap®	ethoprop	Rhone-Poulenc
Monitor®	methamidophos	Mobay, Chevron
Morocide®	binapacryl	Hoechst
Morestan®	oxythioquinox	Mobay
Multicide®	tetramethrin	MGK
Neguvon®	trichlorfon	Mobay
Nemacur®	fenamiphos	Mobay
Nem-A-Tak®	fosthiefane	American Cyanamid
Neo-Pynamin®	tetramethrin	Fairfield Am Corp
Nialate®	ethion	FMC (discontinued)
Niran®	parathion (ethyl)	Monsanto
Noxfish®	rotenone	Prentiss Drug
Nudrin®	methomyl	DuPont
OFF®	deet	Johnson
Oftanol®	isofenphos	Mobay
Omite®	propargite	Uniroyal
Orthene®	acephate	Chevron
Paracide®	para-dichlorobenzene	PPG (discontinued)
Pay-off®	flucythrinate	American Cyanamid

Trade Name	Common Name	Basic Manufacturer
Penncap-M®	methyl parathion	Pennwalt
Pentac®	dienochlor	Hooker
Permectrin®	permethrin	Boehringer Ingelheim
Perthane®	ethylan	Rohm and Haas (discontinued)
Phosdrin®	mevinphos	DuPont
Phostoxin	aluminum phosphide	Phostoxin Sales Inc.
Phosvel®	leptophos	Velsicol (discontinued)
Picfume®	chlorpicrin	Dow
Pirimor®	pirimicarb	ICI
Plictran®	cyhexatin	Dow
Pounce®	permethrin	FMC
Precor®	methoprene	Sandoz
Prolate®	phosmet	ICI
Proxol®	trichlorfon	Upjohn
Pydrin®	fenvalerate	DuPont
Pynamin®	allethrin	MGK
Pyrid®	fenvalerate	Terminix
Rabon®	stirofos	SDS Biotech
Ravap®	mixture-Rabon® & Vapona®	SDS Biotech
Rid-Ezy®	ronnel	Moorman (discontinued)
Roach-Prufe®	boric acid	Copper Brite
Ruelene®	crufomate	Dow (discontinued)
Safrotin®	propertamphos	Sandoz
Scourge®	resmethrin	Penick
Sevin®	carbaryl	Rhone-Poulenc
Silica Aerogel®	silica	W. R. Grace
Spectracide®	diazinon	Ciba-Geigy
Spotton®	fenthion	Mobay
Sumithrin®	d-phenothrin	Sumitomo
Superior oils	petroleum oils	Ortho Agricultural Chemicals
Supracide®	methidathion	Ciba-Geigy
Systox®	demeton	Mobay
Tedion®	tetradifon	FMC
Temik®	aldicarb	Union Carbide
Tetron®	TEPP	Chevron
Thimet®	phorate	American Cyanamid
Thiodan®	endosulfan	FMC
Thuricide®	*Bacillus thuringiensis*	Sandoz
Tiguvon®	fenthion	Mobay
Tiovel®	endosulfan	Velsicol
Torpedo®	permethrin	ICI
Top-Side Dipel®	*Bacillus thuringiensis*	Loveland Industries
Trithion®	carbophenothion	ICI
Trolene®	ronnel	Dow (discontinued)
Vapam®	metam-sodium	ICI
Vapona®	dichlorvos	SDS Biotech
V-C 13®	dichlofenthion	Rhone-Poulenc
Vendex®	fenbutatin-oxide	DuPont
Vikane®	sulfuryl fluoride	Dow
Vydate®	oxamyl	DuPont
Warbex®	famphur	American Cyanamid
Zectran®	mexacarbate	Dow (discontinued)
Zolone®	phosalone	Rhone-Poulenc

Glossary

Abdomen.   The posteriormost of the three main body divisions of insects.

Absolute population estimate.   An estimate of population size expressed as absolute number per ground-surface area (e.g., number per hectare) or per unit of volume (e.g., number per liter of soil).

Acaricide.   A pesticide that kills mites and ticks.

Acceptable daily intake (ADI).   The level of a chemical residue to which daily exposure over a lifetime is not thought to cause appreciable risk.

Accessory gland.   Any secondary gland of a glandular system (Figures 2.44, 2.45).

Acetylcholine.   A synaptic neurotransmitter, $CH_3COOCH_2CH_2N(CH_3)_3^+$.

Acetylcholinesterase.   An enzyme that catalyzes the breakdown of acetylcholine.

Action potential.   A brief electrical depolarization that propagates along an axon or muscle fiber (Figure 2.43).

Active ingredient (AI).   The component of a pesticide formulation responsible for the toxic effect.

Acute poisoning.   Illness or death from a single dose of a toxicant.

Adjuvant.   Any ingredient that improves the properties of a pesticide formulation.

Aedeagus.   The copulatory organ of most male insects (Figure 2.22).

Aerosol.   Pesticide formulation in which propulsion from a pressurized container suspends pesticide particles or droplets in the air.

Aesthetic pest.   A pest whose mere presence is objectionable for psychological reasons.

Aestivation.   Summer dormancy.

African sleeping sickness.   A disease caused by infection with *Trypanosoma* species protozoans; transmitted by the bite of the tsetse fly.

Age distribution.   The proportions of individuals in different age groups of a population at any given time (Figure 5.3).

Aggregation pheromone.   An intraspecific chemical messenger that causes insect congregation at food sites, reproductive habitats, or hibernation sites.

Agroecosystem.   An ecosystem largely created and maintained to satisfy a human want or need (Figure 5.7).

Air sac.   A pouchlike expansion of a trachea (Figure 2.28).

Alarm pheromone.   An intraspecific chemical messenger produced to elicit aggressive or defensive behavior.

Alary muscle.   A wing-shaped muscle that connects the heart to lateral portions of the terga (Figure 2.27).

Alate.   Winged.

Algorithm.   A precisely defined sequence of rules stating how to produce specific outputs from specific inputs through a given number of steps (Figure 8.16).

Alimentary canal. The tube of the digestive system, extending from the mouth to the anus and including the foregut, midgut, and hindgut.

Aliphatic. Organic chemical compound in which the carbon atoms are linked in open chains rather than rings.

Allele. One of two or more alternative forms of a gene at corresponding sites on homologous chromosomes.

Allelochemic. A substance involved in communication between members of different species.

Allethrin. The first pyrethroid insecticide synthesized.

Allomone. An interspecific chemical messenger that benefits the releaser but not the receiver (e.g., repellents, toxicants).

Amino acid. Any organic compound containing an amino group (NH_2) and a carboxylic acid group (COOH).

Amylase. An enzyme that catalyzes the splitting of starch into maltose units.

Anamorphosis. The adding of segments to the body as development progresses.

Anoplura. The order of insects comprising the sucking lice (Figure 3.23).

Antenna (pl. **antennae**). One of a pair of segmented sensory organs located on the head above the mouthparts (Figures 2.5, 2.15).

Anthropocentrism. Regarding humans or human objectives as the principal focus of a system or situation.

Anthropomorphism. The attribution of human characteristics to nonhumans.

Antibiosis. Plant characteristics that affect insects in a negative manner (e.g., increase mortality, reduce fecundity); involves activities of both host plant and insect.

Anticoagulant. Any substance that inhibits, delays, or suppresses blood clotting.

Antixenosis (nonpreference). Plant characteristics that drive insects away from a particular host; involves activities of both host and insect.

Aorta. The anterior portion of the dorsal blood vessel (Figure 2.27).

Aphicide. An insecticide that kills aphids.

Apical. Located at the end or tip.

Apiculture. The management of honey bees.

Apodeme. Invagination of the body wall that strengthens the exoskeleton and provides areas for muscle attachment.

Apolysis. The separation of the old cuticle from the epidermis; an early step in the molting cycle.

Apparent resistance (ecological resistance, pseudoresistance). Temporary host-plant resistance characteristics that rely heavily on environmental conditions; includes host evasion, induced resistance, and host escape.

Approved common name. Generic name of a pesticide (e.g., carbaryl) or common name of an insect as approved by the Entomological Society of America (e.g., European corn borer).

Apterous. Without wings.

Apterygota. The subclass of Insecta that comprises the primitively wingless insects.

Arachnida. The arthropod class that includes the spiders, scorpions, ticks, and mites (Figure 1.7); characterized by having a cephalothorax and an abdomen, and six pairs of appendages including the chelicerae, pedipalps, and four pairs of walking legs.

Arbovirus. A virus transmitted by mosquitoes or other blood-sucking arthropods.

Arolium (pl. **arolia**). An adhesive pad between the claws of the tarsus (Figure 2.18A).

Arthropod. An invertebrate animal with jointed appendages; a member of the phylum Arthropoda.

Atropine. An anticholinergic alkaloid drug used as an antidote for organophosphate and carbamate insecticide poisoning.

Augmentation. Any biological control practice designed to increase the number or effectiveness of existing natural enemies.

Autocidal control. Use of induced sterility for reducing pest population size.

Axon. The part of a neuron that carries impulses away from the cell body (Figure 2.42).

Axonic poison.   A chemical that interrupts normal transmission of impulses along the axon of a neuron (Figure 11.7).

Bait (B).   A pesticide formulation that combines an edible or attractive substance with a pesticide.

Basement membrane.   The noncellular membrane underlying the epidermal cells of the integument (Figure 2.3).

Behavioral resistance.   The ability of an insect population to change its behavior in order to avoid insecticides or other injurious factors.

Benzyl benzoate.   A repellent for use against ticks and chiggers.

Berlese funnel.   A sampling device that uses heat to drive insects from a sample of soil, vegetation, or litter (Figure 6.8).

Binomial.   Having two names, as in a scientific name, composed of a genus name and a trivial name or specific epithet (e.g., *Ostrinia nubilalis*).

Binomial nomenclature.   The system of classification of organisms, as advanced by Carolus Linnaeus.

Biochemical resistance.   The ability of an insect population to develop enzymes that detoxify an insecticide or other injurious material before it can reach its site of action.

Bioeconomics.   The relationship between pest numbers and economic losses.

Biological control (biocontrol).   The employment of any biological agent for control of a pest.

Biomagnification.   The increase in concentration of a substance (usually a pesticide) in animal tissue as related to the animal's higher position in the food chain (Figure 11.4).

Biomass.   The total dry weight or volume of all living organisms in a given area.

Biotic potential.   A measure of the innate ability of a population to survive and reproduce.

Biotype (**race**).   A population of a pest species that differs from other populations of the species in its ability to attack a particular cultivar.

Blastoderm.   The one-cell thick layer of cells that surrounds the yolk of an egg early in embryological development (Figure 4.4).

Blastula.   An early embryonic form that usually consists of a hollow sphere composed of a single layer of cells.

Book lung.   The respiratory cavity of some spiders that contains a series of leaflike folds.

Botanical insecticide.   An insecticide produced from a plant or plant product (e.g., pyrethrum).

Boundary layer.   With respect to insects, the layer of air that extends from ground level upward through increasing wind speeds to a height where wind speed and flight speed are equal.

Brain hormone (AH, activation hormone or PTTH, prothoracicotropic hormone).   A hormone secreted by neurosecretory cells of the insect brain; activates prothoracic glands (Figures 4.10, 4.11).

Brood.   Several cohorts of offspring produced by a parent or parent population at different times or in different places.

Bubonic plague.   See plague.

Bursicon.   A hormone secreted by certain cells of the nervous system that plays a role in the control of cuticular hardening and darkening.

Calyx.   The bulbous junction of the pedicels of the ovarioles.

Campaniform.   Domelike; campaniform sensilla (Figure 2.39B).

Cantharidin.   An irritating substance produced by blister beetles; used medicinally in keratolytic preparations.

Capitate.   Head-shaped or with an apical knob (Figure 2.15).

Carbamate.   An organic compound derived from carbamic acid, NH_2COOH; includes many important synthetic insecticides (e.g., carbaryl).

Carbohydrate.   A compound of carbon, hydrogen, and oxygen that follows the general formula $(CH_2O)_n$.

Carcinogen.   A material that causes cancer.

Cardiac valve.   The valve at the junction of the foregut and the midgut.

Carrying capacity. The maximum population density a given environment will support for a sustained period.

Caste. The type or form of an individual in a colony of social insects (e.g., worker).

Catfacing. Injury resulting from the feeding of piercing-sucking insects on developing fruit; a result of uneven growth and deformation (Figure 7.8).

Cell body. The swollen portion of a neuron that contains the nucleus and is responsible for metabolic and maintenance functions of the neuron (Figure 2.42).

Cellulase. An enzyme that catalyzes the breakdown of cellulose.

Cephalothorax. The united head and thorax of the arachnids and crustaceans.

Cercus (pl. cerci). One of a pair of sensory appendages at the posterior of the abdomen (Figure 2.20).

Cervix. The membranous neck region of an insect.

Chemical name. Name given to a pesticide based on the chemical structure (e.g., 1-naphthyl-N-methylcarbamate); naming follows rules of the International Union of Pure and Applied Chemistry.

Chemigation. Use of irrigation systems for dispensing pesticides (Figure 10.11).

Chemoreceptor. A sense organ involved in the perception of tastes and/or odors (Figure 2.38).

Chemosterilant. A chemical that induces reproductive sterility.

Chilopoda. The arthropod class that includes the centipedes (Figure 1.6); characterized by mandibulate mouthparts and one pair of appendages per body segment, including a pair of venomous claws on the first segment.

Chitin. A polymer of the nitrogenous polysaccharide n-acetyl-D-glucosamine that occurs in the exoskeleton of arthropods.

Chlorinated hydrocarbon. An organic compound containing chlorine, hydrogen, and occasionally oxygen and sulfur; includes the first widely used synthetic organic insecticides (e.g., DDT).

Cholesterol. A fatlike steroid alcohol, $C_{27}H_{45}OH$, which is an important precursor in the formation of other steroids like ecdysone.

Cholinesterase. Shortened term for acetylcholinesterase.

Chordotonal sensillum. A mechanoreceptor that consists of a bundle of bipolar nerve cells stretched between two surfaces of the integument.

Chorion. The outer covering of an insect egg (Figure 4.3).

Chronic poisoning. Illness or death from long-time exposure to low levels of a toxicant.

Chrysalis. The pupal stage in lepidopterans that do not construct cocoons.

Class. A category of the classification scheme that includes one or more orders (e.g., Insecta).

Classification. The ordering of organisms into a hierarchy of categories.

Clavate. Club-shaped or enlarged at the tip; clavate antennae (Figure 2.15).

Coarctate. Compacted and oval in form, concealed by a hardened, thickened, cylindrical case; coarctate pupa.

Cocoon. Case of silk or silk-bound debris in which a pupa develops.

Codlemone. The synthetic sex pheromone of the codling moth.

Coleoptera. The order of insects comprising the beetles (Figure 3.28).

Collembola. The order of insects comprising the springtails (Figure 3.3).

Collophore. Extendable tubelike structure found on the ventral side of the first abdominal segment of collembolans (Figure 3.3).

Comb. The hexagonal wax structure made by honey bees in which young are reared and food is stored.

Community. The interacting "web" of populations in a given area (Figure 5.5).

Compound eye. A visual sensory organ composed of many individual elements, each of which is represented externally by a hexagonal facet (Figure 2.16).

Conditional lethal. An allele that results in mortality of carriers under certain environmental conditions (e.g., cold temperatures).

Conservation. A biological control practice that includes any activity designed to protect and maintain existing populations of natural enemies.

Contact poison. An insecticide that enters the body when the insect walks or crawls over a treated surface; insecticide that is absorbed through the body wall.

Copulation. The joining of male and female genital structures (Figure 4.12); mating.

Cornea. The outer surface or lens of an ommatidium (Figure 2.37).

Cornicle. One of a pair of dorsal tubular structures on the posterior part of the abdomen of aphids.

Corpus allatum (pl. **corpora allata**). One of a pair of neuroendocrine glands just posterior to the brain (Figure 2.41).

Corpus cardiacum (pl. **corpora cardiaca**). One of a pair of neuroendocrine glands adjacent to the corpora allata and posterior to the brain (Figure 2.41).

Coxa (pl. **coxae**). The basal segment of the leg (Figure 2.18).

Cranium. The hardened capsule of the head that bears the mouthparts, antennae, and eyes.

Crawler. Newly emerged, active immature form of a scale insect.

Cremaster. A spine at the tip of the abdomen of a chrysalis.

Crochets. Hooks or claws at the tips of prolegs of Lepidoptera larvae.

Crop. The expanded portion of the foregut just posterior to the esophagus (Figure 2.23).

Crop rotation. Alternating the type of crop planted at a location.

Cross-resistance. The phenomenon of insect resistance to one type of insecticide providing resistance to other insecticides with similar modes of action.

Crustacea. The anthropod class that includes the crayfish, lobsters, pillbugs, crabs, and shrimp (Figures 1.3, 1.4); characterized by having two pairs of antennae, one pair of mandibles, two pairs of maxillae, and gills for respiration.

Crystalline cone. A transparent body that lies beneath the cornea in an ommatidium (Figure 2.37).

Cultural management (ecological management, cultural control). Purposeful manipulation of a cropping environment to reduce rates of pest increase and damage.

Cuneus. A triangular section of a hemelytron, set off by a suture.

Cuticle. The noncellular, multilayered portion of the integument that overlies the epidermis.

Cuticulin. The lipoprotein layer of the epicuticle.

Cyclodiene. A cyclic chlorinated hydrocarbon insecticide (e.g., chlordane).

Cytoplasmic incompatibility. When factors in the egg cytoplasm prevent fusion of sperm and egg nuclei.

Cytoplasmic resistance. Host-plant resistance conferred by mutable substances in cell cytoplasm.

Damage. A measurable loss of host utility, most often including yield quantity, quality, or aesthetic appeal (Figure 7.1).

Damage boundary. The level of injury where damage can be measured (Figure 7.2).

DDT (dichlorodiphenyltrichloroethane). A synthetic chlorinated hydrocarbon insecticide.

DEET (N,N-diethyl-*m*-toluamide). Commonly used biting fly and mosquito repellent.

Degree day. An accumulation of heat units above some threshold temperature for a 24-hour period.

Delayed voltinism. Life cycle requiring more than one year for completion.

Delusory parasitosis. Psychological infliction in which sufferers imagine themselves to be infested with parasites.

Deme. A subpopulation of interbreeding individuals.

Density. The number of individuals per unit of measure (e.g., larvae per square meter).

Density-dependent factor. A population-regulating factor that changes in intensity with changes in population density (e.g., intraspecific competition).

Density-independent factor. A population-regulating factor that causes mortality and is unrelated to a population's density (e.g., weather factors).

Deodorant. A material added to a pesticide formulation for masking unpleasant odors (e.g., cedar oil, pine oil).

Depolarization. The reduction of the electrical potential of a neuron's membrane.

Dermaptera. The order of insects comprising the earwigs (Figure 3.16).

Deterrent. A chemical that prevents feeding or oviposition by insects.

Deutocerebrum. The medial pair of lobes of the brain that innervate the antennae (Figure 2.41).

Developmental threshold. The minimum temperature required for development to proceed.

Diagnosis. The recognition of the presence and extent of a pest infestation and determination of the causal agent responsible.

Diapause. A physiological state of arrested metabolism, growth, and development that occurs at one stage in the life cycle.

Diaphragm. A thin membrane that partitions areas of the body cavity (Figure 2.27).

Diffraction. The splitting of white light into its component spectral colors by a series of fine grooves or ridges.

Diflubenzuron. A chitin synthesis inhibitor for insect control.

Diluent. A substance used as a carrier for concentrated pesticide; can be liquid or solid (e.g., refined oil, organic flour).

DIMBOA (2,4-dihydroxy-7-methoxy-1,4-benzoxazine-3-one). A compound in corn that confers resistance against first-generation European corn borer.

Dinitrophenol. A synthetic organic insecticide characterized by nitro groups (NO_2) attached to a phenol (C_6H_5OH) ring (e.g., dinoseb).

Dioecious. Having male and female elements on separate individuals of the same species.

Diploid. Double; possessing the full complement of maternal and paternal chromosomes.

Diplopoda. The arthropod class that includes the millipedes (Figure 1.5); characterized by cylindrical bodies and two pairs of appendages on each abdominal segment.

Diplura. One of the apterygote orders of insects (Figure 3.4).

Diptera. The order of insects comprising the true flies (Figures 3.44, 3.45).

Direct loss (direct injury). Devaluation of a marketable commodity owing to the presence of insects or insect damage; injury to yield-forming organs.

Dispersal. The process of scattering in various directions.

Dispersion. The spatial arrangement of individuals (Figure 5.1).

Distal. Farthest from the point of attachment or origin.

Dockage. Value depreciation of a product owing to the presence of insects or other foreign matter.

Dormancy. A seasonally recurring period in the life cycle when growth, development, and reproduction are suppressed.

Dormant oil. A petroleum oil applied to trees only when foliage is not present.

Dorsal. Pertaining to the back or upper side of the body.

Dorsal vessel. Primary organ of hemocoel circulation; extends from the abdomen to the head and consists of a posterior heart and anterior aorta (Figure 2.27).

Dorsoventral. From the upper to lower surface; dorsoventral muscle (Figures 2.35, 2.36).

Drone. A male bee.

Dust (D). A dry pesticide formulation prepared by milling the pesticidal compound into a fine powder that is then diluted with a dry material like organic flour.

Ecdysis. The process of shedding the old cuticle.

Ecdysone (molting hormone, prothoracic hormone). A hormone produced in the prothoracic gland; initiates growth and molting activities of epidermal cells (Figures 4.10, 4.11).

Eclosion. The act of an insect's leaving the egg or emergence of the adult insect at the terminal molt.

Eclosion hormone. A substance secreted by certain cells in the nervous system that initiates activities associated with ecdysis.

Ecological backlash. The counter-responses of pest populations or other biotic factors in the environment that devalue the effectiveness of insect management tactics.

Ecological management (cultural management, cultural control). Purposeful manipulation of the cropping environment to reduce rates of pest increase and damage.

Economic damage. The amount of pest-induced injury that justifies the cost of applying pest control measures (Figure 7.2).

Economic-injury level (EIL). The lowest number of insects that will cause economic damage (Figures 7.3, 7.5, 7.6).

Economic poison. Legal classification for a substance used for controlling, preventing, destroying, repelling, or mitigating any pest.

Economic threshold (ET, action threshold). The pest density at which management action should be taken to prevent an increasing pest population from reaching the economic-injury level (Figure 7.5).

Ecosystem. An assemblage of interacting plant and animal communities and nonliving environmental components (Figure 5.6).

Ecotype. A specific strain of a species adapted to a particular set of environmental conditions.

Ectoderm. The outermost of the three primitive germ layers of the developing embryo; gives rise to the body wall, foregut, hindgut, nervous system, respiratory system, and many glands (Figure 4.6).

Ectognathous. Having mouthparts that are exposed, not enclosed within a cavity.

Effective environment. The elements in an ecosystem that have a direct influence on reproduction, survival, and movements of a subject species.

Ejaculatory duct. The terminus of the male sperm duct (Figure 2.45).

Elytron (pl. **elytra**). A thickened, hardened forewing (Figure 2.19F).

Embioptera. The order of insects comprising the webspinners (Figure 3.18).

Emigration. Movement out of an area.

Emulsifiable concentrate (EC, E). A concentrated pesticide formulation containing organic solvent and detergentlike emulsifier to facilitate emulsification with water.

Emulsion. A suspension of microscopic droplets of one liquid in another.

Encephalitis (pl. **encephalitides**). An inflammation of the brain or spinal cord; one type due to infection by mosquito-transmitted viruses.

Endocuticle. The innermost layer of the cuticle (Figure 2.3).

Endoderm. The innermost of the three primitive germ layers of the developing embryo; gives rise to the midgut.

Endopeptidase. An enzyme that catalyzes the splitting of a peptide chain.

Endopterygota. The division of Pterygota comprising insects with wings that develop internally.

Energid. In the fertilized egg, the daughter nuclei that have migrated to the egg periphery and become compartmentalized (Figure 4.4).

Enhanced microbial degradation. Unusually rapid breakdown of soil pesticides by microorganisms.

Entognathous. Having mouthparts housed within a cavity.

Entomophobia. An irrational, pronounced fear of insects.

Envenomation. The poisonous effects caused by the bites, stings, or secretions of insects and other arthropods.

Environmental fate. How a substance (usually a pesticide) behaves in the environment, and what becomes of it and its degradation products in the environment.

Environmental resistance. The physical and biological restraints that prevent a species from realizing its biotic potential.

Enzyme. A protein that regulates the rate of chemical reactions.

Ephemeroptera. The order of insects comprising the mayflies (Figure 3.7).

Epicuticle. The outermost layers of the cuticle (Figure 2.3).

Epideictic pheromone (spacing pheromone). An intraspecific chemical messenger that elicits dispersal from potentially crowded food sources.

Epidermis. The cellular layer of the integument that underlies and secretes the cuticle (Figure 2.3).

Epipharynx (labrum-epipharynx). A fleshy mouthpart structure on the inner surface of the labrum or clypeus.

Epiproct. The dorsal part of the terminal abdominal segment that often appears as a process situated above the anus (Figure 2.20).

Epithelial cell. A cell that covers a body structure or lines a body cavity.

Eradication. Complete and total elimination of a group of organisms from an area.

Esophagus. The narrow portion of the foregut, between the pharynx and crop (Figure 2.23).

Ester. An acid derivative in which the hydrogen of the acid has been replaced by a hydrocarbon group.

Eversible. Capable of being turned outward.

Exocuticle. The middle layer of the cuticle (Figure 2.3).

Exopeptidase. An enzyme that catalyzes the splitting of terminal peptides from a peptide chain.

Exopterygota. The division of Pterygota comprising insects with wings that develop externally.

Exoskeleton. A skeleton or supporting structure on the outside of a body.

Exotic. Introduced from another country or continent.

Exuviae. The cast cuticles of an arthropod.

Facet. The hexagonal external surface of an individual compound eye unit.

Fallow. Land left unseeded and usually clean cultivated during a growing season.

Family. A classification category that includes a number of genera sharing one or a number of characteristics and ending in -idae (e.g., Pyralidae).

Farnesol. An alcohol, $C_{15}H_{25}OH$, with juvenile hormone activity in insects.

Fat body. A meshwork of loosely organized cells in the body that serve as storage depots and function in intermediate metabolism.

Fecundity. The rate at which females produce ova.

Femur (pl. **femora**). The third segment of the leg, between the coxa and the tibia (Figure 2.18).

Feral. Existing in the wild.

Fertility. The rate at which fertilized eggs (zygotes) are produced.

Fidelity. The accuracy with which a population estimate follows actual numbers in a population.

Filariasis. Infection with parasitic nematodes, often transmitted by mosquitoes.

Filiform. Threadlike; filiform antennae (Figure 2.15).

Flagellum (pl. **flagella**). The antennal segments beyond the scape and pedicel.

Flaxseed stage. The puparium of the Hessian fly.

Flowable (F, L). A pesticide formulation in which the active ingredient is wet-milled with a clay diluent and water; has a puddinglike consistency and can be mixed with water for spraying.

Forensic entomology. The employment of entomological information in assessing time, location, or cause of a human death or a criminal act.

Formamidine. A synthetic organic amine insecticide; a recent type of insecticide that appears to mimic some insect neurotransmitters (e.g., chlordimeform).

Formulation (pesticide formulation, insecticide formulation). A mixture of active and inert ingredients for killing pests.

Frass. Insect excrement.

Freezing tolerance. Survival despite freezing of body fluids.

Frontal ganglion. The median ganglion of the stomodaeal nervous system; lies above the esophagus.

Fumigant. A volatile insecticide that enters an insect via the tracheal system.

Fungicide. A substance that kills fungi.

Furcula. A forked abdominal structure used in springing or jumping of Collembola (Figure 3.3).

Gain threshold.   The beginning point of economic damage, expressed in amount of harvestable produce; when cost of suppressing insect injury equals money to be gained from avoiding the damage (Figure 7.2).

Gall.   An abnormal growth of some plant tissue, often induced by the presence or activities of insects or mites.

Ganglion (pl. ganglia).   A bundle of nervous tissue (Figure 2.40).

Gastric caecum (pl. gastric caeca).   A saclike outpocketing of the anterior portion of the midgut (Figure 2.23).

Gastrula.   An embryo in the process of gastrulation, when a single-layered blastula turns into a three-layered embryo consisting of ectoderm, mesoderm, and endoderm (Figure 4.6).

Gene-for-gene relationship.   The concept of a resistant (or susceptible) allele at a gene locus in a particular plant cultivar corresponding to a susceptible (or virulent) allele at the same locus in the pest; a one-to-one relationship between host and pest genotypes.

General equilibrium position (GEP).   A population's long-term average density.

Generation.   A cohort of offspring from a parent population moving through a life cycle together.

Genetic control.   Altering the genetic makeup of organisms to inhibit their reproduction and survival.

Geniculate.   Elbowed or abruptly bent, as with antennae (Figure 2.15).

Genus (pl. genera).   An assemblage of evolutionarily related species sharing one or a number of characteristics.

Geomagnetic receptor.   A sense organ involved in the perception of magnetic fields.

Germ band (germ disk).   A thickened area of the blastoderm that becomes the embryo (Figure 4.5).

Gill.   A respiratory organ in the aquatic immature stage of many insects that obtains dissolved oxygen from water (Figure 2.25).

Granular (G).   A type of pesticide formulation prepared by applying liquid insecticide to coarse particles of porous material like clay, corncobs, or walnut shells.

Growth form (population growth form).   The shape of a population density curve (Figure 5.4).

Gustation.   The sense of taste.

Halter (pl. halteres).   A metathoracic, knoblike guidance organ in Diptera (Figure 2.19H).

Hamuli.   Small hooks used to couple together fore- and hindwings in Hymenoptera.

Haplodiploidy (facultative parthenogenesis).   Reproduction via a combination of sexual and asexual modes resulting in both haploid offspring (from unfertilized eggs) and diploid offspring (from fertilized eggs).

Haploid.   Having only a single (maternal) complement of chromosomes.

Hazard.   The danger that injury will occur with use of a particular pesticide; depends on both toxicity and exposure.

Head.   The anteriormost of the three main body divisions of insects; bears the mouthparts and antennae and houses the brain.

Heart.   The posterior portion of the dorsal blood vessel (Figure 2.27).

Hellgrammite.   A larva of a dobsonfly.

Hemelytron (pl. hemelytra).   The forewing of hemipterans, the basal portion of which is thickened and the apical portion membranous.

Hemiptera.   The order of insects comprising the true bugs (Figure 3.25).

Hemocoel.   The body cavity in which the blood, or hemolymph, flows.

Hemocyte.   A cell of the hemolymph.

Hemolymph.   The blood or lymphlike fluid of invertebrates including insects.

Herbicide.   A substance that kills weeds.

Heterocyclic.   Containing more than one type of atom joined in a ring.

Hibernation.   Winter dormancy.

Histogenesis.   The formation of new body tissues.

Histolysis. A breakdown of body tissues.

Homogeneous. Uniform in structure or composition.

Homoptera. The order of insects comprising the cicadas, leafhoppers, aphids, scale insects, and others (Figure 3.26).

Honeydew. A sugary liquid exuded through the anus of many Homoptera.

Hopperburn. A leaf necrosis resulting from the feeding of leafhoppers.

Horizontal resistance. Host-plant resistance to a broad range of pest genotypes (Figure 12.10).

Hormoligosis. Reproductive stimulation by sublethal doses of pesticides.

Host. An organism in or on which a parasite lives; a plant on which an insect feeds.

Host escape. A type of apparent host-plant resistance in which an unexplained lack of infestation on a susceptible plant occurs in a population of otherwise-infested plants.

Host evasion. A type of apparent host-plant resistance in which the plant passes through a susceptible stage quickly or at a time such that its exposure to potentially injurious pests is reduced.

Host-plant resistance. The relative amount of heritable qualities possessed by a plant that reduces the degree of damage done to the plant by a pest or pests.

Hydroprene. A juvenile hormone analog.

Hygroreceptor. A sense organ involved in the perception of moisture.

Hymenoptera. The order of insects comprising bees, wasps, ants, and sawflies (Figures 3.47–3.54).

Hypermetamorphosis. A type of development in which there are two or more distinct forms of larvae (Figure 4.21).

Hyperparasite. A parasite whose host is another parasite.

Hypognathus. Head positioned vertically and mouthparts pointed downward (Figure 2.6A).

Hypopharynx. A median, tonguelike mouthpart structure anterior to the labium; associated with the ducts from the salivary glands (Figure 2.7).

Imaginal disk. Groups of cells set aside in the embryo and maintained through larval development, as sites from which adult structures arise (Figure 4.19).

Immigration. Movement into an area.

Immunity. Total and complete resistance to a pest or pests.

Imperfectly density-dependent factor. An agent that is usually density-dependent but whose action sometimes fails; includes parasites, predators, and pathogens.

Indigenous. Native to an area.

Indirect loss (indirect injury). Decline in quality or quantity of a marketable commodity (e.g., seeds) owing to the feeding of insects on nonmarketable portions (sometimes roots, stems, leaves) of the plant; injury to nonyield-forming organs.

Induced resistance. A type of apparent host-plant resistance in which a particular plant condition or environmental state (e.g., nutrient level, soil moisture) makes a plant more resistant to pests than under other circumstances.

Inert ingredient. Auxiliary ingredient in a pesticide formulation that has no direct effect on pests.

Injury. The effects of pest activities on host physiology that are usually deleterious (Figure 7.1).

Inoculative release. Release of natural enemies that are expected to colonize and spread throughout an area naturally.

Inorganic. Lacking carbon atoms.

Insecta. The class of arthropods comprising the insects; characterized by three body regions (head, thorax, and abdomen), three pairs of thoracic legs, and one pair of antennae.

Insect equivalent. The amount of injury that could be produced by an individual pest through its complete life cycle.

Insect growth regulator (IGR, biorational, third-generation insecticide). Substance effective in controlling or modifying insect growth processes.

Insecticide. A substance that kills insects.

Insemination. Introduction of sperm into the female reproductive tract.

Instar. The insect between molts (e.g., sixth instar).

Integrated control. Control of pests that emphasizes selective use of insecticides so as to conserve natural enemies in the agroecosystem.

Integument. The outer covering of the body that includes the epidermis and the cuticle (Figure 2.3).

Interference coloration. Coloration resulting from the reflection of light from a series of superimposed surfaces separated by distances comparable with the wavelengths of light (Figure 2.4).

Interneuron. Neuron that serves as a connection between sensory and motor neurons (Figure 2.42).

Interspecific. Between two or more species.

Intestine. The anterior portion of the hindgut (Figure 2.23).

Intima. The cuticular lining of the tracheae (Figure 2.29), foregut, and hindgut.

Intraspecific. Within a species.

Intraspecific competition. Situation in which individuals within a population compete for a limited resource.

Introduction (importation). A biological control practice involving identification of the natural enemies that regulate a pest in its original location, and introduction of these into the pest's new location.

Inundative release. Widespread distribution of natural enemies for the purpose of suppressing a pest population, with little or no impact expected from progeny of the released individuals.

Invertase. An enzyme that catalyzes the breakdown of sucrose into fructose and glucose units.

Isolines. Varieties of a crop plant that differ in only a single gene.

Isoptera. The order of insects comprising the termites (Figure 3.17).

Johnston's organ. A chordotonal organ located in the pedicel of the antenna.

Juvabione. A juvenile hormone analog.

Juvenile. The immature stages in primitively wingless insects with no metamorphosis (Figure 4.14).

Juvenile hormone (JH). Hormone produced by the corpora allata; serves several functions including suppression of adult characteristics (Figure 4.11).

Kairomone. An interspecific messenger substance that benefits the receiver but not the releaser (e.g., attractants, excitants).

Keratolytic. Pertaining to or promoting loosening or separation of the outer layer of the skin.

Key. A tabular arrangement of species, genera, orders, or other classification categories according to characters and traits that serves to identify them (Table 3.1).

Key pest. A perennial, severe pest that causes serious and difficult crop production problems; a pest that dominates cultural activities.

Kinoprene. A juvenile hormone analog.

K strategist. A species characterized by low reproductive rate and high survival rate (Figure 5.11).

Label. The printed information on or attached to a pesticide container; federal regulations stipulate the exact information to be included on it (Figures 11.13, 11.14).

Labium. The lower lip; the fused pair of mouthpart structures just posterior to the maxillae (Figures 2.5, 2.7).

Labrum. The upper lip, lying below the clypeus and in front of the mandibles (Figures 2.5, 2.7).

Lamella (pl. **lamellae**). One in a series of layers or plates.

Lamellate. Platelike; lamellate antennae (Figure 2.15).

Larva (pl. **larvae**). The immature stage between egg and pupa in insects with complete metamorphosis (Figures 4.17, 4.18).

Lateral.   Pertaining to the side of the body.

LC_{50}.   The concentration of a toxicant in some medium (air, water, soil) that will kill 50 percent of the test organisms exposed; generally expressed as mg or cm^3 per animal or as parts per million or billion in the medium.

LD_{50}.   The dose of a toxicant that will kill 50 percent of the test organisms to which it is administered; generally expressed as milligrams of toxicant per kilogram of body weight (Figure 11.9).

Lepidoptera.   The order of insects comprising the butterflies, skippers, and moths (Figures 3.34–3.42).

Life cycle.   The chain or sequence of events that occurs during a lifetime of an individual organism (Figure 4.1).

Life system.   A subdivision of an ecosystem; the subject species and its effective environment (Figure 5.6).

Life table.   A tabulation that accounts for age-specific deaths in a population.

Lipase.   Any enzyme that catalyzes the splitting of fats into glycerol and fatty acids.

Longitudinal.   Lengthwise in the body or an appendage (e.g., dorsal longitudinal muscle) (Figures 2.35, 2.36).

Malaise trap.   A tentlike sampling device that intercepts flying insects (Figure 6.19).

Malaria.   An infectious, fever-inducing disease caused by protozoans of the genus *Plasmodium* and transmitted by *Anopheles* mosquitoes.

Mallophaga.   The order of insects comprising the chewing lice (Figure 3.22).

Malpighian tubules.   Excretory organs that arise near the anterior end of the hindgut and extend into the body cavity (Figure 2.23).

Maltase.   An enzyme that catalyzes the breakdown of the disaccharide maltose into glucose molecules.

Mandible.   The jaw; one of a pair of mouthpart structures located immediately posterior to the labrum (Figures 2.5, 2.7).

Market value.   The amount of money that a seller can expect to obtain for a commodity.

Maxilla (pl. maxillae).   One of the paired mouthpart structures just posterior to the mandibles (Figures 2.5, 2.7).

Mean.   The arithmetic average of a group of samples.

Mechanoreceptor.   A sense organ involved in the perception of tactile stimuli and/or vibration (Figure 2.39).

Mecoptera.   The order of insects comprising the scorpionflies (Figure 3.31).

Median caudal filament.   In Thysanura, a threadlike process at the posterior end of the abdomen that lies between the cerci (Figure 3.5).

Medical pest.   A pest that causes a pathological condition or transmits pathogenic organisms to humans or animals.

Meiosis.   The two successive divisions of a cell's nucleus in which the chromosome number is reduced by one-half.

Meiotic drive.   The unequal recovery of homologous chromosomes during meiosis.

Melanin.   A pigment that produces black, amber, and dark brown colors in the cuticle.

Mesenteron.   The midgut; primary site of digestion and absorption (Figure 2.23).

Mesoderm.   One of the three primitive germ layers in the developing embryo; gives rise to muscles, fat body, heart, blood cells, and reproductive organs (Figure 4.6).

Mesothorax.   The second of the three segments of the thorax (Figure 2.17).

Metamorphosis.   A change in body form during development.

Metathorax.   The third, or posterior, of the three segments of the thorax.

Methoprene.   A juvenile hormone analog.

Microbial insecticide.   A biological preparation of viruses or microorganisms or their products, applied and used in ways similar to conventional chemical insecticides.

Microcoryphia.   The order of insects comprising the jumping bristletails (Figure 3.6).

Microencapsulation.   A pesticide formulation technique in which pesticide-containing microscopic spheres or capsules are created to permit pesticide release at a slow, steady, and effective rate.

Micropyle. A pore in the egg through which sperm may enter (Figure 4.3).

Migration. Movements over great distances, during which locomotion is not inhibited by food, mates, or oviposition sites; an adaptation to periodically transport insects beyond the boundaries of their old reproductive sites and into new ones.

Milky disease. A bacterial disease of Japanese beetle caused by *Bacillus popilliae* and *Bacillus lentimorbus* (Figure 9.14).

Mitochondrion (pl. **mitochondria**). An organelle that is the principal site of energy synthesis in the cell.

Mixed-function oxidase (MFO). An enzyme that metabolizes many foreign substances.

Mode of action. Means by which a toxin affects the anatomy, physiology, or biochemistry of an organism.

Molt. To shed the old cuticle.

Molting fluid. A mixture of enzymes produced by the epidermal cells capable of digesting the endocuticle.

Monogamy. Mating with a single individual.

Monogenic resistance. Resistance resulting through expression of a single gene.

Monophagus. Feeding on one type of food.

Mortality. Death rate or numbers dying per unit of time.

Multivoltine. Having more than one generation per year (Figure 4.23).

Mutagen. A material that induces genetic changes.

Myofibril. A contractile element of a muscle fiber.

Naiad. The immature stage in insects with incomplete metamorphosis; typically feed and develop in water (Figure 4.16).

Narcotic. A chemical that causes reversible depression of the nervous system.

Nasute (pl. **nasuti**). In some termites, a type of soldier that has a snout through which a defensive fluid may be ejected.

Natality. Rate of birth; often measured as total number of eggs or eggs per female per unit of time (Figure 5.2).

Natural control. The suppression of pest populations by naturally occurring biological and other environmental agents.

Natural enemies. Living organisms found in nature that kill insects outright, weaken them, or reduce their reproductive potential.

Negative binomial model. A mathematical model used to describe a clumped, or contagious, dispersion of organisms.

Nematicide. A substance that kills nematodes.

Neoptera. The infraclass of Pterygota comprising insects capable of wing flexion.

Neuron. A nerve cell; the basic unit of nerve impulse transmission.

Neuroptera. The order of insects comprising the lacewings, antlions, and dobsonflies (Figure 3.27).

Neurotoxin. A substance that is poisonous or destructive to nerve tissues.

Neurotransmitter. A substance released from an axon terminal that either excites or inhibits a target cell.

Nicotine. A botanical alkaloid insecticide derived from the tobacco plant.

Nit. The egg of a louse, often attached to a feather or hair.

Nomenclature. The naming of organisms.

Nonpreference (antixenosis). Plant characteristics that lead insects away from a particular host; involves activities of both host plant and insect.

Notum (pl. **nota**). A dorsal surface of a thoracic segment.

Nymph. The immature stage in insects with gradual metamorphosis (Figure 4.15).

Occasional pest. A pest with a general equilibrium position substantially below the economic-injury level; highest pest population fluctuations occasionally and sporadically exceed the economic-injury level (Figure 8.11).

Ocellus (pl. **ocelli**). The small, simple eye of an insect or other arthropod.

Odonata. The order of insects comprising the dragonflies and damselflies (Figure 3.8).

Olfaction. The sense of smell.

Oligogenic resistance (major-gene resistance). Host-plant resistance conferred by one or a few genes.

Oligophagus. Feeding on a few types of food.

Ommatidium (pl. **ommatidia**). An individual sensory unit of a compound eye.

Omnivorous. Feeding on a variety of foods.

Ootheca (pl. **oothecae**). A protective pod, produced by some female insects, that contains several to many eggs.

Opisthognathous. Head positioned obliquely and mouthparts pointed backward (Figure 2.6C).

Order. A category of classification; one of the principal divisions of the class Insecta, based largely on wing structure and usually ending in -ptera (e.g., Lepidoptera).

Organic. Possessing carbon atoms.

Organophosphate. An organic compound containing phosphorus; includes many important insecticides (e.g., malathion).

Organosulfur. A sulfur-containing organic compound used as an acaricide (e.g., tetradifon).

Organotin. A tin-containing organic compound used as an acaricide (e.g., cyhexatin) and/or fungicide.

Orthoptera. The order of insects comprising the grasshoppers, crickets, walkingsticks, mantids, and cockroaches (Figures 3.9–3.15).

Ostium (pl. **ostia**). A slitlike opening in the heart (Figure 2.27).

Ovariole. A tubular division of an ovary (Figure 2.44).

Ovary. The egg-producing organ (Figure 2.44).

Oviduct. The tube leading away from the ovary through which eggs pass (Figure 2.44).

Oviparous. Egg-laying.

Oviposition. The act of egg laying (Figure 4.13).

Ovipositor. The egg-laying structure of female insects (Figure 2.21).

Ovoviviparous. Producing active young immediately following the hatching of the eggs within the female.

Paedogenesis. Reproduction by the juvenile form.

Paleoptera. The infraclass of Pterygota comprising insects incapable of wing flexion.

Palpus (pl. **palpi**). A sensory organ associated with mouthpart structures (Figures 2.5, 2.7).

Paraproct. One of a pair of lobes adjacent to the anus (Figure 2.20).

Parasite. An organism that lives in or on the body of another organism (the host) during some portion of its life cycle (Figure 1.18).

Parasitoid. An arthropod that parasitizes and kills an arthropod host; is parasitic in its immature stages but is free-living as an adult (Figure 9.6).

Parthenogenesis. Reproduction by the development of an egg that has not been fertilized with sperm.

Pasture spelling. Leaving a pasture idle for a period of time by removing livestock from it.

Pathogen. A disease-causing organism.

Pectinate. With processes like the teeth of a comb; pectinate antennae (Figure 2.15).

Pedicel. The second segment of the antenna; the base of the ovariole.

Penis (pl. **penes**). The copulatory organ of the male.

Perennial pest. A pest with a general equilibrium position below but close to the economic-injury level; highest pest population fluctuations regularly reach the economic-injury level (Figure 8.12).

Perfectly density-dependent factor. A factor that never fails to control the increase in a population's numbers (i.e., intraspecific competition).

Peritrophic membrane. A chitinous membrane secreted by cells lining the midgut; forms a semipermeable envelope around the food (Figure 2.24).

Pest. A species that interferes with human activities, property, or health or is objectionable.

Pest control. The application of technology, in the context of biological knowledge, to achieve satisfactory reduction of pest numbers or effects.

Pesticide. Any substance for controlling, preventing, destroying, disabling, or repelling a pest.

Pest management (integrated pest management, IPM). A comprehensive approach to dealing with pests that strives to reduce pest status to tolerable levels by using methods that are effective, economically sound, and ecologically compatible.

Pest management strategy. An overall plan for eliminating or alleviating a pest problem.

Pest management tactic. A method of implementing a pest management strategy.

Pest status. The ranking of a pest relative to the economics of dealing with the species (Figure 1.28).

Petiole. The waistlike constriction in apocritan hymenopterans.

Phagocyte. A blood cell capable of degrading tissues and foreign matter.

Pharynx. The anterior portion of the foregut, between the mouth and the esophagus (Figure 2.23).

Phasic. Characterized by short-duration stimulation; phasic receptors.

Phenology. The periodicity of biological phenomena.

Phenotype. The sum total of the observable or measurable characteristics of an organism without reference to its genetic nature.

Phenyl. An organic radical, C_6H_5, derived from benzene via the removal of a hydrogen atom.

Pheromone. A substance secreted by one animal that causes a specific reaction on reception by another animal of the same species (e.g., sex pheromone).

Photoperiod. Day length.

Photoreceptor. A sense organ involved in the perception of light.

Phylum (pl. **phyla**). One of the major divisions of the animal kingdom.

Physiological resistance. The ability of an insect strain to tolerate insecticides or other factors through physiological change (e.g., via modification of cuticle structure).

Phytoalexins. Phenolic compounds produced by plants when they become diseased or injured by insects; enable plants, once damaged, to resist further attack by insects.

Phytophagous. Feeding on plants.

Phytotoxic. Injures or kills plants.

Piroplasmosis. Infection with tick-transmitted piroplasma protozoans, *Babesia* species, that attack the red blood cells of cattle, dogs, and other animals.

Pitfall trap. A sampling device designed to capture insects moving along the ground (Figure 6.24).

Placoid. Platelike; placoid sensilla.

Plague (bubonic plague, black death). An infectious disease caused by the bacterium *Yersinia pestis* and spread via the oriental rat flea *Xenopsylla cheopis*.

Plasma. The noncellular fluid portion of the hemolymph.

Plasmodium (pl. **plasmodia**). The protozoans that may cause malaria.

Plastron. A film of air over the body of certain aquatic insects that serves as a physical gill for respiration.

Plecoptera. The order of insects comprising the stoneflies (Figure 3.19).

Pleuron (pl. **pleura**). A lateral plate of a thoracic segment.

Plumose. Featherlike; plumose antenna (Figure 2.15).

Poikilothermic. Having body temperatures that fluctuate with environmental temperatures; "cold-blooded."

Poisson model. A mathematical model that describes a random dispersion of organisms.

Polychloroterpene. A chlorinated hydrocarbon insecticide prepared from camphene, a pine tree derivative (e.g., toxaphene).

Polyembryony. More than one embryo formed in a single egg.

Polygenic resistance (minor-gene resistance). Host-plant resistance conferred by many genes.

Polyphagous. Feeding on many types of food.

Population. A group of individuals of the same species within given space and time constraints.

Population dynamics. The interplay between populations and environmental forces that influence them.

Population index. An estimate of population size based on insect effects or products.

Population intensity estimate. An absolute estimate of population size expressed as number per unit habitat (e.g., number per leaf).

Pore canal. A small channel involved in the transport of materials from the epidermis to the body surface (Figure 2.3).

Postdiapause quiescence. A transitional period of dormancy, following diapause, that is not sensitive to token stimuli.

Potential natality. The reproductive rate of individuals in an optimal environment.

Precision. A measure of the degree of error in making estimates of a population's size.

Precocene. An antijuvenile hormone isolated from the plant *Ageratum houstonianum*.

Predator. An organism that attacks and feeds on other animals (usually smaller or less powerful than itself) and consumes more than one animal in its lifetime (Figure 1.24).

Prepupa. The last larval instar.

Prescription. Recommendation of a strategy or course of action.

Preventive practice. A pest management approach that attempts to prevent problems (Figure 16.1).

Prey. An organism caught for food.

Primary metabolite. A metabolic product necessary for growth and reproduction of a plant.

Proboscis. The extended beaklike mouthparts of sucking insects.

Proctodaeum. The hindgut; plays important roles in elimination and water and salt resorption (Figure 2.23).

Prognathous. Head positioned horizontally and mouthparts pointed forward (Figure 2.6B).

Prognosis. Anticipation of pest problems before they occur and prediction of the outcome of these if left unattended.

Proleg. A fleshy, unsegmented abdominal walking appendage of some insect larvae.

Pronotum (pl. **pronota**). The dorsal surface of the prothorax.

Proprioreceptor. A mechanoreceptor that provides information about an insect's position in its environment and/or the position of the insect's body parts.

Proteolysis. The splitting of proteins into smaller components by the breaking of peptide bonds.

Prothoracic gland. An endocrine organ in the prothorax of immature insects that secretes ecdysone (Figures 4.10, 4.11).

Prothorax. The first, or anterior, of the three segments of the thorax.

Protocerebrum. The anterior pair of lobes of the brain that innervate the compound eyes and ocelli (Figure 2.41).

Protura. The order of insects comprising the telsontails or proturans (Figure 3.2).

Proventriculus. The posteriormost portion of the foregut (Figure 2.23).

Proximal. Near the point of attachment or origin.

Psocoptera. The order of insects comprising the booklice and barklice (Figure 3.21).

Pterygota. The subclass of Insecta comprising the winged and secondarily wingless insects.

Pulvillus (pl. **pulvilli**). A pad at the base of each tarsal claw (Figure 2.18B).

Pupa (pl. **pupae**). The stage between the larva and adult in insects with complete metamorphosis (Figure 4.17).

Puparium. A protective enclosure for the pupa of some Diptera formed from the last larval cuticle.

Pyloric valve. The valve at the junction of the midgut and hindgut.

Pyrethroid (synthetic pyrethroid). An organic synthetic insecticide with a structure based on that of pyrethrum (e.g., allethrin).

Pyrethrum. A botanical insecticide derived from *Chrysanthemum* flowers.

Quarantine. A legal instrument to prevent or slow the spread of pests from infested into uninfested areas by restricting movement of infested stock.

Quiescence (torpor). Inactivity induced by unfavorable environmental conditions; activity resumes immediately on return of favorable conditions.

Quinone. A benzene derivative in which two hydrogen atoms are replaced by two oxygen atoms. Quinones are utilized in the hardening and darkening of the insect cuticle.

Range. The difference between the smallest number and the largest number in a sample.

Receptor fibril. The portion of a neuron that transmits impulses toward the cell body.

Rectum. The posterior portion of the hindgut (Figure 2.23).

Reflex. An involuntary movement or other response elicited by a stimulus applied peripherally, conducted through the central nervous system, and conducted back peripherally.

Regression. The amount of change in one variable associated with a unit change in another variable (Figure 7.4).

Regulation (population regulation). The natural phenomenon of population numbers fluctuating within certain bounds for long periods of time.

Relapsing fever. A disease caused by *Borrelia* species of bacteria and transmitted by body lice.

Relative population estimate. An estimate of population size based on the kind of sampling technique used (e.g., number per trap).

Relative variation (RV). Percentage standard error of a mean; used as a measure of sampling precision.

Repellent. A chemical that causes insects to orient their movements away from a source.

Replacement (pest replacement, secondary pest outbreak). When a major pest is suppressed and continues to be suppressed by a tactic but is replaced in importance by another pest, previously with minor status (Figures 15.9, 15.11).

Resistance. The ability of a strain to tolerate or avoid factors that would prove lethal or reproductively degrading to the majority of strains in a normal population (Figures 15.1, 15.2).

Restricted-use pesticide. A pesticide that may be applied only by applicators certified by the state in which they work.

Resurgence (pest resurgence). A situation in which a population, after having been suppressed, rebounds to numbers higher than before suppression occurred (Figures 15.8, 15.10).

Retinula cell. One of a group of cells comprising the receptor unit of an ommatidium (Figure 2.37).

Rhabdom. A rodlike structure formed by the inner surfaces of adjacent retinula cells in an ommatidium (Figure 2.37).

Rickettsia. Obligate intracellular parasitic bacterialike microorganisms that infect arthropods and may also be pathogenic in vertebrates.

Risk. The probability that a pesticide will cause harm.

Rotenone. A botanical insecticide extracted from the roots of *Derris* and *Lonchocarpus* species of plants.

Royal jelly. A glandular secretion of worker bees on which larvae are fed.

r-strategist. A species characterized by high reproductive rate and low survival rate (Figure 5.10).

Ryania. A botanical alkaloid insecticide extracted from the stem and roots of the shrub *Ryania speciosa*.

Sabadilla. A botanical insecticide extracted from seeds of *Schoenocaulon officinale*.

Saprophagus. Feeding on dead and decaying animals or plants.

Sarcoplasm. The cytoplasm of striated muscle fibers.

Scabies. A contagious skin disorder caused by burrowing of the itch mite *Sarcoptes scabiei*.

Scape. The basal segment of the antenna.

Scavenger. An animal that feeds on dead and decaying animals or plants or on animal wastes.

Sclerite. A hardened integumentary plate surrounded by membranous areas or sutures.

Sclerotin. A protein that contributes to the hardness of the cuticle.

Sclerotization. The hardening of the body wall.

Scutellum. A sclerite of the thoracic notum; appears as a triangular plate between the base of the wings in Hemiptera.

Seasonal cycle. The sequence of life cycles of a species that occurs over a one-year period (Figure 4.1).

Secondary metabolite. A plant-produced chemical that is nonessential in primary metabolism but may play a role in defense against herbivory.

Secondary pest. Pest species that are usually present at low levels and are held in check by the action of natural enemies; can assume full pest status when natural enemies are destroyed by a pest management tactic (e.g., insecticide application).

Segment. A subdivision of the body or of an appendage.

Semen. Fluid discharged at ejaculation by the male, consisting of spermatozoa and secretions of glands associated with the genital tract.

Seminal vesicle. A saclike storage structure that holds seminal fluid of the male prior to discharge.

Semiochemical. Any chemical involved in communications among organisms (Figure 12.3).

Sensillum (pl. **sensilla**). A simple sense organ.

Septum (pl. **septa**). A wall or partition.

Sericulture. The propagation of silkworms for silk production.

Seta (pl. **setae**). A bristle or hairlike process.

Severe pest. A pest with a general equilibrium position above the economic-injury level, which makes the pest a constant problem (Figure 8.13).

Sex pheromone. An intraspecific chemical messenger produced for the purpose of bringing the two sexes together.

Sex ratio. Within a population, the proportion of males compared with that of females.

Signal word. A word required on every pesticide label to denote the relative toxicity of the material (e.g., "danger-poison" for highly toxic compounds).

Silviculture. The care and cultivation of forest trees.

Sinus. A partitioned chamber of the body cavity.

Siphonaptera. The order of insects comprising the fleas (Figure 3.46).

Soluble powder (SP). A pesticide formulation consisting of a finely ground solid material that dissolves in water or some other liquid.

Solution. A concentrated liquid pesticide formulation that may be used directly or require diluting.

Species. A group of organisms that are similar in structure and physiology, that actually or potentially interbreed and produce fertile offspring, and that are reproductively isolated from all other such groups.

Spermatheca (pl. **spermathecae**). A saclike structure in which sperm from the male is received and stored in the female (Figure 2.44).

Spermatophore. A packet containing sperm produced by some male insects.

Spiracle. An external opening of the tracheal system (Figures 2.28, 2.29).

Stadium. The time between molts.

Stage. The insect's developmental status (e.g., egg stage).

Standard deviation. The square root of the arithmetic averages of the squares of the deviations from the mean; a measure of variation in a set of samples.

Standard error. The measure of standard deviations of a population of means.

Stem mother. A wingless, parthenogenic, viviparous female aphid that arises from an overwintering aphid egg.

Sternum (pl. **sterna**). The ventral surface of a thoracic segment.

Sticker. An ingredient added to a pesticide formulation to improve its adherence to a surface (e.g., casein).

Stomach poison. An insecticide that is fatal only after being eaten, entering the insect body through the gut (e.g., boric acid).

Stomodaeal nervous system. The main component of the visceral nervous system; controls activities of the anterior gut and dorsal vessel.

Stomodaeum. The foregut, beginning with the mouth and ending with the proventriculus (Figure 2.23).

Strepsiptera. The order of insects comprising the twisted-winged parasites (Figure 3.30).

Stridulation. Sound production accomplished by rubbing body parts together.

Strip harvesting. Harvesting of different areas of a single crop at different times (Figure 10.17).

Stylet. One of the piercing structures in piercing-sucking mouthparts.

Subeconomic pest (noneconomic pest). A pest with a general equilibrium position far below the economic-injury level; highest pest populations do not reach the economic-injury level (Figure 8.10).

Subesophageal ganglion. A swelling of nerve tissue at the anterior end of the ventral nerve cord, below the esophagus (Figures 2.40, 2.41).

Summer oil. A highly refined petroleum oil that can be applied to trees in full foliage without having phytotoxic effects.

Supercooling. Resistance to freezing by lowering the temperature at which freezing of body fluids begins.

Supraesophageal ganglion. A nerve mass that lies in the head above the esophagus; the insect brain.

Surfactant. An agent used to improve emulsifying, wetting, and spreading properties of a pesticide formulation.

Surveillance (scouting). The watch kept on a pest for detection of the species' presence and determination of population density, dispersion, and dynamics.

Survivorship curve. A graphic representation of deaths over time for a given population (Figure 5.13).

Suture. A seam or line indicating the division of distinct parts of the body wall.

Swarm. A concerted departure or association of a large group of insects from a nest or breeding habitat.

Symbiont. One organism living in symbiosis with another.

Symbiosis. A close association between two or more different organisms that benefits one or both of the organisms.

Synapse. The junction between two neurons or between a neuron and another cell (Figure 2.42).

Synaptic poison. A chemical that interrupts normal transmission of impulses across synapses (Figure 11.8).

Synergism. The action of two or more factors together achieving a greater effect than the sum of the individual effects of the factors.

Synergist. A nontoxic chemical used to increase the effectiveness of an insecticide (e.g., piperonyl butoxide).

System. An interrelated set of elements that can be identified and around which a boundary can be drawn (Figure 8.15).

Systematics. The study of the diversity and classification of organisms.

Systemic insecticide. An insecticide that is taken up and translocated within plants and animals (e.g., aldicarb).

Systems technology. The assembly, organization, and employment of an identifiable set of interrelated elements that constitute a system.

Taenidium (pl. **taenidia**). A spiral or circular thickening of the inner surface of the tracheae.

Tagmosis. The differentiation of the body into distinct functional regions, or tagmata (Figure 2.1).

Tarsomeres. The subdivisions of the tarsus (Figure 2.18).

Tarsus (pl. **tarsi**). The terminal leg segment, immediately distal to the tibia; consists of one or more subdivisions or tarsomeres (Figure 2.18).

Taxonomy. The theory and principles of classifying organisms.

Technical-grade material. The relatively pure form of a pesticidal compound that comprises the active ingredient in the final pesticide mixture.

Tegmen (pl. **tegmina**). A leathery or parchmentlike forewing (e.g., Orthoptera).

Telson. The terminal body segment in some primitive arthropods.

Tenaculum. The clasplike structure associated with the furcula of Collembola.

Tentorium. The endoskeleton of the head, formed from invaginations of the body wall (Figure 2.5).

Teratogen. A material that causes birth defects.

Tergum (pl. **terga**). The dorsal surface of a segment.

Termiticide. An insecticide that kills termites.

Terpene. A naturally-occurring compound in which the basic structural component is isoprene, $CH_2C(CH_3)CH_2CH_2$.

Testis (pl. **testes**). The male sex organ that produces sperm (Figure 2.45).

Therapeutic practice. A pest management approach that attempts to cure existing problems (Figure 16.1).

Thermal constant. The number of degree days required for an event (e.g., pupation) to occur.

Thiocyanate. An organic insecticide that contains the SCN group; has a creosotelike odor; interferes with cellular respiration and metabolism.

Thorax. The body region between the head and the abdomen that bears the legs and the wings (Figure 1.2).

Thysanoptera. The order of insects comprising the thrips (Figure 3.24).

Thysanura. The order of insects comprising the silverfish (Figure 3.5).

Tibia (pl. **tibiae**). The fourth segment of the leg, between the femur and the tarsus (Figure 2.18).

Tillage. The cultivation of land.

Token stimuli. Specific environmental cues that induce diapause or elicit a specific behavioral response only at certain times.

Tolerance. The ability of a host to withstand injury by pests; involves a plant response only.

Tonic. Characterized by continuous or long-term stimulation; tonic receptors.

Toxicity. The inherent poisonous potency of a material.

Toxicology. The science or study of poisons.

Trachea (pl. **tracheae**). A tube of the respiratory system that begins with the spiracle and ends with the tracheoles (Figures 2.28, 2.29).

Tracheoles. The minute terminal branches of the tracheal system (Figure 2.29).

Trade name (proprietary name, brand name). Name given to a pesticide by the manufacturer or formulator (e.g., Sevin®).

Trail-marking pheromone. An intraspecific chemical messenger produced by foraging ants and termites to indicate sources of requisites to other colony members.

Trap crop. A small area of a crop used to divert pests from a larger area of the same or another crop (Figure 10.15).

Trench fever. A disease caused by infection with the rickettsia *Rochalimaea quintana* and transmitted via the feces or crushed bodies of the louse *Pediculus humanus humanus*.

Trichoptera. The order of insects comprising the caddisflies (Figure 3.32).

Tritocerebrum. The posterior pair of lobes of the brain that connect to the visceral nervous system (Figure 2.41).

Triungulin. An active, first instar of insects that undergo hypermetamorphosis.

Trivial movement. Displacement of insects within or close to the breeding habitat; nonmigratory movement.

Trochanter.   The second segment of the leg, between the coxa and the femur (Figure 2.18).

Trypanosome.   A protozoan parasite of the blood and lymph.

Tympanum (pl. **tympana**).   An auditory membrane.

Typhus fever.   A human disease caused by the bacteriumlike microorganism *Rickettsia prowazekii* and transmitted by the body louse *Pediculus humanus humanus*.

Ultra low volume (ULV).   Pesticide sprays applied at 0.6 to 4.7 liters per hectare or sprays applied in an undiluted, low-volume formulation.

Univoltine.   Having one generation per year (Figure 4.22).

Urate cell.   A cell of the fat body that stores uric acid.

Uric acid.   The primary nitrogenous waste in insects, $C_5H_4N_4O_3$.

Vagina.   The terminus of the female reproductive tract that opens to the outside.

Vas deferens (pl. **vasa deferentia**).   The sperm duct that leads away from a testis (Figure 2.45).

Vas efferens (pl. **vasa efferentia**).   A short duct connecting a sperm tube in the testis to the vas deferens (Figure 2.45).

Vector.   An organism capable of transmitting pathogens from one host to another.

Vein.   A thickened line in the insect wing.

Ventral.   Pertaining to the underside of the body.

Vertical resistance.   Host-plant resistance to one or a narrow group of pest genotypes (Figure 12.10).

Virulent gene.   A gene that allows a pest species to overcome resistance factors in a plant cultivar.

Visceral.   Pertaining to or affecting the internal organs of the body.

Vitamin.   An organic substance necessary in trace amounts for normal metabolic functioning.

Viviparous.   Giving birth to active young that undergo some growth and development in the mother.

Voltinity.   The number of generations that occur in one year.

Wettable powder (WP, W).   A pesticide formulation of active ingredient mixed with inert dust and a surfactant that mixes readily with water and forms a short-term suspension.

Wing pads.   Externally visible beginnings of wings observed on insect nymphs.

Wireworm.   The larva of a click beetle (family Elateridae).

Yellow fever.   An infectious, mosquito-transmitted viral disease.

Zoophagous.   Feeding on animals.

Zoraptera.   An order of Pterygota that consists of rare, small, and termitelike insects (Figure 3.20).

Zygote.   A fertilized egg.

Index